Ethnoveterinary Botanical Medicine

Herbal Medicines for Animal Health

Ethnoveterinary Botanical Medicine

Herbal Medicines for Animal Health

Edited by

David R. Katerere ▪ Dibungi Luseba

CRC Press
Taylor & Francis Group
Boca Raton London New York

CRC Press is an imprint of the
Taylor & Francis Group, an **informa** business

CRC Press
Taylor & Francis Group
6000 Broken Sound Parkway NW, Suite 300
Boca Raton, FL 33487-2742

Printed in the United States of America on acid-free paper
10 9 8 7 6 5 4 3 2 1

International Standard Book Number: 978-1-4200-4560-4 (Hardback)

Library of Congress Cataloging-in-Publication Data

Ethnoveterinary botanical medicine : herbal medicines for animal health / edited by David R. Katerere and Dibungi Luseba.
p. ; cm.
Includes bibliographical references and index.
ISBN 978-1-4200-4560-4 (alk. paper)
1. Herbs--Therapeutic use. 2. Alternative veterinary medicine. 3. Traditional veterinary medicine. 4. Ethnobotany. I. Katerere, David R. II. Luseba, Dibungi.
[DNLM: 1. Animal Diseases--drug therapy. 2. Phytotherapy--veterinary. 3. Medicine, Traditional. 4. Plants, Medicinal. SF 918.H47 E84 2010]

SF918.H47.E84 2010
636.089'5321--dc22 2009043655

Visit the Taylor & Francis Web site at
http://www.taylorandfrancis.com

and the CRC Press Web site at
http://www.crcpress.com

Dedication

To those who went before us,
their knowledge makes us who we are,
and those who will tread after us
that they may keep and own it, too.

Contents

Foreword

From before written history, plants have been known as sources of medicines for treating human beings. In practically every human culture, there exists a vast treasury of information of this type, and in some civilizations, notably in China and India, this has developed into a sophisticated system of diagnosis, treatment, and preparation of the medicine.

As far as "Western" scientific medicine is concerned, natural substances were, and still are, the source of many isolated chemicals that are incorporated as the active constituents into familiar pharmaceutical dosage forms such as tablets, capsules, injections, and topical applications such as creams and ointments. The rise and application of pharmaceutical chemistry enlarged the potential of naturally occurring compounds for use as drugs since they could also be used as "lead molecules" to improve their efficacy or reduce toxicity by chemical manipulation of the structures. Traditional plant-based extracts became very marginal in most countries where "scientific" medicine became the norm.

However, in spite of the undoubted successes of such a scientific approach to pharmaceuticals, the last few decades have witnessed a spectacular rise in interest and use of "herbal medicinal products" (i.e., plant material or its crude extracts) in those places where sophisticated technologically advanced medicine was common, as well as in countries where it has long formed the mainstay of medicines used by ordinary people. This general interest has been followed by increasing scientific and commercial interest in traditional medicines, and in 1967 the term *ethnopharmacology* was coined to describe the scientific discipline investigating the use of these products.

Since that time, scientific investigations in this area have grown apace, as witnessed by the success of scientific journals such as *Journal of Ethnopharmacology* and *Phytotherapy Research,* as well as by good attendance at international conferences dealing with such matters and introduction of university courses. This has been fueled partly by the pharmaceutical industry seeking new lead compounds but also by an awakened interest and patrimonial pride in their traditions by many countries emerging from their colonial past. At one level, such research gives substance to traditional claims and reinforces the value of the cultural heritage, but if it demonstrates efficacy and safety, it might also lead to substitutes for expensive imported Western drugs. Since Western drugs are often beyond economic or geographical reach of many of the inhabitants of these places anyway, scientific study can also lay the basis for improving the quality of the traditional remedies, thus providing better grounds for efficacy and safety.

As far as the industrially developed countries are concerned, intensive farming has been widely practiced but has raised not only food production, but also many ethical and health concerns, giving rise to the "organic" preference of many consumers. Hence, the investigation of these traditional "natural" medicines might provide alternatives to current treatments of animals that have caused much concern, such as the widespread use of antibiotics in young animals, producing residues in food and thereby the buildup of resistance in humans and other animals. The entry into the food chain of hormones and other steroids for building muscles and stimulating growth may affect human metabolism and health. The study of traditional medicines might provide compounds or extracts with novel structures or different mechanisms of action, with less-acute or chronic side effects, which would make good substitutes for the currently used drugs that are raising concern.

This interesting, and in some ways surprising, explosion of scientific interest and study into traditional materials for helping human health care has only fairly recently been enlarged to encompass use of traditional medicines for treating animals. Urbanized and relatively affluent members of society have a mainly emotional attachment to domestic "companion" animals, on which they are willing to spend considerable amounts of money, and it is not surprising that herbal products to treat domestic animals sell well. However, it should not be forgotten that, for most rural parts of the world, animals are important sources and symbols of wealth and livelihood. When the animal becomes sick or dies, it may mean the loss of transport, aid to farming, dairy products, meat, and other products such as wool or hair, which provide extra income or clothing. The death of a single animal may spell the beginning of poverty for a whole family.

It is therefore not surprising that in many societies there is a pharmacopoeia of substances used to treat livestock and poultry. The growing interest in this from scientists was reflected in a session dedicated to veterinary ethnopharmacology at the Congress of the International Society for Ethnopharmacology in São Paulo, Brazil, in September 2008. This book is another indication and is the first, to my knowledge, to bring together a considerable amount of information about ethnoveterinary medicines from a wide variety of countries. As well as listing plants used, the conditions for which they are used, and the ways that the plant material is treated prior to use, this book covers useful topics such as general research methods for testing claimed effects and safety and the chemical examination of extracts. The inclusion of a chapter on benefit sharing is particularly praiseworthy since it is all too easy for a culture to have its traditional knowledge "stolen" to be used commercially by an outside body that returns no benefits.

Dr. Katerere and Dr. Luseba have been fortunate enough to persuade leading authorities in this field to contribute chapters, and I am sure that it not only will prove to be a valuable reference source but also will stimulate further research into this fascinating area, with the end result of improving not only the health of animals and domestic birds, but also the well-being of their owners and farmers.

Peter Houghton
Emeritus Professor in Pharmacognosy,
Pharmaceutical Sciences Research Division,
King's College London
Editor, Journal of Ethnopharmacology

Acknowledgments

We wish to thank God for seeing this project through and for the support of PROMEC Unit, South African Medical Research Council (MRC) and Tshwane University of Technology (TUT), our employers. We are much obliged to Trish Flaster, Dr. Ronette Gehring, and Nontuthuko Chamane for proofreading edits and providing some much needed peer review and scientific input.

We also want to thank John Sulzycki, our editor, and Pat Roberson and Stephanie Morkert, our project coordinators, and many other dedicated people at Taylor & Francis, none of whom we have met physically, but who have expertly guided this project and brought to our dreams to fruition.

We acknowledge the immense contribution that all the authors who are part of this book have made. Without them, this book would not have known the light of day, and we hope that this project will be the start of further collaborations among them in the future.

Last, we thank our families for the obvious sacrifices they have had to make during the duration of this project. Tinotenda.

Contributors

Joanne Barnes
School of Pharmacy
University of Auckland
Auckland, New Zealand

Hama Benbarek
Department of Veterinary Sciences
Faculty of Agro-Veterinary Sciences
Ibn-Khaldoune University of Tiaret
Tiaret, Algeria

Mokhtar Benhanifia
Department of Veterinary Sciences
Faculty of Agro-Veterinary Sciences
Ibn-Khaldoune University of Tiaret
Tiaret, Algeria

Laïd Boukraa
Department of Veterinary Sciences
Faculty of Agro-Veterinary Sciences
Ibn-Khaldoune University of Tiaret
Tiaret, Algeria

Mary Chikombero
Marine and Coastal Management
Department of Environment and Tourism
Cape Town, Republic of South Africa

Wedson de Medeiros Silva Souto
Bolsista do DAAD (Deutscher Akademischer Austausch Dienst)
Programa Regional de Pós-Graduação em Desenvolvimento e Meio Ambiente (PRODEMA)
Universidade Estadual da Paraíba
Campina Grande
Paraíba, Brazil

Raynner Rilke Duarte Barboza
Pós-Graduação em Ciência e Tecnologia Ambiental
Universidade Estadual da Paraíba
Campina Grande
Paraíba, Brazil

Jacobus N. Eloff
Phytomedicine Programme
Faculty of Veterinary Sciences
University of Pretoria
Onderstepoort, Republic of South Africa

Peter K. Gathumbi
Department of Veterinary Pathology, Microbiology, and Parasitology
University of Nairobi
Nairobi, Kenya

Ronette Gehring
Department of Clinical Sciences (Agricultural Practices)
College of Veterinary Medicine
Kansas State University
Manhattan, Kansas, USA

John B. Githiori
International Livestock Research Institute (ILRI)
Nairobi, Kenya

Mawena Sylvie Hounzangbe-Adote
Animal Production Department
Faculty of Agricultural Sciences
Cotonou University of Agriculture
Cotonou, Benin Republic

Zafar Iqbal
Department of Veterinary Parasitology
University of Agriculture
Faisalabad, Pakistan

Abdul Jabbar
Department of Veterinary Parasitology
University of Agriculture
Faisalabad, Pakistan

Bukirwa Irene Kamara
Sasol Technology (Pty) Limited
Sasolburg, Republic of South Africa

David R. Katerere
PROMEC Unit/SAMRC
Cape Town, Republic of South Africa

Kelly Kindscher
Kansas Biological Survey
College of Veterinary Medicine
Kansas State University
Manhattan, Kansas, USA

Michael I. Lindinger
Department of Human Health and Nutritional Sciences
University of Guelph
Guelph, Canada

Dibungi Luseba
Department of Animal Sciences
Tshwane University of Technology
Pretoria, Republic of South Africa

Viola Maphosa
Agricultural and Rural Development Research Institute (ARDRI)
University of Fort Hare
Alice, Republic of South Africa

Murugaiyah Marimuthu
Faculty of Veterinary Medicine
University Putra Malaysia
Serdang Selangor, Malaysia

Patrick Masika
Agricultural and Rural Development Research Institute (ARDRI)
University of Fort Hare
Alice, Republic of South Africa

Lyndy J. McGaw
Phytomedicine Programme
Faculty of Veterinary Sciences
University of Pretoria
Onderstepoort, Republic of South Africa

Vinny Naidoo
Department of Paraclinical Studies
Faculty of Veterinary Sciences
University of Pretoria
Onderstepoort, Republic of South Africa

M. N. Balakrishnan Nair
Foundation for Revitalisation of Local Health Traditions (FRLHT)
Bangalore, India

Ifeanyi Charles Okoli
Department of Animal Science and Technology
Federal University of Technology
Oweer, Nigeria

Chandrawathani Panchadcharam
Veterinary Research Institute
Ipoh, Perak
Malaysia

Wendy Pearson
CanTox Health Sciences International
Mississauga, Canada

Rômulo Romeu da Nóbrega Alves
Departamento de Biologia
Universidade Estadual da Paraíba
Campina Grande
Paraíba, Brazil

Jurgen Seier
Primate Unit/SAMRC
Cape Town, Republic of South Africa

Debbie Shaw
Traditional and Herbal Medicine
Medical Toxicology Unit
Guy's and St Thomas' Hospital Trust
London, UK

Shanmugavelu Sithambaram
Strategic Livestock Research Centre
MARDI
Kuala Lumpur, Malaysia

Hamidou Hamadou Tamboura
INERA-CNRST
Ouagadougou, Burkina Faso

Lida Teng
Centre for Pharmacognosy and Phytotherapy
School of Pharmacy
University of London
London, UK

Kaizer Thembo
School of Pharmacy
University of Limpopo
Turfloop, Republic of South Africa

Peter Tshisikhawe
Department of Biological Sciences
University of Venda for Science and Technology
Thohoyandou, Republic of South Africa

P. M. Unnikrishnan
Foundation for Revitalisation of Local Health Traditions (FRLHT)
Bangalore, India

Tedjc van Asseldonk
Institute for Ethnobotany and Zoopharmacognosy (IEZ)
DG Beek-Ubbergen, The Netherlands

1 Methods for Evaluating Efficacy of Ethnoveterinary Medicinal Plants

Lyndy J. McGaw and Jacobus N. Eloff

CONTENTS

1.1 INTRODUCTION

In many developing countries with limited access to orthodox health care services, the majority of rural people rely on traditional medicines to alleviate a variety of ailments. Likewise, many pastoralists use customary remedies to treat their sick animals. Commercial pharmaceutical drugs are sometimes available in remote rural areas, but they are often dispensed by untrained vendors or repackaged without printed instructions, which may not be able to be read by illiterate users in any case, leading to drug misuse (Mathias, 2007). In an effort to improve animal health care services in rural areas, it is vital to utilize all available resources, including ethnoveterinary medicine (EVM).

The growing interest in and increasing recognition of the role of EVM has been limited in terms of further development by unavailability of information on the efficacy and safety of these practices. Several research papers in international publications have confirmed a heightened recognition of the need for rigorous scientific investigation of EVM remedies. Studies of biological activity of plants used against veterinary diseases can provide indications of promising leads for extracts that can be developed and used on a commercial basis. Plants with activity may also provide leads for isolation and identification of useful compounds that may be chemically modified to optimize medicinal value and reduce possible toxic effects, in other words, developed into pharmaceuticals.

For common conditions such as coughs, wounds, skin diseases, mild diarrhea, and reproductive disorders, EVM can be a cheap and easily obtainable alternative to expensive orthodox drugs. For epidemic infectious diseases, including anthrax, rinderpest, rabies, and foot-and-mouth disease, modern drugs (mostly vaccines) are preferred. Many drugs in conventional therapy are based on chemical compounds of plant origin or on synthetic derivatives of these chemicals. The search for alternative antibacterials and anthelmintics in particular is intensifying following problems associated with drug resistance and chemical residues in production animals. EVM practices, if proven to be effective and not harmful, may provide answers (not only regarding plant-based remedies but also concerning management customs) to some of these problems currently encountered in conventional veterinary practice.

Much research has been undertaken, especially as reflected by the expanding scientific literature, concerning the ethnopharmacological investigation of plants used to treat humans for various illnesses. This has been followed by an interest in plants used in animal health. It is logical that in evaluating these plants for biological efficacy and safety that similar bioassays are used in a laboratory situation where plants are being investigated for similar types of activity. It is essential to build and maintain standards for assessing the therapeutic potential of plants to facilitate comparison of results between different research groups (Cowan, 1999). In this chapter, methods for evaluating traditional ethnoveterinary plant-based remedies used for treating common diseases are discussed. Some results obtained in these assays with particular reference to the South African context are supplied. Considerations to be taken into account when embarking on research in this field are also presented and discussed.

1.1.1 The Need for Evaluating Traditional Animal Treatments

There is a pressing case for investigating EVM remedies for two major reasons: investigation of their use as effective agents for treatment of particular ailments and to appraise their safety. Undoubtedly, there are dubious, questionable, or even harmful practices forming part of alternative treatment of livestock, and these need to be distinguished from those of value. The profile of EVM would benefit from validation of certain practices in the view of potentially skeptical, conventionally trained veterinarians who may have only been exposed to the results of injurious handling by rural farmers attempting to rid their animals of disease or parasite infestations.

Other advantages to be gleaned from ethnoveterinary research entail prospects for commercial development, whether as refined extracts or lead compounds for the pharmaceutical industry. An important aspect, which falls outside the scope of this chapter, is the necessity for benefit-sharing agreements with owners of the intellectual property concerning EVM (see Chapter 2). This may involve returning value-added knowledge to the community where the information originated if such a community can be identified. Alternatively, they could be included in a profit-sharing arrangement from the outset, without raising initial false hopes of massive profits to be obtained from a product that may display exciting bioactivity in the laboratory, while many obstacles lie in the path to commercial success.

Various approaches have been proposed to validate animal herbal remedies, such as literature reviews, laboratory and clinical studies developed in medicine, social science methods (i.e., ranking of treatments in terms of efficacy by livestock owners), and investigation of the influence of a remedy on animal production and economic considerations (Mathias, 2004). Much work remains to be done in this area. In a review of the current status of published information on the ethnoveterinary use of plants in South Africa and biological activity and toxicity investigations on these plants (McGaw and Eloff, 2008), more than 200 plant species were documented as used in EVM, and of these, a mere 27 species have been tested for bioactivity in targeted ethnoveterinary assays. It was concluded that more plants need to be evaluated, and expanded investigation of those plants already tested in one or two screening systems needs to be carried out (McGaw and Eloff, 2008).

The targeted investigation of plants for commercial application in human medicine has resulted in natural products and their derivatives representing about 50% of all drugs in clinical use, of which higher plants contribute 25% to this figure (Farnsworth, 1984; O'Neill and Lewis, 1993). Plants have a seemingly limitless ability to manufacture unusual and original chemical structures. Also, a major mechanism plants possess to fight infection is the production of chemicals with anti-infective activity (phytoalexins); thus, it is logical and reasonable to explore the potential presence and ability of such compounds to be useful in both human and animal health care, for example, in terms of antibacterial or antifungal activity.

Ethnoveterinary remedies need to be validated before they can be widely promoted. Information on the botany and phytochemistry of particular plants may already be available in the literature, relating to bioactivity and toxicity. There is frequently an overlap between medicinal plants used to treat animals and those used to treat humans. It could be speculated that livestock keepers have over the centuries modified human remedies for use in animals or vice versa. It would make sense that similar treatments are used to treat comparable ailments in humans and their livestock.

Ethnoveterinary medicines can function as leads for drug development, but perhaps a more useful and cost-effective exercise would be to improve a selected preparation by pharmacological research and development, and the resulting remedy can be returned to the community with addition of value. Local farmers can grow the plants and make money from this venture, and such commercialization can also aid in the conservation of useful species. Other issues falling outside the ambit of this chapter include the necessity to conduct research into optimal dosing regimens and effective concentrations of herbal remedies. Potential side effects as well as toxins that may be ingested by the animals and then transferred to humans through milk or meat are further issues to be addressed.

1.2 BIOLOGICAL ACTIVITY SCREENING

It has been estimated that at least 250,000 species of plants inhabit our world (Borris, 1996). A mere 5–15% of higher plants have been systematically investigated for the presence of bioactive compounds, so plant biodiversity is virtually unexplored (Pieters and Vlietinck, 2005). Plants contain an enormous diversity of chemical structures, which are secondary metabolites modulating the relationship of organisms with the environment, for example, as pollinator attractants, signal products, and defensive substances against parasites, predators, and other pathogens (Pieters and Vlietinck, 2005). Such compounds in plants therefore hold much potential for medicinal applications. Following up ethnomedical leads, whether plants are used in human or in animal medicine, is one approach to selecting plants for bioactivity screening. Other methods of bioprospecting for screening studies include random selection and chemotaxonomic selection approaches.

Bioassays used to evaluate plant extracts should meet many criteria, including validity, lack of ambiguity, accuracy, reproducibility, simplicity, and reasonable cost (Pieters and Vlietinck, 2005). More particular considerations are a high selectivity (to limit the number of leads for subsequent evaluation), a high sensitivity (to detect low concentrations of active compounds), and a high specificity (to be insensitive to a variety of inactive compounds and eliminate false positives) (Pieters and Vlietinck, 2005). *In vitro* tests in ethnopharmacological studies are prevalent in the scientific literature even though *in vivo* models supply more accurate evidence for the activity of plant preparations in traditional medicine. This is largely as a result of the fact that the use of *in vivo* models is severely restricted in many countries owing to economic and ethical concerns (Houghton et al., 2007). The smaller amount of plant material needed for an *in vitro* test is often an important consideration.

Methods of extraction and *in vitro* testing need to be standardized so that the evaluation of medicinal plants can be systematic, and comparisons of results obtained by different laboratories may be more useful. In an important review article on the subject of anti-infective potential of natural products, Cos et al. (2006) emphasized that certain pivotal quality standards must be set at the stage of extract processing and primary evaluation in pharmacological screening models. The authors provided extremely useful recommendations to help define a more acceptable "proof of concept" for antibacterial, antifungal, and antiparasitic activity in natural products. They outlined the following requirements for anti-infective screening:

1. Use of reference strains or fully characterized clinical isolates (in the case of microorganisms, the American Type Culture Collection [ATCC] strains are widely used as standards).
2. *In vitro* models on the whole organism, if possible cell based.
3. Selectivity evaluation by parallel cytotoxicity testing or integrated profiling against unrelated microorganisms.

4. Adequate dose range, enabling dose-response curves.
5. Stringent end-point criteria with IC_{50} values generally below 100 μg/mL for extracts and below 25 μM for pure compounds.
6. Proper preparation, storage, and in-test processing of extracts.
7. Inclusion of appropriate controls in each *in vitro* test replicate (negative, positive, and growth controls).
8. Follow-up of *in vitro* activity ("hit" status) in corresponding animal models ("lead" status).

A variety of test systems should be employed in the *in vitro* screens because the use of only one bioassay yields an incomplete picture of the effect of the extract on the whole system involved (Houghton et al., 2007). Many diseases involve more than one factor, and Houghton et al. (2007) cautioned that the use of a single *in vitro* test is generally too simplistic and reductionist to achieve an idea of biological activity. An assortment of chemicals is present in an extract, and these may each have a different biological or pharmacological activity, together contributing to the overall clinical effect (Houghton et al., 2007). It is therefore preferable to use a range of tests for different activities, all related to the particular disease state under investigation.

Pharmacological evaluation of medicinal plants followed by bioassay-guided fractionation can lead to the isolation of pure active compounds with potential for commercialization. An alternative to this route, particularly pertinent to developing countries, is the preparation of standardized, formulated extracts that could contribute to an innovative and successful local pharmaceutical industry (Pieters and Vlietinck, 2005).

A growing number of publications document the use of herbal remedies by small-scale farmers to treat an assortment of livestock diseases, from skin conditions to babesiosis and anaplasmosis (Masika, Sonandi, and Van Averbeke, 1997; Masika, Van Averbeke, and Sonandi, 2000; Dold and Cocks, 2001; Van der Merwe, Swan, and Botha, 2001; Njoroge and Bussmann, 2006; Luseba and Van der Merwe, 2006). Methods of obtaining information range from participatory methods to semistructured interviews, field observations, and questionnaire surveys. This expanding documentation of plants used in EVM is anticipated to precede increased investigation of these plants for bioactivity, reflecting the situation in human ethnomedicine. It is generally accepted that a great deal of work remains to be done on recording the uses of plants in EVM.

Factors to be taken into consideration by researchers aiming to evaluate the biological activity of herbal EVM remedies include whether the plants are used singly or in combination with other plants, the plant part used, method of preparation, dosage, and the way in which the remedy is applied. For example, herbal remedies can be prepared from fresh or dry material in the form of infusions, decoctions, pastes, or expressed juices from fresh plants. Masika, Van Averbeke, and Sonandi (2000) stated that the route and method of application of a remedy depends on the perceived cause of the disease condition. Topical applications are commonly used for skin conditions, powders are rubbed into incisions, drops are placed in the ears and eyes, and drenches are popular in treating systemic conditions. In other studies in South Africa, it was noted that plants are generally not processed or mixed with other materials and are used as single-plant decoctions or infusions for dosing animals or crushed and used topically for wound treatment (Van der Merwe, Swan, and Botha, 2001; Luseba and Van der Merwe, 2006). This is in contrast with traditional medicine intended for human use, for which processing (milling, extracting, etc.) and mixing of two or more plant species (and even animal parts) appears to be common practice.

1.2.1 Limitations of Laboratory Testing of EVM Remedies

As EVM is a complex system of practices involving more than just the application of plant-based remedies to sick animals, it may give rise to misleading expectations about the degree of efficacy of a single plant used as part of a cure. Diagnosis of a disease made by a rural farmer may be inadequate as it is easier to identify symptoms than the actual cause of the disease. Dosages are not precise, contributing to the perception that the remedies are not standardized. The methodology for validating EVM should be scientifically acceptable but also must take into account the probability that EVM remedies might not work as powerfully as orthodox medicines. They may not completely eliminate all microorganisms causing a particular disease, but this could allow the body's immune system to build up immunity against the remaining organisms. It must be kept in mind that ethnoveterinary practices comprise a complex system and isolating one aspect for efficacy studies may not yield the anticipated results, although the system might be adequate for the conditions in the field. Fundamentally, if a plant-based ethnoveterinary remedy is to be deemed suitable for further development, efficacy and toxicity tests must meet certain standards. Legal constraints must be kept in mind if commercial development is anticipated. However, this standard is not so rigorous if the validation and understanding of EVM is the aim of the study. Research can be used to select a particular remedy, improve it through pharmacological and toxicological research, and then return it as a value-added product to the community.

In EVM studies, traditional methods of dosage and preparation of remedies are vital components often neglected for the sake of ease of preparation and standardization of laboratory extracts for testing. For a more true reflection of efficacy of the treatment, it would be advisable to closely follow if possible the preparation and application technique employed by the indigenous user. On the other hand, there is a case to be made for using standard scientific methods to prepare aqueous and organic solvent extracts if extraction of a wide range of chemicals present in the plant is desired for *in vitro* studies.

Traditional remedies may be composed of single plants or mixtures of plants, and if mixtures are employed it is recommended to test not only the individual component plants, but also the mixture in the correct proportions used traditionally. Repeatability of activity is important, and together with well-designed screening programs to elucidate activity using complementary techniques, plant material sourced from different areas could be included in a screening program. The accurate identification of plant material is naturally essential, and numbered voucher specimens should be deposited at a reputable herbarium to allow other researchers to verify the identity of the plants as well as to make allowances for possible future taxonomic revisions.

A further limitation in the laboratory testing of EVM remedies is the difficulty in culturing parasitic nematodes and protozoa, for example. Parasitic infestations of intestinal worms are commonly diagnosed and treated by livestock keepers. Some model systems are available, such as the free-living nematode *Caenorhabditis elegans*, for detecting anthelmintic activity, but sometimes the only true indication of efficacy is activity in an animal model.

It is often not possible to assume efficacy *in vivo* after achieving good results with *in vitro* tests. First, it may not be practical to extrapolate the dose from that which is active *in vitro* to that which would be required to reach adequate plasma concentrations in the target species (Houghton et al., 2007). Bioavailability is an essential consideration if the remedy is orally administered. Second, factors such as absorption and metabolism may be responsible for discrepancies between *in vitro* and *in vivo* tests (Houghton et al., 2007). Even if rat models, for example, are used, differences have been noted between intraperitoneal and oral administration. This can be explained inter alia by metabolic breakdown of compounds in the gastrointestinal tract or by lack of absorption from the gut into the bloodstream (Laupattarakasem et al., 2003). Absorption and metabolism can be affected by other compounds in the extract that may enhance or inhibit absorption, and other compounds may upregulate metabolic enzymes in the liver. Houghton et al. (2007) noted that traditional methods of preparation of plant-based medicines might remove or concentrate such compounds, and if the correct method is not followed in making extracts for pharmacological testing, then the extract may display different activities compared to the extract prepared according to traditional methods.

Most activity investigations in the published scientific literature concentrate on *in vitro* studies for practical, economic, and ethical reasons. Therefore, there exists an unavoidable bias toward *in vitro* tests for evaluation of EVM remedies as this is what is reflected in the available literature. Even in the case of *in vivo* studies, tests of ethnoveterinary remedies in a laboratory where the animals are given fixed diets and kept under controlled conditions, accurate indications of efficacy of the treatment may not be discovered, while in the real-life situation, conditions under which the animals are kept are more varied.

1.2.2 Extract Preparation

The variety of methods by which EVMs are prepared and administered to animals complicates the task of evaluating activity of a particular remedy. The reason for which the screening is being undertaken must be kept in mind when designing the study, including preparation of the extract for pharmacological testing. If the purpose is solely to validate the use of a certain remedy, then it is advisable to closely follow the traditional method of preparation as many factors may influence the activity of the resultant mixture. In addition, the route of administration must somehow be taken into account. In some cases, it is thus possible that *in vivo* tests are the only mechanism by which the efficacy of an EVM medicine can be verified. Alternatively, careful design of *in vitro* screening systems may yield a reasonable idea of the efficacy and nontoxicity of remedies. The selection of a screening system will for the most part depend on the nature of the disease being investigated and the availability of validated laboratory models to identify the relevant biological activity.

For broader screening programs aimed at discovering biological activity in particular plants used in EVM, standardized methods are widely used. Care must be taken to ensure that potentially active compounds are not lost during processing; for example, some constituents may be thermolabile or photosensitive. Plant extracts may be prepared using fresh material or, more commonly, dried powdered material. The plant material can be extracted using water or organic solvents that vary in polarity. For extraction of hydrophilic compounds, polar solvents such as methanol, ethanol, and ethyl acetate can be used, while if lipophilic compounds are being targeted, more nonpolar solvents such as dichloromethane and hexane may be used. Eloff (1998b) examined a spectrum of solvents for their ability to extract antimicrobial compounds from plant material and other factors, including their hazardous nature and ease of removal from the extract. The aim of the study was to identify a more standardized extraction method, and acetone was highlighted as the solvent with the best rating, followed in order by dichloromethane, methanol, ethanol, and water. However, this may vary with the plant species or plant part under investigation (see also Chapter 4).

Following extraction, appropriate handling of the extracts is important to avoid decomposition of active compounds or other changes that may affect biological activity. It is common practice to resuspend dried extract residues in the extracting solvent to a known concentration prior to screening, provided the extract redissolves adequately in the solvent and the solvent is not toxic in the testing system. In quantifying the extract, some researchers dry down only a small aliquot of the extract to determine the original concentration and then use the remaining intact extract for testing. Dimethyl sulfoxide (DMSO) is a popular solvent in which to prepare test compound solutions at a stock concentration (Cos et al., 2006). Test stock solutions in 100% DMSO have the advantages of elimination of microbial contamination, thus obviating the need for filter sterilization, which may lead to loss of compounds, and good compatibility with many test systems as a result of good solubility when diluted to a working concentration in aqueous medium. As DMSO is potentially toxic for cells in tissue culture or microorganisms, the final testing concentration of the solvent should not exceed 1% (Cos et al., 2006). Acetone was put forward as the solvent of choice for use in antibacterial testing systems as, at the concentrations used in a serial broth microdilution assay, it was found to be nontoxic to various species of bacteria tested (Eloff, 1998a). This was also held to be the case for antifungal assays based on a similar method (Masoko, Picard, and Eloff, 2005). Regarding storage issues, Cos et al. (2006) recommended that compounds and extracts should generally be stored without solvent for long-term storage or in 100% DMSO at −20°C with minimal exposure to freeze-thaw cycles or humidity. This is meant to reduce degradation of components. Storage in methylated solvents is not advised because of the possible formation of artifacts.

In the following sections of this chapter, examples of EVM plants that have been screened for biological activity, the methods used to screen them, and indications of activity discovered in the plants are given. Particular emphasis is placed on treatments for those diseases of importance in livestock. Techniques available for bioassaying the plant preparations are described briefly and references given for more detailed information. It is commonly found that there is an overlap between veterinary and human medicine in many communities, but the emphasis here is on the former.

1.2.3 Antibacterial and Antifungal

Different classes of antibacterial assays have been described, and many of these are applicable to antifungal detection as well. Antibacterial assays may be broadly divided into agar diffusion, dilution, and bioautography methods. In agar diffusion, a reservoir containing a known concentration of the test substance is brought into contact with an inoculated medium, and the diameter of the inhibition zone (clear zone) around the reservoir is measured after incubation. Before incubating, the compounds from the reservoir are commonly allowed to diffuse into the agar medium at a lower temperature for a few hours before inoculation with the test bacteria (Cos et al., 2006). The types of reservoirs used can be filter paper disks placed on top of the agar surface or wells punched into the agar, for example. Advantages of the system include small sample requirements and the ability to test up to six extracts per plate against one microorganism (Hadacek and Greger, 2000). However, a major disadvantage is that this method is not suitable for testing nonpolar samples or samples that are unable to diffuse readily through the agar matrix (Cos et al., 2006).

With dilution methods, the test sample is mixed with a medium (liquid broth or solid agar) inoculated with the test microorganism. Growth of the microorganism after incubation can then be monitored in various ways. In agar dilution methods, the minimum inhibitory concentration (MIC) is the lowest concentration of test compound able to inhibit visible microbial growth. In broth dilution methods, turbidity (measured visually or spectrophotometrically) and redox indicators (commonly a tetrazolium salt, e.g., *p*-iodonitrotetrazolium violet [INT], 3-4,5-dimethylthiazol-2-yl-2,5-diphenyl tetrazolium bromide [MTT], or resazurin) are usually used to detect microbial growth. The presence of "cidal" or "static" effects of a certain concentration of compound or extract can be determined using broth dilution methods (Cos et al., 2006). Minimal bactericidal or fungicidal concentrations (MBC or MFC, respectively) can be detected by plating out samples at inhibitory concentrations onto agar and assessing growth (static) or no growth (cidal) after incubation. Dilution methods are useful in testing both polar and nonpolar extracts or compounds. The microdilution assay, using various growth indicators, including tetrazolium salts, has been successfully used with fast-growing species of mycobacteria, including *Mycobacterium smegmatis*, *M. aurum*, and *M. fortuitum*, and with slow-growing species such as *M. bovis* and *M. avium* (Chung et al., 1995; Franzblau et al., 1998; McGaw et al., 2008).

For quantifying antibacterial activity, Eloff (2000) proposed that the quantity of material extracted from 1 g of dried plant material be divided by the MIC value to give the total activity of the plant. This measure, in milliliters per gram, indicates the largest volume to which 1 g of the extract (containing active compounds) can be diluted and still inhibit growth of the bacterial (or fungal) species under investigation and thus the potency of the extract.

Bioautography is a valuable technique that localizes antibacterial or antifungal activity on a thin-layer chromatographic (TLC) plate. Components of an extract are developed on a TLC plate using an appropriate mobile phase (i.e., one that separates the compounds adequately but is also relatively volatile so it evaporates rapidly from the plate). A balance needs to be struck between allowing sufficient time prior to bioautography to pass for the eluting solvent to evaporate completely from the TLC plate, but not too much time for the exposed compounds separated on the TLC plate to decompose as a result of exposure to light and oxygen. In agar overlay bioautography (Hamburger and Cordell, 1987; Rahalison et al., 1991), agar medium mixed with bacterial or fungal culture before it solidifies is poured onto the TLC plate and incubated. In a popular method that avoids the difficulties associated with compounds not being able to diffuse into the agar medium from the TLC plate, a suspension of bacteria or fungi in liquid medium is sprayed onto the developed TLC plate. This is termed direct bioautography (Begue and Kline, 1972). After the plate is sprayed with a suspension of a tetrazolium salt such as INT the presence of clear zones of inhibition are visualized against a purple background to indicate microbial growth. Bioautography facilitates bioassay-guided fractionation for the isolation of antibacterial or antifungal compounds, but its use is restricted to those microorganisms that are able to grow rapidly on a TLC plate with the limited amount of nutrients available for growth in the medium that adheres to the surface of the TLC plate. In this regard, using the technique for filamentous fungi is inappropriate.

Selection of test bacterial species to use in a screening procedure is dependent on the purpose of the study. For routine antibacterial screening, the National Committee for Clinical Laboratory Standards (NCCLS, 1990) (Villanova, Pennsylvania, USA) recommended the Gram-positive *Enterococcus faecalis* (ATCC 29212) and *Staphylococcus aureus* (ATCC 29213) and the Gram-negative *Escherichia coli* (ATCC 25922) and *Pseudomonas aeruginosa* (ATCC 27853). For antifungal screening projects, representatives from the yeasts (e.g., *Candida albicans*), dermatophytic fungi (e.g., *Trichophyton mentagrophytes* and *Epidermophyton floccosum*), and opportunistic filamentous fungi (*Aspergillus niger* and *Fusarium solani*) could be included (Cos et al., 2006). The NCCLS (1990) recommended an inoculum of approximately 10^5 cfu/mL for bacteria, while for yeasts and fungi an inoculum of between 10^3 to 10^4 colony-forming units (CFU)/mL is sufficient for dilution methods (Hadacek and Greger, 2000). An inoculum size that is too low may give false positive results, while a too large inoculum could increase false negatives (Cos et al., 2006).

It has been estimated that about 75% of rural livestock owners in the Eastern Cape province of South Africa use plant-based treatments to treat their livestock (Masika and Afolayan, 2002). When screened against a panel of 10 bacteria and 5 fungi, extracts of *Combretum caffrum*, *Salix capensis,* and *Schotia latifolia* showed good activity against all the Gram-positive bacteria and some antifungal activity (Masika and Afolayan, 2002). Most of the extracts were not active against the Gram-negative bacterial species; interestingly, some water extracts actually promoted fungal growth (Masika and Afolayan, 2002). This may have been due to nutritive sugars, which partition into the aqueous fraction. The organisms used in this study were selected from those generally associated with infections or disease in humans and animals. Different concentrations of each plant extract were mixed with liquid agar at approximately 60°C before being poured into Petri dishes. Solvent was allowed to evaporate overnight from the plates, and bacteria or fungi were inoculated onto the plates and inhibition of growth observed. It was concluded that the inhibition of growth of Gram-positive bacteria, the Gram-negative *Enterobacter cloacae,* and several fungal species by water extracts of the plants indicated possible broad-spectrum antimicrobial effects of the plants, validating to a degree the traditional use of these plants (Masika and Afolayan, 2002).

Ethnoveterinary plants used to treat infectious diseases in cattle were screened in a broth microdilution assay for antibacterial activity (McGaw, Van der Merwe, and Eloff, 2007) against the organisms recommended for antibacterial testing by the NCCLS (1990). Hexane, methanol, and water extracts were found to be most active against the Gram-positive *E. faecalis* and *S. aureus.* Gram-positive species are known to be more susceptible to antimicrobials than are Gram-negative bacteria owing to differences in the bacterial cell wall composition (Vlietinck et al., 1995). It was reported by McGaw, Van der Merwe, and Eloff (2007) that a third of plant extracts tested had MIC values less than 1 mg/mL, and it was largely methanol extracts that displayed activity.

Ziziphus mucronata (Rhamnaceae) demonstrated excellent antibacterial activity in the preliminary assay (McGaw, Van der Merwe, and Eloff, 2007), and the antibacterial compounds 2,3-dihydroxyl-up-20-en-28-oic acid and zizyberanalic acid were subsequently isolated from the leaves (Moloto, 2004). The first compound was very active against *Staphylococcus aureus*, supporting claims of the efficacy of leaf pastes of *Z. mucronata* for the treatment of bacterial infections in animals and humans.

Bizimenyera et al. (2005) identified substantial antibacterial activity against *S. aureus* and *Pseudomonas aeruginosa* in *Peltophorum africanum* (Fabaceae), also using the broth microdilution method. The root and bark extracts are used by farmers to treat stomach ailments such as diarrhea and dysentery in cattle (Bizimenyera et al., 2005), and the antibacterial activity discovered in extracts of the plant may warrant its use against bacterial infections.

Rhizomes and roots of the popular ethnoveterinary plant *Gunnera perpensa* (Gunneraceae) are used to treat endometritis and retained placenta in cattle and women (Hutchings et al., 1996; Van Wyk, Van Oudtshoorn, and Gericke, 1997), and the possibility that antibacterial effects could be responsible for its activity has been investigated (McGaw et al., 2005; Drewes et al., 2005). *Gunnera perpensa* rhizome extracts showed only slight activity against several Gram-negative and Gram-positive bacterial species (McGaw, Jäger, and van Staden, 2000; McGaw et al., 2005), indicating that antibacterial efficacy probably played a mere supporting role to the known uterotonic activity (Kaido et al., 1997) in the reputed medicinal value of the rhizome. Drewes et al. (2005) isolated 1,4-benzoquinones from the stem and leaves of *G. perpensa.* One of the benzoquinones had significant antimicrobial activity, with MIC = 9.8 μg/mL against *Staphylococcus epidermidis* (Drewes et al., 2005).

Noteworthy antifungal activity in several *Terminalia* species (Combretaceae) was reported by Masoko, Picard, and Eloff (2005) against various morphological forms of fungi, including yeasts (*Candida albicans* and *Cryptococcus neoformans*), molds (*Aspergillus fumigatus*), and thermally dimorphic fungi (*Sporothrix schenckii*). These fungal species were carefully selected to represent a spectrum of clinical isolates of the most common and important disease-causing fungi in animals. From extracts of *Terminalia* leaves prepared using several organic solvents, the acetone extracts were most active.

1.2.4 Antiviral

In vitro antiviral assay methods are often based on the abilities of viruses to replicate in cell cultures. Certain viruses cause cytopathic effects (CPEs) or form plaques in lawns of cells, facilitating detection of antiviral effects of a substance. Inhibition of viral replication can also be discovered by monitoring the presence of viral products, such as viral RNA, DNA, or polypeptides. Virucidal substances inactivate the ability of a virus to be infective extracellularly and find application as broad-spectrum biocides. Antiviral agents are more interesting as candidates for clinical use because they may interfere with some aspect of viral biosynthesis (Cos et al., 2006). Vlietinck and Vanden Berghe (1991) supplied a useful outline of cell-based assays that can be used for antiviral or virucidal evaluation of pure compounds or plant extracts.

Toxicity to the host cell system must be assessed as part of the antiviral investigation. The Selectivity Index (SI) is the ratio of the maximum drug concentration causing 50% (or 90%) growth inhibition of normal cells (CC_{50} or CC_{90}, respectively) and the minimum drug concentration at which 50% (or 90%) of the virus is inhibited (IC_{50} or IC_{90}, respectively). It is essential to gain an indication of cytotoxicity of the test substance as, without this, results do not distinguish between antiviral effect and effect against the host cell system.

The choice of viruses to use in a screening panel should include representatives of DNA viruses and RNA viruses and could include criteria such as their ability to replicate in the same cell culture. In the Phytomedicine Program at the University of Pretoria, we have begun investigating antiviral activity of ethnoveterinary plants against feline herpesvirus type 1 (FHV-1) as an enveloped virus relatively sensitive to environmental influences. Plants with good activity in this preliminary screen are then assayed for activity against more resistant viruses, such as the lumpy skin disease virus.

No reports could be found of EVM plants being tested for antiviral activity, although many publications reported on efficacy of ethnobotanically chosen plants against a number of different viruses (e.g., Kudi and Myint, 1999; Asres et al., 2001; Lamien et al., 2005). To screen for antiviral activity, variations on virucidal assays are available in the ethnopharmacological literature, and these mainly focus on inhibition of viral CPE or plaque inhibition. In the virucidal assays, monolayers of the appropriate host cell type are cultured in 96-well microtiter plates. In the cytotoxicity aspect of the assay, serial dilutions of plant extract are exposed to the cells and incubated for a defined period at 37°C in a 5% CO_2 incubator. Following this, the cells are examined using an inverted microscope for signs of damage. Alternatively, a tetrazolium salt or other color indicator of cell viability may be used to detect cytotoxicity compared to untreated cells. In the antiviral test, serial plant extract dilutions are prepared as for the cytotoxicity assay, but virus is added to the cells. In their study, Kudi and Myint (1999) applied tissue culture medium infective dose ($TCID_{50}$) of 10^5 viral particles (100 μL) to each well. The cultures were incubated for an hour to allow adsorption of viral particles, after which 100 μL per well of plant extract dilutions were added to the wells. The plates were incubated for a certain period to allow development of CPEs, if any, and results compared to the controls consisting of only cells and cells with virus only. A range of different viruses was used in this method.

In an example of a plaque inhibition assay, Zhang et al. (2007) infected monolayers of host (Vero) cells grown in six-well culture plates with 100–200 plaque-forming units (PFU) of herpes simplex virus (HSV). After incubating the plates for 1 h to allow adsorption of the virus, the inoculum was aspirated from the cells, and the cultures were overlaid with 0.8% methylcellulose in culture medium containing dilutions of the test plant extract. After 3 days of incubation at 37°C, the plates were fixed with formalin, stained with crystal violet, air dried, and the number of plaques counted. Control plates consisted of those without plant extract, and the percentage of plaque formation inhibition was calculated as follows:

[(Mean number of plaques in control – Mean number of plaques in test)/(Mean number of plaques in control)] × 100

1.2.5 Antiprotozoal and Antirickettsial

Babesiosis is a protozoan, tick-borne disease affecting many vertebrate hosts. The rhizome extract, prepared using acetone, of the popular ethnoveterinary medicinal plant *Elephantorrhiza elephantina* was shown to have *in vitro* antibabesial activity (Naidoo et al., 2005). In this test system, *Babesia caballi* cultures (isolated from a horse) were incubated in 24-well culture plates with plant extracts at varying concentrations, together with uninfected blood. Parasite growth inhibition was monitored initially by a change in the color of the culture medium, where inhibited cultures remained bright red while unaffected protozoal cultures turned a dark coffee color. Culture smears were then evaluated using light microscopy to determine the percentage of infected cells. The registered antibabesial drugs diminazene aceturate (Berenil) and imidocarb diproprionate (Forray-65) were included as positive controls. Acetone extracts of *Urginea sanguinea*, *Rhoicissus tridentata,* and *Aloe marlothii* were not active in this assay (Naidoo et al., 2005).

Another important protozoal disease occurring in domestic livestock and chickens is coccidiosis, which results from infection with *Eimeria* or *Isospora* species. Coccidiosis causes losses worth millions of U.S. dollars annually in the poultry industry, resulting from animal mortality or poor productivity as well as costs of treatment (Williams, 1999). The use of plants to combat coccidiosis is an emerging field of investigation as these remedies may function by mechanisms different from those of conventional therapeutic agents. In one such study, four plant extracts with reported antioxidant activity were screened for their anticoccidial activity against an artificially induced mixed *Eimeria* infection in poultry (Naidoo et al., 2008). Orally administered *Combretum woodii* was toxic to the birds at a concentration of 160 mg/kg, while *Tulbaghia violacea* (35 mg/kg), *Vitis vinifera* (75 mg/kg), and *Artemisia afra* (150 mg/kg) produced feed conversion ratios similar to toltrazuril, the positive control, and higher than the untreated control. *Tulbaghia violacea* significantly decreased the oocyst production in the birds, and it was concluded that antioxidant-rich plant extracts have potential benefits in treating and possibly preventing coccidial infections (Naidoo et al., 2008). The results for extracts of *T. violacea* in particular provide momentum for more detailed investigation of the plant as a potential therapeutic or prophylactic anticoccidial agent.

The antirickettsial activity of *Elephantorrhiza elephantina* and *Aloe marlothii* was evaluated using an *in vitro Ehrlichia ruminantium* culture system (Naidoo, Chikoto, et al., 2006). Acetone extracts of the leaves were incubated with *E. ruminantium* cultures, and their activity was compared to that of oxytetracycline and untreated controls. *Elephantorrhiza elephantina* and *A. marlothii* demonstrated EC_{50} values of 111.4 and 64.5 μg/mL and EC_{90} values of 228.9 and 129.9 μg/mL, respectively, indicating good anti-ehrlichial activity. The EC_{50} and EC_{90} values for oxytetracycline were 0.29 and 0.08 μg/mL, respectively. Naidoo, Chikoto, et al. (2006) surmised that the plant extracts produced their inhibitory activity by a similar mechanism, unrelated to that of the tetracyclines.

1.2.6 Anthelmintic

Helminth parasites of livestock are common in rural areas, and anthelmintic remedies form a major component of EVM, as is the case with human traditional medicine. Laboratory research on anthelmintic activity of plant extracts is constrained by the expense, ethical issues and time associated with performing *in vivo* trials, and the difficulties experienced in maintaining parasitic nematodes in culture systems *in vitro*. A free-living nematode, *Caenorhabditis elegans*, has been used as a model organism in broad screening studies as it is easier and cheaper than using parasitic nematodes (Simpkin and Coles, 1981; Rasoanaivo and Ratsimamanga-Urverg, 1993). Notwithstanding the limitations encountered in extrapolating activity against a free-living nematode to activity against a parasitic species (Geary and Thompson, 2001), most commercially available broad-spectrum anthelmintics demonstrate activity against *C. elegans* (Simpkin and Coles, 1981).

In vitro screening investigations have revealed that many plant extracts show activity against the free-living *C. elegans* nematodes (McGaw, Jäger AK, and van Staden, 2000; McGaw and Eloff, 2005; McGaw, Van der Merwe, and Eloff, 2007). A rapid inhibition assay is easy and simple to perform and entails incubating varying concentrations of plant extracts with nematodes for a defined period of 2 h and scoring the percentage of paralyzed or dead nematodes in comparison to the untreated control (Rasoanaivo and Ratsimamanga-Urverg, 1993). Using this assay, several plant species belonging to the family Combretaceae exhibited interesting anthelmintic activity against *C. elegans* (McGaw et al., 2001). These studies may constitute a first step in validating the use of these plants in treating worm infestations in animals and in humans. In a more complicated screening system that evaluates the ability of the nematodes to grow and reproduce, plant extracts are incubated with nematodes in appropriate culture medium with bacterial and fungal growth inhibitors in 24-well assay plates for 7 days, after which the percentage of surviving nematodes is compared to that of the control wells (Simpkin and Coles, 1981; McGaw, Jäger AK, and van Staden, 2000).

Plant extracts have also been tested using *in vitro* assays with parasitic nematode eggs and larvae. Egg hatch and larval development inhibition against the two most important livestock nematode parasites *Haemonchus contortus* and *Trichostronglyus colubriformis* by various plant extracts have been reported. In these assays, the nematodes are maintained in monospecifically infected lambs, and eggs are collected from the feces. Test substances incubated with the freshly collected eggs may inhibit hatching in the aptly termed egg hatch assay (Coles et al., 1992). The larval development assay (Coles et al., 1988) detects the ability of plant extracts, or other compounds, to retard development of the eggs into infective larvae. The combination of the two assays can provide a practical indication of anthelmintic activity of plant extracts or pure compounds isolated from the extracts.

Peltophorum africanum is a popular plant for use in treating helminthosis, and the acetone extracts of the leaf, bark, and root have been screened for activity against *H. contortus* and *T. colubriformis* in the egg hatch and larval development assays (Bizimenyera, Githiori, Eloff, et al., 2006; Bizimenyera, Swan, et al., 2006). The extracts all showed activity in the assays at a concentration of 0.2 mg/mL, providing some support for the use of this plant in traditional medicine. Further confirmation of nontoxicity and efficacy is required, particularly *in vivo*. Various animal models have been used to detect anthelmintic effects of plant extracts (Kahiya, Mukaratirwa, and Thamsborg, 2003; Iqbal et al., 2006; Jabbar et al., 2007).

1.2.7 Antitick

Tick-borne diseases are a major source of concern for livestock farmers. Research has been undertaken on the repellent and toxic effects of plant extracts against ticks, with promising results thus far. Nchu (2004) analyzed the repellent effects of extracts of *Allium* species, as well as the direct toxicity, against adults of *Hyalomma marginatum rufipes*. Acetone extracts of *A. porrum* revealed a high repellency index (65–79.48%), and the dichloromethane extract of *A. sativum* was toxic to 100% of ticks within an hour of exposure. *Lippia javanica* and *Tagetes minuta* essential oils had a concentration-dependent effect on the ticks (Nchu, 2004), and *T. minuta* delayed molting to adult stage of 60% of engorged nymphs of *H. m. rufipes* in a growth inhibition bioassay. Thembo (2006) showed that *Senna italica* ssp. *arachoides* ethyl acetate extracts had a concentration-dependent acaricidal effect on *H. m. rufipes*. When *S. italica* ssp. a*rachoides* aqueous extracts were fed to guinea pigs and rabbits, the feeding performance of adult *H. m. rufipes* ticks appeared to be impaired (Thembo, 2006).

Plants are used in many African countries as antitick agents on livestock, and 28 of these plants from Ethiopia showed promising repellency activities against adult *Rhipicephalus pulchellus* ticks, with *Calpurnia aurea* displaying the highest toxicity toward the ticks (Zorloni, 2007). Some plants used in South Africa as traditional arthropocides were screened for antitick effects, with *Eucalyptus globoidea* and *Lavendula angustifolia* emerging as effective tick repellents (Mkolo, 2008).

1.2.8 Antioxidant

The current literature reveals a proliferation of recent articles describing antioxidant activity of plant extracts. It should be kept in mind from the outset that most flowering plants contain some antioxidants, and the activity of a plant against a certain disease generally cannot be explained exclusively in terms of its antioxidant activity (Houghton et al., 2007). Oxidative damage caused by free radicals, or reactive oxygen species, has been implicated in contributing to the progression of a number of diseases, such as cardiovascular disease, many cancers, and diabetes. Various *in vitro* screening systems for antioxidant activity have been described, including those based on chemical reactions, for example, the diphenyl–picrylhydrazyl (DPPH) free-radical scavenging test (Mensah et al., 2004), those involving biological models such as liposomes to mimic cell wall lipids (Dickson et al., 2006), and those involving cells challenged with prooxidants (Mensah et al., 2001).

The TEAC (trolox equivalent antioxidant capacity) assay described by Re et al. (1999) has a major advantage in that it is applicable to both aqueous and lipophilic systems. It is a decolorization assay that measures antioxidant activity in comparison to trolox, a water-soluble vitamin E analogue. This assay begins with generation of the radical monocation 2,2′-azinobis-(3-ethylbenzothiazoline-6-sulfonic acid), or $ABTS^{+}$, a blue-green compound produced by reacting ABTS with potassium sulfate. When the free radical is incubated with antioxidants, these compounds reduce the radical to colorless ABTS, and this reaction depends on the concentration of the antioxidant and the time during which the reaction is allowed to occur. The level of decolorization as percentage inhibition of the free radical is calculated relative to the reactivity of trolox under identical conditions (Re et al., 1999).

The widely used ethnoveterinary plant *Peltophorum africanum* is used to promote well-being and resistance to diseases in cattle, in addition to the previously mentioned use in treating stomach upsets (Bizimenyera et al., 2005). On screening the plant for antioxidant activity using the TEAC assay, it was found that root extracts possessed good antioxidant activity, particularly the ethanol extract of the root. The level of polyphenols in the roots was also high, probably contributing in large part to the overall antioxidant activity (Bizimenyera et al., 2005). In the study of Naidoo et al. (2008), plant extracts with high antioxidant activity were shown to be effective in treating coccidiosis infections in chickens.

After Naidoo et al. (2005) reported that extracts of *Rhoicissus tridentata,* a plant used for the treatment of babesiosis in cattle, were not effective in an *in vitro* antibabesial assay, it was proposed that the reputed efficacy of this plant may result from a reduction in antioxidant cellular injury (Naidoo, Zweygarth, and Swan, 2006). Antioxidant evaluation of acetone extracts of different plant parts revealed good activity in the DPPH assay and in the TEAC assay (value of 2.5). The activity was held to owe in part to the presence of catechin, epicatechin, gallic acid, and epigallocatechin-gallate in the tuber acetone extracts as demonstrated by high-performance liquid chromatographic (HPLC) analysis (Naidoo, Zweygarth, and Swan, 2006). It was concluded that *R. tridentata* might be effective in animals infected with babesiosis by limiting the degree of oxidative cellular injury (Naidoo, Zweygarth, and Swan, 2006).

1.2.9 Anti-inflammatory and Wound Healing

Many disease conditions are associated with excessive inflammation (e.g., arthritis, eczema, and asthma), so assays detecting inhibition of one of the many biochemical processes related to inflammation are commonly used as *in vitro* screens for anti-inflammatory activity (Houghton et al., 2007). These include tests for cyclooxygenase inhibition, lipoxygenase inhibition, inhibition of eicosanoid synthesis, or nuclear factor kappa B (NFκB) production.

Traditional EVM makes use of many preparations to treat wounds in livestock animals as such wounds may lead to more serious problems. Wound healing is a complex combination of processes, including inflammation, cell proliferation, collagen formation, and contraction of the collagen lattice (Houghton et al., 2007). Healing of the wound may be complicated by the presence of microbial infection and destruction of cells and tissues by reactive oxygen species. If a livestock owner uses a particular plant medicine to treat wounds, this preparation may affect one or more of the processes described. Therefore, a battery of *in vitro* tests should be conducted to verify the wound-healing activity of the extract, such as stimulation of fibroblast proliferation, antibacterial activity, and free-radical scavenging effects. Anti-inflammatory activity may provide short-term relief. The development of tests for wound healing, from *in vivo* tests to cell-based systems and chemical reactions and further to investigations into effects on secondary messengers and protein expression, has been described by Houghton et al. (2005).

In an efficacy study of South African plants used for wound healing and against retained placenta, Luseba et al. (2007) discovered that several dichloromethane extracts displayed antibacterial and anti-inflammatory activity. Extracts of *Cissus quadrangularis* stem and *Jatropha zeyheri* root showed selective inhibition against cyclooxygenase-2 in the anti-inflammatory experiments. Even though water is traditionally the most commonly used solvent to prepare medicinal extracts, the activity of organic extracts need not be disregarded. According to Luseba et al. (2007), in the treatment of livestock wounds the whole plant material is often locally applied, and in the case of complaints such as retained placenta, for which the treatment mixtures are given orally (and are unlikely to be filtered), active ingredients may be released.

EVM remedies may in some cases be applied topically to wounds or skin infections, and rodent models to evaluate *in vivo* effects of plant extracts and compounds isolated from them with promising *in vitro* activity have been reported (Kruger, 2004; Masoko, 2006). In these methods, a number of small topical wounds are created on the back of a shaved rat, and if investigating antibacterial or antifungal activity as well as wound-healing effects, a bacterial or fungal culture may be used to infect some of the lesions. Following this, preparations of extract or compound in aqueous cream are applied to the wound, and various parameters such as erythema, exudate, and wound diameter are monitored throughout the duration of the study. Signs of toxicity are noted, and gross pathology is performed on necropsy. Owing to the fact that approximately six wounds can be made on each rat, the rats serve as their own controls to reduce the number of rats used in the study. One wound acts as an untreated control, another is treated with a standard antibiotic, and the remaining wounds are treated with test preparations. To allow this reduction in the number of animals used, the assumption is made that systemic effects of the topically applied medication are nonexistent.

1.3 TOXICITY STUDIES

Toxicity investigations on EVM plant extracts are necessary, both to evaluate the potential toxic effects toward the animal being treated and to exclude false-positive activity results in antimicrobial assays arising from nonspecific toxic properties. In southern Africa, there is a rich floral diversity, and approximately 600 toxic species are known to occur in this region (Kellerman, Coetzer, and Naudé, 1992). Van der Merwe, Swan, and Botha (2001) commented that side effects and toxicity associated with the medicinal use of plants in EVM were rarely reported, although several plants used have potentially dangerous toxic effects (e.g., *Boophane disticha, Ricinus communis,* and *Solanum* species). The lack of toxic reports was ascribed to the relatively small quantities used in traditional medicines (van der Merwe, Swan, and Botha, 2001). Toxicity is affected in ruminants by the degradation or binding of toxins by the ruminal microflora or the digestive tract, as harmless precursors can be converted to toxic substances or less toxic substances can be changed to more toxic ones (Naudé, Coetzer, and Kellerman, 1992). It is therefore important to evaluate potential toxic effects of EVM remedies, particularly those preferred for oral dosing. A screening regimen that includes assays to test for genotoxicity, cytotoxicity, and *in vivo* toxicity (acute and chronic) is advised. This section is only touched on as other chapters in this volume deal with this aspect in more detail.

Several approaches have been followed to assess toxicity of natural remedies, including testing for genotoxic effects using *in vitro* bacterial and mammalian cell assays such as the Ames test, micronucleus test, and comet assay (Fennell et al., 2004). Luseba et al. (2007) tested dichloromethane and 90% methanol extracts of 12 South African plants used to treat retained placenta and wounds in livestock for mutagenicity and found that none of the extracts was mutagenic in the Ames test using *Salmonella typhimurium* strain TA98 without metabolic activation.

A quick and easy way to gain a preliminary indication of cytotoxicity is to submit extracts to the brine shrimp assay. This assay has been used to detect *in vitro* cytotoxic or pharmacological effects (Solís et al., 1993) as activity in this assay has been correlated with cytotoxicity in a number of cell lines, including 9KB, P388, L5178Y, and L1210 (Meyer et al., 1982; McLaughlin, 1991; De Rosa, De Giulio, and Iodice, 1994; McLaughlin, Rogers, and Anderson, 1998). The brine shrimp assay involves incubating test substances with freshly hatched brine shrimp larvae and detecting percentage mortality of the larvae. A shortcoming of this technique is that it does not account for metabolic activation of the test extracts or compounds, and it is difficult to extrapolate toxicity against a crustacean to mammalian cytotoxicity even though correlations have been noted with cytotoxicity in some cell lines. As an example, McGaw and Eloff (2005) reported that few extracts of plants known to be toxic to livestock were active in the brine shrimp assay. In a later study of plants used to treat cattle for various ailments (McGaw, Van der Merwe, and Eloff, 2007), the lowest IC_{50} value recorded was 0.6 mg mL^{-1}.

Cytotoxicity assays using cell lines are generally relatively easy to perform, although specialized cell culture facilities are required. A range of cells may be used, including continuous commercially available cell lines as well as primary cells derived from animals. Various indicators of cell viability following incubation with plant extracts may be used, including those to detect mitochondrial activity or cellular integrity.

1.4 CONCLUSION

Research on natural products from plants traditionally used to treat animals as well as humans is a key mechanism for identifying new chemical entities that may have interesting biological activity. Isolation of active compounds is not the only useful pathway in medicinal studies. Optimization of extract preparations by removing bulky inactive constituents while leaving behind compounds that may have a synergistic or additive beneficial effect may also lead to useful medications. This could be of particular value when considering primary health care systems in rural areas as well as assisting rural livestock owners in managing disease in their animals. Standardized and formulated plant extracts may be an initiative for developing countries to follow up in originating successful pharmaceutical industries that can compete with Western pharmaceutical companies for the treatment of various diseases, in both humans and animals (Pieters and Vlietinck, 2005). With the current controversy over antibiotic use as growth promoters in production animals, plants with positive effects on the growth and well-being of animals may provide alternatives to be investigated in this lucrative market.

It should be kept in mind that EVM involves a complex system that merges treatment of diseases with herbal and other remedies with management practices such as disease prevention. To meet the aims of validating the uses of plants in EVM and developing and providing more cost-effective veterinary remedies, other factors such as socioeconomic assessments must take place in conjunction with pharmacological evaluations of efficacy and toxicity. Economic considerations must be taken into account; for example, a particular natural remedy may be less effective than a commercial treatment, but a cost-benefit analysis may reveal that using EVM is economically more beneficial. Management and prevention issues are integral to successful animal health care management, and promotion of the use of pharmacologically proven and nontoxic plant medicines is part of this management system.

In this chapter, the emphasis has of necessity been placed on *in vitro* evaluation of biological activity as a result of the shortage of *in vivo* studies concerning EVM remedies. Ethnopharmacological researchers must assimilate the strengths and weaknesses of *in vitro* tests and take into account pharmacokinetic factors, traditional methods of preparation of medicines, the effect of other added substances or adulterants, and the dose showing activity. The use of accepted laboratory testing equivalents to animal studies is ethically and economically desirable. More than one test system related to the disease under investigation should be employed to evaluate biological activity, appropriate test organisms must be used, and activity should ideally be confirmed by *in vivo* tests if sufficiently justified. Standardization of techniques to improve interlaboratory comparisons of results with different plants analyzed for the same activity is a priority. These methods for evaluating efficacy need to be specific and rapid and not require a large quantity of material. While activity *in vitro* does not necessarily confirm the efficacy and safety of a plant extract, it may provide preliminary indications of the usefulness and potential toxicity of the plant.

Validation is essential to avoid perpetuation of the preconceived notion that traditional medicine is primitive and inherently inferior. Strides have been taken in evaluating the usefulness of human ethnomedicine, and similar action needs to be taken to objectively evaluate EVM to avoid negative stereotypes that obscure the potential benefits to be obtained from traditional ethnoveterinary practices as well as plant-based natural remedies. Investigating the biological activity of ethnoveterinary plants may provide valuable leads for further targeted research that could generate marketable products, whether potentized extracts, isolated compounds, or modified compounds of natural origin. Overall, ethnoveterinary knowledge is an important resource that stands to benefit not only those pastoralists currently making use of traditional remedies to treat their animals, but also researchers seeking innovative and effective treatments for animal and human disease worldwide.

ACKNOWLEDGMENTS

The National Research Foundation (South Africa), University of Pretoria, and the Claude Leon Foundation are gratefully acknowledged for financial support.

REFERENCES

Asres K, Bucar F, Kartnig T, Witvrouw M, Pannecouque C, De Clercq E. 2001. Antiviral activity against human immunodeficiency virus type 1 (HIV-1) and type 2 (HIV-2) of ethnobotanically selected Ethiopian medicinal plants. *Phytotherapy Research* 15, 62–69.

Begue WJ, Kline RM. 1972. The use of tetrazolium salts in bioautographic procedures. *Journal of Chromatography* 64, 182–184.

Bizimenyera ES, Githiori JB, Eloff JN, Swan GE. 2006. In vitro activity of *Peltophorum africanum* Sond. (Fabaceae) extracts on the egg hatching and larval development of the parasitic nematode *Trichostrongylus colubriformis*. *Veterinary Parasitology* 142, 336–343.

Bizimenyera ES, Githiori JB, Swan GE, Eloff JN. 2006. In vitro ovicidal and larvicidal activity of the leaf, bark and root extracts of *Peltophorum africanum* Sond. (Fabaceae) on *Haemonchus contortus*. *Journal of Animal and Veterinary Advances* 5, 608–614.

Bizimenyera ES, Swan GE, Chikoto H, Eloff JN. 2005. Rationale for using *Peltophorum africanum* (Fabaceae) extracts in veterinary medicine. *Journal of the South African Veterinary Association* 76, 54–58.

Borris RP. 1996. Natural products research: perspectives from a major pharmaceutical company. *Journal of Ethnopharmacology* 51, 29–38.

Chung GAC, Aktar Z, Jackson S, Duncan K. 1995. High-throughput screen for detecting antimycobacterial agents. *Antimicrobial Agents and Chemotherapy* 39, 2235–2238.

Coles GC, Bauer C, Borgsteede FHM, Geerts S, Klei TR, Taylor MA, Waller PJ. 1992. World Association for the Advancement of Veterinary Parasitology (WAAVP) methods for detection of anthelmintic resistance in nematodes of veterinary importance. *Veterinary Parasitology* 44, 35–44.

Coles GC, Tritschler JP, Giordano DJ, Laste NJ, Schmidt AL. 1988. Larval development test for detection of anthelmintic resistant nematodes. *Research in Veterinary Science* 45, 50–53.

Cos P, Vlietinck AJ, Vanden Berghe D, Maes L. 2006. Anti-infective potential of natural products: how to develop a stronger in vitro "proof-of-concept." *Journal of Ethnopharmacology* 106, 290–302.

Cowan MM. 1999. Plant products as antimicrobial agents. *Clinical Microbiology Reviews* 12, 564–582.

De Rosa S, De Giulio A, Iodice C. 1994. Biological effects of prenylated hydroquinones: structure–activity relationship studies in antimicrobial, brine shrimp, and fish lethality assays. *Journal of Natural Products* 57, 1711–1716.

Dickson RA, Houghton PJ, Hylands PJ, Gibbons S. 2006. Antimicrobial, resistance-modifying effects, antioxidant and free radical scavenging activities of *Mezoneuron benthamanium* Baill., *Securinega virosa* Roxb. and Wlld. and *Microglossa pyrifolia* Lam. *Phytotherapy Research* 20, 41–45.

Dold AP, Cocks ML. 2001. Traditional veterinary medicine in the Alice district of the Eastern Cape Province, South Africa. *South African Journal of Science* 97, 375–379.

Drewes SE, Khan F, van Vuuren SF, Viljoen AM. 2005. Simple 1,4-benzoquinones with antibacterial activity from stems and leaves of *Gunnera perpensa*. *Phytochemistry* 66, 1812–1816.

Eloff JN. 1998a. A sensitive and quick microplate method to determine the minimal inhibitory concentration of plant extracts for bacteria. *Planta Medica* 64, 711–714.

Eloff JN. 1998b. Which extractant should be used for the screening and isolation of antimicrobial components from plants? *Journal of Ethnopharmacology* 60, 1–8.

Eloff JN. 2000. A proposal on expressing the antibacterial activity of plant extracts—a small first step in applying scientific knowledge to rural primary health care in South Africa. *South African Journal of Science* 96, 116–118.

Farnsworth NF. 1984. The role of medicinal plants in drug development. In: Krogsgaard-Larsen P, Brogger Christensen L, Kofold H. (Eds.), *Natural Products and Drug Development*. Munksgaard, Copenhagen, pp. 17–28.

Fennell CW, Lindsey KL, McGaw LJ, Sparg SG, Stafford GI, Elgorashi EE, Grace OM, van Staden J. 2004. Assessing African medicinal plants for efficacy and safety: pharmacological screening and toxicology. *Journal of Ethnopharmacology* 94, 205–217.

Franzblau SG, Witzig RS, McLaughlin JC, Torres P, Madico G, Hernandez A, Degnan MT, Cook MB, Quenzer VK, Ferguson RM, Gilman RH. 1998. Rapid, low-technology MIC determination with clinical *Mycobacterium tuberculosis* isolates by using the microplate Alamar Blue assay. *Journal of Clinical Microbiology* 36, 362–366.

Geary TG, Thompson DP. 2001. *Caenorhabditis elegans*: how good a model for veterinary parasites? *Veterinary Parasitology* 101, 371–386.

Hadacek F, Greger H. 2000. Testing of antifungal natural products: methodologies, comparability of results and assay choice. *Phytochemical Analysis* 11, 137–147.

Hamburger MO, Cordell GA. 1987. A direct bioautographic TLC assay for compounds possessing antibacterial activity. *Journal of Natural Products* 50, 19–22.

Houghton PJ, Howes M-J, Lee CC, Steventon G. 2007. Uses and abuses of in vitro tests in ethnopharmacology: visualizing an elephant. *Journal of Ethnopharmacology* 110, 391–400.

Houghton PJ, Hylands PJ, Mensah AY, Hensel A, Deters AM. 2005. In vitro tests and ethnopharmacological investigations: wound healing as an example. *Journal of Ethnopharmacology* 100, 100–107.

Hutchings A, Scott AH, Lewis G, Cunningham AB. 1996. *Zulu Medicinal Plants: An Inventory*. Pietermaritzburg, South Africa: University of Natal Press.

Iqbal Z, Lateef M, Akhtar MS, Ghayur MN, Gilani AH. 2006. In vivo anthelmintic activity of ginger against gastrointestinal nematodes of sheep. *Journal of Ethnopharmacology* 106, 285–287.

Jabbar A, Zaman MA, Iqbal Z, Yaseen M, Shamim A. 2007. Anthelmintic activity of *Chenopodium album* (L.) and *Caesalpinia crista* (L.) against trichostrongylid nematodes of sheep. *Journal of Ethnopharmacology* 114, 86–91.

Kahiya C, Mukaratirwa S, Thamsborg SM. 2003. Effects of *Acacia nilotica* and *Acacia karoo* diets on *Haemonchus contortus* infection in goats. *Veterinary Parasitology* 115, 265–274.

Kaido TL, Veale DJH, Havlik I, Rama DBK. 1997. Preliminary screening of plants used in South Africa as traditional herbal remedies during pregnancy and labour. *Journal of Ethnopharmacology* 55, 185–191.

Kellerman TS, Coetzer JAW, Naudé TW. 1992. Plant poisonings and mycotoxicoses of livestock in southern Africa. In: James LF, Keeler RF, Bailey EM, Cheeke PR, Hegarty MP (Eds.), *Poisonous Plants: Proceedings of the Third International Symposium*. Iowa State University Press, Ames, IA, pp. 43–54.

Kruger J. 2004. Isolation and characterization of antibacterial compounds from three *Terminalia* species. PhD thesis, University of Pretoria, South Africa.

Kudi AC, Myint SH. 1999. Antiviral activity of some Nigerian medicinal plant extracts. *Journal of Ethnopharmacology* 68, 289–294.

Lamien CE, Meda A, Mans J, Romito M, Nacoulma OG, Viljoen GJ. 2005. Inhibition of fowlpox virus by an aqueous acetone extract from galls of *Guiera senegalensis* J. F. Gmel (Combretaceae). *Journal of Ethnopharmacology* 96, 249–253.

Laupattarakasem P, Houghton PJ, Hoult JRS, Itharat A. 2003. An evaluation of the activity related to inflammation of four plants used in Thailand to treat arthritis. *Journal of Ethnopharmacology* 85, 207–215.

Luseba D, Elgorashi EE, Ntloedibe NT, Van Staden J. 2007. Antibacterial, anti-inflammatory and mutagenic effects of some medicinal plants used in South Africa for the treatment of wounds and retained placenta in livestock. *South African Journal of Botany* 73, 378–383.

Luseba D, Van der Merwe D. 2006. Ethnoveterinary medicine practices among Tsonga speaking people of South Africa. *Onderstepoort Journal of Veterinary Research* 73, 115–122.

Masika PJ, Afolayan AJ. 2002. Antimicrobial activity of some plants used for the treatment of livestock disease in the Eastern Cape, South Africa. *Journal of Ethnopharmacology* 83, 129–134.

Masika PJ, Sonandi A, Van Averbeke W. 1997. Perceived causes, diagnosis and treatment of babesiosis and anaplasmosis in cattle by livestock farmers in communal areas of the central Eastern Cape Province, South Africa. *Journal of the South African Veterinary Association* 68, 40–44.

Masika PJ, Van Averbeke W, Sonandi A. 2000. Use of herbal remedies by small-scale farmers to treat livestock diseases in central Eastern Cape Province, South Africa. *Journal of the South African Veterinary Association* 71, 87–91.

Masoko P. 2006. Characterisation of antifungal compounds isolated from *Terminalia* and *Combretum* species (Combretaceae). PhD thesis, University of Pretoria, South Africa.

Masoko P, Picard J, Eloff JN. 2005. Antifungal activities of six South African *Terminalia* species (Combretaceae). *Journal of Ethnopharmacology* 99, 301–308.

Mathias E. 2004. Ethnoveterinary medicine: harnessing its potential. *Veterinary Bulletin* 74, 27–37.

Mathias E. 2007. Ethnoveterinary medicine in the era of evidence-based medicine: mumbo-jumbo, or a valuable resource? *The Veterinary Journal* 173, 241–242.

McGaw LJ, Eloff JN. 2005. Screening of sixteen poisonous plants for antibacterial, anthelmintic and cytotoxic activity in vitro. *South African Journal of Botany* 71, 302–306.

McGaw LJ, Eloff JN. 2008. Ethnoveterinary use of southern African plants and scientific evaluation of their medicinal properties. *Journal of Ethnopharmacology* 119, 559–574.

McGaw LJ, Gehring R, Katsoulis L, Eloff JN. 2005. Is the use of *Gunnera perpensa* extracts in endometritis related to antibacterial activity? *Onderstepoort Journal of Veterinary Research* 72, 129–134.

McGaw LJ, Lall N, Hlokwe TM, Michel AL, Meyer JJM, Eloff JN. 2008. Purified compounds and extracts from *Euclea* species with antimycobacterial activity against *Mycobacterium bovis* and fast-growing mycobacteria. *Biological and Pharmaceutical Bulletin*. 31, 1429–1433.

McGaw LJ, Jäger AK, van Staden J. 2000. Antibacterial, anthelmintic and antiamoebic activity in South African medicinal plants. *Journal of Ethnopharmacology* 72, 247–263.

McGaw LJ, Rabe T, Sparg SG, Jäger AK, Eloff JN, van Staden J. 2001. An investigation on the biological activity of *Combretum* species. *Journal of Ethnopharmacology.* 75, 45–50.

McGaw LJ, Van der Merwe D, Eloff JN. 2007. In vitro anthelmintic, antibacterial and cytotoxic effects of extracts from plants used in South African ethnoveterinary medicine. *The Veterinary Journal* 173, 366–372.

McLaughlin JL. 1991. Crown gall tumours on potato discs and brine shrimp lethality: two simple bioassays for higher plant screening. In: Hostettman K (Ed.), *Methods in Biochemistry, Assays for Bioactivity*, Vol. 6. Academic Press, London, pp. 1–32.

McLaughlin JL, Rogers LL, Anderson JE. 1998. The use of biological assays to evaluate botanicals. *Drug Information Journal* 32, 513–524.

Mensah AY, Houghton PJ, Akyirem GNA, Fleischer TC, Mensah MLK, Sarpong K, Adosraku R. 2004. Evaluation of the antioxidant and free radical scavenging properties of *Secamone afzelii* Rhoem. *Phytotherapy Research* 18, 1031–1032.

Mensah AY, Sampson J, Houghton PJ, Hylands PJ, Westbrook J, Dunn M, Hughes MA, Cherry GW. 2001. Effects of *Buddleja globosa* leaf and its constituents relevant to wound healing. *Journal of Ethnopharmacology* 77, 219–226.

Meyer BN, Ferrigni NR, Putnam JE, Jacobsen LB, Nichols DE, McLaughlin JL. 1982. Brine shrimp: a convenient general bioassay for active plant constituents. *Journal of Medicinal Plant Research* 45, 31–34.

Mkolo NM. 2008. Anti-tick properties of some of the traditionally used plant-based products in South Africa. MSc thesis, University of Limpopo, South Africa.

Moloto MP. 2004. Isolation and characterization of antibacterial, anthelmintic, antioxidant and cytotoxic compounds present in *Ziziphus mucronata*. MSc thesis, Medical University of South Africa.

Naidoo V, Zweygarth E, Eloff JN, Swan GE. 2005. Identification of anti-babesial activity for four ethnoveterinary plants in vitro. *Veterinary Parasitology* 130, 9–13.

Naidoo V, Chikoto H, Bekker LC, Eloff JN. 2006. Antioxidant compounds in *Rhoicissus tridentata* extracts may explain their antibabesial activity. *South African Journal of Science* 102, 198–200.

Naidoo V, McGaw LJ, Bisschop SPR, Duncan N, Eloff JN. 2008. The value of plant extracts with antioxidant activity in attenuating coccidiosis in broiler chickens. *Veterinary Parasitology* 153, 214–219.

Naidoo V, Zweygarth E, Swan GE. 2006. Determination and quantification of the *in vitro* activity of *Aloe marlothii* (A. Berger) subsp. *marlothii* and *Elephantorrhiza elephantina* (Burch.) Skeels acetone extracts against *Ehrlichia ruminantium. Onderstepoort Journal of Veterinary Research* 73, 175–178.

National Committee for Clinical Laboratory Standards (NCCLS). 1990. *Performance Standards for Antimicrobial Disc Susceptibility Tests*, 4th ed. Approved Standard. NCCLS Document M2-A4. Villanova, PA: National Committee for Clinical Laboratory Standards.

Naudé TW, Coetzer JAW, Kellerman TS. 1992. In: James LF, Keeler RF, Bailey EM, Cheeke PR, Hegarty MP (Eds.), *Poisonous Plants: Proceedings of the Third International Symposium*. Iowa State University Press, Ames, IA, pp. 11–20.

Nchu F. 2004. Developing methods for the screening of ethnoveterinary plants for tick control. MSc thesis, Medical University of Southern Africa.

Njoroge GN, Bussmann RW. 2006. Herbal usage and informant consensus in ethnoveterinary management of cattle diseases among the Kikuyus (central Kenya). *Journal of Ethnopharmacology* 108, 332–339.

O'Neill MJ, Lewis JA. 1993. The renaissance of plant research in the pharmaceutical industry. In: Kinghorn, AD, Balandrin MF (Eds.), *Human Medicinal Agents from Plants*. ACS Symposium Series 534. American Chemical Society, Washington, DC, pp. 48–55.

Pieters L, Vlietinck AJ. 2005. Bioguided isolation of pharmacologically active plant components, still a valuable strategy for the finding of new lead compounds? *Journal of Ethnopharmacology* 100, 57–60.

Rahalison L, Hamburger M, Hostettman K, Monod M, Frenk E. 1991. A bioautographic agar overlay method for the detection of antifungal compounds from higher plants. *Phytochemical Analysis* 2, 199–203.

Rasoanaivo P, Ratsimamanga-Urverg S. 1993. *Biological Evaluation of Plants with Reference to the Malagasy Flora*. Madagascar: Napreca, pp. 9–43, 72–83.

Re R, Pellegrini N, Proteggente A, Pannala A, Yang M, Rice-Evans C. 1999. Antioxidant activity applying an improved ABTS radical cation decolorization assay. *Free Radical Biology and Medicine* 26, 1231–1237.

Simpkin KG, Coles GC. 1981. The use of *Caenorhabditis elegans* for anthelmintic screening. *Journal of Chemical Technology and Biotechnology* 31, 66–69.

Solís PN, Wright CW, Anderson MM, Gupta MP, Phillipson JD. 1993. A microwell cytotoxicity assay using *Artemia salina* (brine shrimp). *Planta Medica* 59, 250–252.

Thembo MK. 2006. The anti-tick effects of *Senna italica* subsp. a*rachoides* extracts on adults of *Hyalomma marginatum rufipes*. MSc thesis, University of Limpopo, South Africa.

Van der Merwe D, Swan GE, Botha CJ. 2001. Use of ethnoveterinary medicinal plants in cattle by Setswana-speaking people in the Madikwe area of the North West Province of South Africa. *Journal of the South African Veterinary Association* 72, 189–196.

Van Wyk B-E, Van Oudtshoorn B, Gericke N. 1997. *Medicinal Plants of South Africa*. Pretoria, South Africa: Briza.

Vlietinck AJ, Vanden Berghe DA. 1991. Can ethnopharmacology contribute to the development of antiviral drugs? *Journal of Ethnopharmacology* 32, 141–153.

Vlietinck AJ, Van Hoof L, Totté J, Lasure A, Vanden Berghe D, Rwangabo PC, Mvukiyumwami J. 1995. Screening of a hundred Rwandese medicinal plants for antimicrobial and antiviral properties. *Journal of Ethnopharmacology* 46, 31–47.

Williams RB. 1999. A compartmentalized model for the estimation of the cost of coccidiosis to the world's chicken production industry. *International Journal of Parasitology* 29, 1209–1229.

Zhang Y, But PP, Ooi VE, Xu H, Delaney GD, Lee SHS, Lee SF. 2007. Chemical properties, mode of action, and in vivo anti-herpes activities of a lignin-carbohydrate complex from *Prunella vulgaris*. *Antiviral Research* 75, 242–249.

Zorloni A. 2007. Evaluation of plants used for the control of animal ectoparasitoses in southern Ethiopia (Oromiya and Somali regions). MSc thesis, University of Pretoria, South Africa.

2 Logistical and Legal Considerations in Ethnoveterinary Research

Mary Chikombero and Dibungi Luseba

CONTENTS

2.1 INTRODUCTION

Globalization has brought new challenges to the protection of biodiversity. The unprecedented loss of biodiversity in the past few decades due to industrial exploitation and overharvesting has raised global concerns. The manner in which bioprospecting, particularly among indigenous people and the use of traditional knowledge, has been carried out in the past has also been heavily criticized, and such criticisms have centered around allegations of biopiracy, unfair distribution of benefits, illegal appropriation of traditional knowledge, and unethical conduct when it comes to filing and claiming patents on indigenous biological resources.

To tackle these problems, various legal regimes have been instituted at international, regional, and national levels in a bid to promote the sustainable utilization of biodiversity, including medicinal indigenous plants. This chapter discusses plant and information gathering and then outlines the concept of access and benefit sharing (ABS) as one of the approaches adopted at the international level, through the Convention on Biological Diversity (CBD), to protect biodiversity. Some of the challenges of implementing ABS regulations on the local level are also identified. Intellectual property (IP) rights issues and the World Trade Organization (WTO) agreement are also discussed since the proper implementation of the CBD also has an impact on IP rights. Relevant legislation from South Africa, India, and Costa Rica regulating ABS is given as case studies of national laws that have included the CBD provisions/principles.

2.2 MEDICINAL PLANT GATHERING AND STORAGE

People have used ethnoveterinary medicines (EVMs) for generations, and many reasons have been advanced regarding why the practice continues. Farmers claim that medicinal plants are more efficacious than pharmaceuticals for chronic pathologies. They are reputed to have no side effects, and no withdrawal periods for consumption of meat from treated animals are needed since the plants are thought to be nontoxic. In general, ethnoveterinary products are used after conventional pharmaceutical medicines, for chronic cases, have proved ineffective (Luseba and Van der Merwe, 2006).

Traditional knowledge is passed orally from one generation of knowledge holders to the next both formally and informally. In this form of knowledge transmission, plant names that are different from the common names may be used to protect privileged knowledge. Uninformed persons may not be familiar with the nomenclature that traditional healers and knowledge holders use and will therefore not be able to collect medicinal plants by themselves (Reyneke, 1971). Social conventions also control the collection of indigenous plant material to some extent. For example, the felling of fruit trees such as *Sclerocarya birrea* (Marula tree) is often prohibited; the gathering of medicinal plants may be seasonally restricted; and the vegetation around cemeteries may be protected by various taboos (Cunningham, 1988).

In general, traditional healers are not consulted for animal health care. They may be consulted to find lost animals or for advice and medicines to protect animals against witchcraft and ill winds. Most of the farmers learn about traditional treatments from older family members and other farmers. There is little reference to ancestral guidance as seen in human traditional medicine. Farmers are interested in knowing the medicines used by others, and they are willing to share the information. This is contrary to traditional healers, who tend to keep their knowledge to themselves as it may be their only source of livelihood.

Plant parts that are harvested include roots, tubers, bark, leaves, flowers, fruits, seeds, gums, and nectar (Van Wyk, Van Oudtshoorn, and Gericke, 1997). The pharmaceutical and toxicological effects of different parts of a plant may differ substantially (Iwu, 1993; Van Wyk, Van Oudtshoorn, and Gericke, 1997). Plant parts that tend to be constantly available such as roots and bark are used more often than plant parts such as seed or leaves, which may be seasonal (Iwu, 1993). However, the harvesting of roots frequently destroys plants (Van Wyk, Van Oudtshoorn, and Gericke, 1997). The same applies to overharvesting of bark, while collection of leaves, fruits, seeds, and gum is usually less destructive (Cunningham, 1990).

Medicinal plants are used either fresh or after a period of storage. Plant material is usually stored in a dried form. It may be cut into slices to facilitate drying. Dried material is sometimes powdered before storage. Plant material is stored in bags, newspaper, glass jars, and cans. Stored plants should be protected from exposure to sun, water, dust, wind, and contact with strangers. Plant material intended for sale in markets is often tied into bundles (Van Wyk, Van Oudtshoorn, and Gericke, 1997). However, ethnoveterinary herbal materials are often used fresh because they are collected when necessary (Luseba and Van der Merwe, 2006).

Rapid rural appraisal (RRA) methods are preferably used when gathering information from communities (Beebe, 1995). It is important to organize meetings with traditional leaders and the state veterinary service officers at the onset to explain the purpose of the research. Oral interviews can thereafter be conducted with farmers at dipping tanks in groups or individually. It is important to interview traditional healers and herbalists since their knowledge is rarely shared with farmers. In most cases, the interviews are conducted through a translator, who may be one of the local animal health officers. A general feedback session should be organized to correct, harmonize, and share the information among farmers. Key areas of investigation and discussions are the farmer's socioeconomic profile, animal husbandry, and local knowledge of animal health care (ethnoetiology, ethnodiagnostics, treatments, and disease control). Detailed information on plants used is recorded, including local names of the remedies, indications, preparations, routes of administration, and dosage if applicable.

Plants should be collected under the guidance of the respondents. Botanical data should be collected using a collection form. Notes are also taken from discussions with respondents. Pictures of the plants are recorded with a digital camera, and precise coordinates of the locations taken with a global positioning system instrument, if possible. Two to three specimens of each plant species are collected, labeled, and pressed for voucher specimen. If a qualified plant taxonomist is not involved in the collection, a specimen needs to be sent to a specialized botanical institution for identification. This is important because a single plant species may have different common or local names, or the same name is given sometimes to plants of different genera or species. Approximately 2 kg fresh plant materials are usually collected; dried under the shade, ground, and stored in darkened glass jars to prevent oxidation; and kept for subsequent laboratory investigations (Luseba and Van der Merwe, 2006). Special attention should be given to succulent plants because they are prone to contamination by fungi. Freeze-drying or drying in autoclave at 40% has been suggested in the case of succulents (e.g., Aloe species).

2.3 DATA ANALYSIS

Data can be analyzed by pairwise and matrix ranking. Statistical measures of central tendency, dispersion, and percentages are computed using appropriate statistical packages, such as the Statistical Package for Social Sciences (SPSS Inc.). Statistical methods can also be used to identify plants used most frequently in veterinary medicine, based on the frequency of association of a particular or perceived medicinal value (botanical consistency) and the frequency of a particular plant species associated with or used to treat a particular disease (Matekaire and Bwakara, 2004). This is referred to as *consistency of veterinary usage*. There is consistency when the same plant genus or family is mentioned at least twice for treating the same illness.

2.4 THE CONVENTION ON BIOLOGICAL DIVERSITY

At the international regulatory level, countries have negotiated the CBD, which was concluded at Rio de Janeiro on June 5, 1992. The CBD is a remarkable framework instrument that lays down broad goals, key objectives, and general principles to be implemented by contracting parties through measures at the national level. The three objectives of the CBD are stated in Article 1 as the conservation of biological diversity, the sustainable use of its components, and the fair and equitable sharing of the benefits arising from the utilization of genetic resources. The CBD has been ratified by 191 countries; notably, the United States has signed but has not ratified the treaty (Cordell, 2000).

The CBD sought to reconcile the diverging interests between states with rich biological diversity (mainly the developing countries of the global south) and states with advanced technology (industrialized nations) by introducing the concept of ABS. Through the principle of common but differentiated responsibilities, contracting parties (signatories to the convention) agreed to come together to conserve biological diversity and to negotiate its access. In the context of the CBD, *access and benefit sharing* is a phrase used to describe access granted by a contracting party of the CBD to its genetic resources to another contracting party to the CBD, with such access on mutually agreed terms (MAT) and subject to fair and equitable sharing of the results and benefits of any research carried out on the resources. The ABS provisions of the CBD have caused controversial debates between developing and developed countries and among advocacy groups, including indigenous and local communities, business, and industry (Siebenhuner and Suplie, 2005). Further, the progress toward achieving the objectives of the CBD has been painfully slow as many countries still have to pass national legislation to give effect to the spirit of the convention.

Regulating ABS issues is rather complex and challenging, and in most cases it is an area that is not easily understandable as it involves a milieu of socioeconomic, legal, scientific, and environmental issues. Three regional groups—the Association of Southeast Asian Nations (ASEAN), the Andean Pact (comprising Bolivia, Colombia, Ecuador, Peru, and Venezuela), and the Organization of African Unity (OAU)—have taken the approach of preparing guidelines for member states to encourage regional consistency in approaching ABS regulation and to help their member states in preparing the complex legislation required (International Development Research Center [IDRC], 2004).

The ASEAN has the ASEAN Framework Agreement on Access to Biological and Genetic Resources (draft text; ASEAN, 2000) while the Andean Pact has Decision 391 on Common Regime on Access to Genetic Resources (Andean Community Commission, 1996). For its part, the OAU (now the African Union) has the African Model Legislation for the Protection of the Rights of Local Communities, Farmers and Breeders and for the Regulation of Access to Biological Resources (OAU, 2000). The regional guidelines set minimum standards for member state domestic laws and have taken different approaches to regulating the ABS area.

In addition to the regional efforts, there is another international initiative that is aimed at assisting with the implementation of the ABS provisions of the CBD. This is the *Bonn Guidelines on Access to Genetic Resources and Fair and Equitable*

Sharing of Benefits Arising Out of Their Utilization adopted in May 2002 (CBD Secretariat, 2000). These are, however, not binding as they are merely guidelines; they are voluntary and flexible and were designed mainly to facilitate the development process of national ABS laws and policies. The guidelines outline the roles and responsibilities of users and providers of genetic resources and encourage stakeholders to use a bilateral approach to facilitate ABS goals (Carrizosa et al., 2004). The key issues that are contained in the Bonn guidelines include advice on:

1. Involvement of relevant stakeholders and capacity building
2. Steps to be taken in the ABS negotiating process
3. Elements of prior informed consent (PIC)
4. Monetary and nonmonetary benefits and incentives
5. National monitoring and reporting and accountability

2.4.1 Access and Benefit Sharing

Historically, genetic resources were regarded as a "common heritage" of humankind that belonged to the public domain and could not be owned by a single group (Soejator et al., 2005). The advances in biotechnological research and the rise of the concomitant IP over discoveries from living material have contributed to a change in the customary treatment of genetic resources as a common good (Van Overwalle, 2005). Article 15(1) of the CBD recognized the sovereign rights of member states over their natural resources and gives authority to determine access to genetic resources to national governments subject to national legislation (CBD, 1992). Member states are therefore supposed to protect their genetic resources and accompanying traditional or indigenous knowledge by enacting appropriate national ABS legislation. Access to genetic resources is, however, subject to PIC, a fair and equitable sharing of benefits arising from commercial or other utilization of said resources. Access and transfer of relevant technology, which is subject to patents and IP rights, should be provided on terms that recognize and offer adequate and effective protection of such rights.

2.4.1.1 The Principles of ABS

The principles of ABS center around negotiations and agreement on PIC, MAT and benefit sharing (BS). The CBD stipulated that the granting of access to genetic resources shall be on MAT and subject to PIC of the country of origin. The CBD also recognized the important role of indigenous and local communities in conserving and sustainably using biological diversity and stated that the benefits arising from their knowledge and innovation should be equitably shared.

The regional and national procedures are naturally different, but they all aim to achieve the same goals of conservation, sustainable use of resources, and fair compensation to access. For example, the African Model Legislation for the Protection of the Rights of Local Communities, Farmers and Breeders and for the Regulation and Access to Biological Resources of the OAU provides that any access to biological resources and knowledge or technologies of local communities in any part of the country shall be subject to an application for the necessary PIC and written permit. The access permit shall be granted by national governments in terms of their national legislation and through a signed written agreement between the government concerned, the local community or communities, and the applicant or collector.

2.4.1.1.1 *Prior Informed Consent*

The CBD formalized the need to get PIC to access biological material and for utilizing traditional knowledge. It recognized the value of ethnobotany and the wealth of information that traditional or indigenous communities possess. In doing so, the CBD further required member states to respect, preserve, and maintain knowledge, innovations, and practices of indigenous and local communities embodying traditional lifestyles relevant for conservation and sustainable use of biological diversity and promote their wider application with the approval and involvement of the holders of such knowledge, innovations, and practices, as discussed in Articles 8(j), 10(c), 17(2), and 18(4) (CBD, 1992).

PIC was not defined in the CBD, but several authors have identified essential elements as follows:

- Prior: Before access is granted.
- Informed: Access should be based on truthful information about how resources or knowledge will be used. Such information should be adequate for national authorities to understand any implications associated with granting access to the resources or knowledge.
- Consent: There is need for explicit consent of the government and other stakeholders of the country providing the genetic resources (Ten Kate and Laird, 1999, page 27).

The need to get PIC has brought with it a number of challenges for companies or individuals involved in bioprospecting. Bioprospectors are now required to get consent not only from the state but also from other interested stakeholders (e.g., communities, land owners, collaborating institutions, and others; Wynberg and Laird, 2007). This is obviously time consuming and expensive and may be impossible in some cases.

2.4.1.1.2 *Mutually Agreed Terms*

In terms of the CBD, access to genetic resources, traditional knowledge, and the sharing of benefits derived from using such resources and knowledge shall be based on MAT (CBD, 1992). In most cases, national laws of countries that provide biodiversity and traditional knowledge prescribe that contractual arrangements are the legal instrument to regulate access to genetic resources, such as through material transfer agreements (MTAs). The question remains, however, whether these are well-balanced contracts considering that the provider countries or communities are poor and typically lack the necessary negotiation skills (van Overwalle, 2005).

The CBD encourages member states to enter into bilateral agreements, but difficulties may be encountered when using the indigenous knowledge of a community that is considered an enemy by the member state. It is also difficult to determine with whom agreements should be made as it is difficult to identify the official guardians of the knowledge and the exact boundaries of an indigenous culture (Fenwick, 1998).

2.4.1.1.3 Benefit Sharing

As pointed out, one of the objectives of the CBD was the fair and equitable sharing of the benefits arising out of the utilization of genetic resources. CBD stated that BS shall be on MAT. Contracting parties to the CBD have an obligation to take legislative, administrative, or policy measures to ensure the fair and equitable sharing of the results of research and development and the benefits arising from the commercial and other utilization of genetic resources (CBD, 1992). The sharing of benefits is important because community groups must feel that their resources are not being stolen from them, and those making substantial investments in the development of products must feel that they can protect their added value (Cordell, 2000).

National legislation of CBD member states and regional guidelines address the issue of sharing benefits derived from genetic resources. However, it has been argued that the fair and equitable sharing of benefits is a global transnational problem that requires an international response since it involves actors and stakeholders from various countries with highly divergent interests, ranging from development aid agencies and commercial interests in the pharmaceutical and agroindustry to pure academic research (Siebenhuner and Suplie, 2005).

In sharing benefits, the state and the community or communities are recognized to be entitled to a share of the earnings derived from using any biological resources or knowledge collected directly or indirectly. The benefits may be monetary or non-monetary (e.g., technology transfer, training and research capacity building, and IP). Examples of cases in which indigenous people or local communities have benefited from bioprospecting activities include the Kani in India, who receive a proportion of the license fee and royalty from commercialization of Jeevani, a herbal drug that "rejuvenates and builds strength," and the San in southern Africa, who receive a proportion of royalties from the sales of hoodia (a succulent plant used by the San to stave off hunger and thirst) (Wynberg and Laird, 2007). However, it is believed that the main reasons hindering the realization of the CBD objective of "fair and equitable benefit sharing" is the time lag between the collection of samples and the development of a marketable product and the low probability that a sample will succeed in a product. For example, in the pharmaceutical sector it takes 10 to 15 years of research to develop a marketable product from botanical or marine natural products (Richerzhagen and Holm-Mueller, 2005).

2.4.1.2 Problems with ABS Legislation

Many countries find the establishment of ABS regulations containing the ABS principles as enshrined in the CBD challenging as it involves high costs for consultations, employment of more staff, and technical and legal personnel for negotiating contracts and policing. The complexity of the ABS concept and the fear of biopiracy has resulted in some countries overregulating access to biological resources. Many bioprospectors (companies, institutions, and researchers) regard ABS laws in some countries as unclear, bureaucratic, and time consuming and compliance as expensive (Richerzhagen, 2007).

The major problems identified with ABS legislation are bureaucracy, overregulation, negative impact on research, tension and suspicions, biopiracy, community conflict, and commodifying of culture or traditions.

1. *Bureaucracy.* In many developing countries, the process of obtaining access to genetic resources is often met with administrative complexities, such as bureaucratic procedures and a lack of institutional capacity to process applications on time (Richerzhagen, 2007). For example, the process of getting PIC from all interested stakeholders and formalizing this into an agreement may take up to 2 years. In India, it takes about 1–3 years to get a permit (Wynberg and Laird, 2007).
2. *Overregulation.* Some countries have implemented strict and onerous legal regimes to regulate access. This position negatively affects the research conducted by universities and research institutions as researchers are worried about being accused of biopiracy, which has a negative impact on their academic standing and reputation (Richerzhagen, 2007). Overly restrictive rules may also have a negative impact on the pharmaceutical research of that country as its development will lag behind those countries with less-strict rules (Cordell, 2000).
3. *Negative impact on research.* The scientific research community has voiced concerns about national laws, arguing that license fees add to costs of research, and some laws are so onerous that they restrict researchers' academic rights (Cordell, 2000). Furthermore, in negotiating for access there is also a creation of unrealistic expectations that are not likely to be met in most bioprospecting work (which is usually purely for academic and training purposes).
4. *Tension and suspicions.* It is believed that the CBD has done little to ease tensions that exist between scientists searching for potentially valuable therapies and officials in developing countries where most bioprospecting takes place. This tension is growing as the latter have now become more aware of their rights under the CBD. In most instances, bioprospecting is still taking place in an environment of suspicion and mistrust, and accusations of unethical conduct still surface regularly (Dalton, 2004).
5. *Biopiracy.* Other countries have been hesitant about enacting rigorous ABS laws because they fear losing business and potential revenue. This has created a loophole as collectors will simply collect biodiversity in more friendly jurisdictions if the desired species is found there (IDRC, 2004). Because indigenous knowledge or innovations and target flora frequently cross country boundaries, it is particularly difficult to avoid biopiracy (Shiva, 2005). Unscrupulous collectors may also find other ways of smuggling biological resources out of stricter countries because of inadequate policing.
6. *Community conflict.* Although bioprospecting contracts are based on PIC and BS, unlike biopiracy, for which no consent is sought or no benefits are shared, not all owners of traditional knowledge are consulted or compensated pursuant to bioprospecting contracts. It has been argued that such

exclusion not only leads to inequity and injustice but also has the potential to cause conflicts among community members and between communities (Shiva, 2005). An example of a BS arrangement that has caused community conflict is that in India, where many Kani elders believed that the people who revealed the "arogyapacha" plant (*Trichopus zeylanicus* ssp. *travancoricus*), which has antifatigue properties and was used in developing the Jeevani drug, had no right to divulge that information for monetary benefits as their traditional knowledge is sacred (Bijoy, 2007).

7. *Commodifying of culture or traditions.* The concept of bioprospecting has also been criticized as it involves patenting of traditional knowledge that is a product of thousand of years of collective innovation by indigenous cultures; hence, some scholars and activists believe it cannot be called an invention (Shiva, 2005). It is generally believed that many traditional healers and leaders are opposed to the commodifying of their traditions and knowledge as they argue that they are gifts from their forebearers or ancestors that should be used for the greater good of the community rather than for individual commercial gain.

2.4.2 Intellectual Property and the World Trade Organization

Viewed from the point of developing countries that are biologically resource-rich, it is believed that the World Trade Organization's Trade-Related Aspects of Intellectual Property Rights Agreement (TRIPS) of 1994 is almost a counterattack to the spirit of CBD (Cordell, 2000). However, most WTO developed country members such as the United States, Canada, and Japan are of the view that there is no conflict between the CBD and the TRIPS agreement (Lybbert, 2007).

The TRIPS agreement places the protection of IP rights at the heart of international trade (Armour and Harrison, 2007). IP rights generally refer to a bundle of rights aimed at protecting a product of human intellectual innovation. Protection of IP can be done by various legal means (e.g., patents, trade secrets, trademarks, designs, and copyrights). Biotechnological innovations need adequate protection to encourage investors to finance research to create new products or technology. Patents, which cover inventors for a period of 20 years, are the preferred option for most such inventions. Within such a period, creators are given a right to prevent others from using their inventions or creation unless they receive payment.

Under international law, the TRIPS agreement establishes minimum levels of protection that each government has to give to the IP or invention of fellow WTO members (WTO, 1994). The inventor can receive protection by IP rights that are internationally harmonized by the TRIPS agreement in all WTO member countries (Richerzhagen and Holm-Mueller, 2005). Under TRIPS, patents are used to protect any inventions, whether products or processes, in all fields of technology, provided that they are new (i.e., not in the public domain), involve an inventive step, and are capable of industrial application (WTO, 1994; Article 27(1)).

It is important to note that Article 27.3(b) of the TRIPS agreement states that members *may* exclude from patentability "plants and animals other than microorganisms and essentially biological processes for the production of plants or animals other than non-biological and microbiological processes" (WTO, 1994). The flexibility of Article 27.3(b) has led to a variety of approaches in the protection of biological material among provider countries. Van Overwalle (2005) has noted that certain provider countries find patents on biological resources useful and do not opt for the implementation of the exclusionary TRIPS provision in their patent laws. Most developing country WTO members, which are generally biodiversity rich, are concerned about possible conflicts between their national sovereignty over genetic resources mandated by the CBD and the granting of private rights over inventions based on the resources offered under the TRIPS agreement (through national patent legislation) (Lybbert, 2007).

The patenting of biological resources or biotechnology products based on traditional knowledge has raised critical issues. Informal innovations achieved by local communities over generations of years, however, are not covered by the property rights provisions of the TRIPS agreement, particularly with regard to patent protection (Seiler and Dutfield, 2001). There have been suggestions that indigenous communities that are involved in the identification of the therapeutic properties of indigenous plants should be compensated for the exploitation of their medical knowledge by way of patents. However, since the requirements of patent laws have been largely dictated by technologically advanced nations, it is questionable whether the contribution of traditional knowledge from which a pharmaceutical product has been developed can be the sort of contribution that will meet patentability standards (van Overwalle, 2005). For instance, such knowledge cannot be individually owned, and there are various cultural and religious aspects that may be involved. Purists and many indigenous peoples' rights campaigners suggest that historically, traditional knowledge was collectively owned by the community, is intangible, has spiritual dimensions and although worthy of protection, is inalienable and therefore cannot belong to any human being (Posey and Dutfield, 1996).

Furthermore, the main aim of patent law is to provide an incentive for inventiveness and creativity, commercialization, and distribution by offering the patent holder a period of time during which the holder's rights are protected from competition. Traditional knowledge does not fit into this conventional, technically oriented notion of invention and innovation under patent law. Moreover, although there might be diverging views on the value of knowledge in indigenous communities, in general indigenous people have been reported not to be primarily concerned with commercial exploitation of their knowledge and market economic values. The novelty requirement under patent law is also problematic when dealing with traditional knowledge as most of it is transgenerational, communally shared, and considered to be in the public domain and therefore unprotected (van Overwalle, 2005).

Whereas the CBD requires collectors of biological diversity to get PIC and to share benefits derived from the use of such biological resources or traditional knowledge, critics have argued that the TRIPS agreement allows for the granting of patents over genetic materials without addressing the requirements of the CBD (Armour and Harrison, 2007). The Bonn guidelines encourage countries to require patent applicants to disclose the country of origin of any genetic resources or traditional knowledge used in the invention (Lybbert, 2007). In light of this position, some providers of genetic resources, such as South Africa, have amended their patent laws to strengthen the protection of indigenous knowledge. The South African Patent Amendment 2005 (Act No. 20 of 2005) is linked to the South African CBD national laws in that it now requires every applicant who lodges an application for a patent to disclose the role played by an indigenous biological resource, a genetic resource, or traditional knowledge or use and whether the invention is coowned by holders of the relevant traditional knowledge (Department of Trade and Industry, 2005).

2.5 CASE STUDIES OF NATIONAL LAWS

The following are case studies of national laws governing bioprospecting and access to genetic resources in India, Costa Rica, and South Africa. The three countries were selected because they are important sources of genetic resources, they are on three different continents that constitute the global south, and there are interesting elements in the national legislation that point to the difficulties involved in its implementation and enforcements.

2.5.1 An Asian Example: India

In India, the Biodiversity Act of 2002 and the Biological Diversity Rules of 2004 regulate ABS issues. The enactment of the act with rules to regulate ABS in India was a very progressive step as this area was previously unregulated (Kohli, 2007). The Biodiversity Act created the National Biodiversity Fund (NBF), the National Biodiversity Authority (NBA), the state biodiversity boards (SBBs), and the biodiversity management committees (BMC) for the effective implementation of the act. Any monetary benefits, fees, and royalties received as a result of approvals by the NBA will be deposited into the NBF (Biodiversity Act, 2002).

The NBA is mandated to deal with matters relating to requests of access to biological diversity by foreigners (including companies and institutions). The SBBs are in every state in India and are mandated to deal with matters relating to access by Indian nationals for commercial purposes. The BMC is at the municipal level and is consulted by the NBA and SBB on matters related to use of biological resources and associated knowledge within their jurisdiction. Implementing these provisions has also been a challenge because the NBA has granted approvals for accessing biological resources in certain areas where the BMCs have not been in existence. This position is undesirable as it removes the opportunity for public consultation that is needed before approvals are granted. In general, local communities play a limited role in decision making under the act, and there are no clearly defined links provided between the state and the holders of knowledge or biological resources, and the decision lies ultimately in the hands of the state. Article 8(j) of the CBD, which recognizes the claims of indigenous people and communities, is not clearly entrenched in the Biodiversity Act (Bijoy, 2007).

Chapter II of the act deals with the regulation of access to biological diversity. In terms of the act, any foreign companies or foreigners requiring access to any biological resources occurring in India or associated knowledge for research or commercialization should get prior approval through an application to the NBA (Biodiversity Act, 2002). PIC is therefore not required under Indian legislation as it is not explicitly mentioned under the law (Bijoy, 2007). It is important to note that Indian legislation does not specify the time frames within which the NBA should decide on an application to access biodiversity. In general, it can take up to 3 years to get a permit (Wynberg and Laird, 2007).

The NBA ensures that ABS aspects are met in accordance with MAT and conditions between the applicant, local bodies, and the benefit claimants. Benefits may be monetary or material (e.g., royalty, joint ventures, technology transfer, product development, education and awareness-raising activities, institutional capacity building, and venture capital fund).

The NBA, while granting approval to any person for access or application for patent and IP rights of accessed biological resource and associated knowledge may impose terms and conditions to ensure BS. Bijoy (2007) argued that the inconsistency between the CBD provisions and the TRIPS agreement has affected the legislative framework of India as it clearly has been influenced by the monopolistic IP protection of the TRIPS agreement and not by the CBD principles outlined in this chapter.

In terms of the Biodiversity Act, prior approval of the NBA is not required for collaborative research projects involving transfer or exchange of biological resources or related information between institutions of India and institutions of other countries if such projects have been approved by central government and conform to the policy guidelines issued by the central government of India.

Section 4 of the Biodiversity Act prohibits any person from transferring results of any research relating to any biological resources in or obtained from India for financial gain to foreigners or foreign-registered companies without the prior approval of the NBA. This, however, exempts scientific publications and conference presentations.

Whereas other countries such as South Africa require all bioprospecting activities to be approved, Indian citizens or companies registered in India obtain biological resources for commercial utilization only after giving prior intimation to the SBB concerned. However, this requirement does not apply to local people or communities of the area who use indigenous medicine.

2.5.2 Model Law from Latin America: Costa Rica

Costa Rica's ABS system has been regarded as one of the most advanced, transparent, and successful ones. Such success preceded the enactment of the ABS legislation and is attributed to the National Biodiversity Institute (INBio), which entered into bioprospecting contracts with companies. INBio was created as a private, nonprofit organization to coordinate biodiversity activities (Richerzhagen and Holm-Mueller, 2005). However, it should be noted that most bioprospecting in Costa Rica entered into by INBio was conducted in protected areas on state land; property rights were clearly defined, and bioprospectors did not have to waste time trying to identify the interested communities (Richerzhagen and Holm-Mueller, 2005).

The success of bioprospecting activities through INBio seems to portray that the implementation of ABS policies is best achieved in a decentralized system and where the number of parties in the negotiation and permitting process is minimized. However, as discussed in this section, the ABS legislation in Costa Rica seems to have changed this rule as the act introduced an unclear and complex application procedure (Carrizosa et al., 2004). The new application procedure is therefore likely to increase transaction costs and be more time consuming because of the increased participation of other stakeholders (Richerzhagen, 2007).

The CBD provisions were only implemented into law in Costa Rica through the Biodiversity Law of 1998, Law No. 7788 of 1998, and the Rules on Access to Biodiversity, 2003 Presidential Decree No. 31-514, which regulates ABS of natural resources. The Biodiversity Law is regarded as the most detailed national law enacted to implement the CBD provisions and is more restrictive (Carrizosa et al., 2004).

The sovereignty principle of the CBD is enshrined in the Biodiversity Law as it provides that the state exercises total and exclusive sovereignty over the components of biodiversity. The act creates the National Commission for the Management of Biodiversity with defined functions that include formulating and coordinating the policies for access to biodiversity and associated knowledge to ensure adequate scientific and technical transfer and the proper distribution of the benefits.

Stakeholder participation in implementing the act is ensured as the commission is constituted by a wide range of stakeholders, including government departments, the private sector, the indigenous peoples' associations, and the peasants' associations. Whereas PIC is not expressly mentioned under the Indian legislation, Costa Rica's ABS legislation guarantees the participation of local communities in decision-making processes as one of the basic requirements for access is the need to get the PIC of community representatives or land owners. The act adequately addresses the rights of local communities in that it provides that the state recognizes and protects the knowledge, practices, and innovations of indigenous peoples and local communities related to the use of components of biodiversity and associated knowledge. Furthermore, local communities have a right to oppose any access to their resources and associated knowledge for cultural, spiritual, social economic, or other reasons. No form of intellectual or industrial property rights protection regulated in terms of Chapter 5 of the Biodiversity Law, in special laws, and in international law shall affect such historic practices (Biodiversity Law, 1998; Article 82).

All research programs or bioprospecting related to genetic resources meant to be carried out in Costa Rican territory require an access permit. A person who is interested in accessing biodiversity should register at the Technical Office of the commission before applying for any type of access permit and should indicate whether the permit is for basic research, bioprospecting, or economic exploitation. The Technical Office will issue a preliminary identity card to the interested party; it should be used when applying for PIC. To obtain an access permit, there is a need to get PIC and negotiate mutually agreed conditions according to the model contract prepared by the Technical Office. The contract regulates BS (e.g., transfer of technology and equitable distribution of environmental, economic, social, scientific, or spiritual benefits). This should include possible commercial profits of short, medium, or long term of any product or derivative. The interested party should meet with the representatives of the study site to discuss the scope of the access, the terms of protection of the related knowledge, and the practical, economic, and logistical aspects of the granted access. The Technical Office should endorse the PIC and if necessary will make field consultations to verify the agreed-on terms.

An access permit is granted to a researcher or a research center, is valid for 3 years, and is renewable at the discretion of the Technical Office. MTAs, which are authorized by the Technical Office, should be entered between the parties for exchange and transfer resources to be enabled.

The act ensures that immediate benefits flow back to providers of biodiversity resources by requiring the interested party to deposit up to 10% of the research or bioprospection's budget into the bank account indicated by the direct provider of the resources. Such a position is different from South African and Indian legislation, which requires monetary benefits to be deposited into a national fund. For basic research, an interested party should submit three copies of the final results or the scientific papers and publications derived from them, in which the country's contribution and the knowledge holders are acknowledged. For occasional or regular economic exploitation, interested parties may pay up to 50% of the royalties obtained by the interested party on behalf of the conservation areas, local communities or indigenous people, or the landowners.

Registrars of intellectual and industrial property (i.e., patents, trademarks etc) are obliged to consult with the Technical Office before granting protection of intellectual or industrial property to innovations involving components of biodiversity. The individual beneficiaries of intellectual or industrial property protection related to biodiversity will cede, in favor of the state, a compulsory legal license, which will allow it, in cases of declared national emergency, to use such rights for the collective good, with the sole purpose of resolving the emergency, without the need to pay compensation.

Once all the required documents have been submitted, the Technical Office has 30 days to resolve the application. This provision is important as it will ensure that applications are processed in a timely manner, and it is absent from Indian and South African ABS legislation.

To certify the legality of access, the Technical Office will issue to the applicant a certificate of origin, which includes place and date of access, owner, type of material obtained, quantity, and the person or community that have contributed or will contribute with their related knowledge, innovations, and traditional practices.

The act states that the arrangements of the law do not affect university autonomy in the matter of teaching or research in the field of biodiversity, except if the research has commercial purposes. Universities are required to establish internal rules applicable exclusively to academic activity and research that is carried out when it implies access to biodiversity for purposes that are not profit making.

2.5.3 Out of Africa: South Africa

In South Africa, the National Environmental: Biodiversity Act Number 10 of 2004 (the NEMBA) and the bioprospecting and ABS regulations of 2008 regulate bioprospecting and access to genetic resources. Chapter 6 of NEMBA regulates bioprospecting and ABS. The objectives are to regulate bioprospecting of indigenous biological resources and the export from South Africa of such resources for the purpose of bioprospecting or related research and to ensure that there is fair and equitable sharing of benefits (NEMBA, 2004).

The act has a bureaucratic procedure in that applications for bioprospecting permits are submitted to the minister responsible for national environmental management (i.e., the Minister of Environmental Affairs). The discovery phase and commercialization phase of a bioprospecting project may only be carried out with a bioprospecting permit issued by the minister. If the applicant wants to export indigenous biological resources for a bioprospecting project, then the applicant must apply to the minister to obtain an integrated export and bioprospecting permit. These permits are issued when the applicant has obtained the PIC of relevant stakeholders and once the minister has approved the MTA and Benefit Sharing Agreement (BSA). The process of approval of PIC, MTAs, and BSA may take a long time because of the bureaucratic nature of the process, and the act does not prescribe the time period within which an issuing authority should decide on a permit application, although the regulations state that applications must be processed "within a reasonable time." The process of getting a bioprospecting permit is therefore likely to be too costly and onerous for the applicant. The requirement of a BSA in the discovery phase is also problematic as it might be difficult to quantify the financial returns of a bioprospecting project at this stage (Crouch et al., 2008).

The act forbids the issuing of permits to foreign persons or entities unless a South African person or entity is also a coapplicant. It is anticipated that this requirement will lead to technology transfer and information sharing between the foreign persons and the South African citizens (Crouch et al., 2008). A BSA may also not be approved unless it provides for enhancing scientific knowledge and technical capacity to organs of state or national research institutes, universities, or communities to conserve, use, and develop indigenous biological resources.

As with India, the collection and use of biological resources by local healers is exempt from the legislation. However, the threat that the booming trade in medicinal plants poses is real and cause for concern and yet not regulated by any current legislation in South Africa.

Indigenous biological resources for research other than bioprospecting may only be exported with an export permit. The issuing authority for export permits for indigenous biological resources for research other than bioprospecting is the provincial minister of the province in which the biological resources to be exported are collected, gathered, or curated. A permit may only be issued if the applicant has obtained PIC of the stakeholder to access the resources.

The act establishes the Bioprospecting Trust Fund (BTF), to which all money earned from BSAs and MTAs and due to stakeholders must be paid and used in terms of the agreements. The success of this trust fund is still to be tested as the regulations have been in place for a short period of time.

The existence of ABS legislation in South Africa is progressive. However, the current legislation is restrictive and does not provide user certainty in that there is a risk of wasting resources when an applicant goes through the process of obtaining PIC and signing the MTA and BSA and then the collection permit application is refused by a provincial conservation authority on ecological grounds. The need to have collection permits is not specified in the act or regulations, but it is presumed in relation to the Convention on International Trade in Endangered Species of Wild Fauna and Flora (CITES) and threatened or protected species (Crouch et al., 2008). Under the current law, stakeholders are only involved in the granting of PIC and when concluding MTA and BSA but are excluded in decision-making processes, which rest with the issuing authority (i.e., the minister or provincial minister). A lack of administrative capacity might also hamper the implementation of the ABS legislation.

2.6 CONCLUSION AND RECOMMENDATIONS

The current ABS provisions of the CBD have caused heated debates between countries that provide biodiversity and countries that collect biodiversity and between social activists and research scientists. Despite the existence of national and regional legislation governing ABS, most countries have failed to implement their obligations under the CBD as some of the procedures in place are too restrictive, and there is also a belief that the current measures are insufficient to curb the misappropriation and illegal trafficking of resources and knowledge (Wynberg and Laird, 2007). This might emanate from the fact that the ABS principles of PIC, MAT, and ABS are not regulated in more detail under the CBD, and the responsibility for determining the rules was left to member states (Richerzhagen and Holm-Mueller, 2005). The difficulties of implementing the CBD principles may also arise because it is an international agreement that addresses issues relevant to nongovernmental sectors such as business and industry, science, and nongovernmental organizations as well as indigenous and local communities, which are difficult to regulate and police (Siebenhuner and Suplie, 2005).

The current regional and national ABS regimes have taken varied approaches in regulating ABS. Countries should try to enact clear and simple ABS legislation that promotes user certainty and encourages compliance. To prevent collectors of biodiversity approaching certain countries and leaving those with strict regulations requires the harmonization of ABS regimes among member states and codes of ethics among researchers should be put in place. Ethical agreements and codes of conduct offer alternatives to legal protection if legal protection is impractical or incomplete. However, it should be noted that enforcement of ethical codes is limited because they are not legally binding unless they form part of a contractual agreement.

It is recommended that international rules on how benefits should be shared fairly and equitably should be developed as there is currently no international consensus. However, it should be noted that precise quantification of the monetary values of future earnings is extremely difficult because future product development and market potentials remain uncertain (Siebenhuner and Suplie, 2005).

Although there are divergent views on whether there is conflict between the TRIPS agreement and the CBD, it is recommended that the two agreements should be harmonized. TRIPS should be amended to include the CBD provisions that require disclosure of the source and country of origin of biological resources and traditional knowledge used in an invention, evidence of PIC to be produced, and evidence that benefits will be shared fairly and equitably in terms of the legislation of the country where biodiversity or traditional knowledge was collected.

REFERENCES

Andean Community Commission, 1996. Decision 391 on Common Regime on Access to Genetic Resources. (accessed on 30 December 2009 at http://www.cbd.int/doc/measures/abs/msr-abs-acu-en.pdf)

Armour K. J., and Harrison P. S. 2007. Poisons and politics—indigenous rights and IP protection. *World Patent Information* 29: 255–261.

ASEAN, 2000. ASEAN Framework Agreement on Access to Biological and Genetic Resources (draft text). (accessed on 30 December 2009 at http://www.ictsd.org/downloads/2008/04/asean_framework_agreement.pdf)

Beebe J. 1995. Basic concepts and techniques of rapid appraisal. *Human Organization* 54: 42–51.

Bijoy C. R. 2007. Access and benefit sharing from the Indigenous People's Perspective: The TBGRI Kani Model, 3/1. *Law, Environment and Development Journal.* (accessed May 24, 2009 from http://www.lead-journal.org/content/07001.pdf).

Biodiversity Act. 2002. Biodiversity Act, 2002. No. 18 of the Gazette of India Extraordinary, 5 February 2003, Government of India, New Delhi. (accessed on 30 December 2009 at http://www.nbaindia.org/act/act_english.htm).

Biodiversity Law. 1998. Biodiversity Law of 1998, Law No. 7788 of 1998, Official Gazette No. 101 of 27 May 1998, Costa Rica.

Biodiversity Rules. 2004. (with effect from 15 April 2004) Ministry of Environment and Forest, Government of India, New Delhi. (accessed on 30 December 2009 at http://www.nbaindia.org.rules.htm).

Carrizosa S., Brush S., Wright B., and McGuire P. (Eds.). 2004. *Accessing Biodiversity and Sharing the Benefits. Lessons from Implementing the CBD.* IUCN Environmental Policy and Law Paper No. 54. IUCN, Cambridge, UK.

Convention on Biological Diversity (CBD). 1992. *The Convention on Biological Diversity.* United Nations Environment Programme, Geneva.

Convention on Biological Diversity (CBD) Secretariat. 2002. *Bonn Guidelines on Access to Genetic Resources and Fair and Equitable Sharing of the Benefits Arising Out of Their Utilisation.* Secretariat of the Convention on Biological Diversity, Montreal, Canada.

Cordell G. A. 2000. Biodiversity and drug discovery—a symbiotic relationship. *Phytochemistry* 55: 463–480.

Crouch N. R., Douwes E., Wolfson M. M., Smith G. F., and Edwards T. J. 2008. South Africa's bioprospecting, access and benefit-sharing legislation: current realities, future complications and a proposed alternative. *South African Journal of Science* 104(9): 355–366.

Cunningham A. B. 1988. *Development of a Conservation Policy on the Herbal Medicine Trade in Southern Africa: Zulu Medicinal Plants.* Investigational Report 29. Institute of Natural Resources, University of Natal, Pietermaritzburg, South Africa.

Cunningham A. B. 1990. People and medicines: the exploitation and conservation of traditional Zulu medicinal plants. *Proceedings of the Twelfth Plenary Meeting of AETFAT*, Hamburg, 979–990.

Dalton R. 2004. Bioprospects less than golden. *Nature* 429: 596–600.

Department of Environmental Affairs and Tourism (DEAT). 2008. National Environmental Management Act, 2004 (Act No. 10 of 2004): Commencement of Bioprospecting, Access and Benefit Sharing Regulations 2008. Government Gazette No. 30739, 8 February 2008. DEAT, Pretoria, South Africa.

Department of Trade and Industry (DTI). 2005. Patents Amendment Act, 2005 (Act No. 20 of 2005). Government Gazette No. 28319, 9 December 2005. DTI, Pretoria, South Africa.

Fenwick S. 1998. Bioprospecting or biopiracy? *Drug Discovery Today* 3: 399–402.

International Development Research Center (IDRC). 2004. Fine tuning the CBD: the Bonn Guidelines. http://www.idrc.ca/en/ev-67666–201–1-DO_TOPIC.html (accessed April 21, 2009).

Iwu M. M. 1993. *Handbook of African Medicinal Plants.* CRC Press, Boca Raton, FL.

Kohli K. 2007. The Biodiversity Act: A review, http://www.indiatogether.org/2007/jun/env-biofinepr.htm (accessed 30 December 2009).

Luseba D., and Van der Merwe D. 2006. Ethnoveterinary medicine practices among Tsonga speaking people of South Africa. *Onderstepoort Journal of Veterinary Research* 73: 115–122.

Lybbert T. J. 2007. Patent disclosure requirements and benefit sharing: a counterfactual case of Morocco's argan oil. *Ecological Economics* 64: 12–18.

Matekaire T., and Bwakara T. M. 2004. Ethnoveterinary medicine: a potential alternative to orthodox animal health delivery in Zimbabwe. *International Journal of Applied Research in Veterinary Medicine* 2: 269–273.

NEMBA. (2004). National Environmental Management Act, 2004 (Act No. 10 of 2004). Government Gazette No. 26436, 7 June 2004. Department of Environmental Affairs and Tourism, Pretoria.

Organization for African Unity. 2000. African Model Legislation for the Protection of the Rights of Local Communities, Farmers and Breeders and for the Regulation of Access to Biological Resources. (accessed on 30 December 2009 at http://www.cbd.int/doc/measures/abs/msr-abs-oau-en.pdf)

Posey D. A., and Dutfield G. 1996. *Beyond Intellectual Property Rights: Towards Traditional Resource Rights for Indigenous and Local Communities*. IDRC, Ottawa, Canada.

Reyneke J. L. 1971. Towery by die Tswana met besondere verwysing na die Kgatla-Bagakgafela. PhD thesis, Department of Anthropology, University of Pretoria, South Africa.

Richerzhagen C. 2007. Effectiveness and perspectives of access and benefit-sharing regimes in the Convention on Biological Diversity—a comparative analysis of Costa Rica, the Philippines, Ethiopia and the European Union. Dissertation.

Richerzhagen C., and Holm-Mueller K. 2005. The effectiveness of access and benefit sharing in Costa Rica: implications for national and international regimes. *Ecological Economics* 53: 445–460.

Seiler A., and Dutfield G. 2001. *Regulating Access and Benefit Sharing: Basic Issues, Legal Instruments and Policy Proposals*. German Federal Agency for Nature Conservation, Bonn.

Shiva V. 2005. Bioprospecting as sophisticated biopiracy. In: Burrows B (Ed.). *The Catch: Perspectives in Benefit Sharing*. Edmonds Institute, Washington, D.C., 15–35.

Siebenhuner B., and Suplie J. 2005. Implementing the access and benefit-sharing provisions of the CBD: a case for institutional learning. *Ecological Economics* 53: 507–522.

Soejator D., Fong H. H. S., Tan G. T., et al. 2005. Ethnobotany/ethnopharmacology and mass bioprospecting: issues on intellectual property and benefit sharing. *Journal of Ethnopharmacology* 100: 15–22.

Ten Kate K., and Laird S. A. 1999. *The Commercial Use of Biodiversity. Access to Genetic Resources and Benefit-sharing*. Earthscan, London.

Van Overwalle G. 2005. Protecting and sharing biodiversity and traditional knowledge: holder and user tools, *Ecological Economics* 53: 585–607.

Van Wyk B.-E., Van Oudtshoorn B., and Gericke N. 1997. *Medicinal Plants of South Africa*. Briza, Pretoria, South Africa.

WTO. 1994. Agreement on Trade-Related Aspects of Intellectual Property Rights (TRIPS) (Annex 1C of the Marrakesh Agreement). Agreement Establishing the World Trade Organisation. World Trade Organisation, Geneva, Switzerland.

Wynberg R., and Laird S. 2007. Bioprospecting—tracking the policy debate. *Environment*. 49(10): 20–32.

3 Phytochemical Methods

Bukirwa Irene Kamara

CONTENTS

3.1 INTRODUCTION

Natural products chemistry is becoming increasingly interdisciplinary in practice. In the study of natural products chemistry, isolation and purification are mandatory first steps; success and failure are often determined by these steps. Unlike specific protocols followed in synthetic chemistry, which can be remembered and repeated, details of specific isolation steps are usually forgotten, and in most cases isolation is regarded as trial and error. This chapter covers conventional techniques in the separation, purification, and structural elucidation of metabolites from medicinal plants. The theory and practice of natural products isolation techniques that comprehensively serve the natural product scientist in the investigations of plants are outlined. The backdrop of this discussion is the use and knowledge of traditional medicines. It is therefore important to discuss applications of natural products chemistry in the field of herbal medicines and its impact on ethnoveterinary and ethnopharmacological sciences.

Traditional (herbal) medicine is a term loosely used to describe ancient and culture-bound health practices that existed before the application of conventional science to health matters in official, modern, scientific medicine or allopathy. Traditional medicine from natural products (terrestrial and marine environments) still forms the basis for primary health care in 80% of the developing world (Farnsworth and Soejarto, 1985).

In contrast to the primary metabolite compounds such as carbohydrates, the essential amino acids and polymers derived from them that are ubiquitous in living organisms and essential for life, natural products are the secondary constituents of plants. Higher plants have been described as chemical factories that are capable of synthesizing unlimited numbers of highly complex and unusual bioactive chemical substances (Farnsworth, 1988). These bioactive substances are normally present in the plants in quantities that cannot meet or sustain demand. Herbal products, nutraceuticals, and other food supplements are therefore subject to increasingly rigorous investigation and development. Industry standardization, as well as labeling requirements under U.S. federal regulations, have prompted tighter regulation and monitoring of the active ingredients in these products. Extraction and identification of the components in raw and processed plant materials are essential for the quality control of existing products and the development of new ones. Because native plant habitats are destroyed almost daily, sadly many medicinally valuable plants will be gone before scientists can even investigate them. There is a possibility of documentation of traditional knowledge and ethnoveterinary medicine through scientific research. In the past, both conventional and participatory methods have been used to document local knowledge in general. Both approaches have their place, and their results can be complementary and possibly cross-validate each other.

Bioactive principles from medicinal plants are secondary metabolites and not a nutritive element for the plant, so they are not concentrated in the plant. The separation process usually contributes to capturing of the secondary metabolites, which through the efforts of the synthetic chemists may also be synthesized to ensure more abundant supplies for formulation. Previous techniques in the isolation and identification process have indeed had their fair share of flaws. Nevertheless, the practical difficulties of natural products drug discovery are being overcome by advances in separation technologies and in the speed and sensitivity of structure elucidation.

3.2 THE EXTRACTION PROCESS

Although semi- or fully automated purification processes are routinely followed during the final stages of isolation, the initial plant material preparation stages follow a more manual, slow, and tedious protocol. Collection and preparation of plant material (usually a manual process) is subsequently followed by extraction partitioning and purification processes. Method choices for samples and the obtainable analytes are described in Table 3.1. The techniques that utilize pressurized fluid or solvent extraction reduce the amount of solvent needed for extraction and the extraction time when compared to more traditional techniques such as Soxhlet extraction. Solvents, reagents, glassware, and other sample-processing hardware may yield artifacts or interferences in sample analysis. Therefore, factors such as temperature and pressure are well observed.

Regardless of the extraction technique, extracts are concentrated under vacuum using a rotatory evaporator for large volumes of solvents (>5 mL) or nitrogen blown through for small volumes (<5 mL), ensuring that volatile components are not lost. Removal of solvents should be carried out immediately after extraction as natural products may be unstable in the solvent. Aqueous solvents are generally freeze-dried using a freeze dryer. Dried extracts should be stored at –20°C prior to screening for biological activities to decrease the possibility of chemical degradation.

3.2.1 Collection of Material

There are a number of approaches that can be used to collect material needed in the exercise to discover new drugs. The ethnobotanical approach is based on

TABLE 3.1
Extraction Techniques for Different Analytes

Matrix Types	Extraction Technique	Analytes
Aqueous	Liquid-liquid extraction	Semivolatile and nonvolatile organics
	Solid-phase extraction	
Solids	Soxhlet extraction	Semivolatile and nonvolatile organics
	Pressurized fluid extraction (PFE)	Semivolatile and nonvolatile organics
	Microwave extraction	Semivolatile and nonvolatile organics
	Ultrasonic extraction	Semivolatile and nonvolatile organics
	Supercritical fluid extraction (SFE)	Semivolatile petroleum hydrocarbons
		Polynuclear aromatic hydrocarbons
		Polychlorinated biphenyls
		Organochlorine pesticides

observations by ethnobotanists. Valuable local medicinal knowledge is gathered with regard to the use of particular plants for specific ailments by the indigenous people. This knowledge is subsequently used in collection and testing of material for biological activities. When following the chemotaxonomic approach, collection of plant material is based on prediction that taxonomically related plants may contain structurally similar compounds. For example, the plant family Solanaceae is a rich source of alkaloids of the tropane type, while the Combretaceae family is a source of stilbenoids. Last, biomass can be collected randomly, regardless of any previous knowledge of the chemistry or biological activities. This approach relies on the abundance of plants in a geographical area. In all cases, the biomass is collected, dried, and extracted with suitable organic solvents.

3.2.2 Conventional Extraction Processes

The first step of the extraction process is to release and solubilize the smaller secondary metabolites in the matrix, resulting in the initial extract. To produce a good extract, plant material or biomass is usually dried (to avoid water in the extracts), ground, or pulverized prior to addition of the extracting solvent. Most bulk of the biomass, irrespective of whether it is plants or microbes, exists as fairly inert, insoluble, and often polymeric material, such as cellulose of plants or fungi and the microbial cell wall (Cannell, 1998). The highly complex initial extracts can be fractionated using solvents with different polarities. A sample of known volume or weight is repeatedly extracted with different solvents in order of increasing polarities, and the resultant extracts are processed according to the requirements. The value of these polarity-based extractions is that the chemical complexity of the biomass is simplified according to the solubility of the components. Concentration of the compounds in the extracts is also optimized. The initial extractions with low-polarity solvents (e.g., chloroform, dichloromethane, ethyl acetate, or ether) yield the more lipophilic (fat-soluble) components, while the polar alcohols, acetone, and water isolate a broader spectrum of hydrophilic (polar- or water-soluble) compounds from the biomass. For example, when extracting flavonoids from the powdered leaves of herbal teas such as honeybush tea (*Cyclopia* spp.), it is important to extract the chlorophyll at the beginning because the green color may hamper the purification process. In the case of the flavonoids, acetone will mainly extract the less-polar monomeric flavonoids. Subsequent extraction of the same material with methanol will give extracts rich in flavonoid glycosides and dimeric flavonoids. Further extraction with the more polar acetone/water (70:30) will yield very polar extracts containing trimeric and highly glycosylated flavonoids, tannins, and sugars.

The choice of solvent shown in Table 3.2 is dependent on the physicochemical properties of the compounds of interest. These properties include partition coefficients in water or organic solvents, relative polarity of the molecule, and stability of the molecule in light or dark, as well as the temperature employed during the extraction process. Water-soluble compounds and proteins are best extracted in water or buffers, while the organic-soluble compounds are extracted with organic solvents. If a plant is under investigation from an ethnobotanical perspective, the extraction

TABLE 3.2
Different Solvents Used to Extract Different Phytochemicals

Solvent	Fats/Oils	Amino Acids	Chlorophyll	Alkaloids	Phenolics Aglycones	Sugars	Glycosides
Hexane	√						
Cylohexane	√						
Toluene	√			√			
Chloroform			√	√			
Dichloromethane			√	√	√		
Diethylether			√	√	√		
Ethylacetate			√	√	√		√
Acetone			√	√	√		√
Ethanol					√		√
Methanol		√			√	√	√
Water		√			√	√	√
Aqueous acid		√				√	
Aqueous alkali		√				√	

should mimic the traditional use. This means that if the indigenous people use water to extract, then an identical or very similar method should be used in the laboratory to extract the same plant constituents. Poor extraction processes may result in loss of active compounds or degradation of natural products and consequent loss of biological activity.

Several extraction methods are available, the simplest being cold extraction (in a large flask with agitation of biomass using a stirrer), in which the ground dry material is extracted at room temperature with solvents of increasing polarity: first hexane (or petroleum ether) then chloroform (or dichloromethane), ethyl acetate, acetone, methanol, and finally water. The main advantage of this cold extraction is that there is little potential degradation of the natural products. Extraction of chlorophyll with chloroform has been successfully accomplished with minimum interference from the aromatic compounds. Oils and terpenes can easily be extracted with hexane, leaving behind other classes of compounds. Extraction using solvents with different polarities can be used to roughly fractionate a complex mixture of compounds. Reflux extraction (Vogel et al., 1996) is applied in extractions in which continuous application of heat is required. A mantle or water bath is usually a safer source of heat than a direct flame, especially in the case of flammable solvents such as ether. The Soxhlet process (Houghton and Raman, 1998) is useful for the exhaustive extraction of plant material with a single solvent at a time (e.g., when using hexane for defatting, in subsequent extractions with increasing polarities, and when a high yield of a particular component is required). It is, however, unsuitable for neutral, acidified, and basified solvent mixtures.

Other classes of compounds, such as acids or bases, present in the biomass can be extracted using a tailored protocol. The most common group of natural products that are extracted in this manner is the alkaloids, which are often present in plant materials as salts. Extraction of the basic compounds is as follows:

1. Alkaloids can be recovered from their salts by making the dry powdered plant material alkaline with aqueous ammonia. This leaves the alkaloids as free bases that are no longer ionic salts and are more soluble in organic solvents such as ethyl acetate and dichloromethane.
2. This increased solubility in organic solvents allows free partitioning of the free bases into organic solvent, which can then be separated from the aqueous ammonia layer.
3. The organic layer will contain the free bases, which can be extracted with aqueous acid; for example, by extracting three times with 2*M* hydrochloric acid, the alkaloids will transfer from the organic phase to the aqueous phase as hydrochloride salts. The organic layer can be tested using Dragendorff's reagent to ensure total extraction of the alkaloids into the acid layer.
4. Basification of the acid layer results in precipitation of the alkaloids (which are no longer salts and therefore no longer soluble in aqueous media), which can be extracted by organic solvents.

3.2.3 Bioassay-Guided Isolation

In the past, natural products chemists have analyzed plants with the sole aim of establishing their chemical composition and in some cases isolating active compounds. The modern isolation approach, focused on discovery of drugs, has changed the isolation process into a bioassay-guided approach. In the bioassay-guided approach, bioassays for antifungal, anticancer, antimicrobial, antimalarial, and antioxidant activities are performed on the extracts, and the active extracts are investigated further. Bioassay-guided isolation is essential for the drug discovery industry, in which high-throughput work is vital for purposeful results. However, when it comes to chemical analysis of a plant, little is analyzed because the fractions without activity are discarded. In essence, the phytochemistry of the plant is vaguely addressed.

The activity of an extract may change over a period of time, during preparation (e.g., due to fermentation, heat), or after fractionation. Thus, compounds can undergo reactions activated by the surroundings, such as heat, solvents in which they are stored, or light (photochemical reactions). This may cause changes in the activity of the particular extract. Change caused by living organisms (e.g., microbial attacks) is referred to as deterioration, while decomposition is caused by a chemical change. It is therefore important to take precautions when handling extract to minimize deterioration and decomposition.

The general bioassay-guided isolation procedure follows the protocol outlined in Scheme 3.1.

General bioassay-guided isolation procedure

Sample preparation using appropriate solvent

⇩

Bioassay testing of initial extract

⇩

Fractionation by column chromatography

⇩

Bioassay testing of fractions from column

⇩

{ Fractionation by more sensitive HPLC

⇩

Bioassay testing of each HPLC fractions } Repeat these steps to yield pure compounds

⇩

Structural elucidation by spectroscopic and spectrometric methods

⇩

Structure confirmation by synthesis

SCHEME 3.1 Systematic approach to bioassay-guided isolation.

3.2.4 Nonconventional Extraction Processes

The most widely used extraction processes have traditionally been based either on different liquid extraction methods or on vapor-phase extraction methods (Starmans and Nijhuis, 1996). However, there are also a number of nonconventional extraction methods in use that are all, in principle, solid-liquid extractions (SLEs) that introduce some form of additional energy to the process to facilitate the transfer of analytes from sample to solvent. These methods include ultrasonic extraction, microwave-assisted extraction, and pressurized liquid extraction (Huie, 2002; Zygmunt and Namieśnik, 2003), as well as vortical (turbo) extraction. In addition, electrical energy extraction could also be explored (Vinatoru, 2001). Alternatively, forced-flow SLE techniques, such as medium-pressure SLE and rotation planar extraction, which are methods by which the extraction solvent is forced through the sample bed either by pressure or by centrifugal force, respectively, increase the efficiency of the extraction process (Nyiredy, 2001). The main advantage of these nonconventional methods compared to conventional SLE methods is the increased extraction efficiency, which leads to increased yields or shorter extraction times.

3.2.4.1 Supercritical Fluid Extraction

One of the nonconventional methods whose application has steadily increased is supercritical fluid extraction (SFE), which is based on the properties of gases compressed and heated to a state above their critical pressure and temperature (Tservistas, Scheper, and Freitag, 2000).

Essential oils are better extracted using the SFE processes. The characteristic smell of plant materials is usually the result of thc complex interactions occurring among hundreds of compounds. Correct reproduction of the natural fragrance in a concentrated extract is therefore a complex task. The presence of thermolabile compounds, the possibility of hydrolysis, and hydrosolubilization are serious obstacles in the reproduction of natural fragrances. Moreover, severe legislative restrictions are currently being proposed to eliminate solvent residues in these products when used in the food, pharmaceutical, and cosmetic industries. All these problems can be solved easily by employing supercritical fluid in extracting essential oils.

3.2.4.2 Accelerated Solvent Extraction

To screen the vast array of potential product candidates, laboratories must rapidly characterize raw materials for levels of active components. This is normally accomplished by extraction with liquid solvents followed by chromatographic analyses. While chromatographic analysis can be automated, many common extraction techniques are labor intensive and time consuming and require large amounts of costly solvents. Accelerated solvent extraction (ASE) was developed as a solution for laboratories with increased sample throughput needs. By using common organic and aqueous solvents at elevated temperatures and pressures, ASE increases the speed and efficiency of the extraction process compared to the traditional methods described. By means of ASE, sample sizes of 1–10 g are typically extracted in 12–17 min with only 10–15 mL of solvent. Solvents can be selected based on the polarity of the analyte and compatibility with any postextraction processing steps and quantification equipment. Commercially available nutritional supplements, such as hypericin from *Hypericum perforatum* (St. John's wort) and berberine from *Hydrastis canadenis* (goldenseal root) are extracted using ASE.

3.3 ISOLATION METHODS

Often, elegant structure determination of natural products overshadows the most important and often the most difficult part of the work, which is the isolation and purification process. Advances in chromatographic techniques are allowing scientists to venture ever further into hitherto inaccessible areas of natural product isolation. This is particularly true of hydrophilic, nonvolatile, and often labile plant components. But, because of the almost infinite variety of natural products and the diverse, subtle interactions among solvent, solute, and chromatographic substrate, isolation and purification often seem more of an art than a science.

Purification of pure substances from complex extracts and most of the time isolation of the microscale components have spawned development of new separation technologies and upgrading of the conventional methods. Gas chromatography (GC), liquid chromatography (LC), high-performance liquid chromatography (HPLC), and thin-layer chromatography (TLC) are well-known classical techniques, which for the ease and availability of resources to most scientists are discussed in this chapter. Modern countercurrent chromatography (CCC), including rotation locular CCC, droplet CCC, and coil CCC, have been successful in isolation of the molecules that have proven difficult to isolate. Partitioning chromatography and derivatization are procedures that are normally affordable and applicable in most laboratories, and they are still used to deliver good results.

3.3.1 Polarity of Solvents

The polarity can be measured as the dielectric constant or the dipole moment of a compound. The polarity reflects the balance between a polar component (e.g., OH) and a nonpolar hydrocarbon component existing in the same molecule. On an operational basis, solvents that are miscible with water are polar. Making solvent choices for the separation processes is directed by polarities of the components to be isolated. For example, when purifying molecules that are highly hydroxylated, implying that they are highly polar, the solvent used should include acetone and methanol. A mixture of hexane/acetone/methanol in the ratio of 6:3:1 has been used to fractionate underivatized polyphenols using TLC (Ferreira et al., 1998). It must be mentioned that selecting the right solvent is more of a trial-and-error exercise. Most scientists have developed solvent systems in their laboratories; see Table 3.A1 (Appendix) for examples.

3.3.2 TLC: Retention Factor

The retention factor R_f is defined as the distance traveled by the compound divided by the distance traveled by the solvent.

$$R_f = \frac{\text{distance traveled by compound}}{\text{distance traveled by solvent front}}$$

For example, if a compound travels 2.5 cm and the solvent front travels 3.5 cm, the R_f is 0.71, as demonstrated in Figure 3.1.

Since R_f for a compound is dependent on the solvent system, type of adsorbent (stationary phase), thickness of the adsorbent, amount of material loaded, and temperature, and these factors are difficult to keep constant from experiment to experiment, relative R_f values are generally considered. "Relative R_f" means that the values are reported relative to a standard, or it means that you compare the R_f values of compounds run on the same plate at the same time.

If two substances have the same R_f value, they are likely (but not necessarily) the same compound. If they have different R_f values, they are definitely different compounds. Note that this identity check must be performed on a single plate because it is difficult to duplicate all the factors that influence R_f exactly from experiment to experiment. The larger an R_f of a compound, the greater the distance it travels on the TLC plate. When comparing two different compounds run under identical chromatography conditions, the compound with the larger R_f is less polar because it interacts less strongly with the polar adsorbent on the TLC plate. Conversely, if you know the

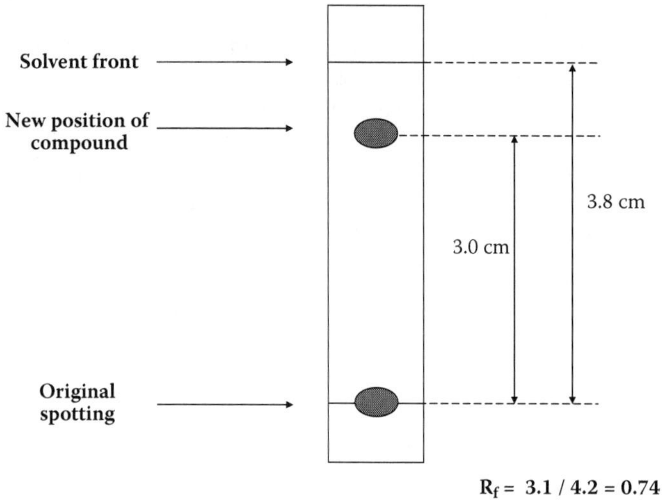

FIGURE 3.1 Calculation of R_f value.

structures of the compounds in a mixture, you can predict that a compound of low polarity will have a larger R_f value than a polar compound run on the same plate.

3.3.3 Types of Chromatography

The commonly used chromatographic methods include adsorption, normal phase, reversed phase, chiral, ion exchange, molecular exclusion, and ion pair/affinity.

3.3.3.1 Adsorption Chromatography

Adsorption chromatography is probably one of the oldest types of chromatography around that is unquestionably the major isolation technique applied in natural products chemistry. It utilizes a mobile liquid or gaseous phase that is adsorbed onto the surface of a stationary solid phase. The equilibration between the mobile and stationary phase accounts for the separation of different solutes.

3.3.3.1.1 Normal-Phase Chromatography

Normal-phase chromatography is a chromatographic technique that uses organic solvents for the mobile phase and a polar stationary phase. Here, the less-polar components elute faster than the more polar components.

3.3.3.1.2 Reversed-Phase Chromatography

Reversed-phase chromatography, a bonded-phase chromatographic technique, uses water as the base solvent. Separation based on solvent strength and selectivity also may be affected by column temperature and pH. In general, the more polar components elute faster than the less-polar components.

3.3.3.1.3 Chiral Chromatography

Separation of the enantiomers can be achieved on chiral stationary phases by formation of diastereomers via derivatizing agents or mobile-phase additives on achiral stationary phases. When used as an impurity test method, the sensitivity is enhanced if the enantiomeric impurity elutes before the enantiomeric drug.

3.3.3.1.4 Ion-Exchange Chromatography

In ion-exchange chromatography, a resin (the stationary solid phase) is used to covalently attach anions or cations onto it. Solute ions of the opposite charge in the mobile liquid phase are attracted to the resin by electrostatic forces. Separation is based on the charge-bearing functional groups, anion exchange for sample negative ion (X^-) or cation exchange for sample positive ion (X^+). The resins are divided into two groups: the cation and anionic exchangers. The cationic exchangers have acidic groups (CO_2H, SO_3H) and are able to exchange their protons with cations of natural products. The anionic exchangers, on the other hand, have basic groups ($^-N^+R_3$) that are incorporated into the resin and can exchange their anions with anions from the natural products. Gradient elution by pH is common. The separation of small polar compounds, in particular ionic natural products, is often problematic. This is because polar compounds are generally strongly adsorbed with normal-phase resins such as silica or alumina and even with the polar solvents and modifiers (e.g., acid and base), so efficient separations may not be achievable. This technique is limited to natural products that carry charge on their functional groups. Separation is achieved by differences in affinity between ionic components (polar natural products) and the stationary phase. The ion-exchange resins may be used in open column chromatography or in closed columns in applications such as HPLC.

3.3.3.1.5 Ion Pair/Affinity Chromatography

Ion pair/affinity chromatography is the most selective type employed. It utilizes the specific interaction between one kind of solute molecule and a second molecule that is immobilized on a stationary phase. The more popular reversed-phase mode uses a buffer and an added counterion of opposite charge to the sample, with separation being influenced by pH, ionic strength, temperature, and concentration and type of organic cosolvents. Affinity chromatography, common for macromolecules, employs a ligand (biologically active molecule bonded covalently to the solid matrix) that interacts with its homologous antigen (analyte) as a reversible complex that can be eluted by changing buffer conditions. For example, the immobilized molecule may be an antibody to some specific protein. When solute containing a mixture of proteins is passed by this molecule, only the specific protein is reacted to this antibody, binding it to the stationary phase. This protein is later extracted by changing the ionic strength or pH.

3.3.3.2 Partition Chromatography

Partition chromatography is based on a thin film formed on the surface of a solid support by a liquid stationary phase. Solute equilibrates between the mobile phase and the stationary liquid.

3.3.3.2.1 Countercurrent Chromatography

Countercurrent chromatography is a type of liquid-liquid chromatography in which both the stationary and mobile phases are liquids. It involves mixing a solution of liquids, allowing them to settle into layers, and then separating the layers. This is a soft separation method and relies on solubility of natural products and not the physical interaction with another medium. Partitioning uses two immiscible solvents to which the extract is added: This can be sequential by using immiscible solvents of increasing polarity, for example, water/hexane (for which a nonpolar fraction is generated in the organic layer) and water/chloroform or water/ethyl acetate (for which a medium-polar fraction is generated in the organic layer). Excellent separations can be obtained from the partitioning process; for example, monoterpenes are easily separated from phenolics such as tannins. Craig countercurrent distribution uses a similar concept to a liquid-liquid extraction in a separatory funnel. A method of multiple liquid-liquid extractions is countercurrent extraction, which permits the separation of substances with different distribution coefficients (ratios) (Craig and Craig, 1950). The Craig apparatus consists of a series of glass tubes (1, 2, 3, …, 20) arranged so that the lighter liquid is transferred from one tube to the next. All extractions take place simultaneously in all tubes of the apparatus, which is usually driven electromechanically.

The lower phase of the heavier solvent (e.g., water) is the "stationary phase," whereas the upper phase of the lighter solvent (e.g., hexane) is the "mobile phase." In the beginning, tube 1 contains the mixture of substances to be separated in the heavier solvent, and all the other tubes contain equal volumes of the same solvent. The lighter solvent is added to tube 1, extraction (equilibration) takes place, and the phases are allowed to separate. The upper phase of tube 1 is then transferred to tube 2; fresh solvent (mobile phase) is added to tube 1, and equilibrium is reached again. The upper layers of tubes 1 and 2 are simultaneously transferred to tubes 2 and 3, respectively, and the cycle is repeated and so on. Obviously, substances with a higher distribution ratio move faster than those with a lower distribution ratio. It is interesting to examine the distribution of a substance A in each tube after a given number of equilibration/transfer cycles. The greater the difference of the distribution ratio of various substances, the better is the separation between each. A much larger number of tubes (more than 20) is required to separate mixtures of substances with almost similar distribution ratios. The Craig apparatus is now only rarely used because modern chromatographic techniques are by far more efficient and convenient.

3.3.3.3 Molecular Exclusion Chromatography

Also known as gel permeation or filtration, separation in molecular exclusion chromatography is based on the molecular size or hydrodynamic volume and to some extent adsorption of the components. This type of chromatography lacks an attractive interaction between the stationary phase and solute. This procedure is widely used as an initial cleanup step. This technique employs a cross-linked dextran (sugar polymer Sephadex) that, when added to a suitable solvent (e.g., chloroform or ethyl acetate, ethanol), swells to form a gel matrix. Molecules that are too large for the pores of the porous packing material on the column elute first, small molecules that enter the pores elute last, and the elution rates of the rest depend on their relative sizes. This is an excellent method for separating out chlorophylls, fatty acids, glycerides, and other large molecules. With the same principle, Sephadex LH-20 is commonly used with ethanol as the eluent to fractionate flavonoids. This is a nondestructive soft method with high recovery, and a large quantity of extract (hundreds of milligrams to grams) may be separated. A further benefit of this technique is that there is a variety of gels available with a variety of pore sizes that can be used to separate compounds from 500 to 250,000 Da. This is the method of choice for large-scale extractions (e.g., when fractionating mixtures for bioassay-guided separations.

3.3.4 Chromatography Techniques

Chromatographic separations can be carried out using a variety of supports, including immobilized silica on glass plates (TLC), volatile gases (GC), paper (paper chromatography), and liquids that may incorporate hydrophilic, insoluble molecules (liquid chromatography). The adaptation of bonded-phase packings to all pressure liquid chromatography, complimented by the diverse chemically bonded phases, including the chiral templates for chiral resolutions, uses separation techniques capable of handling any isolation problem.

3.3.4.1 Thin-Layer Chromatography

Thin-layer chromatography is the simplest of the more common adsorption chromatographic techniques (Vogel et al., 1996). TLC is the most universal chromatographic test method when compared to gas and liquid chromatography as all components are present on the separation plate. With appropriate detection techniques, all components can be seen. However, it normally is not as accurate or sensitive as HPLC. TLC has a higher analytical variation than HPLC. Separation is based on migration of the sample spotted on a coated (stationary-phase) plate with one edge dipped in a mixture of solvents (mobile phase). Adsorption on silica gel in TLC relies on physical attraction of the compounds with another medium. The whole system is contained in an enclosed tank. Detection techniques of the compounds include fluorescence, ultraviolet (UV), and chromogenic sprays (universal and specific) for compounds that are not naturally colored. Location of the analyte on the TLC plate is described by the R_f value, which is the ratio of the migration distance of the compound of interest to the mobile-phase front.

In TLC, there is no direct online coupling between chromatographic development and detection. The chromatogram is freed from the mobile phase prior to the detection step. Thus, different chromatographic conditions, including changing the mobile phase, can be varied during analysis of a TLC chromatogram. For a natural products scientist, this is necessary because extracts and samples usually contain compounds that may have the same R_f when using some mobile phases but may differ in another. In micro-, nano-, and pictogram quantities and for routine use, the TLC technique occupies a firm place. In microchemical detection methods, the TLC plate is sprayed with or dipped into a suitable reagent (Table 3.3) to form absorbing or fluorescent derivatives with the analytes, which can then be detected. Table 3.A1 gives information on different TLC mobile phases and spray reagents that can be used.

Biological detection of separated substances on the TLC plate in situ has been applied to the screening of various extracts, such as antimicrobial and antioxidant activity or

TABLE 3.3
Chemical Tests for Different Classes of Compounds

Chemical Class	Qualitative Test	Observation
Carotenoids in chloroform	Concentrated sulfhuric acid or	Deep blue color
	5% antimony chloride	Deep blue color
Alkaloids	Dragendorff's reagent or[a]	Orange precipitate
	Mayer's reagent[b]	White or buff precipitates
Phenolic compounds	Ferric ions	Dark green, blue, or purple-black
Terpenoids	Liebemann-Burchard reaction[c]	Red-blue
Sugars	Fehling's solution	Red precipitate
Amino acids	Ninhydrin	Reddish or blue color

[a] Potassium iodobismuthate.

[b] Potassium iodomercurate.

[c] Acetic anhydride with a few drops of concentrated sulfuric acid.

Source: From Houghton and Raman, 1998. *Laboratory handbook for the fractionalization of natural extracts*, London, Chapman & Hall.

general toxicity (Eberz et al., 1996; Weins and Jork, 1996; Hostettmann et al., 1997). The most common sorbent of choice in TLC by far is still silica, whereas in HPLC analyses nonpolar reversed-phase sorbents have almost entirely replaced silica and other normal-phase sorbents. The mean particle size and particle size distribution of the silica gel used as adsorbent depends on the nature of the separation task: For high-performance thin-layer chromatography (HPTLC) plates, the mean particle size is approximately 5 μm with a narrow particle size distribution; for TLC, it is approximately 12 μm, and the particle size distribution is wider. The pore diameter for both is approximately 60 Å, and the surface area is approximately 500 m^2/g (Gocan, 2002). Other relatively frequently used polar TLC sorbents include alumina, cellulose, chitin and chitosan, and polar chemically silica-bonded phases such as aminopropyl, cyanopropyl, and diol phases (Geiss, 1987).

3.3.4.2 Preparative Thin-Layer Chromatography

The most conventional and common method of isolation of pure substances at the final stages is to use preparative TLC (PTLC). It is amenable to low-budget conditions and easy to execute. It is labor intensive, and compounds that are in very low concentrations may be missed. The method works on the same principles as TLC (Hostettmann et al., 1997). It involves applying the plant extract (10–15 mg/plate) in the form of a thin line (band) on a TLC plate. The PTLC plates are normally 0.5–1 mm thick. The plate is developed in a solvent system known to separate the components. Although gram quantities of material can be separated by PTLC, most applications involve milligram quantities. Silica gel is the most common adsorbent and is employed for the separation of both lipophilic and hydrophobic substance mixtures. The maximum sample load for a silica layer 1.0 mm thick is about 5 mg/cm^2. PTLC plates can be either self-made or purchased. As the particle sizes of the TLC and PTLC adsorbents are the same, choice of eluent is determined from a preliminary TLC investigation.

Addition of acetic acid or diethylamine in small amounts is useful for the preparation of acidic and basic compounds, respectively. Elution of PTLC usually takes place in glass tanks, which may hold several plates at a time.

3.3.4.3 Paper Chromatography

Paper chromatography is similar to TLC in practice but quite different in principle. Although paper consists mainly of cellulose, the stationary phase is not cellulose but the water that is adsorbed or chemically bound to it. Development is carried out by passing comparatively nonpolar mobile phase through the cellulose fibers, partitioning the solutes between the bound water and the mobile phase. Paper chromatography thus operates by a liquid-liquid partitioning process rather than adsorption on the surface of a solid. Paper chromatography is usually used to separate polar substances such as amino acids and sugars. Paper chromatography has been used extensively in flavonoid chemistry. For preliminary assessment of the extract, it is general practice to run a spotted paper first in an alcoholic solvent, such as saturates secondary butanol, butanol/acetone/water (BAW) or saturated toluene/butanol/acetone (TBA) in the longer dimension (12–30 h) until the solvent line approaches the end of the sheet. The chromatogram is then removed from the tank, air-dried in a fume cupboard, and the paper is run in the second dimension using a more polar solvent, usually an aqueous solvent (2–15% acetic acid in case of flavonoid, 4–6 h).

3.3.4.4 High-Performance Liquid Chromatography

HPLC is a modern development in column chromatography in which a powerful pump is used to force eluents through the column filled with extremely fine particles (3–10 μm) of packing material. HPLC chromatographic separation is based on interaction and differential partition of the sample (partition chromatography) between the mobile liquid phase and the stationary phase. The liquid stationary phase is adsorbed on or chemically bonded to the solid particles of the column packing. Partition chromatography is based on the same principle as extraction, in which a solute is distributed (partitioned) between two liquids. As the molecules of the components elute down the column, they partition between the eluent and the liquid stationery phase according to the partition coefficient ratio.

3.3.4.5 Gas Chromatography

GC is based on the volatilized sample transported by the carrier gas as the moving phase through the stationary phase of the column, where separation takes place by the sorption/desorption process. Samples for gas chromatographic analysis are normally low molecular weight compounds that are volatile and stable at high temperature. In this respect, residual solvents in drug substances and drug products are suitable for gas chromatographic analysis. Chemical derivatives can also be formed to achieve volatility and thermal stability.

Common detectors are flame ionization detectors (FIDs) for carbon-containing compounds, electron capture detectors (ECDs) for halogenated compounds, flame photometric detectors (FPDs) for compounds containing sulfur or phosphorous, and nitrogen-phosphorous detectors (NPDs) for compounds containing nitrogen or phosphorous. Chiral separation also can be achieved by GC. Separation by the packed column is rapidly being replaced by the capillary column, which provides improved resolution and analysis speed. The location of the analyte on the gas chromatogram is described by retention time (R_t), which is similar to HPLC.

3.4 DERIVATIZATION

Although structural elucidation is well mastered and has become quite routine, isolation and purification of specifically natural products that exist only in minute quantities have to be isolated after some derivatization or be monitored by a bioassay. For example, in flavonoid chemistry, in which the occurrence of the compounds is in microgram quantities and the structures are characterized mostly by hydroxylated aromatic rings, derivatization using reactions such as acetylation, benzylation, and methylation (Vogel et al., 1996) is necessary for good separations. Amines are also easily acetylated. Normally, a known amount of extract is derivatized and purified by repeated chromatographic methods, such as preparative chromatography and HPLC. Besides enhancing the purification processes, derivatization also facilitates structural elucidation. It is easy, for example, to quantify the hydroxyl groups present in a compound by counting the number of methyl groups resulting from the methylation experiment. Similarly, acetylation would facilitate good isolation and structural elucidation, except for tertiary alcohols, which are normally resistant to acetylation. Dimethylsulfoxide is usually a good methylating reagent. Benzylation is a valuable protection method for compounds without chromophores (e.g., sugars). However, it should be avoided when derivatizing aromatic compounds (e.g., flavonoids) because the aromatic proton resonances will be buried under the benzyl protons in the nuclear magnetic resonance (NMR) spectra. It is noteworthy to remember that some protective groups are not labile, and it is practically impossible to effect deprotection. The methyl group is usually very stable, and the demethylation process normally requires very harsh conditions, which may consequently destroy the structure.

3.4.1 ACETYLATION

Dry material is dissolved in a minimum volume of pyridine (dried over potassium hydroxide pellets) and twice the amount of dry (distilled) acetic anhydride. After 8–12 h at ambient temperatures (~60°C, normally on top of a hot oven), crushed ice is added to precipitate the acetylated material. The precipitate is filtered under pressure and the excess pyridine washed out with cold water.

3.5 ANALYSIS OF EXTRACTS AND ISOLATED COMPOUNDS

In some cases, it is possible to determine the nature of compounds one is dealing with by following simple qualitative analyses of the extracts. Also, appearances of the extract can provide clues on what the compounds may be. The anthocyanidins, due to the positive charge they carry on the oxygen, are known to have deep red-blue colors, while chalcones and quinones are normally deep yellow. The deep yellow color originates from the conjugate system that exists between the carbonyl and the double bond. Qualitative chemical analyses are normally used as guidelines to assist in determining the solvents, detection method, and route most appropriate for isolation of the components. During the isolation process, it is important to acquire sufficient quantities of the extract or compound because it may not always be easy to reproduce the same compound.

3.5.1 Qualitative Chemical Analysis

Chemical analysis can be used to determine the nature of the chemical group present in the compound (Houghton and Raman, 1998). A color reaction or precipitate in response to specific reagents usually shows the presence of a class of compounds. It should be noted that these tests are seldom specific enough to be unequivocal conclusions, and negative results do not mean that the particular group is not present. Sometimes, concentrations of the particular group are very low, and concentration or purification of the extract is required.

3.5.2 Detection Methods

Detection methods (Table 3.4) are designed to increase sensitivity and selectivity and to provide evidence concerning the quality of the separation. Classes of compounds absorb at different UV wavelengths, and different complementary colors are observed.

3.6 DETERMINING ORGANIC STRUCTURES

Up to the early 1940s, elucidation of complex structures such as terpenoids, alkaloids, and flavonoids was by chemical reactions coupled with skeletal rearrangements and biogenic concepts. Spectroscopy, introduced as early as 1949, has become indispensable for routine research. Organic structures can be determined accurately and quickly by spectroscopy, and molecular formulas, identities of functional groups, carbon connectivities, position of substituents and functional groups on the carbon framework, and stereochemical properties, including dynamic and static aspects, can be derived. The four primary spectroscopic techniques that have been in use since the 1960s are NMR, infrared (IR), ultraviolet-visible (UV-Vis), and mass spectrometry (MS) (Furniss et al., 1989; McMurry, 1996; Fox and Whitesell, 1997). X-ray crystallography is the ultimate tool in structure elucidation, but only for the analysis of well-formed crystals. While X-rays give bond length and angles, NMR tells about the carbon skeleton of the molecule, IR tells us about the types of bond in a molecule (to determine the functional groups in a molecule), and MS determines mass of the molecule and atomic composition. Of these, NMR is more important than all the rest put together, and therefore focus is directed to NMR. Advances in NMR and MS instrumentation along with experimental strategies have made it possible to tackle structural elucidation of very complex compounds made up of nonrepeating subunits.

It is a must to have a fundamental understanding of the principles of organic structures before attempting to analyze organic structures, especially when dealing with exotic compounds. Besides having general knowledge about principles of organic chemistry, there should be clear understanding of the common functional groups, types of structures that are stable (e.g., aromatics), strained systems (e.g., cyclobutane), and moities prone to convert into something else (e.g., tautomerization

TABLE 3.4
Detection Techniques Used in TLC

Detection Technique	Principal Features of the Technique
UV-Vis/densitometry	Free choice of excitation and detection wavelength; possibility to capture UV-Vis spectra of the analytes directly on the plate; fluorescence detection possible; sequential detection of the analytes
UV-Vis/video	Limited number of detection wavelengths; simultaneous densitometry detection of all the analytes on the plate; enables a more convenient analysis of two-dimensional chromatograms
MS	Several ionization methods available; provides structural information about the analytes
FTIR (Fourier transform infrared)	Can be performed directly on the plate or after elution of the analyte from the sorbent; provides structural information about the analytes
Raman spectroscopy	Sensitivity can be considerably increased by treating the plates with silver solution (surface-enhanced resonance Raman spectroscopy); provides structural information about the analytes
Radioactivity-based methods	Applied both for online and offline detection; several methods available; mostly used for metabolic studies
NMR	Cannot be performed directly on the TLC plate; provides structural information about the analytes
FID (flame ionization detection)	Separation and detection performed on a specific sorbent-coated quartz rod; suitable for compounds lacking chromophores
SWASV (square-wave anodic stripping voltammetry)	Electrochemical detection technique; detection directly on the plate; sensitive
Photothermal methods	Several methods available; also provides information about the analytes distributed vertically inside the depth of the layer
AAS (atomic absorption spectroscopy)	Analytes are eluted from the layer before detection; used for the determination of metals

Source: From Crews, Rodríguez, and Jaspars, 1998. *Organic structure analysis.* New York, Oxford University Press. With permission.

of enols into carbonyls). The tentative approach should be organized and systematic, entail understanding of interpretation of a spectral trace, appreciation of the dangers of using negative or unreliable data, and simultaneous use of data from NMR, MS, IR, and UV-Vis methods to support individual conclusions. In contrast to solving a synthetic structure for which only a few pieces of spectroscopic data might be sought, extensive data must be obtained for a newly isolated compound when its structure is presumed to be unknown. The amount of sample at hand, instrument time available, personal experience, and the ability to find appropriate empirical reference data to back up a conclusion are all important factors. Both liquid and solid compounds can easily be examined by NMR, IR, and UV-Vis methods. Alternatively, gases or solids can be analyzed by IR and MS means. In addition, no more than a few milligrams of a compound are needed for any of these techniques. Total syntheses of natural products are the absolute because they can be checked by comparison; structures determined by X-rays are likewise usually final. In contrast, spectroscopic data can be consistent with a structure, but they can never prove it. A single misinterpretation or oversight of an NMR signal can lead to a grossly incorrect structure.

3.6.1 Nuclear Magnetic Resonance

NMR is considered to be the most powerful tool in organic structure analysis because it provides direct insight into the backbone of an organic molecule. It allows detection of atomic nuclei and establishes their environment type within their molecule (Crews, Rodríguez, and Jaspars, 1998). An NMR spectrum arises from transitions between nuclear spin states, and the ^{1}H and ^{13}C isotopes are the most common targets. If 5–50 mg of pure sample are available, then the ^{13}C NMR spectrum can easily provide an analysis of carbon and hydrogen content (two-dimensional [2D] spectra). Only submilligram samples are needed to acquire one-dimensional ^{1}H NMR data to assess the hydrogen content. The identity of various functional groups can often be established entirely by NMR. For example, carbonyl groups (acid, ester, amide, ketone, aldehyde) (Table 3.5) can quickly be identified by analysis of ^{13}C NMR chemical shifts, while benzenoid aromatic rings are revealed by their characteristic ^{1}H NMR resonances (Figure 3.1). The 2D spectroscopic data, including homonuclear ^{1}H-^{1}H COSY (correlation spectroscopy), and heteronuclear ^{1}H-^{13}C COSY or HETCOR NMR experiments, nuclear Overhauser enhancement (nOe, NOESY), together with chemical shifts or coupling constant data from ^{1}H and ^{13}C NMR data, can provide steriochemical insight (Crews, Rodríguez, and Jaspars, 1998). A summary of the most useful 2D methods is shown in Table 3.6.

3.6.2 Deuterated Solvents

A large range of deuterated solvents, such as deuterated chloroform $CDCl_3$ and deuterated dimethyl sulfoxide $(CD_3)_2SO$, exist (Table 3.7), especially for NMR spectra solving. These solvents contain small quantities of undeuterated solvent, which may give rise to a signal; $CHCl_3$ is seen as a single peak at 7.27 ppm. Water may be present as a contaminant; this gives a broad peak whose chemical shift varies greatly with solvent and occurs around 1.6 ppm in $CDCl_3$. Spectra are usually recorded against tertramethyl silane as the internal standard, set as zero. Proton NMR spectra are characterized by chemical shift in the range +12 to −4 ppm using spin-spin coupling between protons.

TABLE 3.5
Typical Chemical Shift Ranges of Carbon Nuclei

Class	Resonances (ppm)
C (alkane)	~0–30
C (alkene)	~110–150
C–N	~50
C–O	~60
C–F	~70
Aromatic	~110–160
Ester, amide, acid	~160–175
Ketone, aldehyde	~200–220

TABLE 3.6
Two-Dimensional NMR Experiments

Information Desired	Normally Detected Experiment	Inverse Detected Experiment	Nucleus
Proton-proton couplings	^{1}H-^{1}H COSY		Proton
Long-range proton-proton couplings	LRCOSY		Proton
Long-range proton-proton couplings via an H–H–H path	^{1}H-^{1}H TOCSY ^{1}H-^{1}H relay		Proton
Unknown proton assignments from that of known carbons or vice versa	^{1}H-^{13}C COSY	Heteronuclear multiple quantum coherence (HMQC) HMBC	Carbon and proton
Long-range proton-carbon couplings an H–C–C path or H–C–C–C path	Long-range ^{1}H-^{13}C COSY COLOC		
Long-range proton-carbon couplings via an H–H–C path (proton to vicinal proton to carbon)	Relay coherence transfer (H–C relay)	HMQC-TOCSY	Carbon and proton
Spatial relationships between protons	^{1}H-^{1}H NOESY ^{1}H-^{1}H ROESY		Proton
Carbon-carbon connectivities	Inadequate		Carbon

TABLE 3.7
Solvents and Quantities for Analytical Physicochemical Measurements

Analytical Chromatography	Common Solvents in Order of Preference	Amount of Sample Needed (mg)
TLC		2
GC		1
HPLC		2
	Physicochemical Measurements	
^{1}H NMR (>400 MHz)	Chloroform,[a] DMSO,[a] pyridine,[a] acetone,[a] methanol[a]	5
^{13}C NMR (>400 MHz)	Same as ^{1}H NMR	20
IR	Chloroform	2
UV/visible	Ethanol methanol (water for UV)	2
Mass spectrometry	Chloroform/methanol	1

[a] Deuterated solvents.

3.6.3 Proton Nuclear Magnetic Resonance

A one-dimensional proton NMR (^{1}H NMR) will display peaks corresponding to different hydrogen atoms, which are characterized by different environments. Chemical shift, usually abbreviated by the symbol δ (in units ppm), is a way of describing a resonance in the NMR spectrum relative to tetramethylsilane (TMS) as an internal standard. Differential decoupling is a valuable technique for uncovering buried multiplets and assigning coupling partners (Crews, Rodríguez, and Jaspars, 1998) as well as revealing information on the axial and equatorial positioning of the protons. It must be noted that the positions of the chemical shifts can to a large degree be influenced by the nature of the solvent (ranging between 0 and 5 ppm). Therefore, one must never compare results of spectra taken with different solvents. The general distribution of proton chemical shifts associated with different functional groups is summarized in Figure 3.2. Bear in mind that these ranges are approximate and may not encompass all compounds of a given class. Note also that the ranges specified for OH and NH protons are wider than those for most CH protons. This is due to hydrogen bonding variations at different sample concentrations. The typical chemical shift ranges of carbon nuclei are available in the literature.

Coupling is when neighboring nuclei might interact through the electrons in the bonds. Coupling is as helpful as chemical shifts in assigning spectra peaks and elucidating structures. The coupling constant J is dependent on three factors: the through bond distance between the protons (H), the angle between the two $C–H$ bonds, and the electronegative substituents. Typical coupling constants are shown in Table 3.A2.

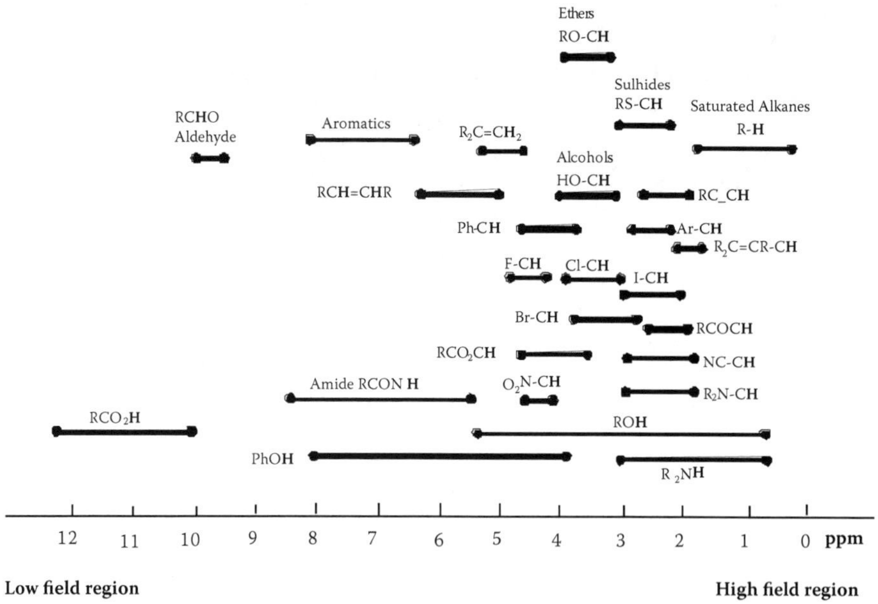

FIGURE 3.2 Proton chemical shift ranges.

3.7 CONCLUSION

Natural products chemistry as an interdisciplinary area can be used to collaborate different fields linked to herbal medicines. It is an important application in ethnoveterinary and ethnopharmacological sciences. If properly utilized, the isolation techniques described in this chapter can provide relief from the laborious protocols that natural products scientists follow when isolating active metabolites from herbal medicines. Application of new experimental procedures in isolation and purification of bioactive products, even those that occur in minute quantities, is feasible. The efficiency and effectiveness of isolation processes can be greatly improved with modern technology.

ACKNOWLEDGMENTS

I would like to extend special thanks to Molahlehi Sonopo, who was my loyal master of science student and friend during my lecturing days, for gathering part of the data presented in this chapter. Many thanks to my children, Joyce, Sam, Edward, and Edmond, all of whom give meaning to my life.

REFERENCES

Cannell, R. J. P. How to approach the isolation of a natural product. In Cannell, R. J. P. (Ed.), *Methods in biotechnology* 4. Natural products isolation. Humana Press, Totowa, NJ, 1998, pp. 1–51.

Craig, L. C., and Craig, D. In Weissberger, A. D., *Technique of organic chemistry*. Vol. 3. New York, 1950, p. 171.

Crews, P., Rodríguez, J., and Jaspars, M. Using spectroscopic data in organic structure analysis, *Organic structure analysis*. Oxford University Press, New York, Chapter 1, 1–14, 1998.

Eberz, G., Rast, H.-G., Burger, K., Kreiss, W., and Weisemann, C. Bioactivity screening by chromatography-bioluminescence coupling. *Chromatographia* 43, 5–9, 1996.

Farnsworth, N. R. Screening plants for new medicines. In Wilson, E. O. (Ed.), *Biodiversity*. National Academy, Washington, DC, 1988, Chap. 9.

Farnsworth, N. R., and D. D. Soejarto. 1985. Potential consequences of plant extinction in the United States on the current and future availability of prescription drugs. *Econ. Bot.* 39(3):231–240.

Ferreira, D., Kamara, B. I., Brandt, E. V., and Joubert, E. Phenolic compounds from *Cyclopia intermedia* (honeybush tea). *Journal of Agricultural and Food Chemistry* 46, 3406–3410, 1998.

Fox, M. A., and Whitesell, J. K. *Organic chemistry*, 2nd ed. Jones and Barlett, Boston, 1997.

Geiss, F. *Fundamentals of thin layer chromatography (planar chromatography)*. Dr. Alfred Hüthig Verlag, Heidelberg, Germany, 1987.

Gocan, S. Stationary phases for thin-layer chromatography. *Journal of Chromatographic Science* 40, 538–549, 2002.

Hostettmann, K., Terreaux, C., Marston A., and Potterat, O. The role of planar chromatography in the rapid screening and isolation of bioactive compounds from medicinal plants. *Journal of Planar Chromatography* 10, 251–257, 1997.

Houghton, P. J., and Raman, A. *Laboratory handbook for the fractionation of natural extracts*. Chapman & Hall, London, 1998.

Huie, C. W. A review of modern sample preparation techniques for the extraction and analysis of medicinal plants. *Analytical and Bioanalytical Chemistry* 373, 23–30, 2002.

Jork, H., Funk, W., Fischer, W., and Wimmer, H. *Thin layer chromatography: reagents and detection methods: physical and chemical methods: fundamental reagents 1*. Vol. 1a, VCH, Weinheim, Germany, 1990.

McMurry, J. *Organic chemistry*, 4th ed. Brooks/Cole, Pacific Grove, CA, 1996.

Nyiredy, Sz. Rotation planar extraction (RPE)—a new exhaustive, preparative forced-flow technique. Part 1: description of the method and practical aspects. *Journal of Planar Chromatography* 14, 393–395, 2001.

Starmans, D. A. J., and Nijhuis, H. H. Extraction of secondary metabolites from plant material: a review. *Trends Food Science and Technology* 7, 191–197, 1996.

Tservistas, M., Scheper, T., and Freitag, R. Supercritical fluid extraction (SFE)—novel strategies in the processing of biomaterials. In Grabley, S., and Thiericke R. (Eds.), *Drug discovery from nature*. Springer-Verlag, Berlin, 2000, pp. 106–113.

Vinatoru, M. An overview of the ultrasonically assisted extraction of bioactive principles from herbs. *Ultrasonics Sonochemistry* 8, 303–313, 2001.

Vogel, A. I., Tatchell A. R., Furnis, B. S., Hannaford, A. J., and Smith, P. W. G. *Vogel's textbook of practical organic chemistry*, 5th ed. Prentice-Hall, Englewood Cliffs, NJ, 1996.

Weins, C., and Jork, H. Toxicological evaluation of harmful substances by in situ enzymatic and biological detection in high-performance thin-layer chromatography. *Journal of Chromatography* A 750, 403–407, 1996.

Zygmunt, B., and Namieśnik, J. Preparation of samples of plant material for chromatographic analysis. *Journal of Chromatographic Science* 41, 109–116, 2003.

SUGGESTED FURTHER READING

Crews, P., Rodríguez, J., and Jaspars, M. Using spectroscopic data in organic structure analysis, *Organic structure analysis*. Oxford University Press, New York, Chapters 2–5, 15–221, 1998.

Sarker, S. D., Latif, Z., and Gray A. I. *Natural products isolation*, 2nd ed. Humana Press, Totowa, NJ, 2006.

Williams, D. H., and Fleming, I. *Spectroscopic methods in organic chemistry*, 5th ed. McGraw-Hill, London, 1991.

TABLE 3.A1
Reagents and Methods of Thin-Layer Chromatography

Type of Compound	Method[a]	Layer (Merk)	Mobile Phase	Wavelength (nm)	Reagent (Jork et al., 1990)[b]	Heat (°C)[c]	Time (min)	Color Developed
Sugars, monosaccharides	HPTLC	Plates, silica gel 60	2-Propanol/0.75% boric acid/glacial acetic acid 40:5:1	365	4-Aminohippuric acid	115–140	8	Orange
Sugars, Carbonyl compounds	HPTLC	Plates, silica gel 60 F_{254}	Toluene/ ethylacetate/ acetone 90:5:5	365	2-Aminodiphynyl-sulfuric acid	105–110	5–10	Green, blue, or purple
Carbohydrates (sugars), mono- and disaccharides, uronic acids	TLC	Plates, silica gel 60	Dichloromethane/ methanol/acetone/ water 50:50:25:10	365	4-Aminobenzoic acid	100	10–15	Reddish-brown
Alkaloids [1,2][d] Mycotoxins [3–5][d] Flavonoids, flavonoid glycosides [6–9][d]	HPTLC	Plates, silica gel 60 F_{254}	Methanol/ chloroform/water 12:8:2	365	Ammonia vapor	110–120	25	Yellow, green, or blue
Antioxidants (e.g., flavonoids) Steroids Prostaglandins Carbohydrates Phenols	HPTLC	Plates, silica gel 60 F_{254}	Toluene/chloroform 10:10 Toluene/acetone/ methanol 6:3:1 (flavonoids)	365	Anisaldehyde (4-methoxybenzaldehyde-sulfuric acid)	90–125	1–15	

Glycosides Sapogenins Terpenes or essential oil components Mycotoxins Antibiotics								
Aromatic hydrocarbons Alkaloids Methyl ethers of fatty acids Tannins	HPTLC	Plates, silica gel 60	Methano/chloroform/water 12:8:2 (alkaloids) Toluene/acetone/methanol 6:3:1 (flavonoids)	365	Formaldehyde-sulfuric acid (Marquis reagent)	110 140-methyl esters	20 10	Variety
Fatty acids, triglycerides, amino acids, sugars, steroids Peptides, nucleosides Alkaloids	HPTLC	Plates, silica gel 60	Methanol/NH_3(0.25%)/acetone 8:1:1		*tert*-Butyl hypochlorite	Room temperature		

[a] Ascending, one-dimensional development in a trough chamber with saturation.

[b] Heating is done after spraying of the reagent.

[c] Reagents preparation.

[d] Chromatogram is heated in the drying cupboard to 110–120°C for 25 min and placed, while still hot, in a trough containing 10 mL ammonia solution for 15 min.

TABLE 3.A2
Typical Coupling Constants

Geminal $^2J_{HH}$ (on the same carbon)	Compound	Coupling Constant between H_A and H_B	Number of Bonds between the Neighboring Hydrogen Atoms
Saturated	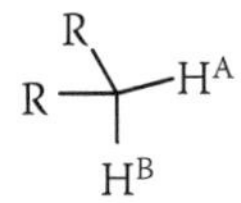	10–16 Hz	Two bonds (H_A–C–H_B)
Unsaturated	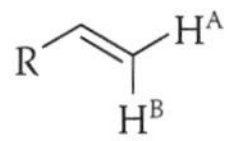	0–3 Hz	Two bonds (H_A–C–H_B)
Vicinal $^3J_{HH}$ (on the adjacent carbon)			
Saturated	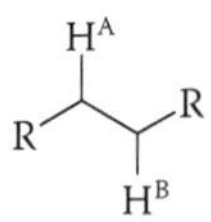	6–8 Hz	Three bonds (H_A–C–C–H_B)
Unsaturated (*trans*-Hs on opposite sides of double bond)	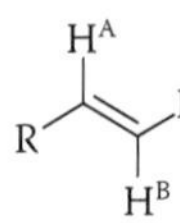	14–16 Hz	Three bonds (H_A–C–C–H_B)
Unsaturated (*cis*-Hs on same sides of double bond)	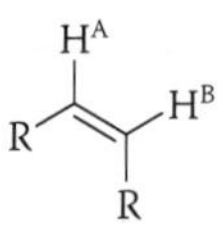	8–11 Hz	Three bonds (H_A–C–C–H_B)
Unsaturated aromatic compound	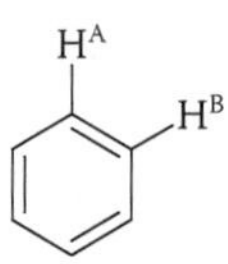	6–9 Hz	Three bonds (H_A–C–C–H_B)
Long-range $^4J_{HH}$			
Meta	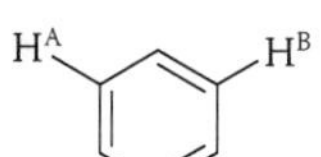	1–3 Hz	Four bonds (H_A–C–C–C–H_B)
Allylic	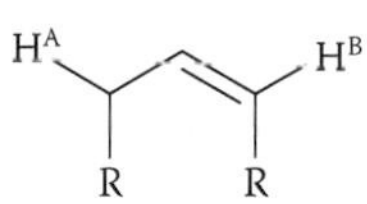	1–2 Hz	Four bonds (H_A–C–C–C–H_B)

$^2J_{HH}$ = 2 bond lengths between the two protons (HH); $^3J_{HH}$ = 3 bond lengths between the two protons (HH); $^4J_{HH}$ = 4 band lengths between the two protons (HH).

4 Preclinical Safety Testing of Herbal Remedies

Vinny Naidoo and Jurgen Seier

CONTENTS

4.1 INTRODUCTION

Is natural safe? This is probably the most important question when considering herbal remedies. Many people are of the mistaken notion that plant or herbal remedies are safe as they are natural, and natural is good for you, that is, "nature knows best." Yet, most toxic substances known to humans are of natural origin, such as tetrodotoxin from the puffer fish (*Fugu rubripes*), black mamba venom (*Dendroaspis polylepis*), aflatoxins (*Aspergillus* spp.), botulinum toxin (*Clostridium botulinum*), and ricins (*Ricinus communis*). From these examples, it is clear that herbal remedies are not inherently safe (Ueno, 1985; Schweitz and Moinier, 1999; Noguchi and Ebesu, 2001; Kellerman et al., 2005; Sobel, 2007). This, of course, does not mean that synthetic pharmaceuticals, on the other hand, are safe, but the more rigorous current legislative framework governing the latter ameliorates the risk associated with their use.

An important factor is that commercial herbal remedies are sometimes processed and concentrated by extraction to maximize the therapeutic effect (percolation or maceration in various solvents). This increases not only the concentration of the beneficial compounds in the extract but also the possible toxins (Foukaridis et al., 1995). If one is to consider the words of Paracelsus, the father of modern toxicology, who essentially stated that all substances are toxic with the dose making the difference, the process of concentration could also increase the concentration of the beneficial agents to a toxic level (Borzelleca, 2000).

Contamination of the formulation due to poor quality control or tampering is possible. In a study in the United Kingdom, many of the topical creams examined were found to be adulterated with steroids (ICH, 2000; Ko, 2006). In another survey on Chinese herbal remedies, formulations were found to be contaminated not only with the steroids, but also nonsteroidal anti-inflammatory drugs as well as diazepenes and other conventional pharmaceutics, such as antibiotics and antidiuretics (Deng, 2002; Ernst, 2002). Therefore, good production quality control systems, as set out in good manufacturing practice (GMP) guidelines, are important.

Another concern is the use of herbal products in animals that are farmed as a source of protein for human nutrition (meat, milk, or eggs). Any substance that gets absorbed into the body also has the potential of becoming incorporated into animal products and can therefore adversely affect human health. At present, residues in animal products are highly controlled around the world by the various regulators (Van Dresser and Wilcke, 1989).

It is clear that every herbal remedy needs to undergo preclinical safety testing to ascertain its potential to cause harm. Preclinical safety testing should not be seen as an obstacle to the development or sale of an herbal remedy but rather as a means to ensure that the end user of the product comes to no harm. However, it should be remembered that most pharmaceutical drugs are researched and developed by commercial companies, many of which are located in highly industrialized nations. Such companies commit considerable financial resources to drug development and in return realize significant profits from their investment (DiMasi, Hansen, and Grawbowski, 2003; Rawlins, 2004). Traditional medicine research, on the other hand, is most often conducted by academic institutions, frequently in less-developed countries. In these cases, it is clear that financial resources and return on investment will be limited, even nonexistent at times. This has implications for toxicity testing since few academic institutions are able to fund the entire range that is conventionally required.

4.2 DOSE-RESPONSE CURVES AND SAFETY

Modern pharmacology has been able to explain the principle of a drug's response or toxicity by means of dose-response curves (Hoel and Portier, 1994; Verlato et al., 1996; Lees, Cunningham, and Elliot, 2004). These curves clearly indicate that the effect achieved is related to the dose administered to a maximum effect (effective dose-response curve). After the point of maximum effect E_{max}, any increase in dose will not result in an increase in the desired effect. Since every product has the potential to induce side effects or to be toxic, this can also be plotted on its own curve (toxic and lethal dose-response curves). In most cases, this curve is slightly to the right of the desired dose-response curve (Timbrell, 1995). The further this curve is to the right of the effective dose-response curve, the safer the product becomes and vice versa (Figure 4.1. The safety of the product (safety indices) is calculated mathematically by comparing the effective dose-response curve to the lethal dose-response curve (a more detailed description of an LD_{50} (median lethal dose) is provided in Section 4.3.1), for example, the therapeutic index (LD_{50}/ED_{50}), the therapeutic ratio ($LD_{25}:ED_{75}$), and the safety factor (LD_1/ED_{99}). The last factor considered is the margin of safety, which looks at the steepness of the slope of the dose-response curve. Steeper curves have a lower margin of safety as a small increase in dose will result in a much larger increase in effect than a product with a more gradual slope (Figure 4.2) (Timbrell, 1995). From an ethical standpoint, the assay determination of an LD_{50} is seen as unacceptable due to the large number of animals used, and the reproducibility of this method has also been questioned (Mutai, 2000; Gad and Chengelis, 2007). As an alternative, we use tolerance testing, in which increasing doses are utilized. This determines the toxicity of a product on a scale system; for example, if a drug is safe at 10 times the recommended dose, then there is little chance of toxicity and no need to expose more animals to the drug (see Section 4.3.8) (Walum, 1998).

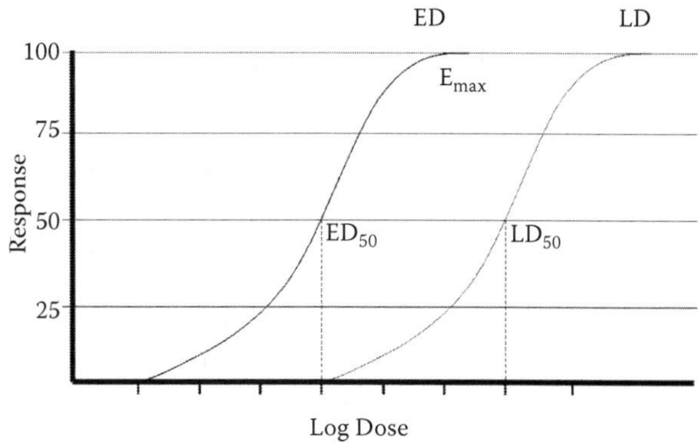

FIGURE 4.1 Illustration of the dose-response relationship for a medicinal product. ED represents the effective dose-response curve; LD represents the lethal dose-response curve. The *x*-axis is plotted on the log scale to linearize the data; the *y*-axis represents the percentage effect. E_{max} represents the maximum therapeutic effect of the product. The ED_{50} and LD_{50} values can be read directly off the *x*-axis. (From Seier et al. 2004. *A toxicity study of IKS/PYAI Consumption*, MRC Press, Cape Town.)

4.3 TOXICITY TEST

Numerous different tests have to be conducted when ascertaining the safety of ethical medical products for registration. These tests have been designed to elucidate all toxic effects at low, medium, and high doses following single or multiple exposures (WHO, 2004). Such tests detect insidious genotoxic as well as carcinogenic effects, while cardiotoxicity is determined *in vitro* by human ether-a-go-go-related gene (hERG) screening (Organization for Economic Cooperation and Development). With the exception of some of the genotoxicity testing and hERG screening, all other safety investigations need to be undertaken in animal models. At present the zebra fish, Chinese hamster ovary (CHO) cells, and human lymphocyte and human lymphoma assays are the more commonly acceptable *in vitro* assays for assessing genotoxicity (OECD, 1997; European

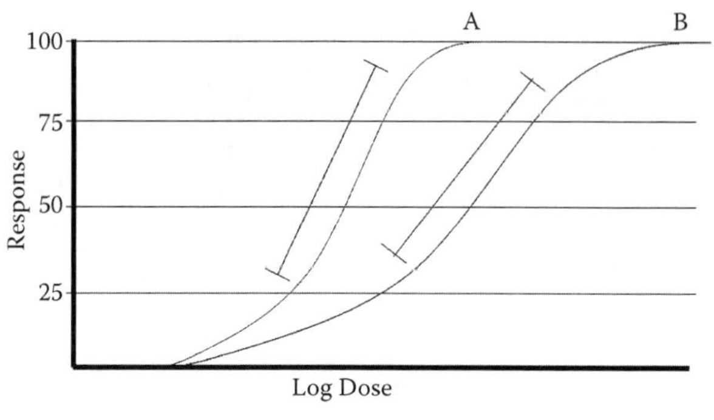

FIGURE 4.2 The importance of the slope of a dose-response curve for drug safety (margin of safety). Product A has a steeper slope than product B. A smaller change in dose for product A will result in a much larger change in response than for product B.

Medicines Agency [EMEA], 1987b). All other *in vitro* tests, such as the brine shrimp and various cell culture assays, are not at present allowed for registration purposes and are not validated due to poor correlation with *in vivo* toxicity in most cases (Toussaint et al., 1995). However, these tests still continue to be widely used in the herbal industry as initial screening for toxicity (Meyer et al., 1982; Solis et al., 1993).

When comparing herbal remedies to ethical medicinal products (pharmaceuticals), some major differences are evident, which may influence registration. The active ingredients in the ethical products are known, while those in the herbal remedies are largely unknown. In addition, the composition of the herbal remedy can change with each batch grown and processed, and this difference can alter the toxicity profile of the herbal product. Therefore, preclinical safety testing for any herbal product is dependent on an adequate quality control strategy during the preparation of said herbal remedy. The results from the preclinical tests can only be applied to production batches that meet the same criteria. At present, this would mean fingerprinting of all batches produced. Any batch that fails to meet the fingerprint requirements should be rejected and destroyed as the toxicity resulting from this difference can be substantial; for example, kava kava extracts appear to be more toxic with an increase in the ethanol quantity in the extraction solvent (Clouatre, 2004).

4.3.1 Single-Dose Acute Toxicity Testing

For single-dose acute toxicity testing, most regulators internationally require use of two mammalian species, with both sexes exposed (EMEA, 1987a; Food and Drug Administration [FDA], 2000; OECD, 2007). The aim of this test is to determine acute toxicity and the mode of death after a single dose as well as provide indications regarding the possible effects of overdosing. The animals are treated with the product by two routes, the first being the same as the recommended route and the second allowing for maximal exposure. The animals are monitored from the time of dosing, and all clinical signs, including food and water intake, are recorded. Animals are monitored until death or to a maximum of 14 days posttreatment, followed by a full necropsy examination at the time of death or euthanasia. Histopathological examinations are only necessary for organ systems showing macroscopic changes (OECD, 2001a, 2007).

The doses selected for this study and the number of animals used per group are defined in the guidelines of the OECD, the FDA, and others for toxicity testing (FDA, 2000; OECD, 2001a). The highest dose should ensure all signs of toxicity become evident, but in some instances a smaller study may need to be undertaken to find the correct dose. When four groups of animals are used, an LD_{50} can be determined by either a probit analysis or a dose-response analysis (Williams, 1986). The LD_{50} is defined as the dose that kills 50% of a uniform population. Although the LD_{50} is still regarded as the best means to compare the toxicity of different compounds, this method is flawed as it differs between laboratories and the strain of animal used (Mutai, 2000). The determination of the LD_{50} also requires a large number of animals, which, in addition to being unacceptable ethically, is expensive.

To overcome this problem, other systems of acute testing have been suggested. One of these systems is the guidelines published by the British Toxicology Society and the OECD (British Toxicology Society, 1984; Bruce, 1985; van den Heuvel et al.,

1990; OECD, 2006). These methods were designed to limit the total number of animals exposed to the potentially toxic compound and the highest dose. The system also avoided death as the end point and relied more on the clinical signs. From these data, compounds can be classified according to their toxicity even though an LD_{50} is not determined. Animals are exposed to 5, 50, and 500 mg/kg of the tested compound, which enables classification as very toxic (LD_{50} < 25 mg/kg), toxic (LD_{50} 25–200 mg/kg), harmful (LD_{50} 200–2,000 mg/kg), and unclassified (>2,000 mg/kg) depending on the number of deaths. The last dose of 2,000 mg/kg also forms part of the tolerance dose set by the OECD (2006). According to this system, a product that fails to show effects at this dose is unlikely to be toxic.

4.3.2 Repeat-Dose Toxicity Testing

The purpose of repeat-dose toxicity testing is to obtain information on the cumulative toxicity of a product during repeat exposure, which might also be representative of the period of exposure in the target animal. The duration of a subchronic study is typically 1–3 months, while a chronic study lasts for about 12 months (OECD, 1981, 1995, 1998, 2001b; EMEA, 1984).

The aim is to reveal physiological or pathological changes by allowing for accumulation of the product so that a steady state can be achieved. From this constant exposure, the effect on the elimination pathways can be determined (i.e., to ascertain if toxic metabolism may result or if the biotransformation becomes saturated). Repeat-dose testing also determines target organ systems when physiological pathways break down or become saturated, establishes the effects of overdose and reversibility, and enables the calculation of the no observed adverse effect level (NOAEL).

The substance should be administered to two animal species (one rodent, one nonrodent) by the same route as for the final product and at a dose selected to demonstrate toxicity. Administration should be for a sufficiently long period, which is determined by the intended duration of exposure during therapeutic use. Blood tests must include the full blood count (FBC) as well as a variety of biochemical parameters in plasma as provided in Table 4.1. For rodents, these blood tests can be performed on days 30 and 60 as well as at the end of a study and for large nonrodent

TABLE 4.1
Blood Tests Typically Required for Repeat Dose Studies

Hematology	Biochemistry
Red blood cell count	Alanine aminotransferase, aspartate aminotransferase, gamma glutamyl transferase, bilirubin, alkaline phosphatase, albumin, ornithine carbamyl transferase
Hemoglobin, hematocrit	Urea, creatinine, Na, K, Cl, Ca, P
Total and differential white blood cell count	Total protein, glucose, total cholesterol
Platelets	Any other tests indicated by specific concerns (e.g., hormones, lipids)

TABLE 4.2
Organ Samples Required for Histopathological Evaluation for a Subacute Test

Any Gross Lesions	Tissue Masses	Blood Smear	Lymph Node	Spinal Cord
Salivary glands	Mammary glands	Pituitary	Thymus	Trachea
Lungs	Heart	Thyroid	Esophagus	Stomach
	Colon	Liver	Gall bladder	Pancreas
Spleen	Kidneys	Adrenals	Bladder	Prostate
Testes	Ovaries	Uterus	Brain	Eye
Sternebrae, femur, or vertebrae (with bone marrow)			Small intestine (Swiss roll method)	

species also before treatment starts (at baseline). It is also acceptable to perform the blood tests only at the end of a study. Microscopic examination of urine for sediment is conducted according to the same schedule.

All animals have to be examined for ophthalmological effects before and at the end of the study and are monitored for any other signs of toxicity. All changes are recorded, and the animals are sacrificed after the last dose. Any death is investigated by full postmortem examination. The organs of either all test animals or only those in the high-dose group as well as the control group have to be evaluated histopathologically (Table 4.2) (OECD, 1995, 2001b).

4.3.3 Reproduction Toxicity Study

4.3.3.1 One-Generational Toxicity Study

The one-generational toxicity studies are designed to determine detrimental effects during the various stages in reproduction, which include reproductive cycles, conception, conception to implantation, implantation to closure of the hard palate, closure of the hard palate to the end of pregnancy, birth to weaning, and weaning to sexual maturity (Kimmel and Makris, 2001). Guidelines are available to define each of these periods (OECD, 1983, 2000a, 2000b). The choice of test species is important since this can have a direct influence on the outcome of the test.

4.3.3.2 Two-Generational Toxicity Study

The two-generational toxicity study is similar to the reproduction toxicity testing but evaluates the effects of the substance on the reproductive potential of the F1 generation and ascertains if deformities are present in the F2 generation (OECD, 2000b).

4.3.4 Mutagenicity Testing

Mutagenesis refers to the changes in the genetic material of a cell, which can be harmful to subsequent generations and may be associated with carcinogenicity. These problems can be related to a single point change in the individual gene, affect an entire chromosome, or even alter the number of chromosomes present. A number of different *in vitro* and *in vivo* tests have been developed to determine the ability of a chemical to induce this type of damage (Kirkland et al., 2005).

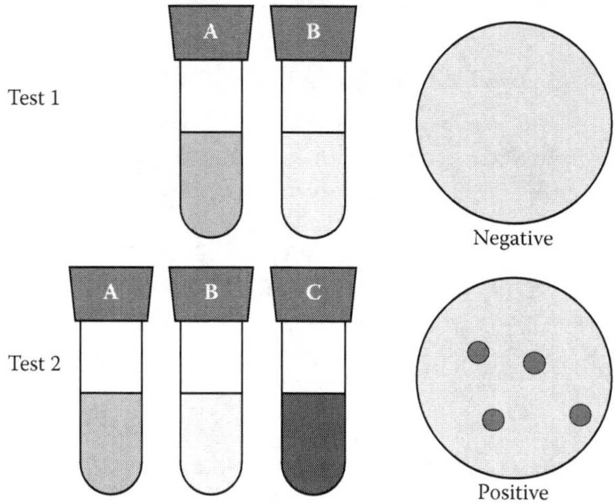

FIGURE 4.3 The Ames test. A is the herbal remedy, B is the bacterial culture, and C is the rat microsomal fraction. In both of the tests, the ingredients are mixed together and plated. The test is positive for mutagenicity if bacterial colonies start to grow.

4.3.4.1 Gene Mutation in Bacteria

The Ames test is most commonly used and carried out with strains of bacteria that have lost the potential to survive without external support (Figure 4.3). In this assay, specific strains of *Salmonella typhimurium* are grown in the presence of the test compound with or without rat microsomal enzymes. The principle of the test is based on the ability of the bacteria to survive on media deprived of histidine, a normal growth nutrient. While the medium does contain the necessary nutrient for the synthesis of histidine, the strains used in the Ames test are genetically deficient in the enzymes necessary to make histidine from this specific media.

The basis of the test is the ability of these bacteria to grow beyond 48 h. Since growth can only occur if the bacteria have been genetically altered, to produce the enzyme necessary for histidine formation, the presence of viable colonies is a direct indication of bacterial mutation and therefore a direct indicator of the potential of a chemical to induce genetic changes in people or animals (i.e., mutagenic potential). A major shortfall of this test is its use of a prokaryotic organism while the mammalian cells are eukaryotic.

4.3.4.2 Chromosomal Aberrations in Mammalian Cells *In Vitro*

The test for chromosomal aberrations in mammalian cells uses an *in vitro* cell culture system to ascertain the ability of the product to damage the chromatid material (EMEA, 1987b; OECD, 1997). In this assay, cell cultures (e.g., lymphocyte cultures) are grown in the presence of the chemical to be tested with and without metabolic activation. The ability of the cell to induce damage to the nucleus is subsequently evaluated microscopically by determining the degree of changes in the nucleus of the exposed cells.

4.3.4.3 Gene Mutations in Eukaryotic Systems

Again, the aim of the test for gene mutations in eukaryotic systems is to determine if a chemical has the ability to induce spontaneous mutation in the cell cultures. While the Ames test used bacteria colonies for the presence of mutagenic potential, this assay uses mammalian cells (EMEA, 1987b) and is usually carried out following a positive result from the Ames test. Since this test uses eukaryotic cells, a more complex system, it is considered to be more representative of mammalian cell physiology. Different types of cultures are available to determine if a compound induces mutations at various specific loci within the cell (e.g., sites coding for the enzymes hypoxanthine-guanine-phosphoribosyl-transferase or thymidine kinase).

4.3.4.4 *In Vivo* Testing for Genetic Damage

The micronucleation test is the most commonly used assay for *in vivo* testing for genetic damage. In this assay, the toxic effect of the compound on the bone marrow is determined as this tissue is highly mitotically active. For this, bone marrow is harvested from the treated animals and evaluated for the degree of micronucleus formation. The micronuclei, when formed, are bits of damaged chromatid material within the cell.

4.3.5 Carcinogenicity Testing

Carcinogenicity testing aims to determine if the material has the ability to induce cancer. While most drugs that are genotoxic may cause cancer, this is only possible if the changes induced do not kill the cell; that is, the changes induced in the nucleus must not kill the cell, thereby allowing the cell to start growing uncontrollably so that it can form a neoplastic condition in the animals. It is important for products in which long-term exposure is possible or, more importantly, when a product shows the potential to be genotoxic. Animals are exposed to the drug for a long period of time, usually 1 to 2 years, and the incidence of cancer is then compared with the control group. The inherent problem of this test is that most of the rodent species only live about 2 years and have a naturally high incidence of cancer when old. There has also been no proper correlation between the occurrence of cancer in humans and in rodents (Lave et al., 1988).

4.3.6 Local Tolerance Testing

The aim of the local tolerance group of studies is to determine the effect a substance has when coming into contact with various parts of the body. This is important for substances destined for human use or that are administered by animal owners. The route tested will be determined by the intended use or exposure in people.

4.3.6.1 Ocular Tolerance Testing

Ocular tolerance testing determines the ability of the product to cause ocular irritation or damage in people following accidental exposure (Environmental Protection Agency [EPA], 1998a). To simulate human ocular exposure, rabbit eyes are exposed to the test compound. In these tests, each animal serves as a self-control, with one eye exposed to the test compound and the other eye used as the comparison. Both single- and repeat-dose testing are conducted, and irritation to the eye and surrounding adnexa is determined. At the termination of dosing, the animals are euthanized for full necropsy. The ethical justification of this test is being questioned despite the use of local anesthetics. At present, this test is not carried out for known corrosive chemical agents (Worth and Cronin, 2001).

4.3.6.2 Dermal Tolerance Testing

Just as for the ocular irritancy test, the aim of dermal tolerance testing is to evaluate the possibility of skin being irritated or damaged following exposure to the chemical while handling a particular medicine (EPA, 1998b). Again, this is important as the package insert may need a warning for the administrator of the risks of handling the said remedy. As for the ocular testing, products are evaluated for irritancy as a single and repeat dose, and rabbits are also used to determine the corrosive effect of medication on both intact and abraded skin.

4.3.6.3 Skin Sensitization

In addition to a compound being directly irritant to skin, sensitization to the particular chemical may result from simple dermal exposure (*sensitization* refers specifically to the ability of the immune system to recognize the molecule) (Buehler, 1965; Magnusson and Kligman, 1969; Magnusson, 1980; Ritz and Buehler, 1980; OECD, 2004). Once the immune system recognizes a molecule, subsequent exposure tends to result in an allergic reaction. While these immune reactions may be mild, such as simple urticaria, they can also be fatal. This test is undertaken in guinea pigs with the animals first exposed to the substance to allow for sensitization, and after a suitable delay, the effect of reexposure is evaluated.

4.3.6.4 Phototoxicity or Photosensitivity

Phototoxicity refers to an increased sensitivity of the skin to ultraviolet (UV) radiation following the exposure to a chemical (Putman, van der Laan, and van Loveren, 2003). This can occur after exposure to a photosensitizing agent that has the unique ability to absorb UV radiation at one wavelength and subsequently emit that radiation at a second wavelength. This change in wavelength makes exposure to UV radiation dangerous, such that direct sunlight can cause severe sunburn and sometimes severe dermal ulceration (doxycycline is the most well-recognized molecule sensitizer in people). At present, this test is dependent on the molecule to be tested and the regulator, with testing undertaken in guinea pigs.

4.3.7 Cardiotoxicity Testing

Drugs that are cardiotoxic may bring about their effect by influencing the opening of potassium channels in the cardiac conduction network; these channels are responsible for the repolarization of the ventricular cardiac action potential, thereby leading to cardiac arrhythmias (Raschi et al., 2008; Fink et al., 2008; Roche et al., 2002). To ascertain this effect, the hERG has been sequenced and cloned into cell lines for further studies.

This gene receives its names from its equivalent *Drosophila* ether-a-go-go gene. In the latter, exposure of the firefly to ether results in the opening of potassium channels and excitation.

4.3.8 Target Species Testing

One of the problems with toxicity testing has been the poor ability for interspecies comparison and extrapolations. This is especially important in people, as was seen with the devastation caused by thalidomide and more recently with the coxib group of nonsteroidal anti-inflammatory drugs. One of the ways in which these comparisons are undertaken in humans is by means of phase testing. In this system, human subjects are slowly exposed to a new chemical entity to first elucidate possible unknown side effects:

- Phase 1: Limited studies in people. A group of healthy adults is exposed to the drug to determine the safety and pharmacokinetics of the drug.
- Phase 2: Limited study in sick people. A selected group of people will be exposed to the drug to determine if the drug is effective in people.
- Phase 3: Wider clinical study. In this phase, a large number of people are exposed to drugs. In this study, one is specifically interested in the effects and side effects of the drug in a wider group of patients (e.g., Chinese, African, Caucasian, etc.). No two persons are alike, and to prevent severe adverse events from occurring, we try to determine the effects of a drug in the widest possible population.
- Phase 4: Postmarketing surveillance (see Section 4.3.10).

In veterinary medicine, the need for phase testing as in humans is not necessary as we have the advantage of being able to test the safety of the product directly in the target species. In practice, this is known as *tolerance testing*. In this system, the target animal is exposed to the product at the expected therapeutic dose, 5 and 10 times the therapeutic dose. This exposure allows the possible toxicity inherent in the final formulation to be determined.

4.3.9 Other Tests

Other toxicity tests are available that may be applicable to the compound under investigation. They include the neurotoxicity and immunotoxicity studies.

Safety pharmacology determines the possible effects on vital functions, including the central nervous, respiratory, and cardiovascular systems. It may be conducted during regular toxicity testing or as a separate study.

4.3.10 Pharmacovigilance

Pharmacovigilance is also referred to as postmarketing surveillance or phase 4 testing. During this stage, the occurrence of adverse reactions following the registration and marketing of a medicinal product are monitored. This is extremely important since in toxicity testing relatively small numbers of animals are used, and results are extrapolated to a large target population. Pharmacogenomics establishes the degree of individual variation after exposure in a large population during therapeutic use.

Pharmacovigilance is based purely on documenting drug reactions and determines the frequency and causality. It allows regulators to react to those side effects occurring frequently. Possible actions include the withdrawal of marketing status, package insert amendments, or limitations on the marketing of a product.

4.4 LIMITATIONS IN COMPLYING WITH CONVENTIONAL STRATEGY

The limitations of the conventional testing strategies are listed in Table 4.3. These may be compounded by a number of other factors:

- Therapeutic doses of herbal medicines: Animals are often treated with very high doses. This may limit the volume that can be administered during toxicity testing, particularly considering that multiples of the therapeutic dose have to be utilized. For example, a 200-g laboratory rat can only be gavaged with a maximum volume of 4 mL (Nebendahl, 2000).
- Consistency: Due to the coarse, bulky consistency of even-milled whole-plant materials, they are difficult to gavage to the test animals and can

TABLE 4.3
Summary of Limitations in Applying Conventional Strategies

Conventional Strategy	In Whole-Plant Testing	In Extract Testing
In vitro (hERG, Ames, etc.)	Not possible as the toxic compound is protected within the whole-plant material	Can be established
Pharmacokinetics, pharmacodynamics	In the absence of lead compound never available	Available if lead compound has been identified
Acute toxicity	Sometimes difficult to establish due to lack of toxicity at any dose	Can be established
Subchronic-chronic toxicity	Can be established	Can be established
Carcinogenicity	Cannot often be established, and dose finding will be difficult	Can be established
Reproductive toxicity	Can be established	Can be established
Developmental toxicity	Can be established	Can be established
High dose should produce toxicity, medium dose lower toxicity	Cannot usually be achieved due to the volume limitation or the acute toxicity test not revealing a toxic dose	Can be achieved
Establish NOAEL	Cannot often be established due to lack of toxicity at any dose	May be possible

sometimes only be administered in the food. Although this does not apply to extracts, alcohol extraction poses its own problem. Due to their low viscosity, they may be dosed into the lungs if the animals struggle during the gavage procedure and cause respiratory tract irritation

- The palatability of the test material: Animals may not like the taste of the test material and refuse to consume it in food or reduce their food consumption.

4.5 SPECIFIC CONSIDERATIONS IN ETHNOBOTANICAL SAFETY TESTING

The process of developing traditional medicines for allopathic use is often reversed compared to that of conventional pharmaceutical development. In the conventional approach, a compound is developed in the laboratory and, after extensive preclinical testing, introduced into clinical practice. Conversely, traditional medicines are usually first used in clinical practice and then brought into the laboratory for testing. Because of this, sometimes extensive, established use in clinical practice, it is occasionally argued that there is no need for further safety testing since this has been established. However, the therapeutic use does not provide information on overdose (acute toxicity), reversibility on drug withdrawal (long-term effects), target organs, chronic and delayed toxicity, or whether there are subclinical side effects. These might not be immediately apparent during clinical use.

4.6 SPECIAL CONSIDERATION FOR PRODUCTION ANIMALS

One important issue when considering animals that provide human nutrition is human safety. According to the general principles of pharmacokinetics, substances entering the animal's body could be distributed to any part of that body and could be present in the food for a period of time after treatment (Figure 4.4). As a general rule, it takes approximately 10 half-lives for the product to be 99% eliminated from the body. Therefore, animal products consumed while containing residue could pose a serious human health risk. Apart from actual ingredients of the herbal remedy, metabolites from processing of such remedy by the animal could be toxic. Since all substances of the herbal remedy are xenobiotic, the animal's body tries to eliminate them as an active drug or as metabolites. Because of the common routes of excretion, people usually get exposed to residues through milk or egg consumption.

To protect the public, withdrawal periods are set for all medicinal products used in production animals and are strictly controlled by the various regulatory authorities across the world. The withdrawal period is determined statistically from the maximum residue levels (MRLs) set by the Joint Expert Committee on Food Additives (JECFA) of the Food and Agriculture Organization (FAO) and World Health Organization (WHO) and are published by Codex Alimentarius. The *withdrawal period* may be defined as the time required from the last dose to the time at which an animal product may be safely consumed or slaughtered for human consumption. This is important as it prevents dangerous drugs from making their way into the human food chain.

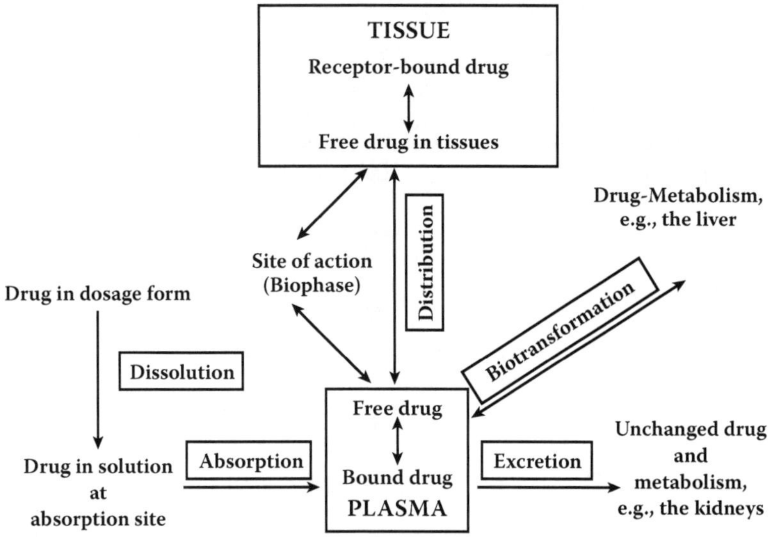

FIGURE 4.4 The normal pharmacokinetics processes every drug undergoes when entering the body.

4.6.1 Determination of Withdrawal Periods

The withdrawal period is based on the *acceptable daily intake* (ADI) of the compound. The ADI is defined as that amount of a product that can be consumed by a person on a daily basis for their entire life without any appreciable risk of harm (Skidmore, Kuiper, and Hamilton, 1997). The ADI is determined from the data derived from the toxicity testing. In each of the toxicity tests, the dose that caused no ill effects, known as the NOAEL, is calculated. Once all the tests are completed, the lowest NOAEL will be selected to determine the ADI by the following equation:

$$\text{ADI} = \frac{\text{NOAEL}}{\text{Safety Factor}}$$

The safety factor is defined by the target population. Usually, this is set at 100 and represents a factor of 10 for extrapolation between species and an additional 10 for differences in sensitivity within animal populations (Skidmore, Kuiper, and Hamilton, 1997). The safety can, at times, be increased depending on the target population (e.g., if the chance of exposure in children is very high, a safety factor may be increased for additional protection of this very susceptible population group). The ADI is used by JECFA and other regulators to set the MRL for the various organ systems and is based on a food basket (the average consumption by a 60-kg person). Usually, only a portion of the ADI is allocated for residues in animal products as exposure is possible via other sources (e.g., via residues on crops).

Once an MRL is set, the withdrawal time can be determined. The usual system requires three groups of animals ($n \geq 3$) to be slaughtered at three different time points. The concentration of drug in each organ system is subsequently measured and plotted, and the time to the MRL is statistically determined. The withdrawal time is taken as the longest time to the MRL for all the tested organ samples (Figure 4.5). The organs and products include muscle, kidneys, fat, injection site, liver, eggs, and

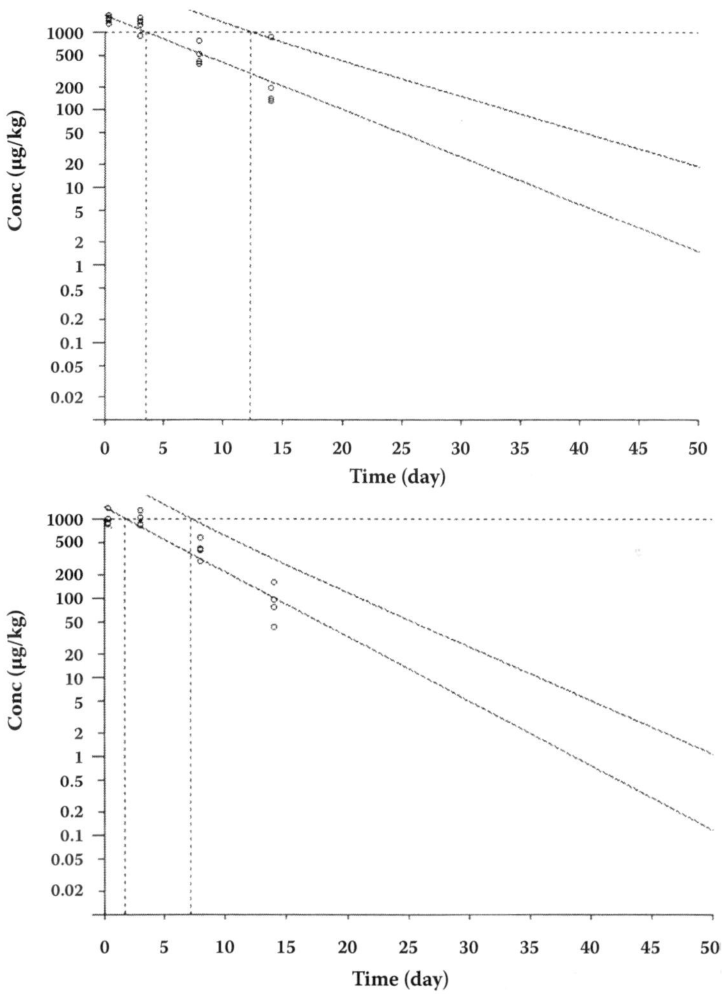

FIGURE 4.5 The statistical determination of the time to MRL for an ethical product in cattle tissue for four groups ($n = 4$). (Top) The kidney drug concentrations of the 16 animals plotted against an MRL of 1,000 µg/kg. From the x-axis, the MRL is reached after 12 days. (Bottom) The liver drug concentrations from the same 16 animals plotted against an MRL of 1,000 µg/kg. From the x-axis, the MRL is reached after 7 days. If the kidneys were the organ system that took the longest to reach the MRL, a withdrawal period of at least 12 days would be required.

milk. (For a more detailed discussion of MRL and withdrawal times, please refer to the guidelines published by your local regulatory authority.)

4.6.2 Dangers of Residues

The presence of residues in animal products may have a negative influence on human health. These include

1. Allergies: Also known as a hypersensitivity reaction. Initial exposure is necessary for the priming of the immune system. *Priming* is the process by which the immune system recognizes that a foreign molecule (antigen) has entered the system and forms a permanent record (antibody) to protect itself thereafter. On further exposure, defenses are activated to fight the invader. With this second or third exposure, one can develop a severe allergic reaction, which may end in death. To prevent allergies, the drug concentrations in the animal product need to be at a level that cannot be recognized by the immune system, which may not be possible to achieve.
2. Enteritis: With their ability to kill pathogenic (disease-causing) bacteria, antibacterial agents may also kill beneficial or commensal bacteria found in the gut. These essential organisms aid normal processes such as digestion. If killed, they could impede these processes and, for example, promote severe diarrhea.
3. Selection and transference of antibiotic resistance: The constant exposure of human gastrointestinal bacteria to low levels of antimicrobial substances in an animal product could promote or select for the development of resistance. Not only does this make the future treatment of infections more difficult and expensive, but this resistance could also be transferred to other pathogens.
4. Other side effects: It is possible that drugs administered to animals may cause other side effects in people by being directly toxic.
5. Other effects of residues may include:
    a. The cheese and yogurt manufacturers make use of bacterial cultures to produce their products. The presence of antimicrobials in the milk could kill these bacteria and prevent efficient food production.
    b. Young animals suckling from their dams may consume drug residues in the milk. This may have an adverse effect on the growth of the young animals by, for example, antimicrobial residues sterilizing the gastrointestinal tract and thereby promoting the development of severe diarrhea and malabsorption.

4.7 SUGGESTIONS FOR TOXICITY TESTING

The degree of testing required for herbal remedies is dependent on the animal species to be treated, on the amount of information already available, and the formulation. The botanical and pharmacognosy or phytochemical literature may provide information on the presence of dangerous compounds. Therefore, it is important to identify knowledge gaps and design a testing strategy around this, considering the limitations mentioned (Table 4.3). The following guide is our recommendation and does not represent the views of any regulatory body. It is therefore incumbent on an applicant to determine the minimum requirements from the local regulatory authorities.

4.7.1 Literature Review

Before any toxicity testing activity, there should be a comprehensive literature review.

This is to establish what is already known about possible harmful effects of the material and to identify gaps in that knowledge. It is these gaps that will guide the testing strategy. The use of therapeutic information to predict toxicity should be considered, and this also applies to herbal medicines already utilized. However, information on overdose and reversibility, organ targets, and indicators such as NOAEL cannot be established during therapeutic use.

It is also important to ascertain the relevance of published data to your extract to ensure the same chemical constituents in the plant, same extraction method, and same plant part in use. As mentioned, any difference in extract could result in possible toxic compounds being introduced or an increase in the concentrations of certain products to toxic levels.

4.7.2 Pharmacovigilance

It has always been argued that due to the established use of many herbal remedies safety is proven. However, evidence for this is often anecdotal, and no documented data are usually available. Without properly kept records describing the use and outcome in animals, it is impossible to ascertain if the information is correct. In one study, a commonly used herbal remedy in South Africa was subjected to a subacute toxicity test. In this study (Figure 4.6, the traditional medicine, consisting of four dried and milled African plants, was tested in vervet monkeys (Seier et al., 2004). Administration of this material was associated with a significant decline in erythrocyte counts (anemia) in all treated groups, and there was a notable 31% loss of body weight in the high-dose group. In addition to clearly establishing toxicity, it was determined that the effects were reversible on cessation of therapy. This study clearly indicated that a history of use was not indicative of safety.

However, if properly recorded safety data, collected in a well-established pharmacovigilance monitoring system is available, these data may provide sufficient and invaluable information in demonstrating safety. For proper interpretation, such data must, however, be evaluated by an expert toxicologist. Any questions or doubt remaining must thereafter be answered by undertaking the relevant toxicity tests. It has to be emphasized that pharmacovigilance data will only be of value for the formulation if history of use is provided; for example, data for dried leaves may not be representative for a decoction from dried leaves as the latter involves a modification and therefore constitutes a new formulation that may have new effects.

4.7.3 Species

In veterinary medicines, the use of a drug can be divided into two broad categories: use in companion animals (dogs, cats, and horses) and use in food-producing animals (cattle, pigs, chickens, sheep, goats, horses, etc.).

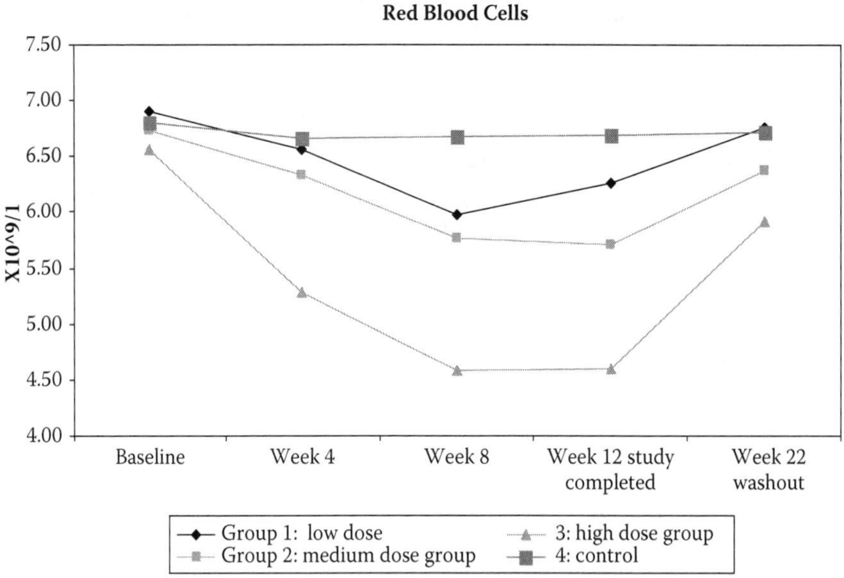

FIGURE 4.6 Changes in erythrocyte counts in vervet monkeys following treatment with a traditional medicine at various doses. This particular product was already used by people and considered safe. (From Seier et al. 2004. MRC, Final Report. With permission.)

4.7.3.1 Testing in Companion Animals

When substances are to be used in companion animals not for food production, we believe that even in the absence of pharmacovigilance or published data not all testing listed need be applied. In these cases, the strategy must be based on the nature of the herbal product.

1. Acute toxicity: This informs the subacute toxicity studies and can be omitted if sufficient information is already available. It is possible that acute toxicity for certain herbal medicines cannot be established as the plant or herbal extract may be devoid of harmful effects even at high doses.
2. Subacute toxicity: These tests are most important, particularly when an animal is to receive multiple doses. They can and should always be undertaken.
3. Reproductive studies: These are important when the product is to be used in animals of reproductive age. Abortions in valuable breeding animals could result in loss of income and prosecution of the manufacturer of the herbal remedy. These should therefore be considered carefully.
4. Genotoxicity studies: Safety from mutagenic effects must be demonstrated.
5. Carcinogenicity: Considering the difficulty in interpreting these tests, it is believed that they can be omitted unless evidence of genotoxicity is present.

6. Local tolerance testing: This is dependent on the nature of the compound or herb being tested. It must also be considered that the person administering the medicine topically can be affected by absorbing it (i.e., dimethyl sulfoxide [DMSO] in formulation) or local irritation. Gloves might have to be worn when applying this treatment, and safety of the medicine in this regard should be demonstrated.
7. Target animal tolerance studies: It is important to establish the safety of the herbal remedy in the intended recipient species. In fact, the entire testing strategy can be conducted in the target species. This has the benefit of decreasing the total number of tests and animals. This method does have the disadvantage of using a larger number of the target animal, which does increase the expense of a study.

It should be borne in mind that the cost of developing herbal remedies for companion animals can be prohibitive, and a cost-benefit analysis prior to the start of testing should be considered. This applies to pharmaceutical remedies as well and has unfortunately resulted in drug development for the companion animals being secondary to human pharmaceutical research and development.

4.7.3.2 Production Animals

Production animals are a more complicated issue as there is the possibility of poisoning not only the animal but also the consumer of such animal or animal products. Another problem is the possible unknown composition of herbal formulations and the resultant inability to identify and quantify active ingredients. This makes it impossible to determine when the withdrawal period will be reached. The use of microbial quantification of tissue levels, fingerprinting, or the use of mass spectroscopy and nuclear magnetic resonance may make this possible, but the onus is on the applicant to prove that the medicinal substance is below MRLs. One should also consider that in many countries, it is illegal to use any medicinal products in a production animal without setting an appropriate withdrawal period.

In production animals, it therefore becomes necessary to undertake all the toxicity tests listed. Exceptions from any of the tests must be based on information in the currently available literature. The use of the same medicines in people is not an accepted reason for exemption from testing since they would be consuming the medicines unknowingly, and no informed consent was given.

4.7.4 Formulation

A number of active and inactive ingredients are present in the formulation of ethical or allopathic medicines. The same applies to herbal remedies, and the amount and ratio of these ingredients are dependent on the preparation of the plant (i.e., the plant part and solvent used). This implies that toxicity testing is dependent on the preparation method. Therefore, each preparation method produces different remedies in terms of toxicity and will require testing unless it can be proven that the ratio and constituents within the various extracts are of an identical chemical nature.

4.8 CONCLUSION

It is clear that in veterinary ethnobotanical development, just as that in humans, establishing toxicity is an important prerequisite for use of all medicines.

Toxicity testing of veterinary herbal remedies is important for the protection of animals and people. Although some believe that such medicines are natural and therefore safe, this is by no means correct. Besides, routine therapeutic use cannot provide answers to the many questions asked during toxicity testing. In the absence of safety information, both the treating veterinarian and the manufacturing company will be liable for any toxicity that results.

The extent of toxicity testing is dependent on the information already available and on finding and filling the gaps. Complete compliance with conventional strategies is often not necessary and at times may not be possible as there are certain practical limitations imposed by the nature of the material to be tested.

During testing, quality control is of critical importance. This involves ensuring that material from plants was harvested at the same time, place, season, and soil condition since plant constituents vary significantly with these factors. This also ensures uniformity of the end product.

There is no doubt that the use of traditional medicine will continue in many countries informally without testing. However, should countries, communities, or organizations wish to enter mainstream drug development with their traditional knowledge, safety will have to be ensured through scientifically obtained data.

APPENDIX A1. CARE AND USE OF LABORATORY ANIMALS

Introduction

In veterinary drug development, substances can be tested in target species, which is an advantage over the human equivalent. However, even in veterinary drug development, preclinical safety testing is often undertaken initially in laboratory animals. The term *laboratory animal* is not an accurate taxonomical term but is often used to refer to animal species commonly used in biomedical research.

Considering this, the following basic insights into laboratory animal science were deemed appropriate for inclusion in this chapter. They are intended to create awareness in readers with little or no knowledge of laboratory animals and stimulate further reading since the literature abounds with information on their care and use.

Choice

Among the rodents, rats (*Rattus norvegicus*) are most commonly utilized for safety testing, while nonrodent species (including dogs, pigs, and nonhuman primates) are also used. In preclinical safety testing for human drugs, it is stipulated that a rodent and nonrodent species be used. Since it is unlikely that drugs for cattle or chickens would be tested in primates, the most likely species utilized would be the rat. Rats have been well characterized and are available in a large variety of strains (breeds). Some of these are produced for particular diseases such as hypertension; others may have been altered genetically as disease models. However, for preclinical safety testing, only normal rats of any strain may be used. While both sexes may be used in testing, it is recommended that female rats be used. This is because through a well-documented history of use, the female of the species has been shown to be more sensitive if sexual sensitivity to toxicity was present.

Care

Laboratory animals, including rats, have highly specific requirements. This is contrary to the popular belief that rats are vermin and are associated with unsanitary conditions. They are, on the contrary, actually prone to a wide variety of diseases, some manifest only in stress situations, and all of them can significantly confound research results. Therefore, to provide high standards of animal welfare and control diseases, rats have to be kept in highly hygienic conditions, and some are even housed in situations that are certified free of certain pathogens, which is called *specific pathogen free* (SPF), or under completely sterile conditions, called *germ free*. Cages must be made from sterilizable materials. Other confounding factors that need to be controlled are environmental conditions and food. Appropriate air conditioning and standardized nutrition consisting of good-quality pelleted diets, produced specifically for rats, will ensure minimal variability and good health.

Environmental Enrichment

Previously, the focus of laboratory animal management tended to be on physical health and disease transmission. However, it is clear that animals also have behavioral needs, and much attention has shifted to address this issue. For example, for hygienic reasons rats were previously housed on grid floors, whereas studies have shown that rats prefer solid ones, which have now become standard in rat housing. The sum of actions taken to enable species-specific behavior is called *environmental enrichment*. This has consequences for cage size, design, and complexity, depending on the animal's needs. Provision of materials for gnawing, hiding, climbing, socializing, and nest making are examples of actions to be taken to address the needs of rats.

Administration of Substances and Collection of Body Fluid

In regulatory safety testing, it will invariably be necessary to administer substances by various routes, which may be gavage, injection, or in food. Collection of body fluids, mainly blood, is another key activity and is required to determine possible hematological and biochemical changes produced by the treatment. All require highly skilled and trained personnel who, in some countries, have to be qualified or registered with relevant authorities.

Ethics

The use of animals in research is an emotive and controversial issue in many societies. Although there is polarization about the right to use animals in research, it is generally agreed that animals are sentient creatures, and that scientists have clear moral obligations to treat them humanely. A commonly used yardstick is that the benefits of the research must outweigh the harm to the research animal. This can be particularly difficult to balance in preclinical safety testing, for which one of the aims is to induce toxicity. However, some of the more severe tests, such as the LD_{50}, are not undertaken routinely, and there is increasing research into finding alternatives such as cell cultures and computer modeling. At present, alternate models are only acceptable for genotoxicity testing. A number of shortcomings of *in vitro* systems still need to be resolved (e.g., cells in culture may lose their natural function and abilities, not being able to test the effects in an interdependent system of organs, and testing for physical symptoms such as ataxia, vomiting, fever, and diarrhea). Reducing safety testing would also require the cooperation of authorities since some tests are legal requirements for product registration.

REFERENCES

Borzelleca, J. F. (2000). Paracelsus: herald of modern toxicology. *Toxicological Sciences* 53, 2–4.

British Toxicology Society. (1984). A new approach to the classification of substances and preparations on the basis of their acute toxicity. A report by the British Toxicology Society Working Party on Toxicity. *Human Toxicology* 3, 85–92.

Bruce, R. D. (1985). An up-and-down procedure for acute toxicity testing. *Toxicological Sciences* 5, 151–157.

Buehler, E. V. (1965). Delayed contact hypersensitivity in the guinea pig. *Archives of Dermatology* 91, 177.

Clouatre, D. L. (2004). Kava kava: examining new reports of toxicity. *Toxicology Letters* 1, 85–96.

DiMasi, J. A., Hansen, R. W., and Grawbowski, J. W. (2003). The price of innovation: new estimates of drug development costs. *Journal of Health Economics* 22, 151–185.

Deng, J. (2002). Clinical and laboratory investigations in herbal poisonings. *Toxicology* 181, 571–576.

Environmental Protection Agency (EPA). (1998a). Health effects test guidelines OPPTS 870.2400 acute eye irritation. EPA 712-C-98-195. Available online: http://iccvam.niehs.nih.gov/SuppDocs/FedDocs/EPA/EPA_870_2400.pdf

Environmental Protection Agency (EPA). (1998b). Health effects test guidelines OPPTS 870.2500 acute dermal irritation. EPA 712-C-98-196. Available online: http://iccvam.niehs.nih.gov/SuppDocs/FedDocs/EPA/EPA_870_2500.pdf

Ernst, E. (2002). Adulteration of Chinese herbal medicines with synthetic drugs: a systematic review. *Journal of Internal Medicine* 252, 107–113.

European Medicines Agency (EMEA). (1984). Repeated Dose Toxicity. Directive 75/318/EEC as amended, 9-20. Available online: http://www.emea.europa.eu/htms/human/humanguidelines/nonclinical.htm

European Medicines Agency (EMEA). (1987a). Single Dose Toxicity. Directive 75/318/EEC as amended, 3-7. Available online: http://www.emea.europa.eu/htms/human/humanguidelines/nonclinical.htm

European Medicines Agency (EMEA). (1987b). Testing of medicinal products for their mutagenic potential. Available online: http://www.emea.europa.eu/htms/human/humanguidelines/nonclinical.htm

Fink, M., Noble, D., Virag, L., Varro, A., and Giles, W. R. (2008). Contributions of HERG K+ current to repolarisation of the human ventricular action potential. *Progress in Biophysics and Molecular Biology* 96, 357–376.

Food and Drug Administration (FDA). (2000). Toxicological Principles. Redbook II. Available online: http://www.fda.gov/Food/GuidanceComplianceRegulatoryInformation/GuidanceDocuments/FoodIngredientsandPackaging/Redbook/default.htm

Foukaridis, G. N., Osuch, E., Mathibe, L., and Tsipa, P. (1995). The ethnopharmacology and toxicology of *Urginea sanguinea* in the Pretoria area. *Journal of Ethnopharmacology* 49, 77–79.

Gad, S. C., and Chengelis, C. P., 1998. *Lethality testing: Acute toxicology testing*. Academic Press, London, pp. 155–196.

Hoel, D. G., and Portier, C. J. (1994). Nonlinearity of dose-response functions for carcinogenicity. *Environmental Health Perspectives* 102, 109–113.

International Conference on Harmonisation (ICH). (2000). Good manufacturing practice guide for active pharmaceutical ingredients. International Conference on Harmonisation of Technical Requirements for Registration of Pharmaceuticals for Human Use. http://www.ICH.org.

Kellerman, T. S., Coetzer, J. A. W., Naude, T. W., and Botha, C. J. (2005). Plant poisonings and mycotoxicoses of livestock in Southern Africa. Oxford University Press, Cape Town, South Africa.

Kimmel, C. A., and Makris, S. L. (2001). Recent developments in regulatory requirements for developmental toxicology. *Toxicology Letters* 1, 73–82.

Kirkland, D., Aardema, M., Henderson, L., and Muller, L. (2005). Evaluation of the ability of a battery of three in vitro genotoxicity tests to discriminate rodent carcinogens and non-carcinogens I. Sensitivity, specificity and relative predictivity. *Mutation Research* 584, 1–256.

Ko, R. (2006). Safety of ethnic and imported herbal and dietary supplements. *Clinical Toxicology* 44, 611–616.

Lave, L. B., Ennever, F. K., Rosenkranz, H. S, and Omenn, G. S. (1988). Information value of the rodent bioassay. *Nature* 336, 631–632.

Lees, P., Cunningham, F. M., and Elliot, J. (2004). Principles of pharmacodynamics and their applications in veterinary pharmacology. *Journal of Veterinary Pharmacology and Therapeutics* 27, 397–414.

Magnusson, B. (1980). Identification of contact sensitizers by animal assay. *Contact Dermatitis* 6, 46–50.

Magnusson, B., and Kligman, A. M. (1969). The identification of contact allergens by animal assay. The guinea pig maximization test. *The Journal of Investigative Dermatology* 52, 268–276.

Meyer, B. N., Ferrigni, N. R., Putnam, J. E., Jacobson, L. B., Nichols, D. E., and McLauglin, J. L. (1982). Brine shrimp: a convenient general bioassay for active plant constituents. *Planta Medica* 45, 31–34.

Mutai, M. (2000). National and international guidelines for the conduct of chemical safety studies: choice of strains. In: Krinke, G. J. (Ed.), *The laboratory rat.* Academic Press, San Diego, CA, pp. 17–27.

Nebendahl, K. (2000). Routes of administration. In: Krinke, G. J. (Ed.), *The laboratory rat.* Academic Press, San Diego, CA, pp. 463–483.

Noguchi, T., and Ebesu, J. S. M. (2001). Puffer poisoning: epidemiology and treatment. *Toxins Review* 201, 1–10.

Organization for Economic Cooperation and Development (OECD). (1981). OECD Guideline for testing of chemicals. Chronic Toxicity Testing. Guideline 452. Available online: http://lysander.sourceoecd.org/vl=5947981/cl=12/nw=1/rpsv/cw/vhosts/oecdjournals/1607310x/v1n4/contp1-1.htm

Organization for Economic Cooperation and Development (OECD). (1983). OECD Guideline for testing of chemicals. Two generation reproduction toxicity. Guideline 426. Available online: http://lysander.sourceoecd.org/vl=5947981/cl=12/nw=1/rpsv/cw/vhosts/oecdjournals/1607310x/v1n4/contp1-1.htm

Organization for Economic Cooperation and Development (OECD). (1995). OECD Guideline for testing of chemicals. Repeated Dose 28-day Oral Toxicity Study in Rodents. Guideline 407. Available online: http://lysander.sourceoecd.org/vl=5947981/cl=12/nw=1/rpsv/cw/vhosts/oecdjournals/1607310x/v1n4/contp1-1.htm

Organization for Economic Cooperation and Development (OECD). (1997). OECD Guideline for testing of chemicals. In vitro mammalian chromosomal aberration test. Guideline 473. Available online: http://lysander.sourceoecd.org/vl=5947981/cl=12/nw=1/rpsv/cw/vhosts/oecdjournals/1607310x/v1n4/contp1-1.htm

Organization for Economic Cooperation and Development (OECD). (1998). OECD Guideline for testing of chemicals. Repeated Dose 90-day Oral Toxicity Study in Non-Rodents. Guideline 409. Available online: http://lysander.sourceoecd.org/vl=5947981/cl=12/nw=1/rpsv/cw/vhosts/oecdjournals/1607310x/v1n4/contp1-1.htm

Organization for Economic Cooperation and Development (OECD). (2000a). OECD Guideline for testing of chemicals. Prenatal developmental toxicity study. Test no 414. Available online: http://lysander.sourceoecd.org/vl=5947981/cl=12/nw=1/rpsv/cw/vhosts/oecdjournals/1607310x/v1n4/contp1-1.htm

Organization for Economic Cooperation and Development (OECD). (2000b). OECD Guideline for testing of chemicals. Two-generation reproduction toxicity study. Guideline 416. Available online: http://lysander.sourceoecd.org/vl=5947981/cl=12/nw=1/rpsv/cw/vhosts/oecdjournals/1607310x/v1n4/contp1-1.htm

Organization for Economic Cooperation and Development (OECD). (2001a). OECD Guideline for testing of chemicals. Acute Oral Toxicity - Acute Toxic Class Method. Guideline. Guideline 423. Available online: http://lysander.sourceoecd.org/vl=5947981/cl=12/nw=1/rpsv/cw/vhosts/oecdjournals/1607310x/v1n4/contp1-1.htm

Organization for Economic Cooperation and Development (OECD). (2001b). OECD Guideline for testing of chemicals. Repeated Dose 90-day Oral Toxicity Study in Rodents. Guideline 408. Available online: http://lysander.sourceoecd.org/vl=5947981/cl=12/nw=1/rpsv/cw/vhosts/oecdjournals/1607310x/v1n4/contp1-1.htm

Organization for Economic Cooperation and Development (OECD). (2004). OECD Guideline for testing of chemicals. Acute dermal toxicity, Draft guideline. Guideline 434. Available online: http://lysander.sourceoecd.org/vl=5947981/cl=12/nw=1/rpsv/cw/vhosts/oecdjournals/1607310x/v1n4/contp1-1.htm

Organization for Economic Cooperation and Development (OECD). (2006). OECD Guideline for testing of chemicals. Acute Oral Toxicity - Up-and-Down-Procedure (UDP). Guideline 425. Available online: http://lysander.sourceoecd.org/vl=5947981/cl=12/nw=1/rpsv/cw/vhosts/oecdjournals/1607310x/v1n4/contp1-1.htm

Putman, E., van der Laan, J. W., and van Loveren, H. (2003). Assessing immunotoxicity: guidelines. *Fundamental and Clinical Pharmacology* 17, 615–626.

Raschi, E., Vasina, V., Poluzzi, E., and De Ponti, F. (2008). The hERG K^+ channel: target and antitarget strategies in drug development. *Pharmacological Research* 57, 181–195.

Rawlins, M. D. (2004). Cutting the cost of drug development? *Nature Reviews* 3, 360–364.

Ritz, H. L., and Buehler, E. V. (1980). Planning, conduct, and interpretation of guinea pig sensitization patch tests. In: Drill, V. A. and Lazar, P. (Eds.) *Current Concepts in Cutaneous Toxicity,* Academic Press, New York, 25–42.

Roche, O., Trube, G., Zuegge, J., Pflimlin, P., Anlanine, A., and Schneider, G (2002). A virtual screening method for prediction of the hERG Potassium channel liability of compound libraries. *Chembiochem: A European Journal of Chemical Biology* 3, 455–459.

Schweitz, H., and Moinier, D. (1999). Mamba toxins. *Perspectives in Drug Discovery and Design* 15–16, 83–110.

Seier, J. V., Mdhluli, M., Matsabisa, M., Dhansay, M. A., Laubscher, R., Mbewu, A., and Dhai, A. (2004). A toxicity study of IKS/PYAI Consumption. MRC, Final Report. May 17, pp. 1–97, MRC Press, Cape Town.

Skidmore, M. W., Kuiper, H. A., and Hamilton, D. (1997). Safety evaluation of pesticide residue in food. *Chemistry International* 19, 119–126.

Sobel, J. (2007). Botulism. *Clinical Infectious Diseases* 41, 1167–1173.

Solis, P. N., Wright, C. W., Anderson, M. M., Gupta, M. P., and Phillipson, J. D. (1993). A microwell cytotoxicity assay using *Artemia salina* (brine shrimp). *Planta Medica* 59, 250–252.

Timbrell, J. A. (1995). Introduction to toxicology. Taylor and Francis, London.

Toussaint, M. W., Shedd, T. R., van der Schalie, W. H., and Leather, G. R. (1995). A comparison of standard acute toxicity tests with rapid-screening toxicity tests. *Environmental Toxicology and Chemistry* 14, 907–915.

Ueno, Y. (1985). The toxicology of mycotoxins. *Critical Reviews in Toxicology* 14, 99–132.

Van den Heuvel, M. J., Clark, D. G., Fielder, R. J., Koundakjien, P. P., Oliver, G. J. A., Pelling, D., Tomlinson, N. J., and Walker, A. P. (1990). The international validation of a fixed-dose procedure as an alternate to the classical LD_{50} test. *Food and Chemical Toxicology* 28, 469–482.

Van Dresser, W. R., and Wilcke, J. R. (1989). Drug residues in food animals. *Journal of the American Veterinary Medical Association* 194, 1700–1710.

Verlato, G., Cerveri, I., Villani, A., Pasquetto, M., Ferrari, M., Fanfulla, F., Zanolin, E., Rijken, B., and de Marco, R. (1996). Evaluation of methacholine dose-response curves by linear and exponential mathematical models: goodness-of-fit and validity of extrapolation. *European Respiratory Journal* 9, 506–511.

Walum, E. (1998). Acute oral toxicity. *Environmental Health Perspectives*, 106, 497–503.

World Health Organization (WHO). (2004). Consideration in the non-clinical development of pharmaceuticals. In Kioy, D. (Ed.), *Non-clinical safety testing. Special Programme for Research and Training in Tropical Diseases,* pp. 11–23.

Williams, D. A. (1986). Median lethal dose. *Biometrics* 42, 641–645.

Worth, P. A., and Cronin, T. D. (2001). The use of pH measurements to predict the potential of chemicals to cause acute dermal and ocular irritation. *Toxicology* 169, 119–131.

FURTHER READING

Association for Assessment and Accreditation of Laboratory Animal Care. Home page. http://www.aaalac.org/. Provides information on animal care standards.

Institute of Laboratory Animal Resources, NCR. (1996). *Guide for the care and use of laboratory animals,* National Academy Press, Washington, DC.

Poole, T. (Ed.). (1999). *UFAW handbook on the care and management of laboratory animals.* 7th ed., vol. 1. Blackwell, Malden, MA. Note: The UFAW (University Federation for Animal Welfare) publishes extensively in the field of laboratory animal science, including principles of humane experimentation and *in vitro* alternatives.

World Health Organization. Good laboratory practice. *WHO/TDR handbook.* (2001), Geneva.

5 Revitalizing Ethnoveterinary Medical Traditions

A Perspective from India

M. N. Balakrishnan Nair and P. M. Unnikrishnan

CONTENTS

5.1 INTRODUCTION

Nearly 70% of the world's rural poor depend on livestock as a critical component of their livelihood (640 million poor farmers in rain-fed areas, 190 million pastoralists in the arid and mountain zones, and more than 100 million people in landless households; Swarup and Patra, 2005). Prevention, control, and eradication of diseases among domesticated animals are major concerns as diseases in animals lead to economic losses and possible transmission of the causative agents to humans. Livestock are the foundation of agrarian economy, contributing about 32% of the total share from agriculture to the national gross domestic product (GDP; Swarup and Patra, 2005). Livestock rearing is considered a supplementary occupation and a source of additional income for those engaged in agricultural operations.

People use meat, eggs, and dairy products for their food, wool and leather for clothing, and animal power to till their fields. Livestock also provide manure to fertilize the crop (Food and Agriculture Organization [FAO] *Livestock Policy Brief No. 3*). The Indian livestock sector today has 185 million cattle, 98 million buffalo, 124.5 million goats, 60 million sheep, 343 million poultry, 13.5 million pigs, 1.6 million equines, and 6.3 million camels. The world share is 16% of cattle and 57% of buffalo (ranked first), 18% of goats (ranked second), and 6% of sheep (ranked third). The contribution of livestock to GDP is 5.38% (valued Rs 120,938 cr (about US$26121 million) in 2003) (Anonymous, 2006).

It is estimated that approximately Rs 50 billion (US$1 = about Rs 46.3) annually are lost on account of livestock diseases in India. The high treatment cost, inaccessibility, and indiscriminate use of antibiotics and hormones, which leads to user-unfriendly effects such as high antibiotic and hormone residues in the milk and other animal products, are serious limitations of modern veterinary management (Khoda, 2005; Merai and Boghra, 2004). Decline in the animal husbandry budget and privatization has led to the scanty veterinary services provided by the government to the poor in rural areas (Anonymous, 2004). Veterinary services have a crucial role in controlling highly contagious diseases and zoonotic infections, which have implications for human health as well as livestock health.

5.2 MODERN VETERINARY SERVICES

The veterinary service in India under the government sector is solely based on allopathic veterinary infrastructure. There were about 310,269 million adult units per the 1992 livestock census (assuming that either 1 cattle, 1 buffalo, 10 sheep, 10 goats, 5 pigs, or 100 poultry are equivalent to 1 adult unit) (Ravindra and Rao, 1999). There are 45,760 veterinary institutions in India, comprised of 7,415 veterinary hospitals, 14,573 veterinary dispensaries, and 23,682 veterinary aid centers. There is one veterinary institution for every 7,052 adult units, and one veterinarian serves around 10,000 adult units. Common problems in cattle are given in Table 5.1. The budget for veterinary medicine and biologicals is less than Rs 13 (US$1 = about Rs 46.3) per adult unit per annum. The existing infrastructure is inadequate (Ravindra and Rao, 1999).

5.3 A HISTORICAL PERSPECTIVE OF ETHNOVETERINARY MEDICINE

Traditional veterinary medicine has evolved simultaneously with the evolution of human beings specifically to take care of the health of animals, which were being domesticated. A historical record of treatment of diseased animals is present in the river valley civilizations. Ancient Egyptians used various methods, including application of herbs for treatment of animal diseases, and it is believed that during that period the physicians had knowledge of more than 250 medicinal plants and 120 mineral salts (Swarup and Patra, 2005). Veterinary science in India has a documented history of around 5,000 years. The veterinary and animal husbandry practices are mentioned in *Rigveda,* one of the oldest vedas (4000–1500 BC)—the primary scripture of Hinduism, and *Atharvaveda,* one of the four vedas. The codified knowledge existed in the form of texts and manuscripts on various aspects of veterinary care of livestock. Sage Shalihotra's (1800 BC) *asvasatra*

TABLE 5.1
Common Problems in Cattle

Serial No.	Modern Veterinary Science	Ayurveda (ISM)
1	Abortion	
2	Black quarter	*Pittajasopha/Visarpa*
3	Bloat	*Adhmana, Gulma, Vistabdha ajeerna*
4	Bronchitis	*Kasa*
5	Dermatitis	*Kusta*
6	Enteritis/diarrhea	*Atisara*
7	Ephemeral fever	*Trithiyaka Jvara*
8	Foot-and-mouth disease	*Sannipata jvara/Agni visarpa*
9	Fracture/dislocation	*Asthi bhagna/sandhi mukta*
10	Helminthiasis	*Krimi*
11	Hypogalactia	*Stanya kshaya*
12	Impaction of rumen	
13	Indigestion	*Ajeerna (vistabdha)*
14	Mastitis	*Sthana vidrahi*
15	Prolapses of uterus	*Yoni bhramsha*
16	Pyrexia	*Jvara*
17	Repeat breeder/habitual abortion	*Apatyaya/Garbha srava*
18	Retention of placenta	*Apathitha apara*
19	Sterility	
20	Vitamin deficiency	
21	Yoke gall	

Source: From Raneesh et al., 2005. *A user's guide on ethnoveterinary health practices*, FRLHT, Bangalore, India. With permission.

(Asva Shastra translated by R. Vijayaraghavacharya, 1927, Sri Venkaleswara Oriental Library, Thirupathi) is considered the first work on veterinary science. *Hastyayurveda* (1000 BC) by sage Palakapya is the oldest text on elephants (Palakapya Maharshi Hastyayurvedam translated in Malayalam by Vaidyamadham Cheria Narayanan Namboodiri Mathrubhoomi Printing and Publishing Co. Ltd. Kozhikode, 2006). Treatises such as *Asva Vaidyaka* by Nakula Garuda Purana, *Asvayur Veda Sarasindu* by Vysampayana are based on the principles of the humoral theory of the ayurvedic system. A few vernacular scriptures like *Sahadeva Pasu Vaidya Sastramu* (Telugu) and *Mattu Vaidya Bodhini* (Tamil) dealt with traditional veterinary medicine (Ravindra and Rao, 1999). *Pashu Vaidya Mattuvagadam,* a Tamil book based on ancient Tamil palm leaves manuscript, discusses over 250 diseases in cattle and their management. *Matsyapurana, Garudapurana, Agnipurana, Brahmanandapurana, Lingapuranah, Caraka Samhita,* and *Shusruta Samhita* (Trikamgi 1954, 1992) have veterinary information. *Arthasatra* by Kautilya described rearing of cattle, buffalo, sheep, goats, horses, elephants, and other animals. It also gave a detailed account of welfare practices of livestock and regulation for the protection of wildlife. King Ashoka created animal hospitals between 269 and 232 BC (documentation found in Sagari and Nitya, 2004).

5.4 COMMUNITY-BASED ANIMAL HEALTH CARE

Ethnoveterinary or folk medicine pertaining to animal health care is as old as the domestication of various livestock species. There are rich and efficient ethnoveterinary traditions existing in the villages of India that form an integral part of the village life and play an important social, religious, and economic role. They consist of belief, knowledge, practices, and skills pertaining to health care and management of livestock. Table 5.2 shows the common problems in cattle and their management by folk healers.

TABLE 5.2
Common Problems in Cattle and Their Management

Serial No.	Health Conditions	Number of Practices	Folk Remedies
1	Black quarter	1	*Holoptelea integrifolia* (bark) 50 g is pounded and boiled in a liter of water and is reduced to 500 mL to make a decoction. 500 mL of the decoction are administered orally twice a day before feed on alternate days until the condition is relieved. Vd. Siddharamappa, Benegere, Shimoga District
		2	Paste of *Brassica nigra* (seeds) in sufficient quantity is applied over the affected area until the condition is relieved. Vd. Siddharamappa, Benegere, Shimoga District
		3	Rice cooked with palm jaggery should be given orally as feed until the condition is relieved. Vd. Guruswami, Desiya Nendhal, Virudhunagar District
		4	*Cuminum cyminum* (seeds) 10 g and *Ferula asafoetida* (resin) 5 g are ground to a fine paste. The paste is administered orally before feed along with lukewarm water until the condition is relieved. Vd. Thangaswami Thevar, Virudhunagar District
		5	*Curcuma longa* (rhizome) is mixed with sufficient quantity of coconut oil to form a fine paste, which is used as an external application until the condition is relieved. Vd. Thangaswami Thevar, Virudhunagar District
2	Bloat	6	*Ricinus communis* (seed oil) 100 mL along with 100 mL of water are administered orally twice a day for 2 days. Vd. Appaji, Hassan
		7	*Zingiber officinale* (rhizome) 5 g, *Acorus calamus* (rhizome) 5 g, *Syzygium aromaticum* (flower buds) 10 nos., *Piper betel* (leaves) 10 nos., and *Piper nigrum* (fruit) 10 nos. are boiled in 500 mL of water and reduced to 250 mL. Add 25 g of jaggery to the prepared lukewarm decoction. This is administered orally twice a day for 2 days. Vd. Shivanna, Tumkur

TABLE 5.2 (CONTINUED)
Common Problems in Cattle and Their Management

Serial No.	Health Conditions	Number of Practices	Folk Remedies
		8	*Pongamia pinnata* (bark) 200 g is boiled in a liter of water and reduced to 500 mL. Add the juice of *Citrus limon* (fruit) 1–2 nos., *Ricinus communis* (seed oil) 50–100 mL, and baking soda 5 g to the prepared decoction. The decoction is administered orally once a day. After 6 h of administration of the decoction, 3 teaspoons of the fine powder of the compound mixture of *Piper nigrum* (fruit) 50 g, *Embelia ribes* (seed) 200 g, *Trachyspermum ammi* (seed) 200 g, *Zingiber officinale* (rhizome) 25 g, *Butea monosperma* (seed) 25 g, *Holoptelea integrifolia* (bark) 200 g, and *Ailanthes excelsa* (bark) 200 g are added to *Pongamia pinnata* decoction and are administered orally three times daily for a day. Vd. Siddabasavaraju, Bimangala, Siddlagatta, Kolar District
		9	*Holoptelea integrifolia* (leaves) 1 handful, *Aegle marmelos* (leaves) 1 handful, *Zingiber officinale* (rhizome) 5 g, and *Allium sativum* (bulb) 1 no. are ground to a fine paste and administered orally in a liter of water twice a day for 2 to 3 days. Vd. Ramesh, Hassan
		10	*Sapindus laurifolia* (fruit) 1 nos. (The fruit peel is squeezed in water.) *Zingiber officinale* (rhizome) 10 g is ground to a fine paste and mixed with water. This is administered orally once a day for 2 days. Vd. Siddharamappa, Benegere, Shimoga District
		11	*Dolichos biflorus* (seed) 50 g, *Acorus calamus* (rhizome) 5 g, *Ferula asafoetida* (resin) 5 g, and *Ricinus communis* (seed oil) 50 mL are ground to a fine paste. The paste is administered orally once with 50 mL of water. Vd. Shekarappa, Shimoga
3	Bronchitis	12	*Allium sativum* (bulb) 4 nos., jaggery 50 g, *Cinnamomum camphora* (resin) 5 g, *Piper nigrum* (fruit) 5 nos., and *Coriandrum sativum* (seeds) 250 g are ground to a fine paste. The paste is administered orally in two divided doses for a day. Vd. Shivaswami Gowda, Tumkur
		13	*Piper nigrum* (fruit) 20 nos. and *Zingiber officinale* (rhizome) 10 g are boiled in 500 mL of water and reduced to 250 mL. Then 250 mL of this decoction is administered orally twice a day for 3 days. Vd. Saraswathi Bhatta, Sutla Taluk, Udupi

(continued)

TABLE 5.2 (CONTINUED)
Common Problems in Cattle and Their Management

Serial No.	Health Conditions	Number of Practices	Folk Remedies
		14	10 g of each of the stem barks of *Tabernaemontana heyneana, Zanthoxylum rhetsa, Pongamia pinnata, Holarrhena pubescens,* and *Moringa oleifera* are ground along with *Capsicum annum* (fruit) 10 g, *Piper nigrum* (fruit) 10 g, *Brassica nigra* (seeds) 10 g, *Trachyspermum ammi* (seeds) 10 g, and *Allium sativum* (bulb) 10 g to a fine paste. The paste is rolled in pills of betel nut size and is administered orally twice a day along with 500 mL of *Trachyspermum ammi* decoction until the condition is relieved. Vd. Kunjeeramulya, Karkala
4	Dermatitis	15	*Plumbago zeylanica* (roots) 5 g, *Allium sativum* (bulb) 2 nos., and *Piper nigrum* (fruits) 1–2 nos. are ground in the juice of *Citrus limon* to make a fine paste. The paste is applied over the affected part once a day until the condition is relieved. Vd. Gangadarappa, Hassan
		16	Finely powder 100 g of each of *Toddalia asiatica* (fruit), *Aristolochia indica* (root), *Vitex negundo* (leaf), *Clerodendrum inerme* (leaf), *Argemone mexicana* (leaf), *Cipadessa baccifera* (leaf), *Solanum nigrum* (leaf), and *Leucas aspera* (leaf). The powder is made into paste with water and is applied over the affected part once a day for a week. Vd. Anantharamiah, Bangalore Rural
		17	Mix a kilogram of dried leaves of *Streblus asper* and 250 g of charcoal. Make a fine paste with *Pongamia pinnata* seed oil. This is applied over the affected part. Vd. Siddhabasavaraju, Kolar
		18	A handful of dried *Evolvulus alsionoides* (whole plant) is mixed with 500 mL cow's milk. This is administered orally once a day for 3 days. Vd. Siddharamappa, Shimoga
		19	*Scleropyrum pentandrum* (seed oil) is applied over the affected part for a week. Vd. Kunjeeramulya, Karkala
		20	*Curcuma longa* rhizome powder (5 g), is mixed with the paste of *Mimosa pudica* and 50 mL of neem oil. The oil is heated until bubbles appear on top of the oil and is devoid of moisture content. The oil is applied over the affected part once a day until the condition is relieved. Vd. Shivaswami Gowda, Tumkur

TABLE 5.2 (CONTINUED)
Common Problems in Cattle and Their Management

Serial No.	Health Conditions	Number of Practices	Folk Remedies
5	Enteritis	21	*Aerva lanata:* A handful of the whole plant is ground with rice-washed water to a fine paste. This is administered orally twice a day until the animal gets cured. Ms. Eliyamma, Waynad
		22	*Glycosmis mauritiana:* A handful of root bark is ground with rice-washed water to a fine paste. The paste is administered orally twice a day until the animal gets cured. Ms. Eliyamma, Waynad
		23	*Glycosmis mauritiana:* A handful of root bark is ground with rice-washed water to a fine paste. The paste is administered orally twice a day until the animal gets cured. Ms. Eliyamma, Waynad
		24	The pulp juice of *Garcinia gummi gutta* and juice of *Zingiber officinale* are mixed with a sufficient quantity of honey. 100 mL of this mixture is administered orally twice a day until the condition is relieved. Ms. Moli Jose Pullikal, Waynad
		25	A handful of each of *Punica granatum* (tender leaves) and *Syzygium cumini* (bark) are ground to a fine paste in a liter of tender coconut water. This paste is administered orally three times a day for 3 days. Vd. Ramesh, Hassan
		26	Mix 2 teaspoons of *Wattakaka volubilis* (stem powder) and 500 g of *Eleusine coracana* flour along with buttermilk. This mixture is administered orally in three divided doses for a day. Vd. Siddhabasavaraju, Kolar
		27	*Caesalpinia crista:* A handful of leaves is fed to the animal orally three times a day until the condition is relieved. Vd. Appaji, Hassan
6	Ephemeral fever	28	10 g of each of *Bauhinia racemosa* (bark), *Buchanania axillaris* (bark), *Ailanthes excelsa* (bark), *Capparis sepiaria* (bark), *Balanites aegyptica* (bark), *Piper nigrum* (fruit), *Trachyspermum ammi* (seed), *Butea monosperma* (seed), *Ferula asafoetida* (resin), *Allium sativum* (bulb), *Brassica nigra* (seed), and *Toddalia asiatica* (bark) are finely powdered. Orally, 50 g of the powder along with a liter of boiled and cooled water are administered twice a day for 5 days. Vd. Narayanappa, Kolar

(*continued*)

TABLE 5.2 (CONTINUED)
Common Problems in Cattle and Their Management

Serial No.	Health Conditions	Number of Practices	Folk Remedies
		29	Decoction is prepared by boiling *Toddalia asiatica* (root) 25 g, *Balanites aegyptica* (root) 25 g, *Capparis sepiaria* (bark) 25 g, and *Capparis zeylanica* (bark) 25 g in 2 L of water and reduce to 1 L. To this decoction add 2 tablespoons of the powder prepared from the following ingredients: *Embelia ribes* (seed) 250 g, *Trachyspermum ammi* (seed) 250 g, *Curcuma aromatica* (rhizome) 250 g, *Acorus calamus* (rhizome) 20 g, *Butea monosperma* (flower and seed) 20 g, *Ferula asafoetida* (resin) 20 g, *Zingiber officinale (*rhizome) 25 g, *Ailanthes excelsa* (bark) 250 g, and *Terminalia chebula* (fruit) 50 g. This medicine is administered orally twice a day for a fortnight. Note: The bark of *Ailanthes excelsa* should be collected either on Sunday (Hasta nakshatra) or Thursday (Pushya nakshatra). Vd. Siddhabasavaraju, Kolar
7	Foot-and-mouth disease	30	Warm 200 g of *Calotropis gigantea* (flowers) and *Syzygium aromaticum* (flower buds) 4 nos. in 500 mL of tender coconut water. To this add 50 g of jaggery; this is administered orally once a day for 5–9 days. Vd. Shivanna, Tumkur
		31	*Mesua ferrea* (fruit) 50 g, *Strychnos nux-vomica* (bark) 25 g, *Curcuma longa* (rhizome) 10 g, and *Vernonia anthelmintica* (seed) 5 g are heated in 300 mL of coconut oil. The oil is applied over the affected area three times a day until the condition is relieved. Vd. Kunjeeramulya, Karkala
		32	*Vigna radiata* is boiled in water to which jaggery is added to form a pudding. A ripe banana fruit is given separately prior to feeding the pudding. The pudding is administered orally twice a day for 3 days. Vd. Karibasappa, Shimoga
		33	*Cissus quadrangularis* (stem) 1 kg, *Capsicum annum* (fruit) 50 g, *Piper nigrum* (fruit) 10 g, *Zingiber officinale* (rhizome) 50 g, *Allium sativum* (bulb) 25 g, *Curcuma longa* (rhizome) 25 g, and jaggery 25 g are dried and powdered. 50 g of the powder are given orally with water twice a day for a period of 6 days. Vd. Shekarappa, Shimoga

TABLE 5.2 (CONTINUED)
Common Problems in Cattle and Their Management

Serial No.	Health Conditions	Number of Practices	Folk Remedies
		34	*Sesamum indicum* (seed oil) 100 mL, *Zingiber officinale* (rhizome) 100 mL of juice, and *Citrus limon* (fruit juice) 100 mL are mixed and administered orally once a day until the condition is relieved. Vd. Saraswathi Bhatta, Sutla Taluk, Udupi
8	Helminthiasis	35	A handful of *Carica papaya* (seeds) is ground in water to form a paste. The paste is administered orally on an empty stomach or after evening feed for 2 days. The same treatment is repeated after 6 months. Ms. Sisily Mathew, Monipillil, Waynad
		36	*Myristica fragrans* (fruit) is rubbed in honey to form a paste and is administered orally on an empty stomach or after evening feed for 2 days. Ms. Sisily Mathew, Monipillil, Waynad
		37	A handful of each of *Embelia ribes* (seeds), *Kaempferia galanga* (rhizome), *Allium sativum* (bulb), and *Vitex negundo* (leaves) are ground into a fine paste, and pills of coffee seed-size are dried in shade. This is administered orally, 3–4 pills once a day for 3 days consecutively. Ms. Sisily Mathew, Monipillil, Waynad
		38	Take equal amounts of *Trachyspermum ammi* (seeds), *Allium sativum* (bulb), *Kaempferia galanga* (rhizome), *Zingiber officinale* (rhizome), *Myristica fragrans* (fruit), and *Acorus calamus* (rhizome) and grind into a fine paste. The paste is rolled into pills of gooseberry size, which are administered orally once daily for 7 days. Ms. Eliyamma, Vadakkayil Veedu, Waynad
		39	Pastes of *Piper nigrum* (fruit) and *Allium sativum* (bulb) 10 g each are heated in neem oil. The oil is administered orally once a month. Adult dosage is 30 mL, and dosage for young ones is 10 g. Mr. E. V. Antony, Impalil, Waynad
		40	Mix juice of one *Citrus limon* (fruit) with 25 mL of *Sesamum indicum* (seed oil). No heating of oil is done. The oil is administered orally once a day for 3 days. Vd. Kunjeeramulya, Karkala
		41	*Citrullus colocynthis* (fruit) is burned over red-hot coal and mixed with 500 mL of buttermilk. This is administered orally only once. Vd. Vishveswara, Shimoga

(*continued*)

TABLE 5.2 (CONTINUED)
Common Problems in Cattle and Their Management

Serial No.	Health Conditions	Number of Practices	Folk Remedies
		42	Boil 50 g of *Cassia italica* (leaves) in 250 mL of water and reduce it to 125 mL. This is administered orally once a day for 2 days. This kashaya acts as a purgative. Vd. Siddharamappa, Shimoga
		43	*Momordica charantia* (leaves) 2 handfuls are finely ground to form a paste. The paste is mixed with water and administered orally. This medicine should not be given to pregnant cows. Vd. Saraswathi Bhatta, Sutla Taluk, Udupi
		44	Equal quantities of the leaves of *Delonix elata, Clerodendrum inerme, Tylophora indica, Azadirachta indica,* and *Capparis zeylanica* are dried under sun and finely powdered. 150 g of the powder are administered orally with lukewarm water once a day for a week. Vd. Anantharamiah, Bangalore Rural
9	Hypogalactia	45	A handful of *Asparagus racemosus* (tubers) is ground in the water in which *Cuminum cyminum* seeds are kept soaked overnight to form a fine paste. The paste is administered orally once a day for a week. Vd. Siddharamappa, Shimoga
		46	A handful of *Sesamum indicum* (seeds) is ground to a fine paste. The paste is administered orally twice a day for a week. Ms. Mary Thomas, Neenduthalakal, Waynad
		47	The fruit of *Carica papaya* is boiled and smashed to a paste. 250 g of the paste are administered orally twice day for a week. This formulation is contraindicated in pregnant animals. Ms. Sisily Jose, Kadakkayan House, Waynad
		48	250 g of *Aerva lanata* (entire plant) is boiled in 800 mL of water and is reduced to 200 mL. Then 200 mL of the decoction is administered orally once a day for a period of 5 days. This formulation is to be administered only for 15 days after delivery. Ms. Moli Jose Pullikal, Ammarakuni, Waynad
		49	*Ipomea mauritiana* (tuber and leaves) (250 g) is dried and finely powdered. Two teaspoons of the powder are administered after milking the cow. Avoid a dose exceeding 2 teaspoons. Ms. Thangamma Mathai, Mattathil House, Waynad

TABLE 5.2 (CONTINUED)
Common Problems in Cattle and Their Management

Serial No.	Health Conditions	Number of Practices	Folk Remedies
		50	Paste of 5 g of *Glycyrrhiza glabra* (roots) is administered once a day before milking the cow regularly. Ms. Eliyamma, Vadakkayil Veedu, Waynad
		51	50 g of the fine paste of *Gossipium arborium* (seeds) is used orally frequently. Ms. Eliyamma, Vadakkayil Veedu, Waynad
10	Impaction of rumen	52	Fine powder of *Holoptelia integrifolia* (leaves) (two handfuls) is mixed with a liter of boiled water. This mixture is administered orally twice a day on an empty stomach until the condition is relieved. Vd. Palani, Pappampallam, Dharmapuri
		53	A handful of leaves of each of *Cocculus hirsutus, Aristolochia indica, Waltheria indica, Hibiscus rosa-sinensis,* and *Holoptelia integrifolia* is ground to a fine paste. The paste is mixed with 1,500 mL of boiled water and administered orally three times a day until the condition is relieved. Vd. Chenrayan, Adhagapadi, Dharmapuri
		54	150 g of *Merremia tridentata* (leaves and stem) are mixed with 600 mL water. 200 mL of this mixture (water together with leaves and stem) are administered orally once a day for 3 days. Vd. Kunjeeramulya, Karkala
11	Indigestion	55	The fine powder from the seeds of *Coriandrum sativum* 50 g along with the bark of *Holoptelea integrifolia* and *Ailanthes exelsa* 150 g each, 25 g of *Zingiber officinale* (rhizome), 100 g of *Withania somnifera* (roots), 10 g of *Acorus calamus* (rhizome), 5 g of *Ferula asafoetida* (resin), 15 g of flower and seed of *Butea monosperma,* 10 g of *Piper nigrum* (fruit), and 20 g of *Terminalia chebula* (fruit rind) are mixed thoroughly. Three teaspoons of the powder are mixed with 50 mL of castor oil and are administered orally three times a day for 1 to 2 days. Vd. Siddhabasavaraju, Kolar
		56	A handful of *Cassia fistula* (bark) is finely powdered and is administered orally with a liter of boiled water twice a day for 1 day. Vd. Anataramiah, Bangalore

(*continued*)

TABLE 5.2 (CONTINUED)
Common Problems in Cattle and Their Management

Serial No.	Health Conditions	Number of Practices	Folk Remedies
		57	A paste from the rhizome of *Acorus calamus* and *Zingiber officinale* 5 g each, 5 g of *Ocimum basilicum* (leaves), *Piper nigrum* (fruit) 5 nos., *Allium sativum* (bulb) 5 flakes, along with 10 g of jaggery is rolled into a pill. Two pills are administered orally twice a day until the condition is relieved. Vd. Shivanna, Tumkur
		58	*Phyllanthus emblica* fruits are crushed mildly and filtered in a mesh to separate the pulp. A small amount of rock salt is added to the pulp and dried. 50 g of this pickle is administered orally along with water twice a day until the condition is relieved. Vd. Saraswathi Bhatt, Sutla Taluk, Udupi District
		59	Take 25 g of each of *Syzygium aromaticum* (flower bud), *Myristica fragrans* (fruit), *Piper longum* (fruit), *Allium sativum* (bulb), *Brassica nigra* (seed) along with 100 g of bark powders of *Moringa oleifera, Zanthoxylum rhetsa, Holarrhena pubescens, Tabernaemontana heyneana, Oroxylum indicum,* and *Pongamia pinnata.* All the ingredients are made into paste form and rolled into pills of betel nut size. Each pill is taken along with 600 mL of *Trachyspermum ammi* water (25 g of *Trachyspermum ammi* is kept in a vessel to which 600 mL of boiled water is added and keep a lid closed over the vessel) twice a day until the condition is relieved. Vd. Kunjeeramulya, Karkala
		60	Decoction made from rhizome of *Zingiber officinale* and *Piper nigrum* (fruit) 50 g each is administered orally twice a day on empty stomach until the condition is relieved. Ms. Mary Sebastian, Madiyakkankal, Waynad
		61	A handful of each of *Zingiber officinale* (rhizome), *Piper nigrum* (fruit), *Allium sativum* (bulb), and *Psidium guajava* (Tender leaves) are ground to a fine paste. The paste is mixed in water and filtered. The filtered water is administered orally three times a day until the condition is relieved. Ms. Sisily Jose, Kadakkayan House, Waynad
		62	Take 50 g each of *Trachyspermum ammi* (seeds), *Allium sativum* (bulb), and *Piper nigrum* (fruit). All these ingredients are ground to a fine paste, which is rolled into pills of gooseberry size. The pill is administered orally twice a day until the condition is relieved. Ms. Sisily Mathew, Monipillil, Waynad

TABLE 5.2 (CONTINUED)
Common Problems in Cattle and Their Management

Serial No.	Health Conditions	Number of Practices	Folk Remedies
		63	The bark of *Jatropha curcas* 100 g, *Zanthoxylum rhesta* 50 g, and *Erythrina indica* 50 g along with *Piper nigrum* 2 teaspoons, *Trachyspermum ammi* 2 teaspoons, *Allium sativum* bulb 4 flakes, and 50 g of jaggery are ground to a fine paste and administered orally as a single dose. Vaidyas of Dakshina Kannada region
		64	*Piper nigrum* (fruits) 100 g and 50 g of each of *Trachyspermum ammi* (seed), *Allium sativum* (bulb), *Moringa oleifera* (bark), and *Punica granatum* (fruit rind) are boiled in a liter of water, which is reduced to 500 mL. This decoction is administered orally before feed until the condition is relieved. Vaidyas of Dakshina Kannada region
		65	A handful of *Vitex negundo* (leaves), *Piper nigrum* (fruit) 5 teaspoons, *Leucas aspera* (leaves) 20 nos., *Allium sativum* (bulb) 5 flakes, *Tabernaemontana heyneana* (bark) 20 g, *Allium cepa* (bulb) 1 no., *Wattakaka volubilis* (leaves) 4 nos., *Aristolochia indica* (root/stem) 10 g, *Trachyspermum ammi* (seeds) 3 teaspoons, *Zingiber officinale* (rhizome), and *Myristica fragrans* (fruit) 1 no. are ground to a fine paste and administered orally once a day until the condition is relieved. Vaidyas of Dakshina Kannada region
12	Mastitis	66	Seeds of *Cajanus cajan* and the leaves of *Piper betel* are ground to a fine paste. The paste is applied over the udder once a day for 3 days. Vd. Gandhi, Virudhunagar District
		67	A handful of *Ocimum basilicum* (leaves) and *Curcuma longa* (rhizome) 20 g is ground into a fine paste. The paste is applied over the udder twice a day for 4 days. Vd. Thangaswami Thevar, Virudhunagar District
		68	Add a pinch of *Calcium hydroxide* to leaf paste of *Pergularia daemia* and apply over the udder once a day for 2 days. Vd. Govindhan, Nathapatti, Dharmapuri
		69	Prepare a paste from a handful of leaves of *Pergularia daemia, Phyllanthus reticulatus, Tylophora indica,* and *Acalypha indica,* a handful of *Curcuma longa* rhizome, a leaf of *Piper betle,* and a pinch of calcium hydroxide. The paste is applied over the udder three times a day for 2 days. Vd. Muthu Gounder, Muthappanagaram, Dharmapuri

(continued)

TABLE 5.2 (CONTINUED)
Common Problems in Cattle and Their Management

Serial No.	Health Conditions	Number of Practices	Folk Remedies
		70	Five handfuls of *Pisonia aculeata* (leaves) are boiled in a liter of water, which is reduced to 500 mL. 500 mL of the decoction is administered orally twice a day until the condition is relieved. Vd. Sevatha Gounder, Chinna Peramanur, Dharmapuri
		71	The leaves and stems of *Wattakaka volubilis* are dried in shade and powdered well. This powder is made into a paste by adding boiled water, and the paste is applied over the udder every half hour for nearly 12 times a day until the condition is relieved. Vd. Siddhabasavaraju, Kolar
		72	The leaves of *Commelina benghalensis* and the stem of *Wattakaka volubilis* are dried well and powdered. This mixture of powders is made into a paste by adding boiled water, and the paste is applied over the udder every half an hour for nearly 12 times a day for 2 to 3 days. Vd. Siddhabasavaraju, Kolar
		73	The leaves and roots of *Andrographis serpyllifolia* 500 g each, 15 flakes of *Allium sativum,* and 9 nos. of *Piper nigrum* fruit are ground to a fine paste and rolled into boluses of egg size. Three boluses are administered orally three times a day for a period of 9–21 days. Vd. Shivanna, Tumkur
		74	A handful of leaves of each of *Mimosa pudica* and *Datura metel*, 150 g of *Sorghum vulgare* flour, along with 2 inches of *Aloe vera* leaf pulp are made into a paste, which is applied over the udder three times a day for 3 days. Vd. Vishweswar, Shimoga
		75	Take 10 leaves of *Vitex negundo, Aegle marmelos,* and *Azadirachta indica,* along with a teaspoon of *Zingiber officinale* rhizome powder, 9 nos. of *Piper nigrum* fruits, and jaggery of gooseberry size. Grind well to make a fine paste. This paste is divided into six equal parts and made into six boluses. One bolus is given twice a day for 3 days. Vd. V. Shekarappa, Yaraganalu, Davanagere
		76	The leaves of *Phyllanthus reticulatus and Curcuma longa* in equal quantities are ground into a fine paste and applied over the udder for 2 to 3 days. Ms. Mary John, Cherriyambanadu, Waynad

TABLE 5.2 (CONTINUED)
Common Problems in Cattle and Their Management

Serial No.	Health Conditions	Number of Practices	Folk Remedies
		77	The leaf pulp of *Aloe vera,* a teaspoon of *Curcuma longa* rhizome powder, and a pinch of calcium hydroxide are ground to a paste. This paste is applied over the udder three times a day for 3 days. Vd. Ramesh, Hassan
		78	A handful of *Wattakaka volubilis* (root) and 50 g of *Cuminum cyminum* (seeds) are ground to a fine paste along with sour buttermilk. This paste is applied over the udder twice a day for a period of 4–7 days. Vd. Kunjeeramulya, Karkala
		79	Grind 50 g of *Cassia alata* (leaves) and 10 g of *Cuminum cyminum* (seeds) to a fine paste. The paste is applied over the udder three times a day until the condition is relieved. Vd. Shankar Shetty, Hosamane, Udupi District
		80	Powder the shade-dried leaves of *Aegle marmelos, Delonix elata, Azadirachta indica, Capparis zeylanica,* and *Leucas aspera* taken in equal quantities. A handful of the powder is given with lukewarm water on an empty stomach twice daily for 3 days. Vd. Anantharamiah, Bangalore Rural
		81	The fruit rind of *Datura metel* and seeds of *Sesamum indicum* and *Brassica nigra* are boiled in cow's milk in equal quantities to make a fine paste, which is applied over the udder until the condition is relieved. Ms. Mary Sebastian, Madiyakkankal, Waynad
		82	Grind a handful of *Asparagus gonocladus* tubers into a fine paste and add a pinch of common salt to it. Apply the paste over the udder until the condition is relieved. Ms. Thangamma Mathai, Mattathil House, Waynad
		83	The leaves of *Gymnema sylvestris* and *Curcuma longa* in equal quantities are ground into a fine paste. The paste is applied over the udder until the condition is relieved. Ms. Mary Thomas, Neenduthalakal, Waynad
		84	*Aloe vera* (leaf pulp) 1 foot long, *Curcuma longa* (rhizome) 2 inches long, and *Citrus limon* (fruit juice) 2 nos. are ground to a fine paste and administered orally until the condition is relieved. Vaidyas of Dakshina Kannada region

(*continued*)

TABLE 5.2 (CONTINUED)
Common Problems in Cattle and Their Management

Serial No.	Health Conditions	Number of Practices	Folk Remedies
		85	A handful of each of *Benincasa hispida* (leaves), *Cocos nucifera* (inflorescence), *and Indigofera tinctoria* (leaves), *Entada pursaetha* (fruit) 3 nos., *Tabernaemontana divaricata* (leaves) 3 nos., and *Vitex negundo* (leaves) 3 nos. is ground to a fine paste, which is applied over the udder until the condition is relieved. Vaidyas of Dakshina Kannada region
13	Prolapse of the uterus	86	*Diplocyclos palmatus* (entire plant) two handsful and *Mimosa pudica* (leaves) 1 handful are ground to a fine paste. The paste is administered orally with buttermilk once a day until the condition is relieved. Vd. Shekarappa, Shimoga
		87	*Pongamia pinnata* (bark) 1 handful is finely powdered and administered orally along with rice-washed water three times a day until the condition is relieved. Vd. Siddharamappa, Shimoga
		88	*Andrographis panniculata* (leaves) 2 teaspoons is finely powdered and administered orally along with rice-washed water three times a day until the condition is relieved. Vd. Siddharamappa, Shimoga
14	Pyrexia	89	Take 20 g of fine bark powder of each of *Zanthoxylum rhetsa, Tabernaemontana heyneana, Pongamia pinnata, Holarrhena pubescens,* and *Moringa oleifera* along with the fine powders of 10 g of each of *Capsicum annum* (fruits), *Piper nigrum* (fruits), *Trachyspermum ammi* (seeds), *Brassica nigra* (seeds), and *Allium sativum* (bulb) and grind well to a paste. Pills of betel nut size are made, and a pill is administered orally along with *Trachyspermum ammi* decoction (25 g of *Trachyspermum ammi* is kept in a vessel to which 600 mL of boiled water is added with a lid kept closed over the vessel for 15 minutes). Vd. Kunjeeramulya, Karkala
		90	Take equal amounts of dried leaves of *Delonix elata, Aegle marmelos, Capparis zeylanica, Azadirachta indica,* and *Leucas aspera.* The ingredients are powdered, and 50 g are administered with lukewarm water twice a day for 2 days. Vd. Anantharamiah, Bangalore Rural

TABLE 5.2 (CONTINUED)
Common Problems in Cattle and Their Management

Serial No.	Health Conditions	Number of Practices	Folk Remedies
		91	Take a handful leaves of each of *Vitex negundo, Aegle marmelos,* and *Azadirachta* indica along with *Piper nigrum* (fruit) 9 nos., *Allium sativum* (bulb) 9 flakes, *Zingiber officinale* (rhizome powder) 10 g, and *Piper betle* (leaves) 9 nos. The ingredients are ground to a fine paste, and six pills with jaggery are made. A pill is administered once a day. Vd. Shekarappa, Shimoga
		92	Prepare a decoction of *Careya arborea* by boiling 500 g of bark powder (coarse) in 5 L of water and reduce it to 3 L. Crush 250 g of *Allium sativum* (bulb) and 150 g of jaggery. Add them to the prepared kashaya. Administer a liter as a single dose three times a day for 2 days. Vd. Saraswathi Bhatt, Sutla Taluk, Udupi District
		93	Take 10 g of each of *Tinospora cordifolia* (stem and leaves), *Ocimum sanctum* (leaves) *Ocimum basilicum* (leaves and stem) along with *Allium sativum* (bulb) 10 flakes, *Piper nigrum* (fruit) 10 nos., and *Syzygium aromaticum* (dried flower buds) 10 nos. Grind the ingredients to a fine paste and make a bolus of hen egg size. Administer a bolus before feed three times a day for 3 days along with boiled water. Vd. Shivanna, Tumkur
		94	Boil 100 g of each of *Carvum carvi* (seeds), *Brassica nigra* (seeds), *Acorus calamus* (rhizome), *Andrographis paniculata* (leaves), *Ocimum sanctum* (leaves), *Cuminum cyminum* (seeds), and *Tinospora cordifolia* (stem) in 1,500 mL of water and reduce it to 400 mL. Add 10 g of rock salt to the decoction and administer twice a day before feed until the condition is relieved. Ms. Eliyamma, Vadakkayil Veedu, Waynad
		95	100 g of each of *Coleus zeylanicus* (whole plant) and *Vetivera zizanioides* (root) are mildly heated in 500 mL of water and administered twice a day before feed until the condition is relieved. Note: During the administration of these medications, avoid feeding with fresh grass. Ms. Moli Jose Pullikal, Ammarakuni, Waynad

(*continued*)

TABLE 5.2 (CONTINUED)
Common Problems in Cattle and Their Management

Serial No.	Health Conditions	Number of Practices	Folk Remedies
15	Retention of placenta	96	Equal quantities of the leaves of *Ficus hispida, Bambusa arundinaceae,* and *Saccharum officinarum* along with the entire plant of *Cyathula prostata* and the inflorescence of *Musa paradisiaca* are crushed to extract fresh juice. The fresh juice is administered only once. Ms. Mary John, Cherriyambanadu, Waynad
		97	50 g of the roots of *Jasminum angustifolium* along with a pinch of *Ferula asafoetida* and rock salt (saindhava in Sanskrit) are ground to a fine paste. The paste is administered only once. Ms. Mary John, Cherriyambanadu, Waynad
		98	A handful of fresh leaves of *Saccharum officinarum* is administered only once. Vd. Kunjeeramulya, Karkala
		99	A handful of fresh leaves of *Ficus hispida* is administered only once. Vd. Kunjeeramulya, Karkala
16	Sterility	100	Gruel is prepared by boiling a handful of *Pterocarpus marsupium* (bark powder) and 500 g of rice grain. This gruel is administered once a day on an empty stomach for 5 consecutive days a month for 3 consecutive months. Thus, the medicine is administered a total of 15 times. Vd. Kunjeeramulya, Karkala
		101	A handful of bark powder of each of *Pterocarpus marsupium, Ficus racemosa, Spondias pinnata,* and *Madhuca latifolia* are mixed thoroughly, and one-eighth of the total quantity is cooked with 500 g of rice grains to prepare gruel. Black gram is fed for 5 consecutive days prior to administering the medicated gruel. Vd. Kunjeeramulya, Karkala
		102	Two handfuls inflorescence of *Cocos nucifera* are ground in warm water to yield a paste, which is squeezed in a clean cloth to collect the fresh juice. 25 g *Cuminum cyminum* seeds are powdered and added to the juice. The fresh juice is administered once a day for 3 days. Each month, the procedure is repeated for 3 days. Vd. Saraswathi Bhatt, Sutla Taluk, Udupi District

TABLE 5.2 (CONTINUED)
Common Problems in Cattle and Their Management

Serial No.	Health Conditions	Number of Practices	Folk Remedies
		103	The crushed barks of *Erythrina indica and Adenanthera pavonina* along with the stem of *Coccinia indica,* each 100 g are kept in a cloth and are immersed in a vessel containing broken rice (250 g) and water for gruel preparation. Gruel is prepared and fed to the cattle for 3 days. Vd. Saraswathi Bhatt, Sutla Taluk, Udupi District
		104	100 g each of the barks of *Calophyllum inophyllum, Pterocarpus marsupium,* and *Madhuca latifolia* are added to the rice gruel and further boiled and filtered. The medicine is administered once a day for 10 days before the estrus cycle of the cow. Vd. Shankara Shetty, Udupi
		105	A handful each of *Biophytum reinwardtii* (entire plant) and *Mimosa pudica* (leaves) are ground to a fine paste and administered orally along with a liter of buttermilk. On the 21st day after consuming the medication, the cow enters the estrus cycle. During estrus, a handful of *Biophytum reinwardtii* (entire plant), *Piper nigrum* (fruit) 9 nos., and *Allium sativum* (bulb) 7 flakes are administered orally with buttermilk. Vd. Shekarappa, Shimoga
		106	The roots of *Tribulus terrestris* 100 g are boiled in 2 L of water and reduced to 500 mL. This decoction is administered once a day for 5 days. Mr. E. V. Antony, Impalil, Waynad
		107	Mix thoroughly the fine powders of the barks each 20 g of *Fagraea ceilanica, Putranjiva roxburghii, Erythrina indica,* and *Oroxylum indicum.* The powder is administered orally by mixing with a liter of water twice a week for 4 consecutive weeks. Vd. Vishveswara, Shimoga
		108	100 g of the bark of each of *Madhuca indica, Albizia lebbeck* and *Breynia* spp., 50 g of *Moringa oleifera* (bark), and *Carica papaya* root 50 g are crushed and tied in a thick cloth suspended in a bowl in which rice gruel is cooked. The juices from the herbs are thoroughly mixed with 1 liter gruel, which is administered orally. No feed or water is given for 3 hours after the administration of the medicine. Vaidyas of Dakshina Kannada region

(*continued*)

TABLE 5.2 (CONTINUED)
Common Problems in Cattle and Their Management

Serial No.	Health Conditions	Number of Practices	Folk Remedies
		109	A handful of leaves of each of *Biophytum sensitivum* and *Premna meruneta* are mixed with rice gruel and administered orally for 3 days. Vaidyas of Dakshina Kannada region
		110	*Albizia chinensis* (bark) 200 g is powdered and kept inside a thick cloth, which is suspended into a bowl that contains about a liter of rice for gruel preparation. This preparation is administered orally for 3 days. Vaidyas of Dakshina Kannada region
17	Yoke gall	111	Initially, sesamum oil is applied over the wound followed by application of the leaf paste of *Albizia amara*. The paste is applied over the affected area until the condition is relieved. Vd. Poongavanam, Periyapangunatham, Pangutham Post, Dharmapuri
		112	*Justicia glauca* (entire plant) is ground with groundnut oil to form a fine paste. The paste is applied over the affected part until the condition is relieved. Vd. Thathan, E. K. Pudur, Dharmapuri
18	Vitamin A deficiency	113	*Flacourtia ramontchi* (leaves) 100 g, *Anacardium occidentale* (leaves) 5 nos., and *Oryza sativa* (boiled rice) 50 g are cooked and administered orally early in the morning. Vaidyas of Dakshina Kannada region
		114	*Calophyllum inophyllum* (seed oil) is mixed with a teaspoon of lemon juice and is administered orally. Vaidyas of Dakshina Kannada region

Source: From Raneesh et al., 2005. *A user's guide on ethnoveterinary health practices*, FRLHT, Bangalore, India. With permission.

Note: These folk remedies are the property of folk healers; therefore, any further use of the knowledge must be after prior consent from the healers.

There are local healers and livestock farmers (both settled and nomadic) who are knowledgeable and experienced in traditional veterinary health care. The folk health practices largely remain undocumented and are passed on from one generation to another by word of mouth. They use the locally available medicinal plants for treatment of animals. The local healers are popular in their communities, and farmers generally consult them. The ethnoveterinary medicines used are ecosystem and ethnic-community specific; therefore, the characteristics, sophistication, and intensity of these practices differ greatly among individuals, societies, and regions. These local veterinary health practices are accessible, affordable, and culturally acceptable.

The use of antibiotics and other chemical products have in fact been banned for animal health care in many countries, and the world is looking for safer herbal alternatives. Ethnoveterinary medicine has great potential to address current challenges faced by veterinary medicine as it has decentralized ecosystem-specific, resource-based applications that are safe and efficacious and create little or no adverse effects in the target animals. However, they are facing the threat of rapid erosion due to urbanization and westernization. The urgent revival of these traditional veterinary practices is a high priority in the light of the constraints of modern medicine and the benefits of these practices in terms of their accessibility, affordability, and acceptability.

5.5 REVIVAL OF ETHNOVETERINARY MEDICINE

Mathias-Mundy and McCorkle (1989) compiled an annotated bibliography with emphasis on the ethnoveterinary practices of the African continent and Latin America and, to a lesser extent, Asia. The authors opined that the gap between contemporary allopathic veterinarians and the traditional practitioners can be reduced if the two can work together for the betterment of livestock owners. Ethnoveterinary medicine was not recognized and appreciated by researchers, scholars, planners, developers, and veterinarians (McCorkle, 1986; Mathias-Mundy, 1989; Mathias-Mundy and McCorkle, 1989; McCorkle et al., 1999), even though conventional veterinary medicine has its origins in aspects of traditional medicine (Wanzala et al., 2005). Presently, traditional veterinary health techniques and practices, particularly the use of herbal and holistic medicines, are increasingly being accepted in Western societies and becoming mainstream (Mugera, 1970a, 1970b; Mathias-Mundy and McCorkle, 1989; Bizimana, 1994; Farah, Ngatia, and Munyua, 1996; Intermediate Technology Development Group and International Institute of Rural Reconstruction [ITDG and IIRR], 1996; Schillhorn van Veen, 1997; Munyua et al., 1998). It is now recognized that a complementary medical approach is crucial and necessary to boost livestock production at the community level (Toyang et al., 1995). Because of their holistic nature, traditional remedies may offer efficacy combined with safety more often than single cosmopolitan or conventional drugs (Varier, 1996; Wynn, 2001). There are no harmful effects in most cases with the use of traditional medicine (Rangnekar, 1999). The Indian Council of Agricultural Research in 2000 documented 595 veterinary traditions from different sources (Swarup and Patra, 2005). About 48 of them were recommended for scientific validation, and some have shown therapeutic and ameliorative potential. In another study, 158 plants were catalogued and 50 were evaluated for antiparasitic activity (Anonymous, 2004).

The commercialization of herbal medicines for animals in India is much more advanced than in other countries, with the exception of China. The total market of animal health care products in India is estimated at about Rs 1,600 cr (about US$400 million), of which about Rs 200 cr (about US$50 million) is spent on herbal products. An estimate showed that the gap in demand and supply of veterinary health care products is Rs 7,600 to 10,500 million (Swarup and Patra, 2005). Modern veterinary care reaches only 20% of livestock owners. Hence, there is enormous scope to develop standardized herbal products for veterinary health care.

The key issues expected to be resolved are as follows:

1. Reduction in the cost of health care among the dairy animals.
2. Reduction in the antibiotic and hormone residues in the milk and other animal products by using safe, effective, and standardized products based on time-tested local traditions.
3. Provision of timely veterinary service to farmers in response to the primary health care needs of the livestock.
4. Contribution to the economy of the local producers.

A participatory rapid assessment program was designed to study and promote ethnoveterinary traditions and to establish safe and efficacious ethnoveterinary health practices in select locations of south India. The program has the following components:

- Prioritize and document local ethnoveterinary practices.
- Assess this knowledge and practices for their efficacy and safety based on evidence from both codified traditional and scientific literature.
- Promote positively assessed practices through various extension programs, such as training, establishment of home herbal gardens, publications, and product development through local enterprises.

5.6 THE INDIAN SYSTEMS OF MEDICINE AND RAPID ASSESSMENT OF ETHNOVETERINARY HEALTH TRADITIONS

Traditional systems of medicine in India include codified systems such as ayurveda, siddha, unani, and Tibetan and the noncodified oral or folk traditions. The codified systems are based on the theory of physiological functioning, disease etiology, and clinical practices. They have formal traditions of training and possess an extensive collection of written documents, the materia medica, specialized subjects related to medicine and surgery, clinical procedure, and medical ethics. The noncodified folk medicines are practiced by local healers and have been transferred from generation to generation orally or by demonstration. These Indian systems derive from a worldview that living beings share all its elements with the world outside and vice versa, thus showing oneness. The philosophical foundations of ayurveda are derived from *sad-darshanas*. Its logical system is called *nyaya vaisesika*. The means of knowledge are *apta* (verbal testimony), *pratyaksa* (direct perception), *anumana* (inference), and *upamana* (analogy). For example, in indigenous pharmacology (*Dravya guna sastra*) the entire plant or its parts (leaves, stem, roots, bark, flowers, fruits, and seeds) are studied as a whole in terms of their *in vivo* systemic effects on parameters such as six *Rasa* (six tastes), each suggestive of the composition, properties, and biological activity; *Virya* (the potency of the substance immediately after ingestion); *Vipaka* (the postdigestion state of a substance); and *Prabhava* (unique biological activity of a substance) (Tripati, 1982).

Even though the logic of folk knowledge lacks the theoretical rigor of ayurveda or modern medicine, there is an inherent relationship between classical textual

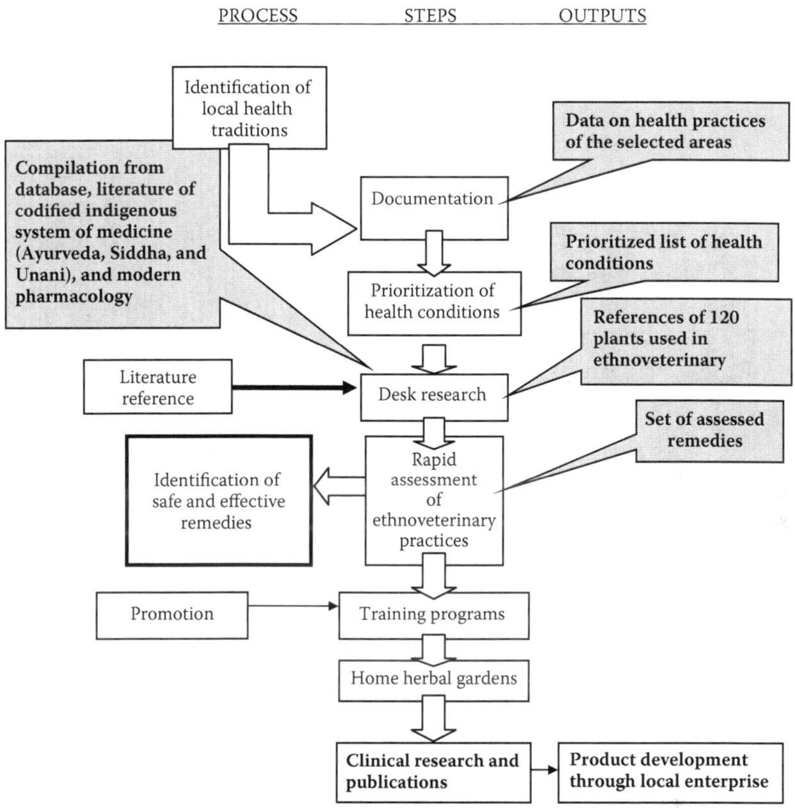

FIGURE 5.1 Process of rapid assessment of local health traditions.

knowledge and folk knowledge (Unnikrishnan and Darshan Shankar, 2005). The worldview of the theoretical foundation of modern science and traditional knowledge are completely different. The codified Indian medical traditions share a similar worldview as that of the oral folk traditions' therefore, they are (ayurveda/mrugayurveda) better tools for assessment of the folk health traditions (Unnikrishnan and Darshan Shankar, 2005).

It is necessary to find out the effectiveness and contemporary relevance of the ethnoveterinary knowledge and practices. This involves critical and comprehensive assessment of these practices using the Indian systems of medicine. We developed such a method, called the rapid assessment of local ethnoveterinary health traditions (RALHT, see Figure 5.1).

RALHT is a participatory method developed to document and validate ethnoveterinary knowledge in a rapid and cost-efficient way. In this process, the traditional health practices are assessed through a method of dialogue and consensus; local healers, veterinary doctors, researchers, community members, and other ethnoveterinary experts take part. The process involves comprehensive documentation of health practices, desk research for finding out and compiling scientific data on these practices, and assessment workshops for prioritizing and selecting practices for promotion. The assessment workshops form a pluralistic medicine platform for a cross-cultural dialogue between traditional and the contemporary medical sciences. The participants in this program are traditional healers, veterinary doctors, ayurvedic doctors, ethnobotanists, documenters, and community members.

The RALHT process consists of the following steps:

- Ethnographic documentation and identification of important ethnoveterinary health traditions
- Community validation of these identified traditions along with multidisciplinary assessment (using comprehensive ayurveda, siddha, unani, and pharmacology databases) for encouraging the best practices, adding to the incomplete and discouraging distorted or harmful ones
- Promotion of the use of the validated ethnoveterinary health traditions at household and community levels

The documentation of local health traditions is aimed at systematic recording of local health traditions and the resources used in them. The RALHT process will help select the best ethnoveterinary practices for their promotion. This form of assessment is termed *rapid* as it does not involve laboratory or clinical studies. The herbarium records of the plants used by the community are sent to competent local taxonomists for identification.

5.6.1 Desk Research Report

The desk research report consists of assessment of the local health practices based on the three Indian systems of medicine and the pharmacology report. An ethnographically sensitive approach is used to draw correlation of the locally described health problem with traditional diagnosis or modern health condition. In this process, the documented health conditions and practices were rigorously analyzed and correlated with literature on ayurveda or mrugayurveda and modern veterinary science and on plant pharmacological studies. This is a prerequisite for assessing the local remedies. The details of the locally used formulation, such as the ingredients in local name, botanical names, part used, proportions used, method of preparation, form of medicine, and the dosage used, are conveyed to the evaluators.

5.6.2 RALHT Workshop

The RALHT workshop is an essentially participatory method of assessment that involves the community, vaidyas (local healers), medical practitioners from various systems of medicine, pharmacologists, botanists, and the facilitators (nongovernmental organizations [NGOs], peoples' organizations, etc.). The prioritized list of health conditions or illnesses and the causes and symptoms are presented to the panel of practitioners. The panel discusses with the community (in the case of

RALHT for household remedies) and the folk healers (in the case of RALHT for folk remedies) and seeks clarifications if required. During the analysis, the total number of formulas having consensus from the community and all the experts of the Indian systems of medicine and modern medicine, the number of formulas having consensus from the community and one of the Indian Systems of Medicine (ISM) experts, formulas suggested for modification or additions, and formulas to be discouraged as per consensus of the participants are listed.

5.6.3 Promotion of the Best Traditional Health Practices

The positively assessed remedies are promoted among the communities in the same region. This is taken up through promotional workshops and awareness programs for the households. Following the assessment, clinical evidence is collected on a particular practice based on pilot clinical studies. Selected best remedies are made into suitable product through local enterprises or the program can be used to develop medicinal plant kits for the home herbal gardens. The data collected from the community are kept in a place decided by the community where the community can access it. This serves as a register of the health traditions of the area.

The first level of assessment is purely based on consensus among the healers and other practitioners and is supported with literature and available pharmacological studies. These assessments are not based on clinical study in the field. Therefore, many of the practices do not carry the tag of any clinically proven records or observational data from the field.

5.7 CONSERVATION OF ETHNOMEDICO KNOWLEDGE

In India, ethnomedico-botanical surveys and inventories receive little attention, and detailed information and documentation on the uses of medicinal plants in indigenous communities are lacking. This disinterest in the existing knowledge and wanton neglect by authorities and public slowly result in the loss of this knowledge. It is high time that effective measures and programs are adopted to save this information before it is lost forever. Ethnomedico-botanical surveys in each district are needed. Comparing and cross-checking the results will provide information about herbal medicines presently in use. Furthermore, concerted efforts are needed toward the usage of plant resources, linking conservation strongly with utilization.

5.8 COLLABORATION BETWEEN INDIGENOUS MEDICAL SYSTEMS

The lack of communication between indigenous medical systems of different countries is a major reason why the systems are not able to convince the world of their rich traditions and sound principles. For example, there are many commonalties between India and China. Both countries have cultural histories dating uninterrupted thousands of years. Actually, if the countries of South and Southeast Asia that have common knowledge systems come together to discuss these aspects, much hidden information on indigenous systems could be consolidated.

5.9 CONCLUSION

Ethnoveterinary practices have immense contemporary relevance. A rapid participatory assessment model for finding out the best practices was developed in FRLHT and tested in four geographical locations in southern India. Of the documented folk remedies, about 70% had positive evidence from various systems of medicine and practical experience. The medicinal plants used in the remedies can be easily grown in home herbal gardens and are locally available. Some of the remedies have gone through pilot clinical studies and have been made into products, which are now being made through local enterprises. It is suggested that this model, if promoted widely, can be of immense use for rural communities and dairy farmers. We believe revitalization of ethnoveterinary theory and practice holds the key to better animal health and hence to the prosperity of animal farmers in rural India. It also ensures reduction of antibiotic and hormone residue from the animal products and thus promotes green health.

ACKNOWLEDGMENTS

We are grateful to the community, folk healers, ISM doctors, botanists, and all others who are directly or indirectly associated with this program.

Aegle marmelos

Asperagus racemosus

Benincasa hispida

Hibiscus rosasinensis

REFERENCES

Anonymous. 2004. Identification and evaluation of medicinal plants for control of parasitic diseases of livestock. Technical report. CIRG, Makhdoom (Madhura).

Anonymous. 2006. Basic animal husbandry statistics. Government of India, Ministry of Agriculture, Department of Animal Husbandry, Dairying and Fisheries, Krishibhavan, New Delhi.

Bizimana N. 1994. Traditional veterinary practice in Africa. Deutsche Gesellschaft für Technische Zusammenarbeit (GTZ), Eschborn, Germany, 917 pp.

FAO. 2006. Livestock policy Brief 3. Cattle ranching and deforestation, 1–6.

Farah OI, Ngatia TA, and Munyua WK. 1996. *Ethnoveterinary techniques practiced by the Maasai pastoralist of Kajiado District, Kenya.* The African Pastoral Forum, Pastoral Information Network Program. Working Paper Series No. 8, June.

Intermediate Technology Development Group and International Institute of Rural Reconstruction (ITDG and IIRR). 1996. *Ethnoveterinary medicine in Kenya: a field manual of traditional animal healthcare practices.* Intermediate Technology Development Group and International Institute of Rural Reconstruction, Nairobi, Kenya, 226 pp.

Khoda VK. 2005. Current priorities for the use of ethnoveterinary in production of quality milk. In: *Proceedings of the National Conference on Contemporary Relevance of Ethnoveterinary Medical Traditions of India* (Nair MNB, Ed.). FRLHT, Bangalore, India.

Mathias-Mundy E. 1989. Indigenous knowledge on veterinary medicine. Paper presented at the 66th annual meeting of the Central States Anthropological Society, University of Notre Dame, South Bend, IN, March 10–12, 1989.

Mathias-Mundy E and McCorkle CM. 1989. *Ethnoveterinary medicine: An annotated bibliography.* Bibliographies in technology and social change, no. 6. Iowa State University, Ames, Iowa.

McCorkle CM. 1986. An introduction to ethnoveterinary research and development. *Journal of Ethnobiology* 6: 129–149.

McCorkle CM, Rangnekar VD, and Mathias-Mundy E. 1997. Introduction: whence and whither ER and D? In: Ethnoveterinary medicine: alternatives for livestock development. *Proceedings of an international conference held in Pune, India*, November 4–6, 1997 (Mathias E, Rangnekar VD, McCorkle CM, and Martin M, Eds.), Volume 1: selected papers. File 2 of 9: part 4: introduction and part 1: applied studies of ethnoveterinary systems. BAIF Development Research Foundation, Pune, India.

Merai M, and Boghra VR. 2004. Chemical pollutants—their occurrence, source of contamination in milk and milk products and methods of removal—a review. *Indian Journal of Dairy Science* 57: 291–305.

Mugera GM. 1970a. Toxic and medicinal plants of east Africa, part I. *Bulletin of Epizootic Diseases of Africa* 18: 377–387.

Mugera GM. 1970b. Toxic and medicinal plants of east Africa, part II. *Bulletin of Epizootic Diseases of Africa* 18: 389–403.

Munyua MJS, Mbai MK, Karioki ID, and Chibeu MD. 1998. Indigenous knowledge and the use of medicinal plants in small ruminant theriogenology in arid and semi-arid area of Kenya. The African Pastoral Forum, Pastoral Information Network Program. Working Paper Series No. 17, December.

Rangnekar VD. 1999. Participatory studies with women on ethnoveterinary practices for livestock health management. In: *Ethnoveterinary medicine: alternatives for livestock development. Proceedings of an international conference held in Pune, India*, November 4–6, 1997 (Mathias E, Rangnekar VD, McCorkle CM, and Martin M, Eds.), Volume 1: selected papers. File 5 of 9: part 4: application of ethnoveterinary medicine (pp. 162–170). BAIF Development Research Foundation, Pune, India.

Raneesh S, Hafeel A, Hariramamurthi G, and Unnikrishnan PM. 2005. *A user's guide on ethnoveterinary health practices*, FRLHT, Bangalore, India, pp. 31–41.

Ravindra D, and Rao RK. 1999. Ethnoveterinary medicine—a boon for improving the productivity of livestock in rural India. In: *Ethnoveterinary medicine: Alternatives for livestock development. Proceedings of an international conference held in Pune, India*, November 4–6, 1997 (Mathias E, Rangnekar VD, McCorkle CM, and Martin M, Eds.), Volume 1: selected papers. File 2 of 9: Part 4: application of ethnoveterinary medicine. BAIF Development Research Foundation, Pune, India.

Sagari RR, and Nitya SG. 2004. *Ethnoveterinary research in India: An annotated bibliography*. Mudra. 383 Narayan Peth, Pune, India.

Schillhorn van Veen TW. 1997. Sense or nonsense? Traditional methods of animal parasitic disease control. *Veterinary Parasitology* 71: 177–194.

Swarup D, and Patra RC. 2005. Perspectives of ethnoveterinary medicine in veterinary practice. *Proceedings of the National Conference on Contemporary Relevance of Ethnoveterinary Medical Traditions of India* (Nair MNB, Ed.). FRLHT, Bangalore, India.

Toyang NJ, Nuwanyakpa M, Ndi C, Django S, and Kinyuy CW. 1995. Ethnoveterinary medicine practices in the northwest province of Cameroon. *Indigenous Knowledge and Development Monitor* 3: 20–22.

Varier NVK. 1996. Repossession of our ancient wisdom. Symposium on Traditional Veterinary Medicine, Malabar Regional Co-operative Milk Producers Union, Calicut, Kerala, India, May 15.

Trikamji YA (Ed.). 1992. *Caraka samhita*, 5th ed. Munshiram Manoharlal, New Delhi.

Trikamji YA (Ed.). 1954. Susruta samhita with dalhana commentary. Chaukhambha, Varanasi, India.

Tripati I (Ed.). 1982. Raja nighantu with dravyaguna prakashika hindi commentary. Krishandas Academy, Varanasi, India.

Unnikrishnan PM, and Shankar D. 2005. Ethnoveterinary medicine in ayurvedic perpective. *Proceedings of the National Conference on Contemporary Relevance of Ethnoveterinary Medical Traditions of India* (Nair MNB, Ed.). FRLHT, Bangalore, India.

Wanzala W et al. 2005. Ethnoveterinary medicine: A critical review of its evolution, perception, understanding and the way forward. *Livestock Research for Rural Development* 17(11).

Wynn GS. 2001. Herbs in veterinary medicine. Alternative veterinary medicine. http://www.altvetmed.com/articles/herbs.html (accessed on October 9, 2001).

6 Inventory of Traditional Veterinary Botanicals from around the World

Zafar Iqbal and Abdul Jabbar

CONTENTS

6.1 INTRODUCTION

There has been a resurgence of interest in traditional health care practices in the Western as well as in the developing world. In animal health, this has led to further attention in ethnoveterinary research and development.[1–16] This is a relatively new field of study that covers traditional practices, ethnobotany, and application of animal care practices embedded in local tradition. During its annual convention (1996), the American Veterinary Medical Association[17] recognized veterinary acupuncture and acutherapy, veterinary chiropractic, physical therapy, massage, homeopathic, botanical, nutraceutical, and holistic veterinary medicine as "modalities to be offered in the context of a valid veterinarian/client/patient relationship" (http://www.veterinarywatch.com/guidelines.htm). These developments have coincided with the emergence of a consensus on the validity and importance of indigenous knowledge and traditional ecological knowledge and the risk that such knowledge, including traditional veterinary practice as well as medicine, is to be lost.[18,19]

Traditional practices were commonly used in many parts of the world but were rarely recorded in mainstream literature except in anthropology,[13, 20] in early colonial reports in the context of broader narratives about livestock,[21,22] or if related to ethnobotany.

Plants constitute a major part of traditional veterinary practices throughout the world. Farmers and traditional healers have acquired the empirical knowledge of medicinal and toxic plants and the medicinal prescriptions based on these plant products passed down to them from generation to generation. Natural (plant) products can also be important sources for new pharmaceuticals for both animals and humans.[23] There is significant evidence that plants are an important source of medicines for indigenous people.[24] Even so-called weeds have a significant representation in the flora from which drugs are derived. Therefore, undervaluing and even destruction of weeds could have an impact on availability of certain medicinal plants. Also, the realization that medicinal plants are readily available in a "living pharmacy" right outside the door and along trailsides rather than deep in the forest could lead governments and nongovernmental organizations (NGOs) to encourage and promote traditional medical practices rather than discourage them.[25] Based on a World Health Organization survey[26] and more recent review articles based on literature searches for drugs of natural plant product origin,[27,28] there are more than 121 pharmaceutical compounds used as medicine worldwide that are derived from plants.

Therefore, key themes of the emerging debate surrounding the status and future of medicinal plants are the recognition that medicinal plants constitute a vast, undocumented, and overexploited economic resource, and that they are the principal health care resource for the majority of the world's human and animal populations. This chapter serves as an inventory of the plants used in traditional veterinary practices in different parts of the world.

6.2 INVENTORY OF PLANTS USED IN ANIMAL HEALTH AND PRODUCTION

A total of 451 plants representing 308 genera and 116 families have been listed for the treatment of different health and production problems in animals (Tables 6.1 through 6.15). The most frequently used plants were *Malva sylvestris* L. (12), *Sambucus nigra* L. (11), *Khaya senegalensis* A. Juss (9), *Linum usitatissimum* L. (9), *Olea europaea* L. (8), *Allium sativum* Linn. (7), *Foeniculum vulgare* Mill. (7), *Fraxinus ornus* L. (7), *Ruta chalepensis* L. (7), *Urtica dioica* L. (7), *Allium cepa* L. (6), *Butyrospermum parkii* (6), *Helleborus viridis* L. (6), *Nicotiana tabaccum* L. (6), *Apium nodiflorum* (L.) Lag. (5), *Arachis hypogea* Linn. (5), *Elaeis guineensis* Jacq. (5), *Ficus thonningii* (5), *Hedera helix* L. (5), *Juniperus communis* L. (5), *Mercurialis annua* L. (5), *Triticum aestivum* L. (5), *Vitis vinifera* L. (5), and *Zea mays* L. (5). The frequency of usage for the remaining listed plants was 15 ($n = 4$), 49 ($n = 3$), 62 ($n = 2$), and 300 ($n = 1$). The values in parentheses indicate the number of conditions or diseases for which the respective number of plants (outside the parentheses) was used.

The most frequently represented families were Asteraceae ($n = 67$ plants), Lamiaceae ($n = 45$), Fabaceae ($n = 41$), Poaceae ($n = 35$), Solanaceae ($n = 29$), Apiaceae ($n = 27$), Euphorbiaceae ($n = 23$), Malvaceae ($n = 22$), Urticaceae ($n = 20$), Liliaceae ($n = 17$), Oleaceae ($n = 17$), Moraceae ($n = 16$), Boraginaceae ($n = 14$), Ranunculaceae ($n = 14$), Rosaceae ($n = 14$), Rutaceae ($n = 14$), Scrophulariaceae ($n = 14$), Alliaceae ($n = 13$), Meliaceae ($n = 13$), Caprifoliaceae ($n = 12$), Cupressaceae ($n = 12$), Araceae ($n = 11$), Crassulaceae ($n = 10$), Chenopodiaceae ($n = 9$), Plantaginaceae ($n = 8$), Polygalaceae ($n = 8$), Salicaceae ($n = 8$), Sapotaceae ($n = 8$), Verbenaceae ($n = 8$), Brassicaceae ($n = 8$), Combretaceae ($n = 7$), Cucurbitaceae ($n = 7$), Papaveraceae ($n = 7$), Pinaceae ($n = 7$), Vitaceae ($n = 7$), Acanthaceae ($n = 6$), Aspleniaceae ($n = 6$), Melanthiaceae/Liliaceae

TABLE 6.1
Plants Used in the Treatment of General Gastrointestinal Complaints in Animals

Plants Used	Plant Family	References
Allium cepa L.	Alliaceae	144–146
Allium sativum L.	Alliaceae	144–147
Apium nodiflorum L.	Apiaceae	148
Cuminum cyminum L.	Apiaceae	149
Pimpinella anisum L.	Apiaceae	150
Hedera helix L.	Araliaceae	151
Asplenium trichomanes L.	Aspleniaceae	148
Artemisia absinthium L.	Asteraceae	150–153
Cynara cardunculus ssp. *Scolymus* (L.) Hayek	Asteraceae	146
Chamomilla recutita (L.) Rauschert	Asteraceae	145, 154
Santolina marchii Arrigoni	Asteraceae	153
Calendula officinalis L.	Asteraceae	155
Borago officinalis L.	Boraginaceae	156
Sambucus nigra L.	Caprifoliaceae	147, 157, 158
Sempervivum tectorum L.	Crassulaceae	150, 159–162
Equisetum arvense L.	Equisteraceae	163
Euphorbia lathyris L.	Euphorbiaceae	148, 164
Mercurialis annua L.	Euphorbiaceae	146, 150, 151, 155, 156, 161, 164, 165
Glycyrrhiza glabra L.	Fabaceae	156
Lotus corniculatus L.	Fabaceae	155
Vicia faba L.	Fabaceae	156
Quercus sp.	Fagaceae	144
Geranium rotundifolium L.	Geraniaceae	166
Geranium sanguineum L.	Geraniaceae	166
Aesculus hippocastanum L.	Hippocastanaceae	146
Marrubium vulgare L.	Lamiaceae	160, 167
Mentha sp. pl.	Lamiaceae	146
Rosmarinus officinalis L.	Lamiaceae	148, 167
Teucrium chamaedrys L.	Lamiaceae	159
Laurus nobilis L.	Lauraceae	163
Linum usitatissimum L.	Linaceae	147, 148, 151, 153, 156, 159, 161, 162, 168, 169
Lavatera cretica L.	Malvaceae	165, 170
Malva sp. pl.	Malvaceae	146
Malva sylvestris L.	Malvaceae	144, 145, 147, 148, 150, 158, 169
Ficus carica L.	Moraceae	146, 150, 160, 163, 170
Fraxinus excelsior L.	Oleaceae	171, 148, 1712, –173
Fraxinus ornus L.	Oleaceae	146, 150, 157, 159–161, 171, 174–177
Olea europaea L.	Oleaceae	158

(*continued*)

TABLE 6.1 (CONTINUED)
Plants Used in the Treatment of General Gastrointestinal Complaints in Animals

Plants Used	Plant Family	References
Agropyron repens (L.) Beauv.	Poaceae	146
Cynodon dactylon Pers.	Poaceae	178
Triticum aestivum L.	Poaceae	148, 162
Triticum durum Desf.	Poaceae	156
Zea mays L.	Poaceae	153
Rumex sp.	Polygonaceae	146
Clematis vitalba L.	Ranunculaceae	179
Crataegus laevigata L.	Rosaceae	180
Rubus fruticosus L.	Rosaceae	154
Rubus sp. pl.	Rosaceae	156
Sorbus domestica L.	Rosaceae	156
Ruta chalepensis L.	Rutaceae	180
Ruta graveolens L.	Rutaceae	146, 161, 181, 182
Saccharomyces cerevisiae Rees	Saccharomycetaceae	162
Populus sp.	Salicaceae	153, 162
Salix alba L. subsp. *Alba*	Salicaceae	146, 148, 150, 156, 159, 183, 184
Salix sp.	Salicaceae	144, 160, 162
Salix viminalis L.	Salicaceae	167
Solanum tuberosum L.	Solanaceae	146
Ulmus minor Mill.	Ulmaceae	156, 162
Parietaria judaica L.	Urticaceae	146
Urtica dioica L.	Urticaceae	148
Vitis vinifera L.	Vitaceae	159, 160

TABLE 6.2
Plants Used as Antidiarrheals or Antidysentrics in Animals

Plants Used	Plant Family	References
Blepharis ciliaris	Acanthaceae	185
Anacardium occidentale	Anacardiaceae	186, 187
Foeniculum vulgare Mill.	Apiaceae	188
Trachyspermum ammi L.	Apiaceae	188
Elaeis guineensis Jacq.	Arecaceae	189
Gomphocarpus fruticosus (L.) Ait. F.	Asclepiadaceae	190
Carduus nyassanus (S. Moore) R. E. Fries	Asteraceae	190
Crepis rueppellii Sch. Bip.	Asteraceae	190
Sphaeranthus gomphrenoides O. Hoffm.	Asteraceae	191
Vernonia amygdalina Delile	Asteraceae	192

TABLE 6.2 (CONTINUED)
Plants Used as Antidiarrheals or Antidysentrics in Animals

Plants Used	Plant Family	References
Adansonia digitata Linn.	Bombacaceae	189
Cynoglossum coeruleum Hochst.	Boraginaceae	190
Opuntia maxima Mill.	Cactaceae	193
Deterium microcarpum	Caesalpiniaceae	189
Herniaria glabra L.	Caryophyllaceae	193
Chenopodium opulifolium Koch & Ziz	Chenopodiaceae	192
Guiera senegalensis Lam	Combretaceae	189
Terminalia macroptera Guill & Perr.	Combretaceae	189
Cupressus lusitanica Mill.	Cupressaceae	190
Juniperus procera L.	Cupressaceae	190
Clutia abyssinica Jaub. & Spach.	Euphorbiaceae	190
Ceratonia siliqua L.	Fabaceae	193
Millettia ferruginea (Hochst.) Bak.	Fabaceae	190
Senna occidentalis L.	Fabaceae	192
Acacia polyacantha	Fabaceae	194
Piliostigma thonningii Schum	Fabaceae	189
Quercus ilex L.	Fagaceae	193
Quercus pubescens Willd.	Fagaceae	193
Mentha pulegium L.	Lamiaceae	193
Mentha suaveolens Ehrh.	Lamiaceae	193
Ocimum lam Hochst.	Lamiaceae	190
Thymus vulgaris L.	Lamiaceae	193
Cordyla africana	Leguminosae	194
Aloe secundiflora Engl.	Liliaceae	191
Linum usitatissimum L.	Linaceae	195
Lythrum salicaria L.	Lythraceae	193
Corchorus olitorius	Malvaceae	194
Malva sylvestris L.	Malvaceae	193
Khaya senegalensis A. Juss	Meliaceae	189
Ficus thonningii B.	Moraceae	191
Psidium guajava L.	Myrtaceae	186, 187, 191
Pinus patula Schlect. & Charm.	Pinaceae	191
Avena sativa L.	Poaceae	193
Oryza sativa L.	Poaceae	193, 195
Phragmites australis Cav.	Poaceae	189, 195
Sorghum bicolor Linn. Moench	Poaceae	189
Zea mays L.	Poaceae	191
Zizyphus nummularia	Rhamnaceae	185
Agrimonia eupatoria L.	Rosaceae	193
Rubus ulmifolius Schott.	Rosaceae	193
Galium verum L.	Rubiaceae	195
Harrisonia abyssinica Oliv.	Rutaceae	192

TABLE 6.3
Plants Used in Gastroenteritis or Abdominal Pain and as a Laxative or Purgative

Plants Used	Plant Family	References
Gastroenteritis/Abdominal Pain		
Apium nodiflorum (L.) Lag.	Apiaceae	195
Sphaeranthus gomphrenoides O. Hoffm.	Asteraceae	191
Lepidium sativum	Brassicaceae	185, 196
Sambucus nigra L.	Caprifoliaceae	195
Cupressus sempervirens L.	Cupressaceae	195
Khaya senegalensis A. Juss	Meliaceae	189
Ficus thonningii B.	Moraceae	191
Psidium guajava L.	Myrtaceae	191
Fraxinus excelsior L.	Oleaceae	195
Fraxinus ornus L.	Oleaceae	195
Pinus patula Schlect. & Charm.	Pinaceae	191
Zea mays L.	Poaceae	191
Solanum incanum Linn.	Solanaceae	189
Laxative/Purgative		
Rhazya stricta	Apocynaceae	185, 196
Artemesia abyssinica	Asteraceae	185, 197
Artemesia inculata	Asteraceae	185
Beta vulgaris L. ssp. *vulgaris* var. *Vulgaris*	Chenopodiaceae	193
Citrullus colocynthis	Cucurbitaceae	185
Euphorbia cocuneata	Euphorbiaceae	185, 198
Euphorbia lathyris L.	Euphorbiaceae	195
Linum usitatissimum L.	Linaceae	195
Papaver rhoeas L.	Papaveraceae	195
Malva sylvestris L.	Pinaceae	193, 199
Cymbopogon schoenanthus	Poaceae	185, 196
Triticum aestivum L.	Poaceae	195
Agrimonia eupatoria L.	Rosaceae	193
Solanum nigrum L.	Solanaceae	195

($n=6$), Apocynaceae ($n=5$), Asclepiadaceae ($n=5$), Canellaceae ($n=5$), Fagaceae ($n=5$), and Myrtaceae ($n = 5$). The frequency of distribution of the remaining plants among different families was four plants each in Basellaceae, Caryophyllaceae, Rhamnaceae, Rubiaceae, and Zingiberaceae; three each in Araceae, Balanitaceae/Zygophyllaceae, Convallariaceae, Hypericaceae/Clusiaceae, Lauraceae, Leguminosae, Sapindaceae, Ulmaceae, and Zygophyllaceae; two each in Amaranthaceae, Anacardiaceae, Aristolochiaceae, Betulaceae, Cactaceae, Caesalpinaceae/Fabaceae, Capparaceae/Capparidaceae, Geraniaceae, Hippocastanaceae, Lythraceae, Phytolaccaceae, and Tamaricaceae.

TABLE 6.4
Plants Used for Rumination, Bloat, and Colic and as Detoxicants

Plants Used	Plant Family	References
Rumination		
Foeniculum vulgare Mill.	Apiaceae	195
Anchomanes difformis Engl.	Araceae	189
Asplenium trichomanes L.	Aspleniaceae	195
Artemisia abisinthium L.	Asteraceae	195
Helichrysum italicum (Roth) G. Don fil.	Asteraceae	199
Sempervivum tectorum L.	Crassulaceae	195
Juniperus communis L.	Cupressaceae	195
Euphorbia hirta Linn.	Euphorbiaceae	189
Spartium junceum L.	Fabaceae	195
Marrubium vulgare L.	Lamiaceae	195
Rosmarinus officinalis L.	Lamiaceae	195
Satureja montana L.	Lamiaceae	199
Linum usitatissimum L.	Linaceae	195
Gossypium barbadens Linn.	Malvaceae	189
Malva sylvestris L.	Malvaceae	195
Khaya senegalensis A. Juss	Meliaceae	189
Triticum aestivum L.	Poaceae	195
Salix alba L.	Salicaceae	195
Ulmus minor Miller	Ulmaceae	195
Vitis vinifera L.	Vitaceae	195
Bloat		
Sambucus nigra L.	Caprifoliaceae	193
Momordica balasamia Linn.	Cucurbitaceae	189
Arachis hypogea Linn.	Fabaceae	189
Cymbopogon schoenanthus	Poaceae	185, 196
Salix aiha L.	Salicaceae	199
Schwenkia americana Linn.	Solanaceae	189
Colic		
Brassica campestris L. var. *sarson* Prain	Brassicaceae	188
Cymbopogon schoenanthus	Poaceae	196
Lycium barbarum	Solanaceae	196
Detoxicants		
Nelsonia canescens	Acanthaceae	189
Elaeis guineensis Jacq.	Arecaceae	189
Arachis hypogea Linn.	Fabaceae	189
Khaya senegalensis A. Juss	Meliaceae	189
Olea europaea L.	Oleaceae	195
Striga hermontheca (Del.) Benth	Scrophulariaceae	189

TABLE 6.5
Plants Used for Anorexia/ or Indigestion/ or as a Carminative

Plants Used	Plant Family	References
Allium cepa L.	Alliaceae	188
Allium sativum L.	Alliaceae	188
Anethum graveolens L.	Apiaceae	188
Foeniculum vulgare Mill.	Apiaceae	188
Trachyspermum ammi L.	Apiaceae	188
Calotropis procera	Asclepiadaceae	200, 201
Schkuria pinnata (Lam.) Thell.	Asteraceae	191
Tithonia diversifolia (Hemsl.) Gray	Asteraceae	191
Matricaria chamomilla L.	Asteraceae	199
Vernonia anthelmintica Willd.	Asteraceae	188
Lepidium sativum L.	Brassicaceae	188
Eruca sativa Mill.	Brassicaceae	188
Capparis deciduas (Forssk.) Edgew.	Capparaceae	188
Sempervivum tectorum L.	Crassulaceae	199
Citrullus colocynthis (L.) Kuntze	Cucurbitaceae	188
Melilotus alba	Fabaceae	202
Trigonella foenum-gracum L.	Fabaceae	188
Malva sylvestris L.	Malvaceae	199
Piper nigrum L.	Piperaceae	188
Punica granatum L.	Punicaceae	188
Picrorhiza kurroa Royle ex Benth	Scrophulariaceae	188
Capsicum annuum L.	Solanaceae	188
Lycium barbarum	Solanaceae	196
Withania coagulans (Stocks) Dund.	Solanaceae	188
Amomum subulatum Roxb.	Zingiberaceae	188
Zingiber officinale Rosc.	Zingiberaceae	188

Maximum number of plants were used as antiseptic ($n = 67$) followed by reproductive and metabolic problems ($n = 60$), skin conditions ($n = 58$), gastrointestinal disorders ($n = 54$), respiratory and viral problems ($n = 44$), antidiarrhoeal/ antidysentric ($n = 43$), ectoparasitic problems ($n = 42$), blood protozoal diseases ($n = 28$), anthelmintic ($n = 22$), anorexia, indigestion and carminative ($n = 21$), rumination ($n = 18$), laxative/purgative ($n = 14$), and gastroenteritis ($n = 13$).

The most diversified use in animal health and production problems was that of plants belonging to the family Asteraceae ($n = 14$; value indicates the number of ailments or conditions for which the plants of this family are used) followed by Fabaceae ($n = 12$), Apiaceae (11), Euphorbiaceae (11), Lamiaceae (10), Poaceae (10), Malvaceae (10), Solanaceae (9), Meliaceae (9), Moraceae (8), Boraginaceae (7), Rosaceae (7), Ranunculaceae (7), Caprifoliaceae (7), Linaceae (7), Oleaceae (6), Alliaceae (6), Brassicaceae (6), Crassulaceae (6), Cupressaceae (6), Polygonaceae (6),

TABLE 6.6
Plants Used as Anthelmintics

Plants Used	Plant Family	References
Ruellia tuberosa	Acanthaceae	187
Achryanthes indica	Amaranthaceae	186
Rhazya stricta	Apocynaceae	185, 196
Tanacetum balsamita L.	Asteraceae	193
Artemesia inculata	Asteraceae	185
Artemesia abisinthium L.	Asteraceae	195
Balanites aegyptiaca L.	Balanitaceae	192
Heliotropium strigosum	Boraginaceae	185, 203
Cassia alata	Caesalpinaceae	186
Carica papaya	Caricaceae	186
Chenopodium ambrosiodes	Chenopodiaceae	186
Anogeissus leicarpa D.C. Guill & Perr	Combretaceae	189
Guiera senegalensis Lam.	Combretaceae	194
Citrullus colocynthis	Cucurbitaceae	185
Cucurbita maxima Lam.	Cucurbitaceae	191
Juniperus communis L.	Cupressaceae	193
Cajanus cajan	Fabaceae	186
Senna didymobotrya (Fresen.) Irwin & Barneby	Fabaceae	191
Cordyla africana	Leguminosae	194
Gossypium spp.	Malvaceae	186
Azadirachta indica	Meliaceae	187
Khaya senegalensis A. Juss	Meliaceae	189
Ficus sycomorus Linn.	Moraceae	189
Fraxinus ornus L.	Oleaceae	195
Olea europaea L.	Oleaceae	191
Petiveria alliacea	Phytolaccaceae	187
Sorghum bicolor Linn. Moench	Poaceae	189
Zizyphus mauritania	Rhamnaceae	194
Stachytarpheta jamaicensis	Verbenaceae	187
Zygophylum album	Zygophyllaceae	185, 196
Zygophylum coccineum	Zygophyllaceae	185, 196

Apocynaceae (6), Vitaceae (6), Scrophulariaceae (5), Rutaceae (5), Chenopodiaceae (5), Urticaceae (4), Combretaceae (4), Plantaginaceae (4), Salicaceae (4), Verbenaceae (4), Acanthaceae (4), Liliaceae (4), Papaveraceae (4), Pinaceae (4), Asclepiadaceae (4), Myrtaceae (4), Melanthiaceae (3), Aspleniaceae (3), Cucurbitaceae (3), Fagaceae (3), Sapotaceae (3), Araliaceae (3), and Arecaceae (3)

Some plants used in poultry have also been documented (Table 6.15). Six plants representing six families were used for egg production, one plant to improve plumage, and eight plants representing seven families as parasiticides.

TABLE 6.7
Plants Used in Some Protozoal Diseases

Plants Used	Plant Family	References
Trypanosomiasis		
Terminalia superba	Combretaceae	204
Terminalia vorensi	Combretaceae	204
Alchornea cordifolia	Euphorbiaceae	204
Acacia artaxacantha	Fabaceae	204
Eugenia uniflora	Myrtaceae	204
Adenia cissampeloides	Passifloraceae	204
Murraya koenigii	Rutaceae	204
Clerodendron capitata	Verbenaceae	204
Theileriosis		
Thunbergia alata Sims	Acanthaceae	191
Solanecio mannii (Hook. f.) C. Jeffrey	Asteraceae	191
Sonchus oleraceus L.	Asteraceae	191
lpomoea batatas (L.) Lam.	Convolvulaceae	191
Kalanchoe densiflora Rolfe	Crassulaceae	191
Ricinus communis L.	Euphorbiaceae	191
Synadenium compactum N. E. Br.	Euphorbiaceae	191
Ajuga remota Benth	Lamiaceae	191
Plectranthus barbatus Andr.	Lamiaceae	191
Dodonaea angustifolia L. f.	Sapindaceae	191
East Coast Fever		
Synadeniuni grantii Hook.		192
Steganotaenia araliacea Hochst.	Apiaceae	192
Carissa edulis (Forssk.)	Apocynaceae	192
Pistia stratiotes L.	Araceae	192
Aristolochia elegans Mast.	Aristolochiaceae	192
Ananas comosus L.	Bromeliaceae	192
Cannabis sativa L.	Cannabaceae	192
Maytenus senegalensis Lam.	Celastraceae	192
Euphorbia tirucalli L.	Euphorbiaceae	192
Jatropha curcas L.	Euphorbiaceae	192
Albizia coriaria Oliv.	Fabaceae	192
Oncoba spinosa Forssk.	Flacourtiaceae	192
Tetradenia riparia Hochst.	Lamiaceae	192
Asparagus racemosus (Wild)	Liliaceae	192
Strychnos innocua Del.	Loganiaceae	192
Milicia excelsa Welw. C. Berg.	Moraceae	192
Musa paradisiaca L. var. *paradisiaca*	Musaceae	192
Securidaca longipedunculata Fres.	Polygalaceae	192
Oxygonurn sinuatum Meisn.	Polygonaceae	192
Gardenia tern Schumach. & Thonn.	Rubiaceae	192

TABLE 6.7 (CONTINUED)
Plants Used in Some Protozoal Diseases

Plants Used	Plant Family	References
Citrus limon L.	Rutaceae	192
Citrus sp., *Buniimu,* NC	Rutaceae	192
Harrisonia abyssinica Oliv.	Rutaceae	192
Physalis peruviana L.	Solanaceae	192
Clerodendrum myricoides Hochst.	Verbenaceae	192
Anaplasmosis		
Thunbergia alata Sims	Acanthaceae	191
Helichrysum odoratissimum (L.) Less.	Asteraceae	191
Solanecio mannii (Hook. f.) C. Jeffrey	Asteraceae	191
Sonchus oleraceus L.	Asteraceae	191
Kalanchoe densiflora Rolfe	Crassulaceae	191
Euclea divinorum Hiern	Ebenaceae	191
Synadenium compactum N. E. Br.	Euphorbiaceae	191
Senna didymobotrya (Fresen.) Irwin & Barneby	Fabaceae	191
Achyrospermum schimperi (Hochst.) Perkins	Lamiaceae	191
Ficus thonningii BI.	Moraceae	191
Dodonaea angustifolia L. f.	Sapindaceae	191
Cyphostemma maranguense (Gilg) Desc.	Vitaceae	191

TABLE 6.8
Plants Used for Ectoparasitic Problems

Plants Used	Plant Family	References
Scabies/Mange		
Apium nodiflorum (L.) Lag.	Apiaceae	195
Rhazya stricta	Apocynaceae	196
Eclipta alba	Asteraceae	186
Sonchus bipontini Asch.	Asteraceae	190
Crescentia. Cujete	Bignoniaceae	186
Bixa orellana	Bixaceae	186
Eruca sativa Mill.	Brassicaceae	188
Musa spp.	Musaceae	186
Discopodium eremanthum Chiov.	Solanaceae	190
Fagonia bruguieri	Zygophyllaceae	202
Myiasis		
Anthostema senegalense	Euphorbiaceae	194
Manilkara zapota	Sapotaceae	186

(continued)

TABLE 6.8 (CONTINUED)
Plants Used for Ectoparasitic Problems

Plants Used	Plant Family	References
Ectoparasite		
Tithonia diversifolia (Hemsl.) Gray	Asteraceae	191
Cordia curassavica	Boraginaceae	186
Warburgia ugandensis Sprague	Canellaceae	191
Tagetes minuta L.	Compositae	191
Synadenium compactum N. E. Br.	Euphorbiaceae	191
Caesalpinia volkensii Harms	Fabaceae	191
Ajugu remota Benth	Lamiaceae	191
Mentha aquatica L.	Lamiaceae	195
Melia volkensii Gurke	Meliaceae	191
Pouteria sapota	Sapotaceae	186
Nicotiana tabacum	Solanaceae	186
Insecticidal		
Allium cepa L.	Alliaceae	195
Sambucus ebulus L.	Caprifoliaceae	195
Sempervivum tectorum L.	Crassulaceae	195
Lupinus albus L.	Fabaceae	195
Ticks, Fleas, Lice		
Allium sativum Linn.	Alliaceae	189
Nicotiana tobbacum Linn.	Solanaceae	188, 189
Antiparasitics/or Repellents		
Allium cepa L.	Alliaceae	146
Allium sativum L.	Alliaceae	144
Apium nodiflorum L.	Apiaceae	148
Achillea ligustica All.	Asteraceae	205
Artemisia absinthium L.	Asteraceae	145–147, 153
Tanacetum vulgare L.	Asteraceae	149, 182, 206
Alnus glutinosa L.	Betulaceae	148
Echium vulgare L.	Boraginaceae	162
Sambucus nigra L.	Caprifoliaceae	206
Mercurialis annua L.	Euphorbiaceae	156
Lupinus albus L.	Fabaceae	144, 148, 152, 156, 159
Juglans regia L.	Juglandaceae	146, 205, 207
Glechoma hederacea L.	Lamiaceae	157
Laurus nobilis L.	Lauraceae	146, 205
Veratrum album L.	Melanthiaceae	154, 207
Veratrum nigrum L.	Melanthiaceae	146
Veratrum sp.	Melanthiaceae	208
Fraxinus ornus L.	Oleaceae	153, 208
Ruta graveolens L.	Rutaceae	146, 153, 163, 168, 208
Salix sp.	Salicaceae	144
Datura stramonium L.	Solanaceae	145, 155, 179

TABLE 6.9
Plants Used in the Treatment of Reproductive and Metabolic Ailments

Plants Used	Plant Family	References
General		
Allium cepa L.	Alliaceae	146
Heracleum sphondilium L.	Apiaceae	207
Ceterach officinarum Lam. et DC.	Aspleniaceae	148
Helianthus annuus L.	Asteraceae	146
Picris echioides L.	Asteraceae	156
Carduus pycnocephalus L.	Asteraceae	156
Chamomilla recutita (L.) Rauschert.	Asteraceae	144, 146, 208
Sonchus oleraceus L.	Asteraceae	156
Nasturtium officinale R. Br.	Brassicaceae	146
Sambucus nigra L.	Caprifoliaceae	148
Stellaria media (L.) Vill.	Caryophyllaceae	159
Juniperus sabina L.	Cupressaceae	156, 207
Pteridium aquilinum (L.) Kuhn	Dennstaedtiaceae	146, 153, 156
Onobrychis viciifolia Scop.	Fabaceae	153
Galega officinalis L.	Fabaceae	160, 179
Melissa officinalis L.	Labiatae	146
Teucrium chamaedrys L.	Lamiaceae	156
Scorpiurus subvillosus L.	Leguminosae	155
Malva neglecta Wallr.	Malvaceae	209
Papaver rhoeas L.	Papaveraceae	146
Plantago lanceolata L.	Plantaginaceae	161
Plantago major L.	Plantaginaceae	161
Zea mays L.	Poaceae	156
Helleborus viridis L.	Ranunculaceae	148
Prunus avium L.	Rosaceae	148
Rubia tinctorum L.	Rubiaceae	210
Scrophularia canina L.	Scrophulariaceae	156
Capsicum annum L.	Solanaceae	146, 208
Solanum tuberosum L.	Solanaceae	144
Parietaria judaica L.	Urticaceae	148
Parietaria sp. pl.	Urticaceae	146, 208
Urtica dioica L.	Urticaceae	147, 153, 155, 156, 211
Urtica sp. pl.	Urticaceae	146, 212
Abortive		
Hedera helix L.	Araliaceae	195
Postpartum Problems		
Adiantum capillus-veneris L.	Adiantaceae	195
Apium nodiflorum L.	Apiaceae	195
Ceterach officinarum DC.	Aspleniaceae	195

(continued)

TABLE 6.9 (CONTINUED)
Plants Used in the Treatment of Reproductive and Metabolic Ailments

Plants Used	Plant Family	References
Mercurialis annua L	Euphorbiaceae	195
Laurus nobilis L.	Lauraceae	195
Linum usitatissimum L.	Linaceae	195
Rubia peregrine L.	Rubiaceae	195
Parietaria diffusa M. et K.	Urticaceae	195
Depurative		
Angelica sylvestris L.	Apiaceae	195
Ceterach officinarum DC.	Aspleniaceae	195
Linum usitatissimum L.	Linaceae	195
Retained Placenta		
Balanites aegyptiaca Del.	Balanitaceae	189
Hibiscus sabdariffa Linn.	Malvaceae	189
Ficus thonningii Blume	Moraceae	189
Sorghum bicolor Linn. Moench	Poaceae	189
Infertility/Fertility Enhancer		
Guiera senegalensis Lam	Combretaceae	189
Tamarindus indica Linn.	Fabaceae	189
Khaya senegalensis A. Juss	Meliaceae	189
Avena sativa L.	Poaceae	195
Prunus avium L.	Rosaceae	195
Striga hermontheca Del. Benth	Scrophulariaceae	189
Galactogogue		
Foeniculum vulgare Mill.	Apiaceae	195
Euphorbia hirta Linn.	Euphorbiaceae	189
Arachis hypogea Linn.	Fabaceae	189
Melilotus alba	Fabaceae	202
Galega officinalis L.	Fabaceae	195, 199
Gossypium barbadens Linn.	Malvaceae	189
Malva sylvestris L.	Malvaceae	195
Ficus carica L.	Moraceae	195
Ficus sycomorus Linn.	Moraceae	189
Schwenkia aritime Linn.	Solanaceae	189
Urtica dioica L.	Urticaceae	195
Stachytarpheta jamaicensis	Verbenaceae	186
Acetonaemia		
Arachis hypogea Linn.	Fabaceae	189
Mineral Deficiencies		
Euphorbia aegyptica Boiss	Euphorbiaceae	189

TABLE 6.10
Plants Used for Wound Healing and as Antiseptics

Plants Used	Plant Family	References
Blepharis ciliaris	Acanthaceae	202
Allium sativum L.	Alliaceae	193, 199
Angelica sylvestris L.	Apiaceae	193
Conium maculatum L.	Apiaceae	146, 156
Eryngium campestre L.	Apiaceae	193
Foeniculum vulgare Mill.	Apiaceae	193
Hedera helix L.	Araliaceae	199
Achillea millefolium L.	Asteraceae	195
Inula aritim L.	Asteraceae	154, 162
Tanacetum parthenium (L.) Schultz-Bip.	Asteraceae	193
Borago officinalis L.	Boraginaceae	207
Cynoglossum creticum Miller	Boraginaceae	195
Heliotropium strigosum	Boraginaceae	203
Brassica oleracea L.	Brassicaceae	156
Opuntia ficus-indica (L.) Mill.	Cactaceae	213
Cleome arabica	Capparaceae	202
Sambucus nigra L.	Caprifoliaceae	195
Beta vulgaris L. ssp. *vulgaris* var. *Vulgaris*	Chenopodiaceae	193
Hamada elegans	Chenopodiaceae	202
Cistus laurifolius L.	Cistaceae	193
Sedun telephiuin L. ssp. *Maximum* (L.) Krocker	Crassulaceae	193
Umbilicus rupestris (Salisb.) Dandy	Crassulaceae	193
Juniperus oxycedrus L. subsp. *Oxycedrus*	Cupressaceae	211
Equisetum telmateja Ehrh.	Equisteraceae	195
Euphorbia cocuneata	Euphorbiaceae	198
Mercurialis annua L.	Euphorbiaceae	193
Cytisus scoparius (L.) Link	Fabaceae	195
Tephrosia vogelii Hook.	Fabaceae	192
Quercus pubescens L.	Fagaceae	199
Quercus suber L.	Fagaceae	214
Hypericum perforatum L.	Hypericaceae	162, 195
Crocus sativus L.	Iridaceae	193
Marrubium vulgare L.	Lamiaceae	145, 154, 164, 193
Teucrium polium L. ssp. *Polium*	Lamiaceae	193
Thymus vulgaris L.	Lamiaceae	193
Lavandula latifolia Medik.	Lamiaceae	193

(continued)

TABLE 6.10 (CONTINUED)
Plants Used for Wound Healing and as Antiseptics

Plants Used	Plant Family	References
Salvia officinalis L. ssp. *Lavandulifolia* (Vahl) Gams	Lamiaceae	193
Stachys germanica L.	Lamiaceae	156
Stachys officinalis (L.) Trevisan	Lamiaceae	195
Asphodelus microcarpus Salzm. et Viv.	Liliaceae	213
Urginea aritime (L.) Baker	Liliaceae	193
Malva sylvestris L.	Malvaceae	193
Veratrum album L. subsp. *Lobelianum* (Bernh.) Arcang.	Melanthiaceae	146
Veratrum nigrum L.	Melanthiaceae	199
Fraxinus ornus L.	Oleaceae	146
Olea europaea L.	Oleaceae	193
Chelidonium majus L.	Papaveraceae	182
Papaver rhoeas L.	Papaveraceae	193
Pinus halepensis Miller	Pinaceae	195
Pinus sylvestris L.	Pinaceae	193
Plantago lanceolata L.	Plantaginaceae	193
Plantago major L.	Plantaginaceae	144, 148, 162, 168, 181, 195
Agropyron repens (L.) Beauv.	Poaceae	146
Arundo donax L.	Poaceae	193
Avena sativa L.	Poaceae	162
Oryza sativa L.	Poaceae	193
Triticum aestivum L.	Poaceae	193, 199
Zea mays L.	Poaceae	195
Polygonum persicaria L.	Polygonaceae	193
Helleborus viridis L.	Ranunculaceae	176
Zizyphus nummularia	Rhamnaceae	202
Agrimonia eupatoria L.	Rosaceae	193
Rubus fruticosus L.	Rosaceae	145
Ruta chalepensis L.	Rutaceae	193
Saxifraga paniculata Mill.	Saxifragaceae	193
Kichxia elatine (L.) Dumort.	Scrophulariaceae	154, 184
Scrophularia canina L.	Scrophulariaceae	146, 153, 156, 184, 215
Verbascum densiflorum Bert.	Scrophulariaceae	216
Verbascum sinuatum L.	Scrophulariaceae	154, 217
Verbascum sp. pl.	Scrophulariaceae	184
Parietaria judaica L.	Urticaceae	147
Parietaria officinalis L. ssp. *judaica* (L.) Beg.	Urticaceae	193
Urtica dioica L.	Urticaceae	193
Vitis vinifera L.	Vitaceae	162, 193

TABLE 6.11
Plants Used for Skin Problems

Plants Used	Plant Family	References
Dermatosis/Skin Disorders		
Conium maculatum L.	Apiaceae	146, 156
Hedera helix L.	Araliaceae	155, 180
Hedera sp.	Araliaceae	167
Achillea millefolium L.	Asteraceae	160, 168, 218
Cirsium arvense (L.)	Asteraceae	168
Inula viscosa L.	Asteraceae	154, 162
Sonchus oleraceus L.	Asteraceae	156
Ecballium elaterium (L.) A. Rich	Benincaseae	164
Cordia curassavica	Boraginaceae	186
Cynoglossum creticum Miller (syn. *Cynoglossum pictum* Aiton)	Boraginaceae	160, 167, 218
Cynoglossum officinale L.	Boraginaceae	160, 167
Echium vulgare L.	Boraginaceae	162, 195
Brassica oleracea L.	Brassicaceae	156
Lepidium sativum	Brassicaceae	185, 196
Sambucus nigra L.	Caprifoliaceae	146, 148, 153, 162, 219
Cupressus sempervirens L.	Cupressaceae	195
Tamus communis L.	Dioscoreaceae	146, 159
Euphorbia cocuneata	Euphorbiaceae	185, 198
Hypericum perforatum L.	Hypericaceae	162
Marrubium vulgare L.	Lamiaceae	145, 154, 164
Satureja montana L.	Lamiaceae	154
Linum usitatissimum L.	Linaceae	144, 156, 157, 161
Lycoperdon sp.	Lycoperdaceae	156
Althaea officinalis L.	Malvaceae	146, 156, 183
Malva sylvestris L.	Malvaceae	144, 153, 159, 218, 220
Veratrum album L. subsp. *lobelianum* (Bernh.) Arcang	Melanthiaceae	146
Morus nigra L.	Moraceae	162
Olea europaea L.	Oleaceae	144
Chelidonium majus L.	Papaveraceae	182
Plantago lanceolata L.	Plantaginaceae	158
Plantago major L.	Plantaginaceae	144, 148, 162, 168, 181
Rumex crispus L.	Polygonaceae	151
Anagallis arvensis L.	Primulaceae	156
Helleborus viridis L.	Ranunculaceae	176
Ranunculus ficaria L.	Ranunculaceae	146
Salix alba L. subsp. *Alba*	Salicaceae	159
Scrophularia canina L.	Scrophulariaceae	153, 156, 184, 215
Verbascum thapsus L.	Scrophulariaceae	159

(*continued*)

TABLE 6.11 (CONTINUED)
Plants Used for Skin Problems

Plants Used	Plant Family	References
Tamarix aphylla	Tamaricaceae	221
Vitis vinifera L.	Vitaceae	195
Ringworm		
Arachis hypogea Linn.	Fabaceae	189
Khaya sengegalensis A. Juss	Meliaceae	189
Ficus thonningii Blume	Moraceae	189
Butyrospermum parkii Kotsky	Sapotaceae	189
Fungal Skin Infections		
Pistacia lentiscus L.	Anacardiaceae	193
Conium maculatum L.	Apiaceae	193
Nerium oleander L.	Apocynaceae	195
Eclipta alba	Asteraceae	186
Calendula arvensis L.	Asteraceae	193
Heliotropium europaeum L.	Boraginaceae	193
Raphanus raphanistrum L. ssp. *sativus* (L.) Dornin	Brassicaceae	193
Calpurnia aurea (Ait.) Benth.	Fabaceae	190
Bersama abyssinica Fresen.	Melianthaceae	190
Olea europaea L.	Oleaceae	193
Helleborus foetidus L.	Ranunculaceae	193
Nicotiana tabacum L.	Solanaceae	193
Solanum marginatum L. f.	Solanaceae	190
Footrot		
Elaeis guineensis Jacq.	Arecaceae	189
Acacia nilotica Del.	Fabaceae	189
Khaya senegalensis A. Juss	Meliaceae	189
Butyrospermum parkii	Sapotaceae	189
Nicotiana tobbacum Linn.	Solanaceae	189
Snakebite		
Calotropis procera	Asclepiadaceae	200, 201
Euphorbia cocuneata	Euphorbiaceae	198
Lycium barbarum	Solanaceae	196

TABLE 6.12
Plants Used in the Treatment of Some Bacterial Diseases

Plants Used	Plant Family	References
Anthrax		
Euphorbia schimperiana Scheele	Euphorbiaceae	190
Leonotis ocymnifera (Burm. f.) Iwarsson	Lamiaceae	190
Khaya senegalensis A. Juss	Meliaceae	189
Antimastitic		
Buxus semperirens L.	Buxaceae	195
Sambucus nigra L.	Caprifoliaceae	195
Malva sylvestris L.	Malvaceae	195
Avena sativa L.	Poaceae	195
Butyrospermum parkii Kotsky	Sapotaceae	189
Schwenkia americana Linn.	Solanaceae	189
Vitis vinifera L.	Vitaceae	195
Blackleg		
Crepis rueppellii Sch. Bip.	Asteraceae	190
Sonchus bipontini Asch.	Asteraceae	190
Vernonia myrantha Hook. f.	Asteraceae	190
Cucumis ficifolius A. Rich.	Cucurbitaceae	190
Erythrina brucei Schweinf.	Fabaceae	190
Salvia merjamie Forssk.	Lamiaceae	190
Salvia nilotica Jacq.	Lamiaceae	190
Sida schimperiana Hochst ex A. Rich.	Malvaceae	190
Cymbopogon citrates (DC.) Stapf	Poaceae	190
Rumex nepalensis Spreng.	Polygonaceae	190
Clematis hirsute Perr. & Guill.	Ranunculaceae	190
Nigella sativa L.	Ranunculaceae	190
Alchemilla abyssinica Fresen.	Rosaceae	190
Ruta chalepensis L.	Rutaceae	190
Discopodium eremanthum Chiov.	Solanaceae	190
Nicotiana tabaccum L.	Solanaceae	190
Solanum incanum L.	Solanaceae	190
Urinary Infections		
Petroselinum sativum Hoffm.	Apiaceae	195
Potentilla reptans L.	Rosaceae	162, 195

TABLE 6.13
Plants Used for the Treatment of Respiratory and Viral Infections

Plants Used	Plant Family	References
Respiratory Problems/Pneumonia		
Gomphocarpus fruticosus (L.) Ait. f.	Asclepiadaceae	190
Helichrysum italicum (Roth) Don	Asteraceae	195
Inula sp.	Asteraceae	182
Solanecio mannii (Hook. f.) C. Jeffrey	Asteraceae	191
Vernonia anthelmintica Willd.	Asteraceae	188
Alnus glutinosa L.	Betulaceae	153
Borago officinalis L.	Boraginaceae	146
Polygonatum multiflorum All.	Convallariaceae	157
Polygonatum officinale All.	Convallariaceae	157
Mercurialis annua L.	Euphorbiaceae	156
Senna didymobotrya (Fresen.) Irwin & Barneby	Fabaceae	191
Aesculus hippocastanum L.	Hippocastanaceae	157
Mentha pulegium L.	Lamiaceae	195
Thymus sp.	Lamiaceae	
Linum usitatissimum L.	Linaceae	148
Malva sylvestris L.	Malvaceae	153
Ficus carica L.	Moraceae	156
Ficus sycomorus Linn.	Moraceae	189
Eucalyptus saligna	Myrtaceae	191
Plantago major L.	Plantaginaceae	154
Rumex abyssinicus	Polygonaceae	190
Helleborus viridis L.	Ranunculaceae	176, 195
Eryobotrya japonica (Thunb.) Lindley	Rosaceae	161
Dodonaea angustifolia L. f.	Sapindaceae	191
Striga hermontheca Del. Benth	Scrophulariaceae	189
Ulmus minor Mill.	Ulmaceae	156
Parietaria diffusa Mert. et Koch	Urticaceae	178
Colds and Coughs		
Foeniculum vulgare Mill.	Apiaceae	188
Vernonia amygdnlina Delile	Asteraceae	192
Vernonia anthelmintica Willd.	Asteraceae	188
Chenopodium opulifolium Koch & Ziz.	Chenopodiaceae	192
Haloxylon ammodendron	Chenopodiaceae	185
Acacia seyal Del. var. *fistula*	Fabaceae	192
Mentha pulegium L.	Lamiaceae	195
Helleborus viridis L.	Ranunculaceae	195
Measles		
Vernonia amygdnlina Delile	Asteraceae	192
Balanites aegyptiaca L.	Balanitaceae	192

TABLE 6.13 (CONTINUED)
Plants Used for the Treatment of Respiratory and Viral Infections

Plants Used	Plant Family	References
Cannabis sativa L.	Cannabaceae	192
Chenopodium opulifolium Koch & Ziz.	Chenopodiaceae	192
Azadirachta indica A. Juss.	Meliaceae	192
Lantana camara L.	Verbenaceae	192
Hepatitis		
Allium sativum L.	Alliaceae	190
Kniphofia foliosa Hochst.	Asphodelaceae	190
Carduus nyassanus (S. Moore) R. E. Fries	Asteraceae	190
Crepis rueppellii Sch. Bip.	Asteraceae	190
Solanecio gigas (Vatke) C. Jeffrey	Asteraceae	190
Sonchus bipontini Asch.	Asteraceae	190
Vernonia amygdalina Del.	Asteraceae	190
Silene macrosolen A. Rich.	Caryophyllaceae	190
Lotus corniculatus L.	Fabaceae	190
Cymbopogon citrates (DC.) Stapf	Poaceae	190
Cheilanthes farinose (Forssk.) Kaulf.	Pteridaceae	190
Rhamnus prinoides L'Herit.	Rhamnaceae	190
Cycniopsis hum (Forssk.) Sengl	Scrophulariaceae	190
Rabies		
Carduus nyassanus (S. Moore) R. E. Fries	Asteraceae	190
Asparagus africanus Lam.	Liliaceae	190

TABLE 6.14
Plants Used in the Treatment of Some Miscellanous Conditions

Plants Used	Plant Family	References
Anti-inflammatory		
Allium cepa L.	Alliaceae	193
Hedera helix L.	Araliaceae	195
Aristolochia rotunda L.	Aristolochiaceae	193
Phyllitis scolopendrium (L.) Newm.	Aspleniaceae	193
Helichrysum italicum (Roth) Don	Asteraceae	195
Inula viscosa (L.) Aiton	Asteraceae	193
Sambucus nigra L.	Caprifoliaceae	193
Beta vulgaris L. ssp. *vulgaris* var. *Vulgaris*	Chenopodiaceae	193
Hrpericum perforatum L.	Clusiaceae	193
Teucrium chamaedrys L.	Lamiaceae	195
Thymus vulgaris L.	Lamiaceae	193
Lavandula latifolia Medik.	Lamiaceae	193
Althaea officinalis L.	Malvaceae	195

(*continued*)

TABLE 6.14 (CONTINUED)
Plants Used in the Treatment of Some Miscellanous Conditions

Plants Used	Plant Family	References
Malva sylvestris L.	Malvaceae	193
Marchantia paleacea Bertol.	Marchantiaceae	193
Morus nigra L.	Moraceae	195
Olea europaea L.	Oleaceae	193
Oryza sativa L.	Poaceae	193
Triticum aestivum L.	Poaceae	193
Anemone hepatica L.	Ranunculaceae	193
Parietaria officinalis L. ssp. *judaica* (L.) Beg.	Urticaceae	193
Urtica dioica L.	Urticaceae	193
Lippia triphylla (L'Hér.) Kuntze	Verbenaceae	193
Analgesic/Antipyretic		
Juniperus communis L.	Cupressaceae	195
Hypericum perforatum L.	Hypericaceae	195
Helleborus odorus W. et K.	Ranunculaceae	195
Helleborus viridis L.	Ranunculaceae	195
Newborn Navel Diseases Protection		
Elaeis guineensis Jacq.	Arecaceae	189
Butyrospermum parkii Kotschy	Sapotaceae	189
Sprain		
Elaeis guineensis Jacq.	Arecaceae	189
Butyrospermum parkii Kotschy	Sapotaceae	189
Emollients		
Umbilicus rupestris (Salisb.) Dandy	Crassulaceae	193
Olea europaea L.	Oleaceae	193
Strengthens Joints		
Juniperus communis L.	Cupressaceae	195
Swollen Joints or Feet		
Colocasia esculentum (L.) Schott Var.	Araceae	189
Lawsonia inermis Linn.	Lythraceae	189
Butyrospermum parkii Kotschy oil	Sapotaceae	189
Antirheumatic		
Cupressus sempervirens L.	Cupressaceae	195
Fraxinus ornus L.	Oleaceae	195
Myositis and Arthritis		
Calotropis procera	Asclepiadaceae	200, 201
Lycium barbarum	Solanaceae	196

TABLE 6.14 (CONTINUED)
Plants Used in the Treatment of Some Miscellanous Conditions

Plants Used	Plant Family	References
Eye/Ocular Problem		
Calendula arvensis L.	Asteraceae	193
Acacia girardil Benth.	Fabaceae	222
Ruta chalepensis L.	Rutaceae	193
Solanum incanum L.	Solanaceae	192
Tamarix aphylla	Tamaricaceae	185, 221
Zérbád (Dropsy)		
Vernonia anthelmintica Willd.	Asteraceae	188
Piper nigrum L.	Piperaceae	188
Picrorhiza kurroa Royle ex Benth	Scrophulariaceae	188
Capsicum annuum L.	Solanaceae	188
Amomum subulatum Roxb.	Zingiberaceae	188
Anhidrosis		
Trachyspermum ammi L.	Apiaceae	188
Brassica campestris L. var. *sarson* Prain	Brassicaceae	188
Camellia sinensis (L.) O. Kuntze	Theaceae	188
Amomum subulatum Roxb.	Zingiberaceae	188
Salutiferous		
Foeniculum vulgare Mill.	Apiaceae	193
Sambucus nigra L.	Caprifoliaceae	193
Juniperus communis L.	Cupressaceae	193
Mentha suaveolens Ehrh.	Lamiaceae	193
Hyparrhenia hirta (L.) Stapf in Oliver	Poaceae	193
Ruta chalepensis L.	Rutaceae	193
Smilax aspera L.	Smilacaceae	193
Parietaria officinalis L. ssp. *judaica* (L.) Beg.	Urticaceae	193
Urtica dioica L.	Urticaceae	193
Diuretic		
Scabiosa atropurpurea L.	Dipsacaceae	193
Arctostaphylos uvaursi (L.) Spreng.	Ericaceae	193
Arundo donax L.	Poaceae	193
Abortion		
Acacia polyacantha Willd.	Fabaceae	191
Aloe secundiflora Engl.	Liliaceae	191
Ekebergia capensis Sparrm.	Meliaceae	191
General Weakness		
Basella alba L.	Basellaceae	191
Cucumis aculeatus Cogn.	Cucurbitaceae	191

(continued)

TABLE 6.14 (CONTINUED)
Plants Used in the Treatment of Some Miscellanous Conditions

Plants Used	Plant Family	References
Acacia mearnsii De Wild.	Fabaceae	191
Ajuga remota Benth	Lamiaceae	191
Aloe secundiflora Engl.	Liliaceae	191
Urtica massaica Mildbr.	Urticaceae	191
Clerodendrum johnstonii Oliv.	Verbenaceae	191
Cayratia ibuensis (Hook. f.) Suesseng.	Vitaceae	191
Weakening of Bones		
Cyathula polycephala Bak.	Amaranthaceae	191
Synadenium compactum N. E. Br.	Euphorbiaceae	191
Aloe secundiflora Engl.	Liliaceae	191
Urtica massaica Mildbr.	Urticaceae	191
Dietary Deficiencies		
Warburgia ugandensis Spragueb	Canellaceae	191
Plectranthus barbatus Andr.	Lamiaceae	191
Urtica massaica Mildbr.	Urticaceae	191
Nervous/Sensory/Locomotory Ailments		
Allium sp. pl.	Alliaceae	156
Inula viscosa (L.) Aiton	Asteraceae	162
Brassica nigra (L.) Koch	Brassicaceae	177
Cannabis sativa L.	Cannabaceae	182
Sambucus nigra L.	Caprifoliaceae	147
Satureja montana L.	Lamiaceae	181
Teucrium chamaedrys L.	Lamiaceae	160
Linum usitatissimum L.	Linaceae	144
Papaver somniferum L.	Papaveraceae	156
Lolium temulentum L.	Poaceae	156, 217
Clematis vitalba L.	Ranunculaceae	162
Ruta chalepensis L.	Rutaceae	161
Hyoscyamus niger L.	Solanaceae	155
Valeriana officinalis L.	Valerianaceae	153, 157

6.3 VALIDATION OF THE TRADITIONAL USE OF PLANTS

Although validation of safety of traditional practices and drugs is universally accepted, the need for efficacy testing, using Western standards, is still debated. Agarwal[29] noted that indigenous knowledge differs from Western scientific knowledge on substantive, methodological or epistemological, as well as contextual grounds, arguing that indigenous knowledge is more deeply rooted in its environment, is based on different values, and is assessed by different methods. Accordingly, validation and preservation of indigenous knowledge should be done in situ. Ex situ preservation

TABLE 6.15
Plants Used in Different Poultry Conditions

Plants Used	Plant Family	References
Egg Production		
Stellaria media (L.)	Caryophyllaceae	195
Fraxinus ornus L.	Oleaceae	195
Popover rhoeas L.	Papaveraceae	199
Echinochloa crus-galli Beauv.	Poaceae	195
Rumex acetosa L.	Polygonaceae	195
Urtica dioica L.	Urticaceae	195
To Improve Plumage		
Phytolacca americana L.	Phytolaccaceae	195
Parasiticide		
Umbilicus rupestris (Salisb.) Dandy	Crassulaceae	193
Psoralea bituminosa L.	Fabaceae	193
Mentha suaveolens Ehrh.	Lamiaceae	193
Lavandula latifolia Medik.	Lamiaceae	193
Eucalyptus globulus Labill	Myrtaceae	193
Plantago coronopus L.	Plantaginaceae	193
Asplenium adiantum-nigrum L. ssp. *onopteris* (L.) Heufl.	Polypodiaceae	193

creates a museum for knowledge and is likely to benefit outsiders. As some practices are risky or incorrect dosing of medicinal extracts can lead to poisoning, validation of safety is a minimal requirement. Efficacy may be more difficult to demonstrate ex situ, and guidelines such as those by the World Health Organization (WHO)[30] may be helpful.

In the validation process, recognition that certain plants are used in combinations is important as there are complex chemical interactions among constituents of a single plant and with mixtures of plants.[31] Other considerations are whether some plants mixed together increase availability of bioactive compounds or if preparations diminish toxicity while retaining therapeutic actions.

Various ailments, in humans or their animals, are the result of microbial infections, nutritional deficiencies, and genetic disorders. Traditional medicine mainly constitutes the use of plants. Very few plants emprically used in veterinary medicine have been validated. However, there is no dearth of literature on the phytochemical aspects of several plants. Although not specifically targeted at veterinary medicine, antimicrobial and antifugal activities of several plants have been reported in the literature. An increasing number of researchers have demonstrated that commonly used herbs and spices such as garlic, black cumin, cloves, cinnamon, thyme, oregano, allspice, bay leaves, mustard, and rosemary possess antimicrobial properties.[32] The activity is mainly due to the presence of essential oils.[33] The use of plants in different animal ailments has been rarely validated except as anthelmintics.

As far as could be ascertained, anthelmintic activity of plants is one of the most important areas subjected to scientific validation in veterinary medicine. There are more than 150 plants that have been validated for their anthelmintic effects in animals using standard parasitological procedures. Some of the examples include *Acacia albida*;[34] *Albizia anthelmintica*;[35,36] *Albizia coriavera*;[36] *Albizia lebbek*;[37] *Allium sativum*;[36] *Aloe barteri*;[37] *Ananas comosus*;[38] *Anogeissus leiocarpus*;[39] *Khaya senegalansis*;[40] *Artemisia annua*;[41] *Artemisia brevifolia*;[42] *Artemisia herba-alba*;[43] *Artemisia inflorescence*;[44] *Artemisia maritime*;[38,45–47] *Artemisia senna*;[48,49] *Azadirachta indica*;[50,51] *Azadirachta indica*;[52] *Vernonia anthelmintica* and *Bixa orellana*;[38] *Butea* spp.;[53–57] *Caesalpina crista*;[58–60] *Calliandra portoricensis*;[61–64] *Calotropis procera*;[65] *Carissa edulis*;[36] *Cassia alata*;[38] *Chenopodium album*;[66,67] *Caesalpinia crista*;[67] *Cucurbita maxima*;[68] *Cucurbita pepo*;[60] *Cymbopogon nardus* and *Cymbopogon citrates*;[69] *Hyoscyamus niger*;[70] *Juglans regia, Musa paradisaca,* and *Scindapsus officinalis*;[60] *Melia azedarach*;[71–74] *Nicotiana tabacum*;[75] *Nigella sativa*;[76] *Peganum harmala*;[77,78] *Punica granatum*;[79] *Swertia chirata*;[80] *Tamarindus indica*;[36] *Trachyspermum ammi*;[81,82] *Vernonia anthelmintica*;[83,84] *Withania coagulans*;[85] and *Zingiber officinale*.[86] Githiori et al.[87] and Jackson and Miller[88] claimed that plants with anthelmintic properties typically contain saponins, alkaloids, nonprotein amino acids, tannins and other polyphenols, lignins, alkaloids, terpenes, lactones, glycosides, and phenolic compounds. Other active anthelmintic compounds include cysteine proteinases, which digest the protective cuticle of the rodent gastrointestinal nematode *Heligmosomoides polygyrus*;[89] anthraquinones, which are active against *Schistosoma mansoni*;[90] and condensed tannins, which are active against gastrointestinal nematodes.[91]

Onions (*Allium cepa*) and garlic (*Allium sativum*) have a wide array of veterinary uses, ranging from treating gastrointestinal complaints (i.e., anorexia, indigestion, or as a carminative) to proven insecticidal, antiparasitic, repellant, and antiseptic actions. Sufficient scientific evidence is available regarding the use of *Allium* species as antiprotozoal agents against *Giardia lamblia, G. intestinalis, Entamoeba histolytica, Trichomonas vaginalis,* and different strains of *Leishmania*;[92,93] antifungals[94–96] with activity against *Candida, Trichophyton, Torulopsis, Rhodotorula, Cryptococcus, Aspergillus,* and *Trichosporon*;[97] and as an antibacterial against *Pseudomonas, Proteus, Escherichia coli, Staphylococcus aureus, Klebsiella, Salmonella, Micrococcus, Bacillus subtilis, Mycobacterium*, and *Clostridium*.[98] The antiviral activities of various commercial garlic products, including garlic powder tablets and capsules, oil-macerated garlic, steam-distilled garlic oils, garlic aged in aqueous alcohol and fermented garlic oil, against herpes simplex virus types 1 and 2, influenza A and B viruses,[99] human cytomegalovirus,[100] vesicular stomatitis virus, rhinovirus, human immunodeficiency virus (HIV), viral pneumonia, and rotavirus have also been studied. Evidence from several investigations suggested that the biological and medical functions of garlic and onions are mainly due to the high organo-sulfur compound content.[101] The biological effects of additional constituents of intact garlic and onion, such as lectins (the most abundant proteins in garlic and onion); prostaglandins; fructan; pectin; adenosine; vitamins B_1, B_2, B_6, C, and E; biotin; nicotinic acid; fatty acids; glycolipids; phospholipids; and essential amino acids, have been studied for over several decades.[99] Some proteins, saponins, and phenolic compounds of onions and garlic can also contribute to their pharmacological activity.[102]

Artemesia spp. are used as laxatives, purgatives, and anthelmintics in traditional veterinary medicine. Several essential oils from *Artemesia* species have shown important antimicrobial activity against bacteria, yeasts, dermatophytes, and *Aspergillus* strains[103–105] and have therapeutic potential, mainly in diseases involving mucosal, cutaneous, and respiratory tract infections. The major constituents of many of these oils are phenolic compounds (terpenoids and phenylpropanoids) like thymol, carvacrol, or eugenol, for which antimicrobial and antioxidant activities are well documented.[106]

Anacardium occidentale has been reported for its antimicrobial property[107] and is used as an antidiarrheal and antidysentric in veterinary medicine. It has been reported to contain cardol, methylcardol, cardanol (alkylphenol), and anacardic acid.[108] Cardanol derivatives were found to show antibacterial, antifungal, antioxidant, and antitumor activities.[109]

Ruta spp. are used for gastrointestinal complaints; blackleg; and nervous, sensory, or locomotory ailments and as an antiparasitic, repellent, and antiseptic in veterinary medicine. *Ruta* spp. leaves have been reported to possess antibacterial activity against *Bacillus subtilis* and *Staphylococcus aureus*, whereas, essential oil has a slight anthelmintic effect attributed to nonylmethyl ketone, and its odor is effective in inhibiting the fifth instar of the triatomine *Rhodnius prolixus*.[110–112]

Aloe secundiflora is used as an antidiarrheal and antidysentric and in abortion, general weakness, and weakness of bones in veterinary medicine. Emodin, an anthraquinone, is the virucidal agent (enveloped viruses) in *Aloe* spp. (*A. vera*) and possesses antibacterial, diuretic, vasorelaxant, anti-inflammatory, antiproliferative, and anticarcinogenic properties.[113] Dhananjeyan et al.[90] successfully tested other anthraquinones against the human filarial parasite *Brugia malayi* and the pathogenic trematode *Schistosoma mansoni*.

Mentha spp. are used in gastrointestinal complaints and respiratory and viral problems, against ectoparasites, and as an antidiarrheal or antidysentric in veterinary medicine. *Mentha* spp. oil has been found effective against *Staphylococcus aureus, Salmonella enteritidis, Enterococcus faecium, Shigella sonei,* and *Micrococcus flavus*.[114]

Flaxseed (*Linum usitatissimum*) is used in gastrointestinal complaints, reproductive disorders, respiratory ailments, and skin problems. Flaxseed flour has been reported to possess antifungal activity.[115] Health benefits of flaxseed are associated with α-linolenic acid (ALA), lignan, and flax dietary fiber.[116]

6.4 PARTS OF PLANTS USED

The commonly used parts of the plants or plant products are leaves, bark, oil, shoots, whole plant, root, flowers, cobs, fruit, sap, gum, seeds, pulp, aerial parts, latex, stem, pods, galls, wood, buds, bulb, rhizomes, cake, cladode, inflorescence, infructescence, stigmas, floral tops, gametophyte, petals, grain, frond, vinegar, and wine.

6.4.1 Modes of Preparation and Administration

The plant parts to be used are crushed and powdered, boiled with water, beaten, burned to smoke or ash, chewed, ground, brewed, soaked in solvent, added in feed, used as forage, macerated, embrocated, and used as concoction, decoction, infusion, poultice, smoke, distillate, ash powder, juice, syrup, tisane, lotion, wine, sap, and so on. The common modes of administration are ingestion, oral and nasal administration, drinking, licking, enema, stinging, and topical application such as painting, rubbing, ointment, poultice, massage, emulsion, liniment, and so on.

Different vehicles are also used frequently for administration of plants. For liquids these include water, milk whey, oils, wine, vinegar, and the like. Jaggery, salt, butter, and so on are used for making physic drench balls, and oils, butter, Vaseline, and so on are used for making preparations for topical application of plants.

6.5 DISCUSSION

As stated by Schillhorn van Veen,[19] farmers and herders have historically relied on empirically derived practices in management and health. Veterinary medicine as practiced today has its origins in traditional medicine as practiced in prehistory in China, India, and the Middle East. King Hammurabi of Babylon, for example, laid out laws concerning the fees veterinarians could charge for treatment of cattle and donkeys as early as 1800 BC. Early Buddhist Indian society already worshipped cattle and other animals, and animal health was mentioned in the Rock Edict II of King Ashoka (269–232 BC). It is suggested that animal hospitals were created during his reign (Smith, 1924, as cited by Lodrick[20]). China, Egypt, and later Arabia developed into centers of veterinary and other medical practice knowledge. Schwabe[1] compared the beliefs and practices of past and present herders in the Nile valley and suggested a long tradition behind contemporary ethnoveterinary practice. Throughout much of history, however, medical knowledge was subordinate to religious certainties and seen as an empirical art rather than a science. This was also the case with veterinary medicine, and the distinction between traditional practice and "studied" empirical knowledge was vague, although early veterinary schools tried to distinguish between quackery and rational explanation of diseases and their treatment. These attitudes changed drastically in the 20th century, especially after World War II, when chemotherapeutic control of disease became predominant and when diseases were explained and treated in ways based on the understanding of pathophysiology and immunology. This change was dominated by a rational Western society and its beliefs and rarely reached the non-Western world, apart from mass vaccinations, insecticides, and some (sometimes outdated) antibiotics and anthelmintics. Indeed, the Western developments in the elucidation and treatment of animal disease bypassed many livestock owners in poorer developing countries, who continued to rely on their age-old methods in disease control and often considered Western animal health care expensive, not embedded in local beliefs and concerns, and not always in tune with animal welfare.

A compendium of ethnoveterinary medical practices in parts of Africa has been published.[36] In other studies,[117–119] the active principles as well as the mechanisms of action of some plant extracts that are used in ethnoveterinary medicine have been discussed. In some of the studies cited here, the appropriate dosages of the plant extracts required to suppress the growth of causative organisms of some diseases have been given, thus suggesting the potential of traditional drugs in primary animal health care.

Mainstream medicine is increasingly receptive to the use of antimicrobial and other drugs derived from plants as traditional antibiotics (products of microorganisms or their synthesized derivatives) become ineffective and as new, particularly viral, diseases remain intractable to this type of drug. Another driving factor for the renewed interest in plant antimicrobials in the past 20 years has been the rapid rate of (plant) species extinction.[120] There is a feeling among natural products chemists and microbiologists alike that the multitude of potentially useful phytochemical structures that could be synthesized chemically is at risk of being lost irretrievably.[121] There is a scientific discipline known as *ethnobotany* (or *ethnopharmacology*) whose goal is to utilize the impressive array of knowledge assembled by indigenous peoples about the plant and animal products they have used to maintain health.[122–125]

Plants have an almost limitless ability to synthesize aromatic substances, most of which are phenols or their oxygen-substituted derivatives.[126] There is increasing evidence to support the hypothesis that plants are relatively high in bioactive secondary compounds and are thus likely to hold promise for drug discovery. Secondary compounds in weeds are important for a variety of ecological functions. Chief among these are allelopathy, for which secondary compounds inhibit germination and growth of other plants and as chemical defense against herbivory.[127] At least 50 species of weeds have been shown to interfere with crops through allopathic secondary compounds.[128] However, because allelopathy usually occurs through the complex chemical matrix of the soil, it is difficult to conclusively show a causal relationship.[129] Investigations into plant antiherbivore defense are perhaps further developed. The two major antiherbivory chemical defense strategies for plants are metabolically inactive immobile (or quantitative, as defined by Feeny[130]) defenses such as tannins and lignins that reduce digestibility and mobile (or qualitative) defenses such as alkaloids, cardiac glycosides, or terpenoids.[130,131] It is the latter types of compounds that are the basis for plant-derived pharmaceuticals. Ephemeral, successional, or *r*-selected species (all common characteristics of weeds) tend to rely on these sorts of toxic chemical defenses.[132,133] Most of the plant-derived chemicals are secondary metabolites, of which at least 12,000 have been isolated, a number estimated to be less than 10% of the total.[134] Some of these secondary metabolites, such as terpenoids, give plants their odors; others (quinones and tannins) are responsible for plant pigment. Many compounds are responsible for plant flavor (e.g., the terpenoid capsaicin from chili peppers). Useful antimicrobial phytochemicals can be divided into phenolics and polyphenols (e.g., simple phenols and phenolic acids, quinones, flavones, flavonoids, and flavonols, tannins, coumarins, etc.); terpenoids and essential oils; alkaloids; lectins and polypeptides; mixtures; and other compounds (reviewed by Cowan[135]).

The efficacy of different plants based on their traditional or empirical use is attributed to the compounds or chemical groups given in this section. For example, phenolics possess a wide spectrum of biochemical activities, such as antioxidant, antimutagenic, and anticarcinogenic properties, as well as the ability to modify gene expression.[136,137] There is, however, variation in the chemistry of plants, which leads to differences in the efficacy even within same species of plants. It has been reported that plant genotype[138] and cultivation[139] affect total phenolic and flavonoid contents in fruit. The variation in phenolic compounds depends on many factors, such as degree of maturity at harvest, genetic differences, environmental conditions, development stages of the plant at harvesting, drying process, and storage technique.[140,141] Moreover, the type of solvent can also influence the amount and spectrum of the active components in the final extract of a plant.[140]

6.6 CONCLUSIONS

1. Although ethnomedicine is a readily available alternative to modern veterinary inputs and services, there is a lack of information on the efficiency and mode of action of many traditional drugs. Standard drugs are often more effective and convenient to use than traditional remedies. Furthermore, traditional drugs have not been tested extensively for their effects on organ integrity. There is also the need to determine the active ingredients and their amounts in the plant. Such information is important in dosages. Biochemical studies also would reveal any poisons (such as alkaloids and cardiac glycosides) that may exist in the plants.
2. The treatments and practices reported in this chapter are based on selected reviews and papers and need to be validated to identify those that can be of practical advantage in agricultural development.[14] Issues that should be addressed are efficacy, quality, safety, and standardization of doses. Models and guidelines to validate and develop human medicines have been developed, and these could probably be modified and applied to animals.[142,143]

REFERENCES

1. Schwabe, C., *Cattle, priests and progress in medicine,* University of Minnesota Press, Minneapolis, 1978.
2. Ba, A.S., *L'art veterinaire en milieu traditionnel africain*, Agence de cooperation culturelle et technique, Paris, 1982.
3. Stem, C., Ethnoveterinary R and D in production systems, in *Ethnoveterinary research and development,* McCorkle, C.M., Mathias, E., and Schillhorn van Veen, T.W., Eds., Intermediate Technology, London, 1996, 193.
4. Food and Agriculture Organization-Regional Office for Asia and the Pacific (FAO-RAPA), *Preliminary study of traditional systems of veterinary medicine*, FAO-RAPA, Bangkok, 1980.
5. Food and Agriculture Organization-Regional Office for Asia and the Pacific (FAO-RAPA), *Traditional (indigenous) systems of veterinary medicine for small farmers in India*, FAO-RAPA, Bangkok, 1984a.
6. Food and Agriculture Organization-Regional Office for Asia and the Pacific (FAO-RAPA), *Traditional (indigenous) systems of veterinary medicine lot small farmers in Thailand,* FAO-RAPA, Bangkok, 1984b.

7. Food and Agriculture Organization-Regional Office for Asia and the Pacific (FAO-RAPA), *Traditional (indigenous) systems of veterinary medicine for small farmers in Nepal*, FAO-RAPA, Bangkok, 1984c.
8. Food and Agriculture Organization-Regional Office for Asia and the Pacific (FAO-RAPA), *Traditional (indigenous) systems of veterinary medicine for small farmers in Pakistan*, FAO-RAPA, Bangkok, 1986.
9. Food and Agriculture Organization-Regional Office for Asia and the Pacific (FAO-RAPA), *Traditional veterinary medicine in Sri Lanka*, FAO-RAPA, Bangkok, 1991a.
10. Food and Agriculture Organization-Regional Office for Asia and the Pacific (FAO-RAPA), *Traditional veterinary medicine in Indonesia*, FAO-RAPA, Bangkok, 1991b.
11. Food and Agriculture Organization-Regional Office for Asia and the Pacific (FAO-RAPA), *Traditional veterinary medicine in the Philippines*, FAO-RAPA, Bangkok, 1992.
12. Office of International des Epizootics (OIE), Early methods of animal disease control, *Rev. Sci. Techn. Off. Int. Epizoot.*, Paris, 1994.
13. McCorkle, C.M., An introduction to ethnoveterinary research and development, *J. Ethnobiology,* 6, 129, 1986.
14. McCorkle, C.M., Veterinary anthropology, *Hum. Org.,* 48, 156, 1989.
15. McCorkle, C.M., and Mathias-Mundy, E., Ethnoveterinary medicine in Africa, *J. Int. African Inst.*, 62, 59, 1992.
16. Iqbal, Z., Akhtar, M.S., Sandhu, Z.D. et al., Herbal dewormers in livestock—a traditional therapy, *Int. J. Agric. Biol.,* 5, 199, 2003.
17. American Veterinary Medical Association (AVMA), *133rd AVMA Convention Notes*, July 20–24, Louisville, KY, AVMA, Schaumburg, IL.
18. Inglis, J.T. (Editor), *Traditional ecological knowledge: concept and cases,* International Development Research Center, Ottawa, 1995, 141.
19. Schillhorn van Veen, T.W., Sense or nonsense? Traditional methods of animal parasitic disease control, *Vet. Parasitol.,* 71, 177, 1997.
20. Lodrick, D.O., *Sacred cows, sacred places. Origin and survivals of animal homes in India*, University of California Press, Berkeley, CA, 1981, 51.
21. Larrat, R., Medecine et pharmacie indigenes: trypanosomiases et piroplasmoses. Bull. Serv. Zootechn. t~pizoot, *Afr. Occid. Francoph.,* 2, 55, 1939.
22. Nigeria, *Annual report of the Veterinary Department, northern provinces for the year 1928*, Government Printer, Lagos, 1929.
23. Abelson, P.H., Medicine from plants, *Science,* 247, 513, 1990.
24. Stepp, J.R., and Moerman, D.E., The importance of weeds in ethnopharmacology, *J. Ethnopharmacol.,* 75, 25, 2001.
25. Stepp, J.R., The role of weeds as sources of pharmaceuticals, *J. Ethnopharmacol.,* 92, 163, 2004.
26. Farnsworth, N.R., and Soejarto, D.D., Potential consequence of plant extinction in the United States on the current and future availability of prescription drugs, *Econ. Bot.,* 39, 231, 1985.
27. Farnsworth, N.R., and Soejarto, D.D. Global importance of medicinal plants, in *Conservation of medicinal plants*, Akerele, O, Heywood, V., and Synge, H., Eds., Cambridge University Press, Cambridge, UK, 1991, pp. 25–52.
28. Cragg, G.M., Newman, D.J., and Snader, K.M., Natural products in drug discovery and development, *J. Nat. Prod.,* 60, 52,1997.
29. Agarwal, A., Dismantling the divide between indigenous and scientific knowledge, *Dev. Change,* 26, 413, 1995.
30. World Health Organization (WHO), *Research guidelines for evaluating the safety and efficacy of herbal medicines*, World Health Organization, Regional Office for the Western Pacific, Manila, 1993.
31. Etkin, N.L., Anthropological methods in ethnopharmacology, *J. Ethnopharmacol.*, 38, 93, 1993.

32. Lai, P.K., and Roy, J., Antimicrobial and chemopreventive properties of herbs and spices, *Curr. Med. Chem.*, 11, 1451, 2004.
33. Singh, G., and Upadhyay, R.K., Essential oils—a potent source of natural pesticides, *J. Sci. Ind. Res.,* 52, 676, 1993.
34. Nwude, N., and Ibrahim, M.A., Plants used in traditional veterinary medical practice in Nigeria, *J. Vet. Pharmacol. Ther.*, 3, 261, 1980.
35. Minja, M.M.J., Collection of Tanzanian medicinal plants for biological activity studies, *Proc. 7th Tanzania Vet. Assoc. Sci. Conf.,* Arusha, 7, 67, 1989.
36. Intermediate Technology Development Group and International Institute of Rural Reconstruction (ITDG and IIRR), *Ethnoveterinary medicine in Kenya. A field manual of traditional health care practices*, Nairobi, Kenya, 1996.
37. Kalesaraj, R., Screening of some indigenous plants for anthelmintic action against human *Ascaris lumbricoides, Indian J. Physiol. Pharmacol.*, 19, 47, 1975.
38. Fernandez, T.J., Local plants having anthelmintic activity, *ASEAN J. Sci. Technol. Dev.*, 8, 115, 1991.
39. Ibrahim, M.A., Nwude, N., Ogunsusi, R.A. et al., Screening West African plants for anthelmintic activity, *ILCA Bull.*, 17, 19, 1984.
40. Bizimana, N., *Traditional veterinary practices in Africa*, Deutsche Gesellschaft für Technische Zusammenarbeit (GTZ), Eschborn, Germany, 1994.
41. Shuhua, X., Jiqing, Y., Jinying, M. et al., Preventive effect of artemether in experimental animals infected with *Schistosoma mansoni, Parasitol. Int.*, 49, 19, 2000.
42. Iqbal, Z., Lateef, M., Ashraf, M. et al., Anthelmintic activity of *Artemisia brevifolia* in sheep, *J. Ethnopharmacol.,* 93, 265, 2004.
43. Idris, A.A., Adam, S.E., Tortour, G. et al., The anthelmintic efficacy of *Artemisia herba-alba* against *Haemonchus contortus* infection in goats, *Nat. Inst. Anim. Health Q.*, 22, 138, 1982.
44. Slepnev, N.K., Study of the anthelmintic activity of some plants, *Nauchnye-Trudy-Nauchno-Issledovatel'skogo-Veterinarnogo-Instituta,-Minsk*, 8, 107, 1970.
45. Akhtar, M.S., Chattha, M.S., Chaudhry, A.H. et al., Comparative efficacy of santonin and piperazine against *Neoascaris vitulorum* in buffalo calves, *J. Vet. Pharmacol. Ther.*, 5, 71, 1982.
46. Farnsworth, N.R., Akerele, Q., Bingen, A. et al., Medicinal plants in therapy, *Bull. WHO*, 63, 965, 1985.
47. Sherif, A., Hall, R.G., El-Amamy, M. et al., Drugs, insecticides and other agents from Artemisia, *Med. Hypotheses*, 23, 187, 1987.
48. Francois, M.T., Examination of a few plant species with anthelmintic properties by thin layer chromatography in conjunction with the T.A.S. micro-oven, *Plantes Med. Phytother.,* 8, 109, 1974.
49. Narayana, K., Setty, P.R.L., Rao, H.S. et al., A clinical and experimental study of "Wopell" in canine cestodiasis, *Mysore J. Agric. Sci.*, 10, 98, 1976.
50. Gowda, S.K., Biological effects of neem (*Azadirachta indica*) derivatives in animals, in *Ethnoveterinary medicine: alternatives for livestock development. Vol. 2. Abst. Proc. Int. Conf. Pune,* India, Nov. 4–6, 1997.
51. Mostofa, M., McKellar, Q.A., Alam, M.N. et al., Epidemiology of gastrointestinal helminth parasites in small ruminants in Bangladesh and their anthelmintic therapy. Sustainable parasite control in small ruminants, *Int. workshop sponsored by ACIAR, Bogor, Indonesia*, 22–25 April 1996, pp. 105–108.
52. Hördegen, P., Hertzberg, H., Neilmann, J. et al., The anthelmintic efficacy of five plant products against gastronintestinal trichostrongylids in artificially infected lambs, *Vet. Parasitol.*, 117, 51, 2003.
53. Kalesaraj, R., and Kurup, P.A., Anthelmintic activity, toxicity and other pharmacological properties of palasonin, the active principle of *Butea frondosa* seeds and its piperazine salt, *Indian J. Med. Res.*, 56, 1818, 1968.

54. Lal, J., Chandra, S., and Sabir, M., Modified method for isolation of Palasonin—the anthelmintic principle of *Butea frondosa* seeds, *Indian J. Pharm. Sci.*, 40, 97, 1978.
55. Akhtar, M.S., Iqbal, Z., Khan, M.N. et al., Anthelmintic activity of medicinal plants with particular reference to their use in animals in Indo-Pakistan sub-continent. *Small Ruminant Res.*, 38, 99, 2000.
56. Prashanth, D., Asha, M.K., Amit, A. et al., Anthelmintic activity of *Butea monosperma, Fitoterapia*, 72, 421, 2001.
57. Iqbal, Z., Lateef, M., Jabbar, A., et al., In vivo anthelmintic activity of *Butea monosperma* against trichostrongylid nematodes in sheep, *Fitoterapia,* 77, 137, 2006.
58. Akhtar, M.S., Javed, I., Hayat, C.S. et al., Efficacy and safety of *Caesalpinia crista* Linn. seeds and its extracts in water and methanol against natural *Neoascaris vitulorum* infection in buffalo-calves, *Pak. Vet. J.*, 5, 192, 1985.
59. Javed, I., Akhtar, M.S., Rehman, Z.U. et al., Comparative anthelmintic efficacy and safety of *Caesalpinia crista* seed and piperazine adipate in chickens with artificially induced *Ascaridia galli* infection, *Acta Vet. Hungarica,* 42, 103, 1994.
60. Sharma, L.D., Bhaga, H.S., and Srivastava, P.S., In vitro anthelmintic screening of indigenous medicinal plants against *Haemonchus contortus* (Rudolphi, 1803) Cobbold, 1898 of sheep and goats, *Indian J. Anim. Res.*, 5, 33, 1971.
61. Adewunmi, C.O., and Akubue, P.I., Preliminary studies on the anthelmintic properties of the aqueous extract of *Calliandra portoricensis* (Jacq.) Benth., *Bull. Anim. Health Prod. Africa*, 29, 171, 1981.
62. Garg, L.C., and Atal, C.K., Anthelmintic activity of Calotropain and Bromelain, *Indian J. Pharmacol.*, 25, 422, 1963.
63. Jain, S.C., Sharma, R., Jain, R. et al., Antimicrobial activity of *Calotropis procera, Fitoterapia,* 67, 275, 1996.
64. Al-Qarawi, A.A., Mahmoud, O.M., Sobaih, M.A. et al., A preliminary study on the anthelmintic activity of *Calotropis procera* latex against *Haemonchus contortus* infection in Najdi sheep, *Vet. Res. Commun.*, 25, 61, 2001.
65. Iqbal, Z., Lateef, M., Jabbar, A. et al., Anthelmintic activity of *Calotropis procera* (Ait.) Ait. F. flowers in sheep, *J. Ethnopharmacol.,* 102, 256, 2005.
66. Akhtar, M.S., Iqbal, Z., and Khan, M.N., Evaluation of anthelmintic activity *Chenopodium album* (Bathu) against nematodes in sheep, *Int. J. Agric. Biol.*, 1, 121, 1999.
67. Jabbar, A., Zaman, M.A., Iqbal, Z. et al., Anthelmintic activity of *Chenopodium album* (L.) and *Caesalpinia crista* (L.) against trichostrongylid nematodes of sheep, *J. Ethnopharmacol.,* 114, 86, 2007.
68. Diaz-Obregon, D., Lloja-Lozano, L., and Carbajal-Zuniga, V., Preclinical studies of *Cucurbita maxima* (pumpkin seeds) a traditional intestinal antiparasitic in rural urban areas, *Rev. Gastroenterol.,* 24, 323, 2004.
69. Kokate, C.K., and Varma, K.C., Anthelmintic activity of some essential oils, *Indian J. Hosp. Pharm.*, 8, 150, 1971.
70. Akhtar, M.S., and Ahmad, I., Anthelmintic and phytochemical studies on *Hyoscyamus niger* Linn. (Ajwain Khurasani) seeds and *Morringa oleifera* Linn. (Sohanjna) roots, *J. Pharm. Punjab Univ. Lahore, Pakistan*, 3, 75, 1990.
71. Neogi, N.C., Baliga, P.A.C., and Srivastava, R.K., Anthelmintic activity of some indigenous drugs, *Indian J. Pharmacol.*, 26, 37, 1964.
72. Sangwan, S.A.K.N., and Sangwan, N., In vitro effects of leaf extracts of *Melia azedarach* on mortality of *Haemonchus contortus, Indian J. Anim. Res.*, 32, 70, 1998.
73. Akhtar, M.S., and Riffat, S., Evaluation of *Melia azedarach* Linn. seeds (Bakain) and piperazine against *Ascaridia galli* infection in chickens, *Pak. Vet. J.*, 5, 34, 1985.
74. Akhtar, M.S., and Riffat, S., Efficacy of *Melia azedarach* Linn. (Bakain) and morantel against naturally acquired gastrointestinal nematodes in goats, *Pak. Vet. J.*, 4, 176, 1984.

75. Iqbal, Z., Lateef, M., Jabbar, A., In vitro and in vivo anthelmintic activity of *Nicotiana tabacum* L. leaves against gastrointestinal nematodes of sheep, *Phytother. Res.,* 20, 46, 2006.
76. Kailani, S.U.R., Akhtar, M.S., and Ashraf, M., Antifasciolic efficacy of indigenous plant drugs: Kalonji, Shahterah and Karanjwa in buffaloes, *Pak. J. Pharmacol. Sci.*, 8, 17, 1995.
77. Akhtar, M.S., and Ahmad, I., Evaluation of antinematodal efficacy of tetrahydroharmine in goats, *Vet. Arhiv.*, 6, 307, 1991.
78. Akhtar, M.S., and Riffat, S., A field trial of *Paganum harmala* Linn. seeds (Harmal) against natural cestodal infection in Beetal goats, *J. Pharm. Univ. Karachi, Pakistan,* 4, 79, 1986.
79. Akhtar, M.S., and Riffat, S., Efficacy of *Punica granatum* Linn. (Anar) fruit-rinds against naturally acquired nematodal and cestodal infections, *J. Pharm. Punjab Univ. Lahore, Pakistan*, 6, 17, 1985.
80. Iqbal, Z., Lateef, M., Khan, M.N. et al., Anthelmintic activity of *Swertia chirata* Buch-Ham. whole plant against gastrointestinal nematodes of sheep, *Fitoterapia,* 77, 463, 2006.
81. Jabbar, A., Iqbal, Z., and Khan, M.N., In vitro anthelmintic activity of *Trachyspermum ammi* seeds, *Pharmacognosy Mag.,* 2, 126, 2006.
82. Lateef, M., Iqbal, Z., Akhtar, M.S. et al., Preliminary screening of *Trachyspermum ammi* (L.) seed for anthelmintic activity in sheep, *Trop. Anim. Health Prod.,* 38, 491, 2006.
83. Javed, I., and Akhtar, M.S., Screening of *Veronia anthelmintica* seed and *Embelia ribes* fruit mixed in equal parts against gastrointestinal nematodes, *Pak. J. Pharmacol. Sci.*, 3, 69, 1990.
84. Iqbal, Z., Lateef M., Jabbar, A. et al., Anthelmintic activity of *Vernonia anthelmintica* (L.) Willd. seeds against trichostrongylid nematodes of sheep, *Pharm. Biol.,* 44, 563, 2006.
85. Gaind, K.N., and Budhiraja, R.D., Antibacterial and anthelmintic activity of *Withania coagulans* Dunal., *Indian J. Pharmacol.*, 29, 185, 1967.
86. Iqbal, Z., Lateef, M., Akhtar, M.S., In vivo anthelmintic activity of ginger against gastrointestinal nematodes of sheep, *J. Ethnopharmacol.*, 106, 285, 2006.
87. Githiori, J.B., Athanasadou, S., and Thomsberg, S.M., Use of plants in novel approaches for control of gastrointestinal helminths in livestock with emphasis on small ruminants, *Vet. Parasitol.,* 139, 308, 2006.
88. Jackson, F., and Miller, J., Alternative approaches to control—Quo vadit? *Vet. Parasitol.,* 139, 371, 2006.
89. Stepek, G., Lowe, E.A., Buttle, D.J., The anthelmintic efficacy of plant-derived cysteine proteinases against the rodent gastrointestinal nematode, *Heligmosomoides polygyrus*, in vivo, *Parasitology*, 134, 1409, 2007.
90. Dhananjeyan, M.R., Milev, Y.P., Kron, M.A. et al., Synthesis and activity of substituted anthraquinones against a human filarial parasite, *Brugia malayi*, *J. Med. Chem.,* 48, 2822, 2005.
91. Iqbal, Z., Sarwar, M., Jabbar, A. et al., Direct and indirect anthelmintic effects of condensed tannins in sheep, *Vet. Parasitol.,* 144, 125, 2007.
92. Lun, Z., Burri, C., Menzinger, M. et al., Antiparasitic activity of diallyl trisulfide (Dasuansu) on human and animal pathogenic protozoa (*Trypanosoma* sp., *Entamoeba histolytica* and *Giardia lamblia*) in vitro, *Ann. Soc. Belge Med. Trop.,* 74, 51, 1994.
93. Saleheen, D., Ali, S.A., and Yasinzai, M.M., Antileishmanial activity of aqueous onion extract in vitro, *Fitoterapia,* 75, 9, 2004.
94. Schmidt, P. W., and Marquardt, U., Über den antimykotischen Effekt ätherischer Öle von lauchgewächsen und kreuzblülern auf pathogene Hautpilze, *Zentralblatt für Bakteriologie, Parasitenkunde, Infektionskrankheiten und Hygiene. Abteilung,* 138, 104, 1936.

95. Zohri, A.N., Abdel-Gawad, K., and Saber, S., Antibacterial, antidermatophytic and antioxigenic activities of onion (*Allium cepa* L.) oil, *Microbiol. Res.,* 150, 167, 1995.
96. Shen, J.K., Davis, L.E., Wallace, J.M. et al., Enhanced diallyl trisulfide has in vitro synergy with amphotericin B against *Cryptococcus neoformans*, *Planta Med.,* 62, 415, 1996.
97. Davis, S.R., and Perrie, R., The in-vitro susceptibility of *Cryptococcus neoformans* to allitridium. In *Program and Abstracts of the 15th Congress of ISHAM* (abstract 113), San Antonio, TX, May 25–29, 2003.
98. Delaha, E.C., and Garagusi, V.F., Inhibition of mycobacterial by garlic extract (*Allium sativum*), *Antimicrob. Agents Chemother.,* 27, 485, 1985.
99. Fenwick, G.R., and Hanley, A.B., The genus *Allium*, *CRC Crit. Rev. Food Sci. Nutr.,* 22, 199, 1985.
100. Meng, Y., Guo, N., Zhang, L. et al., Anti-HCMV effect of garlic components, *Virol. Sin.,* 8, 147, 1993.
101. Augusti, K.T., and Mathew, P.T., Lipid lowering effect of allicin (diallyl disulfide oxide) on long-term feeding in normal rats, *Experientia,* 30, 468, 1974.
102. Griffiths, G., Trueman, L., Crowther, T. et al., Onions—a global benefit to health, *Phytother. Res.,* 16, 603, 2002.
103. Griffin, G.E., Leung, K., Folks, T.M. et al., Activation of HIV gene expression during monocyte differentiation by induction of NF-κB, *Nature (London),* 339, 70, 1989.
104. Rios, J.L., Recio, M.C., and Villar, A., Screening methods for natural products with antimicrobial activity: a review of the literature, *J. Ethnopharmacol.,* 23, 127, 1988.
105. Janssen, A.M., Scheffer, J.J.C., and Svendsen, A.B., Antimicrobial activity of essential oils: a 1976–1986 literature review: aspects of the test methods, *Planta Med.,* 5, 395, 1987.
106. Lawrence, B.M., *Antimicrobial/biological activity of essential oils*, Allured, Carol Stream, IL, 2005.
107. Muroi, H., and Kubo, I., Antibacterial activity of anacardic acid and toatarol, alone and in combination with methicillin, against methicillin-resistant *Staphylococcus aureus*, *J. Appl. Bacteriol.,* 80, 387, 1996.
108. Stasiuk, M., Bartosiewicz, D., and Kozubek, A., Inhibitory effect of some natural and semisynthetic phenolic lipids upon acetylcholinesterase activity, *Food Chem.,* 108, 996, 2008.
109. Chen, J., Zhang, Y.H., Wang, L.K. et al., Inhibitors of DNA polymerase b from *Schoepfia californica*, *J. Chem. Soc. Chem. Commun.,* 24, 2769, 1998.
110. Atta, A.H., and Alkofahi, A., Anti-nociceptive and anti-inflammatory effects of some Jordanian medicinal plant extracts, *J. Ethnopharmacol.,* 60, 117, 1998.
111. Anon., *Ruta graveolens*. Committee for Veterinary Medicinal Products. The European Agency for the Evaluation of Medicinal Products, 1999, http:*www.emea.eu.int/pdfs/vet/mrls/054298en.pdf.*
112. Abramson, C.I., Aldana, E., and Sulbaran, E., Exposure to citral, cinnamon and ruda disrupts the life cycle of a vector of Chagas disease, *Am. J. Environ. Sci.,* 3, 7, 2007.
113. Alves, D.S., Perez-Fons, L., and Estepa A., Membrane related effects underlying the biological activity of the anthraquinones emodin and barbaloin, *Biochem. Pharmacol.,* 68, 549, 2004.
114. McKay, D.L., and Blumberg, J.B., A review of the bioactivity and potential health benefits of peppermint tea (*Mentha piperita* L.), *Phytother. Res.,* 20, 619, 2006.
115. Yingying Xu, Hall III, C., Wolf-Hall, C. et al., Fungistatic activity of flaxseed in potato dextrose agar and a fresh noodle system, *Int. J. Food Microbiol.,* 121, 262, 2008.
116. Payne, T.J., Promoting better health with flaxseed in bread, *Cereal Foods World,* 45, 102, 2000.
117. Nok, A.J., King, A.N.E., Adaudi, A. et al., Trypanocidal activity of an organotin compound (tri-n-butyltin oxide) toward *Trypanosoma brucei*, *J. Clin. Biochem. Nutr.,* 13, 81, 1992.

118. Nok, A.J., Esievo, K.A.N., Hongdet, I. et al., Trypanocidal potentials of *Azadirachta indica*: in vivo activity of leaf extract against *Trypanosoma brucei*, *J. Clin. Biochem. Nutr.*, 15, 113, 1993.
119. Nok, A.J., and Williams, S., *Allium sativum*—induced death of African trypanosomes, *Parasitol. Res.*, 82, 634, 1996.
120. Lewis, W.H., and Elvin-Lewis, M.P., Medicinal plants as sources of new therapeutics, *Ann. Missouri Bot. Gard.*, 82, 16, 1995.
121. Borris, R.P., Natural products research: perspectives from a major pharmaceutical company, *J. Ethnopharmacol.*, 51, 29, 1996.
122. Georges, M., and Pandelai, K.M., Investigations on plant antibiotics. IV. Further search for antibiotic substances in Indian medicinal plants, *Indian J. Med. Res.*, 37, 169, 1949.
123. Rojas, A., Hernandez, L., Pereda-Miranda, R. et al., Screening for antimicrobial activity of crude drug extracts and pure natural products from Mexican medicinal plants, *J. Ethnopharmacol.*, 35, 275, 1992.
124. Silva, O., Duarte, A., Cabrita, J. et al., Antimicrobial activity of Guinea-Bissau traditional remedies, *J. Ethnopharmacol.*, 50, 55, 1996.
125. Berghe, V., D.A., Vlietinck, A.J., and Van Hoof, L., Plant products as potential antiviral agents, *Bull. Inst. Pasteur*, 84, 101, 1986.
126. Geissman, T.A., Flavonoid compounds, tannins, lignins and related compounds, in *Pyrrole pigments, isoprenoid compounds and phenolic plant constituents,* vol. 9, Florkin, M., and Stotz, E. H., Eds., Elsevier, New York, 1963, 265.
127. Harborne, J.B., *Introduction to ecological biochemistry,* 4th ed., Harcourt Brace, London, 1993.
128. Putnam, A.R., Weed allelopathy, in *Weed physiology, vol. 1. Reproduction and ecophysiology*, Duke, S.O., Ed., CRC Press, Boca Raton, FL, 1994, 131–155.
129. Zimdahl, R.L., *Fundamentals of weed science*, 2nd ed., Academic Press, San Diego, CA, 1999.
130. Feeny, P.P., Plant apparency and chemical defense, in *Recent advances in phytochemistry,* vol. 10, Wallace, J.W., and Mansell, R.L., Eds., Plenum, New York, 1976, 1–40.
131. Coley, P.D., Bryant, J.P., and Chapin, F.S., Resource availability and plant antiherbivore defense, *Science,* 230, 895, 1985.
132. Rhoades, D.F., and Cates, R.G., A general theory of plant antiherbivore chemistry, in *Recent advances in phytochemistry*, vol. 10, Wallace, J.W., and Mansell, R.L., Eds., Plenum, New York, 1976, 168–213.
133. Abe, T., and Higashi, M., Cellulose centered perspective on terrestrial community structure, *Oikos,* 60, 127, 1991.
134. Schultes, R.E., The kingdom of plants, in *Medicines from the earth,* Thomson, W.A.R., Ed., McGraw-Hill, New York, 1978, 208.
135. Cowan, M.M., Plant products as antimicrobial agents, *Clin. Microbiol. Rev.,* 12, 564, 1999.
136. Nakamura, Y., Watanabe, S., Miyake, N. et al., Dihydrochalcones: evaluation as novel radical scavenging antioxidants, *J. Agric. Food Chem.*, 51, 3309, 2003.
137. Tapiero, H., Tew, K.D., Nguyen, Ba G. et al., Polyphenols: do they play a role in the prevention of human pathologies? *Biomedicine & Pharmacotherapy*, 56, 200, 2002
138. Scalzo, J., Politi, A., Pellegrini, N. et al., Plant genotype affects total antioxidant capacity and phenolic contents in fruit, *Nutrition,* 21, 207, 2005.
139. Hakkinen, S.H., and Torronen, A.R., Content of flavonols and selected phenolic acids in strawberries and *Vaccinium* species: influence of cultivar, cultivation site and technique, *Food Res. Int.,* 33, 517, 2000.
140. Croom, E.M., Documenting and evaluating herbal remedies, *Econ. Bot.,* 37, 13, 1983.
141. Zadernowski, R., Naczk, M., and Nesterowicz, J., Phenolic acid profiles in some small berries, *J. Agric. Food Chem.,* 53, 2118, 2005.

142. World Health Organization (WHO), *General guidelines for methodologies on research and evaluation of traditional medicine*, WHO, Geneva, 2000.
143. Diallo, D., and Paulsen, B.S., Pharmaceutical research and traditional practitioners in Mali: experiences with benefit sharing, in *Responding to bioprospecting: from biodiversity in the south to medicines in the north*, Svarstad, H., and Dhillion, S.S., Eds., Spartacus Forlag, Ås, Oslo, 2000.
144. Corrain, C., Zampini, P., and Note di, P., Veterinaria popolare riguardantiil polesine, *Arch. l'Antropologia e la Etnologia,* 91, 249, 1961.
145. De Simoni, E., and Guarrera, P.M., Indagine etnobotanica nella Provinciadi Teramo, *Quaderni Botanica Ambientale Applicata*, 15, 3, 1994.
146. Guarrera, P.M., *Il patrimonio etnobotanico del Lazio*, Regione Lazioe Dipartimento di Biologia Vegetale, Roma, 301, 1994.
147. Nardelli, G.M., *Cultura e tradizione*, Demomedicina nell'alta Umbria, Provincia di Perugia, 1987.
148. Corsi, G., Gaspari, G., and Pagni, A.M., L'uso delle piantenell'economia domestica della Versilia collinare e montana, *Atti Soc. tosc. Sci. Nat. Mem. Ser.,* 87, 309, 1981.
149. Zampiva, F., Erbe e piante della Lessinia, *Erboristeria Domani,* 9, 17, 1981.
150. Guarrera, P.M., Ricerche etnobotaniche nelle Province di Macerata e di Ancona, EPPOS, 2, 99, *Rivista Italiana* EPPOS, 4, 220, 1981.
151. Bandini, A., Le piante della medicina tradizionale nell'alta Val diVara (*Liguria orientale), Webbia*, 16, 143, 1961.
152. Bellomaria, B., and Della Mora, L., Novità nell'uso delle piante officinali per la zona di Matelica (Macerata) anche in confronto con altre zonedelle Marche, *Archivio Botanico e Biogeografico Italiano,* 61, 51, 1985.
153. Guarrera, P.M., Usi tradizionali delle piante nel territorio della Majella, In: *Fasc. Monograf ed., Erbe e piante medicinali nella storia e nelletradizioni popolari abruzzesi*, Reg. Abruzzo, Chieti, Tip. Anxanum, Lanciano, 1987, 17.
154. Tammaro, F., *Flora officinale d'Abruzzo*, Giunta Reg. d'Abruzzo, Chieti, Italy, 1984.
155. Tammaro, F., and Pietrocola, L., Piante nella medicina popolare d'Abruzzo, *Ann. Bot.,* 34, 269, 1975.
156. Manzi, A., Piante utilizzate nella veterinaria popolare a Gessopalena (CH), *Rivista Abruzzese,* 3, 253, 1989.
157. De Capite, L., and Menghini, A., Le piante medicinali in Umbrianell'uso della veterinaria popolare, *Annali Facoltà di Agraria Universitàdi Perugia*, 28, 589, 1973.
158. Corrain, C., Ricordi di Folklore Polesano. Minelliana, Rovigo, 47, 1977.
159. Bellomaria, B., Le piante di uso popolare nel territorio di Camerino (Marche), *Archivio Botanico e Biogeografico Italiano,* 58, 1, 1982.
160. De Bellis, A., *Erbe di val d'Orcia*, Editori del Grifo, Montepulciano, Siena, 1988.
161. Maccioni, S., and Marchini, G., *La bassa val di magra*, Collana "Liguria in parole povere," Sagep Eds., Genova, 1999.
162. Viegi, L., Bioli, A., Vangelisti, R. et al., Prima indagine sulle piante utilizzate in medicina veterinaria popolare in alcune località dell'alta Val di Cecina, *Atti della Societa Toscana di Scienze Naturali, Memorie, Serie B,* 106, 1, 1999.
163. Ciccodicola, F., Pratiche di guarigione e memoria collettiva—considerazioni relative ad una ricerca sul campo, in *Le piante magiche—una ricerca storico-antropologica*, Giusti, S., Ed., Domograf, Rome, 1995, 241.
164. Pieroni, A.C., Quave, S.N., and Heinrich, M., Ethnopharmacy of the Arbëreshë in Lucania (southern Italy), *Fitoterapia*, 73, 217, 2002.
165. De Feo, V., and Senatore, F., Medicinal plants and phytotherapy in the Amalfitan Coast, Salerno province, Campania, Southern Italy, *J. Ethnopharmacol.,* 39, 39, 1993.

166. Gastaldo, P., Barberis G., and Fossati, F., Le piante della medicinatradizionale nei dintorni di Praglia (Appennino ligure-piemontese), *Atti Accademia Ligure di Scienze e Lettere,* 35, 125, 1978.
167. De Bellis, A., *Erbe di val d'Orcia*, Pienza, 1978.
168. Amici, L., *Medicina popolare della Teverina*, Regione Lazio (Assessorato alla Cultura) e Associazione Intercomunale della Teverina per la Cultura, 1992.
169. Uncini Manganelli, R.E., and Tomei, P.E., Indagini farmaco-botaniche in Garfagnana (Lucca): il versante appenninico, *Atti della Societa Toscana di Scienze Naturali, Memorie, Serie B,* 102, 1995.
170. De Feo, V., Aquino, A., Menghini, E. et al., Traditional phytotherapy in the Peninsula Sorrentina, Campania, Southern Italy, *J. Ethnopharmacol.,* 36, 113, 1992.
171. Ferri, S., Piante medicinali e fitoterapia nel territorio di Cetona eSarteano, *Webbia*, 31, 105, 1977.
172. Martini, E., La fitoterapia popolare nell'alta Valle dell'Orba (Appenninoligure), *Atti Società Ligure di Scienze e Lettere,* 39, 1, 1982.
173. Chiavoni, M., and Raffo, E.S., *Ricerca etnobotanica nella Provincia diGrosseto*, Istituto Tecnico Agrario Statale "Leopoldo II di Lorena," Grosseto, Italy, 1994.
174. Cappelletti, E.M., Trevisan, R., Foletto, A. et al., Lepiante utilizzate in medicina popolare in due vallate trentine: Val di Ledro e Val dei Mocheni, *Studi Trentini di Scienze Naturali,* 58, 119, 1981.
175. Uncini Manganelli, R.E., and Tomei, P.E., Indagini farmaco-botaniche in Garfagnana (Lucca): il versante apuano, *Atti della Societa Toscana di Scienze Naturali, Memorie, Serie B,* 103, 63,1996.
176. Camangi, F., and Uncini Manganelli, R.E., L'etnobotanica nel territorio di Capannori: stato delle conoscenze e nuove acquisizioni, *Studi Capannoresi,* 3, 179, 1999.
177. Uncini Manganelli, R.E., and Tomei, P.E., Documenti per la conoscenza delle tradizioni etno-farmacobotaniche in Toscana. Accademia Lucchese di Scienze, *Lettere e Arti.,* 1999.
178. Lentini, F., Catanzaro, F., and Aleo, M., Indagini etnobotaniche in Sicilia. III. L'uso tradizionale delle piante nel territorio di Mazara delVallo (Trapani), *Atti Accademia di Scienze, Lettere e Arti di Palermo,* 1, 1988.
179. Guarrera, P.M., Usi tradizionali delle piante in alcune aree marchigiane, *Informatore Botanico Italiano,* 22, 155, 1990.
180. Bellomaria, B., and Lattanzi E., Le piante del territorio di Cupra Marittima (Marche) attualmente usate nella medicina popolare, *Archivio Botanico e Biogeografico Italiano,* 58, 1, 1982.
181. Tammaro, F., Piante officinali e pratica della fitoterapia nel territoriodel Gran Sasso d'Italia, *Rivista Italiana EPPOS,* 58, 593, 1976.
182. Picchi, G., Erbe ed animali nelle società agricole tradizionali, in *Herbs, humans and animals/Erbe, uomini e bestie*, Pieroni, A., Ed., Experiences Verlag, Cologne, Germany, 1999, 94.
183. Tomei, P.E., and Gaspari, G., Indagine sulle zone umide della Toscana, XVI. Le piante officinali dei bacini palustri della Toscana settentrionale, in *Il Bacino di Massaciuccoli*, Grafiche Pacini, Pisa, 1981, 13.
184. Guarrera, P.M., and Tammaro, F., Ethnobotanical research in the Abruzzo and Latium (central Italy), in *Ethnopharmacologie: sources, méthodes, objectifs*, Actes du 1er Colloque Européen d'Ethnopharmacologie, Orstom, Parigi, 1991, 168.
185. Abbas, B., Al-Qarawi, A.A., and Al-Hawas, A., The ethnoveterinary knowledge and practice of traditional healers in Qassim Region, Saudi Arabia, *J. Arid Environ.,* 50, 367, 2002.
186. Lans, C., Harper, T., Georges, K. et al., Medicinal plants used for dogs in Trinidad and Tobago, *Prev. Vet. Med.,* 50, 45, 201, 2000.
187. Bowman, D.D., Successful and currently ongoing parasite eradication programs. *Vet. Parasitol.,* 139, 293, 2006.

188. Muhammad, G., Khan, M.Z., Hussain, M.H. et al., Ethnoveterinary practices of owners of pneumatic-cart pulling camels in Faisalabad (Pakistan), *J. Ethnopharmacol.,* 97, 241, 2005.
189. Alawa, J.P., Jokthan, G.E., and Akut, K., Ethnoveterinary medical practice for ruminants in the subhumid zone of northern Nigeria, *Prev. Vet. Med.,* 54, 79, 2002.
190. Yineger, H., Kelbessa, E., Bekele, T. et al., Ethnoveterinary medicinal plants at Bale Mountains National Park, Ethiopia, *J. Ethnopharmacol.,* 112, 55, 2007.
191. Njoroge, G.N., and Bussmann, R.W., Herbal usage and informant consensus in ethnoveterinary management of cattle diseases among the Kikuyus (Central Kenya), *J. Ethnopharmacol.,* 108, 332, 2006.
192. Tabuti, J.R., Dhillion, S.S., and Lye, K.A., Ethnoveterinary medicines for cattle (*Bos indicus*) in Bolamugi country, Uganda: plant species and mode of use, *J. Ethnopharmacol.,* 88, 279, 2003.
193. Bonet, M.A., and Valles, J., Ethnobotany of Montseny biosphere reserve (Catalonia, Iberian Penninsula): plants used in veterinary medicine, *J. Ethnopharmacol.,* 110, 130, 2007.
194. Madge, C., Therapeutic landscapes of the Jola, The Gambia, West Africa, *Health and Place*, 4, 293, 1998.
195. Uncini Manganelli, R.E., Camangi, F., and Tomei, P.E., Curing animals with plants: traditional usage in Tuscany (Italy), *J. Ethnopharmacol.,* 78, 171, 2001.
196. Al-Yahya, M.A., Al-Meshal, J.S., Mossa, A.B. et al., *Saudi plants: a phytochemical and biological approach*, KACST, King Saud University Press, Riyadh, Saudi Arabia, 1990, 123.
197. Seif-Elnasr, M.M., Yousef, M.M., and Yousef, M.A.M., GLC/MS analysis of the essential oil of *Artemesia abyssinica, Herba Hungarica,* 22, 89,1983.
198. Albadr, A.A., Almeshal, A.I., Tariq, M. et al., Phytochemical and pharmacological studies of some Saudi medicinal plants, in *Proceedings of the Seventh Symposium of Biological Aspects of Saudi Arabia*, Al-Hassa, Saudi Arabia, 1985, 250.
199. Guarrera, P.M., Traditional phytotherapy in central Italy (Marche, Abruzzo, and Latium), *Fitoterapia,* 76, 1, 2005.
200. Watt, J.M., and Breyer-Brandwijk, N.G., *Medicinal and toxic plants of southern and eastern Africa*, Livingstone, Edinborough, 1962, 1108.
201. Abbas, B., Eltayeb, A.E., and Sullieman, S.R., *Calotropis procera* feed potential for arid zones, *Vet. Rec.,* 5, 132, 1993.
202. El-Shanawani, M.A.A., *Plants used in Saudi traditional medicine,* KACST, King Saud University Press, Riyadh, Saudi Arabia, 1996, 126.
203. Zalkow, L.H., Bonetti, S., Gelbaum, L. et al., Pyrroliziondine alkaloids from Middle Eastern plants, *J. Nat. Prod.,* 42, 612, 1979.
204. Adewunmi, C.O., Agbedahunsi, J.M., Adebajo, A.C. et al. Ethno-veterinary medicine: screening of Nigerian medicinal plants for trypanocidal properties, *J. Ethnopharmacol.,* 77, 19, 2001.
205. Guarrera, P.M., Castellacci A.M., and Tacconi, M., Gli usi tradizionalidelle piante, *Agricoltura e Ambiente*, 24, 47, 1984.
206. Cappelletti, E.M., Antifungal, parasiticide, insecticide, and anthelmintic herbal remedies in the traditional medicine of north-eastern Italy, *Curare*, 3, 39, 1985.
207. D'Andrea, M., Le piante officinali del parco Nazionale d'Abruzzoe gli usi popolari di esse nell'Alta Valle del Sangro, *Rivista Abruzzese,* 3, 157, 1982.
208. Guarrera, P.M., Fitoterapia e uso tradizionale delle piante nel territorio della Valle di Comino (Frosinone), in *Lepiante magiche—una ricerca storico-antropologica*, Giusti, S., Ed., Domograf, Rome, 1995, 121.
209. Chimenti Signorini, R., and Fumagalli, M., Indagine etnofarmacobotanicanella Valtournanche (Val d'Aosta), *Webbia,* 37, 69, 1983.

210. Bertagnon, E., Sulla flora medicinale della Liguria. Usi tradizionali dell'Alta Fontanabuona, *Atti Accademia Ligure di Scienze e Lettere,* 11, 201, 1955.
211. Bruni, A., Ballero, M., and Poli, F., Quantitative ethnopharmacological study of the Campidano Valley and Urzulei district, Sardinia, Italy, *J. Ethnopharmacol.,* 57, 97, 1997.
212. Catanzaro, F., Piante officinali dell'Isola di Pantelleria, *Webbia,* 23, 135, 1968.
213. Amico, F.P., and Sorge, E.G., Medicinal plants and phytotherapy in Mussomeli area (Caltanissetta, Sicily, Italy), *Fitoterapia,* 68, 143, 1997.
214. Mearelli, F., and Tardelli, C., Maremma Mediterranea, *Erboristeria Domani,* 7/8, 45, 1995.
215. Chichiriccò, G., Cifani, M.P., Frizzi, G. et al., Phytotherapy in the Subequana Valley, Abruzzo, central Italy, *J. Ethnopharmacol.,* 2, 247, 1980.
216. Barbagallo, C., Grillo M., and Meli, R., Nota sulle piante officinali spontanee e coltivate del territorio di Cesarò (Messina), *Fitoterapia*, 50, 57, 1979.
217. Catanzaro, F., Le piante officinali del territorio di Bivona (Agrigento) nella tradizione popolare, *Fitoterapia,* 41, 66, 1970.
218. De Bellis, A., *Uomini e piante dell'Amiata*, Editori del Grifo, Montepulciano, Siena, 1986.
219. Guarrera, P.M., Primo contributo allo studio del patrimonio etnobotanico del Molise, in Università degli Studi del Molise, Di.ST.e.B.A., 97 Congresso della Società Botanica Italiana, Lecce, September 24, 2002, 209.
220. Fossati, F., Bianchi, A., and Favali, M.A., Farmacopea popolare del parmense: passato e presente, *Informatore Botanico Italiano,* 31, 171, 1999.
221. Nawar, M.A.M., Elsherbeiny, A.E.A., and Elansari, M.A., Chemical constituents of *Tamarix aphylla* flowers (Tamariaceae), *Experientia,* 31, 1118, 1975.
222. Ole-Miaron, J.O., The Maasai ethnodiagnostic skill of livestock diseases: a lead to traditional bioprospecting, *J. Ethnopharmacol.,* 84, 79, 2003.

7 The Current Status and Future Prospects of Medicinal and Aromatic Plants in Veterinary Health Care in Southeast Asia

Shanmugavelu Sithambaram, Murugaiyah Marimuthu, and Chandrawathani Panchadcharam

CONTENTS

7.1 INTRODUCTION

Southeast Asia (SEA) is geographically divided into two regions: mainland SEA (or Indochina) and the Maritime SEA (or the Malay Archipelago). The subcontinent (Figure 7.1) includes Brunei Darussalam, Cambodia, Democratic Republic of Timor Leste (East Timor), Indonesia, Lao People's Democratic Republic (PDR), Malaysia, Myanmar, Philippines, Singapore, Thailand, and Vietnam. Four of the world's 25 main biodiversity hot spots lie in SEA (Myers et al., 2000). The major hot spots are Indonesia, Malaysia, Thailand, and Vietnam with 29,000, 15,500, 11,625, and 10,500 plant species, respectively (Anon., 2004). Based on traditional and cultural practice, medicinal and aromatic plants (MAPs) constitute a major portion of health care in these parts of the world. For instance, it was estimated that in Lao PDR, about 77% of the population uses traditional treatments (Sydara et al., 2005). In Indonesia, on

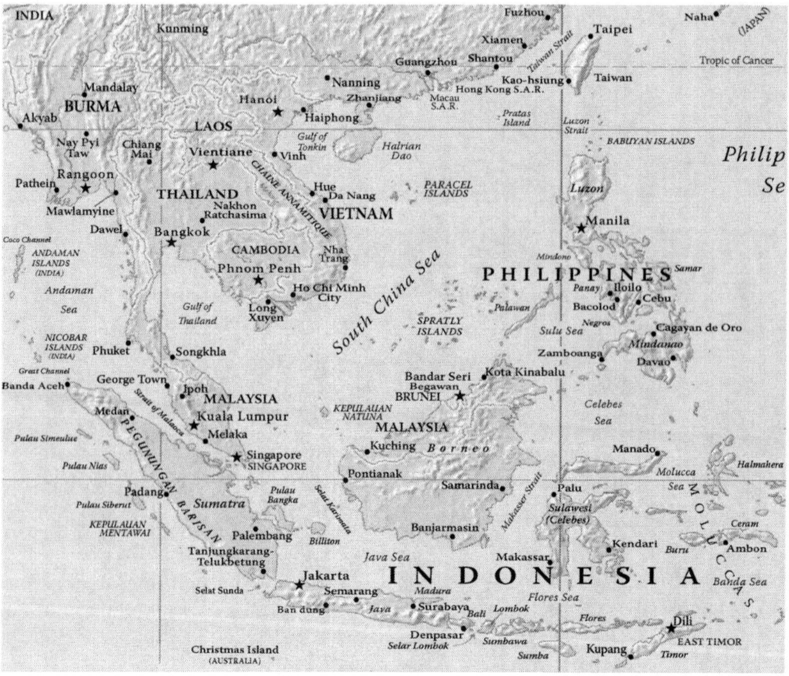

FIGURE 7.1 Countries of Southeast Asia. (*Source:* Factbook 2008. Central Intelligence Agency Office of Public Affairs, Washington, DC. https://www.cia.gov/about-cia/site-policies/index.html#copy)

the other hand, 40% of the population uses traditional medicine, with 28% from the in rural areas (World Health Organization [WHO], 2001). The status of countries of SEA and their ethnic makeup are discussed briefly in the following section.

7.2 COUNTRIES OF SEA

7.2.1 Brunei

The source of herbal and medicinal plants in Brunei is its natural forest, which covers almost 80% of its land area of 5,770 km^2. There is no commercial-scale cultivation in the country, which has about 5,000 native species of plants. Brunei has also been known for its superior camphor, with medicinal properties derived from the trunk of *Dryobalanops aromatica*. Traditional medicine practice is based on the ethnic makeup of Brunei, which includes Malay (66.3%), Chinese (11.2%), indigenous (3.4%), and others (19.1%). Its official religion is Muslim (66.3%), followed by 13% Buddhist, 10% Christian, and 10% others. Most of its traditional medicines are imported, such as the *Jamu* medicines from Indonesia.

7.2.2 Cambodia

Cambodia covers an area of 181,035 km^2 with a population of more than 14 million people. The ethnic makeup of Cambodia includes about 90% Khmer followed by 5% Vietnamese, 1% Chinese, and 4% of others. About 95% of Cambodians are Theravada Buddhist, with 5% of other religions. Traditional medicine has been practiced for thousands of years, and the traditional practitioners are known as *Kru Khmer* since only the rich can afford modern medicine. The research and documentation of MAP in Cambodia is very limited. Some of the important plants are listed in Appendix 7.1.

7.2.3 Democratic Republic of Timor Leste (East Timor)

The ethnic groups of Timor Leste are Austronesian (Malayo-Polynesian), Papuan, and a small Chinese minority. Their religious makeup is Roman Catholic (98%), Muslim (1%), and Protestant (1%). Their official languages are Tetum and Portuguese, with some Indonesian and English. There are about 16 indigenous languages spoken, with the main being Tetum, Galole, Mambae, and Kemak. Many locals rely on traditional medicine for their health care needs. However, there is almost negligible information on the use of MAP resources.

7.2.4 Indonesia

Indonesia consists of 17,000 islands and is the world's largest archipelago (Figure 7.1). It is separated from the peninsula of Malaysia and Singapore by the Straits of Malacca and the South China Sea. The ethnic groups of Indonesia are Javanese (40.6%), Sundanese (15%), Madurese (3.3%), Minangkabau (2.7%), Betawi (2.4%), Bugis (2.4%), Banten (2%), Banjar (1.7%), and others (29.9%). The main religions are Muslim (86.1%), Protestant (5.7%), Roman Catholic (3%), Hindu

(1.8%), and others (3.4%). The official language is Bahasa Indonesia, but other commonly used languages include English and Dutch, while the local dialect is mainly Javanese. About 60% of Indonesia is under forest cover, which represents about 10% of all tropical forests of the world. It is second largest in biodiversity after Brazil, and about 90% of medicinal plant species of Asia can be found there. About 40% of Indonesia's population uses traditional medicine, with 70% of users from rural areas (WHO, 2001). The well-known traditional medicine is called *Jamu,* and it is reported to be preferred over other "available over-the-counter" Western drugs (Stevensen, 1999).

7.2.5 Lao People's Democratic Republic

Lao PDR has tropical rain forests covering almost 47% of its total land area of 236,800 km^2. These serve as a reserve for natural resources, including MAPs. The ethnic groups of Lao PDR are Lao (55%), Khmou (11%), Hmong (8%), and over 100 minor ethnic groups (26%). Their main religion is Buddhism (67%), followed by Christianity (1.5%) and others (31.5%). The official language is Lao, with some French, English, and various ethnic languages. The traditional medicine in Lao PDR is known as ya phurn meung, which means "medicine from the foundation of the country." The origin of this medicine goes back to the 12th century and is a combination of Buddhist and Indian medical systems, resulting in traditional Laotian medicine. However, the invasion of the French in 1893 brought about the introduction of allopathic medicine in Lao PDR (WHO, 2001). There are an estimated 4,000 plant species found in Lao PDR, of which a number have important medicinal and aromatic values (Kraisintu, 2003).

7.2.6 Malaysia

Malaysia is listed as the 14th most biologically diverse country in the world (Anon., 2006) and ranks fourth in Asia, with over 15,000 flowering plants. Of these, only about 50 species are used commercially and researched scientifically for their medicinal properties, although about 2,000 plant species have been reported to have medicinal values (Soepadmo, 1999). Malaysia is rich in its biological and cultural heritage, which consists of three main communities: Malay, Indian, and Chinese. These three major ethnic groups have traditions and knowledge of herbal use for health and healing spanning generations. The Malays comprise about 50.4% of the total population, followed by Chinese (23.7%), indigenous (11%), Indians (7.1%), and others (7.8%). There are about 60.4% Muslims in Malaysia, followed by Buddhists (19.2%), Christians (9.1%), Hindus (6.3%), traditional Chinese religions (Confucianism, Taoism, 2.6%), other or unknown (1.5%), and none (0.8%). The official language is Bahasa Malaysia with other spoken languages, such as English, Chinese (Cantonese, Mandarin, Hokkien, Hakka, Hainan, Foochow), Tamil, Telugu, Malayalam, Panjabi, and Thai. In East Malaysia, the most widely spoken indigenous languages are Iban and Kadazan. The traditional medicine system in Malaysia follows its ethnic makeup and includes ayurveda, siddha, traditional Chinese, traditional Malay, and unani. The indigenous Malay medicinal system has been influenced by those from Arabia, India, Java, and aboriginal races. Based on a survey by

Aziz (2004), it was observed that about 19% of physicians recommended some form of herbal remedies for their patients. At present, the Malaysian market for herbal and natural products has been estimated to be worth more than US$1.5 billion, of which 90% of the raw materials are imported. However, the use of MAP products in veterinary health care in Malaysia has been limited.

7.2.7 Myanmar

Myanmar, the largest country in SEA, has an area of 676,553 km^2 with about 43% of its total area under forest. The Myanmar traditional medicine system is generally based on Buddhist philosophy and ayurveda. Allopathic medicine was introduced during the colonial period (1885), but its shortage during World War II drove the people toward traditional medicine (WHO, 2001). The ethnic groups of Myanmar include Burman (68%), Shan (9%), Karen (7%), Rakhine (4%), Chinese (3%), Indian (2%), Mon (2%), and others (5%). The Buddhists in Myanmar comprise about 89%, followed by Christians (4%), Muslims (4%), animist (1%), and others (2%). About 7,000 plant species have been recorded in Myanmar, of which 1,071 species are endemic (Food and Agriculture Organization [FAO], 2002). The number of traditional medicine practitioners registered in Myanmar exceeds 8,000 (WHO, 2001).

7.2.8 The Philippines

The Philippines is made up of 7,100 islands, with forest covering only about 20% of the total land area. The ethnic groups of the Philippines are Tagalog (28.1%), Cebuano (13.1%), Ilocano (9%), Bisaya/Binisaya (7.6%), Hiligaynon Ilonggo (7.5%), Bikol (6%), Waray (3.4%), and others (25.3%). The main religious groups are Roman Catholics (80.9%), Muslims (5%), Evangelicals (2.8%), Iglesia ni Kristos (2.3%), Aglipayans (2%), other Christians 4.5%, others 2.4%, and none (0.1%). The official languages are Tagalog (which has eight major dialects) and English. The traditional medicine practitioners in the Philippines are known as arbolario and have common roots with other traditional healing methods in SEA countries, including Myanmar, Cambodia, and Indonesia (Apostol, 2003). There are about 13,500 plant species in the Philippines, of which 1,500 have medicinal value (Tan, 2003). Herbal medicines in the Philippines are regulated as over-the-counter medicines, while medical claims have to be scientifically supported (WHO, 2005). The use of traditional medicine is more common in rural areas despite the availability of allopathic medicine, with about 80% of the rural population utilizing herbal remedies (FAO, 2002).

7.2.9 Singapore

The ethnic breakdown of Singapore has hardly changed since the 1980s and now stands at 76% Chinese, 15% Malays, 8% Indians, and 1.4% others. The 400,000 people who make up the Malay community today have roots throughout the Malayan archipelago (especially Bali, Java, Sulawesi, Bawean, and Malaysia), with strong cultural links (Tuschinsky, 1995). The religious makeup is Buddhist (42.5%), Muslim (14.9%), Taoist (8.5%), Hindu (4%), Catholic (4.8%), other Christian (9.8%), other (0.7%), and none (14.8%). The major languages are Mandarin (35%), English (23%), Malay (14.1%), Hokkien (11.4%), Cantonese (5.7%), Teochew (4.9%), Tamil (3.2%), other Chinese dialects (1.8%), and others (0.9%). About 45% of Singaporeans have consulted traditional medicine practitioners at one time (WHO, 2001), while *Jamu*, the traditional Indonesian medicine, is widely practiced, especially among the Malay community (Tuschinsky, 1995).

7.2.10 Thailand

Thailand covers an area of 514,000 km^2, of which less than 33% is forest cover. The ethnic groups of Thailand include Thai (75%), Chinese (14%), and others (11%). The major religions are Buddhism (94.6%), Muslim (4.6%), Christianity (0.7%), and others (0.1%). The official language is Thai with some English and ethnic regional dialects. Thailand's traditional medicine is famous for its herbal saunas, steam baths, hot compresses, traditional massages, acupressure, and reflexology, and traditional medicine practitioners are an important part of country's health care system (WHO, 2001). Thailand has about 11,000 species of vascular plants, of which 10,200 are flowering plants. The *Thai Traditional Materia Medica* lists about 1,400 plants with medicinal and aromatic properties (Handa, Rakesh, and Vasisht, 2006).

7.2.11 Vietnam

Vietnam covers an area of 330,000 km^2, of which 60% is forest. The ethnic groups of Vietnam are Kinh or Viet (86.2%), Tay (1.9%), Thai (1.7%), Muong (1.5%), Khome (1.4%), Hoa (1.1%), Nun (1.1%), Hmong (1%), and about 4.1% others. The majority of the Vietnamese (80.8%) do not have any religious beliefs, with some Buddhists (9.3%), Catholics (6.7%), Hoa Haos (1.5%), Cao Dais (1.1%), Protestants (0.5%), and Muslims (0.1%). The official language is Vietnamese, followed by English (a favored second language), French, Chinese, Khmer, and mountain area languages (Mon-Khmer and Malayo-Polynesian). Vietnamese medicine is a merger of Chinese, Vietnamese, and Western medical systems (Ladinsky, Volk, and Robinson, 1987). Western medicine was introduced when the French arrived in Vietnam in the 19th century. Vietnam is the 16th most biologically diverse country in the world, with some 12,000 plant species, of which 40% of the total flora has been identified as endemic (An, 2000).

7.3 CURRENT RESEARCH AND APPLICATION OF MEDICINAL AND AROMATIC PLANTS

Although traditional or complementary medicine has been practiced for a long time, it was only recently that the efficacy and safety of a small number of MAPs have been subjected to scientific investigations. Most MAPs investigated are for human medicine, and very limited ethnoveterinary studies have been done. Based on their widespread traditional use, the plants shown in Table 7.1 are cultivated commercially both for local and export markets. The relative importance of these cultivated plants differs among the different SEA countries.

TABLE 7.1
Major Commercially Cultivated Medicinal and Aromatic Plants (for Both Humans and Animals)

Family	Botanic Name	Country of Cultivation
Acanthaceae	*Adhatoda vasica*	Vietnam
	Andrographis paniculata	Thailand, Vietnam
Apiaceae	*Ammi majus*	Vietnam
	Angelica acutiloba	Vietnam
Apocynaceae	*Catharanthus roseus*	Philippines, Vietnam
	Rauvolfia serpentina	Vietnam
Araliaceae	*Panax pseudoginseng*	Vietnam
	Panax vietnamensi	Vietnam
Arecaceae	*Areca catechu*	Thailand, Vietnam
Asphodelaceae	*Aloe barbadensis*	Thailand
Asteraceae	*Artemisia annua*	Thailand, Vietnam
	Chrysanthemum morifolium	Thailand, Vietnam
Convallariaceae	*Ophiopogon japonicus*	Vietnam
Fabaceae	*Cassia angustifolia*	Thailand, Vietnam
Lamiaceae	*Mentha arvensis*	Thailand, Vietnam
	Vitex negundo	Philippines
Lauraceae	*Cinnamomum camphora*	Thailand, Vietnam
Malvaceae	*Hibiscus sabdariffa*	Thailand, Vietnam
Menispermaceae	*Tinospora crispa*	Philippines
Myrtaceae	*Syzygium aromaticum*	Indonesia, Malaysia
Paeoniaceae	*Paeonia lactiflora* Pall.	Vietnam
Piperaceae	*Piper betle*	Vietnam
	Piper nigrum	Indonesia, Malaysia, Thailand, Vietnam
	Piper retrofractum	Indonesia, Thailand
Poaceae	*Cymbopogon winterianus*	Indonesia, Thailand
Rubiaceae	*Morinda officinalis*	Vietnam
Solanaceae	*Atropa belladonna*	Vietnam
Zingiberaceae	*Curcuma domestica*	Indonesia, Thailand, Vietnam
	Kaempferia galanga	Indonesia, Vietnam
	Zingiber officinale	Indonesia, Thailand

Source: Adapted from Handa, Rakesh, and Vasisht, *Compendium of medicinal and aromatic plants: Asia* ICS-UNIDO, 2006.

In Malaysia, for instance, 10 major plant species were identified for the evaluation of their efficacy, safety, and cultivation and propagation methodologies. These 10 plants are *Andrographis paniculata* (Hempedu bumi), *Centella asiatica* (Pegaga), *Citrus hystrix* (Limau purut), *Eurycoma longifolia* (Tongkat Ali), *Ficus deltoidea* (Emas cotek), *Labisia pumila* (Kacip Fatimah), *Morinda citrifolia* (Mengkudu or Noni), *Orthosiphon stamineus* (Misai kucing), *Phyllanthus niruri* (Dukung anak), and *Zingiber officinale* (Halia). Although there are a number of published reports

TABLE 7.2
Herbal Medicines Used for Different Disorders in Animals

Disorders/Use	Herbs Used
Degenerative joint disease	*Picrorrhiza kurroa, Zingiber officinale, Allium sativum*
Liver tonic and performance enhancer	*Andrographis paniculata, Eclipta alba, Phyllanthus niruri*
Ectoparasities: antidandruff, antilice	*Cedrus deodara, Azadiracta indica*
Respiratory conditions and allergy	*Adhatoda vasica, Solanum xanthocarpum, Glycyrrhiza glabra, Albezia lebbek, Alpinia galanga, Piper longum, Hedychium spicatum*
Antidiarrheal and gut function stabilizer	*Aegle marmelos, Holorrhoena antidysentrica, Punica granatum, Woodfordia fruticosa, Areca catechu, Symplocos racemosa, Cyperus rotundus, Salmalia malabarica, Zingiber officinale, Berberis aristata, Tinospora cordifolia*
Renal protection, kidney tonic, and diuretic	*Tinospora cordifolia, Boerhaavia diffusa, Tribulus terrestris, Bergenia ligulata, Crataeva nurvala, Andrographis paniculata, Solanum nigrum, Eclipta alba, Terminalia arjuna*
Immune modulator	*Ocimum sanctum, Withania somnifera, Tinospora cordifolia*

Source: Adapted from Handa, Rakesh, and Vasisht, *Compendium of medicinal and aromatic plants: Asia* ICS-UNIDO, 2006.

regarding the safety, efficacy, and phytochemical properties of these plants, a detailed study of these species has been initiated as the phytochemical properties are known to vary between subspecies and geographical locations. For instance, there are at least four different local subspecies of *Phyllanthus,* which vary in phytochemical profiles. Most of the studies on MAPs have been geared toward treatment of human diseases (e.g., bacterial and viral infections, diabetes mellitus, cancer, and hypertension). Besides these studies, other investigations are ongoing to identify herbs commonly used by traditional communities. This is being done for the preservation of local knowledge and identification of new resources (Andersen, Nilsson, and De Richelieu, 2003). The common MAPs traditionally used to treat various animal ailments in SEA are listed in Table 7.2. Some of the chemical constituents and use of these MAPs that have ethnoveterinary applications are described briefly next.

7.3.1 *Andrographis paniculata* (Hempedu bumi)

Phytochemistry: Several diterpenoids and diterpene glycosides have been isolated from the aerial parts. Its bitter taste was attributed to its andrographolide constituent (Yanfang et al., 2006). The leaves were observed to contain flavone derivatives such as oroxylin and wogonin.

Biological studies: Diterpene andrographolide has been shown to have protective effects on hepatotoxicity induced in mice by carbon tetrachloride. *Andrographis paniculata* has been traditionally known to have an antifertility effect and has been demonstrated to have these effects in both male and female mice. It has also been used traditionally for diabetes mellitus, but it has been demonstrated in experimental studies that these effects were only apparent in the chloroform extracts of *A. paniculata*. Leaf extracts were found to have *in vitro* activity against *Heterometrus laoticus* scorpion venom with low cytotoxicity (Uawonggul et al., 2006).

Veterinary uses: Uses include reduction of flatulence, abdominal pain, and cough and as a scorpion venom antidote in Thailand (Uawonggul et al., 2006).

7.3.2 *Centella asiatica* (Pegaga)

Phytochemistry: The main components are triterpenoid saponins, including asiaticoside, centelloside, madecassoside, and asiatic acid. It also contains other components, including volatile oils, flavonoids, tannins, phytosterols, amino acids, and sugars (Anon., 2008a; Ling, 2004).

Biological studies: Traditionally, this is used for wound-healing properties with its ability to relieve skin conditions such as lupus, varicose ulcers, eczema, psoriasis, and wounds. Asiaticoside, the main active constituent, exhibited significant wound-healing activity in normal as well as delayed-healing models (Shukla et al., 1999). It has also been used for the treatment of diarrhea, fever, amenorrhea, and genitourinary tract diseases. *Centella asiatica* has also been demonstrated to have cardioprotective as well as cognitive-enhancing effects.

Veterinary uses: It is used as a health tonic.

7.3.3 *Citrus hystrix* (Limau purut)

Phytochemistry: The leaves of *Citrus hystrix* have been reported to contain up to 400 mg/kg of alpha-tocopherol besides glyceroglycolipids, coumarins, flavonoids (principally flavanone), and glycosides (e.g., narirutin, naringin, hesperidin, and neohesperidin), and flavone aglycons such as nobiletin and tangeretin. The oil of *C. hystrix* contains citronellal, geranial, and d-limonene (Lawrence et al., 1971).

Biological studies: The leaves of *C. hystrix* have traditionally been used for the relief of flu, fever, hypertension, abdominal pains, and diarrhea in children. The extracts of *C. hystrix* have demonstrated anti-tumor-promoting activity, antiviral activity against herpes simplex virus 1 (HSV-1) but not poliovirus and antibacterial activity, especially against gram-positive bacteria. *Citrus hystrix* extracts have also shown anti-inflammatory effects, probably due to its d-limonene content. Antifertility activities such as the inhibition of embryo implantation and abortion have also been reported with *C. hystrix* extracts. The volatile oils of *C. hystrix* have demonstrated mosquito-repelling potential.

Veterinary uses: Although not specific to this species, other *Citrus* spp. such as *C. junos* have been used against poultry viral respiratory infections (Kim, Jeon, and Ko, 2001).

7.3.4 *Eurycoma longifolia* (Tongkat Ali)

Phytochemistry: Eurycoma longifolia has been studied extensively and consists of a number of phytochemical components, such as terpenoids, steroids, benzoquinones, coumarins, and alkaloids.

Biological studies: The reported aphrodisiac effect of *E. longifolia* has made it much sought after in Asia. Extracts of *E. longifolia* were observed to promote the growth of rat prostate and seminal vesicles (Ang and Sim, 1998). The extracts of *E. longifolia* have shown dose-dependent antimalarial activities against six Malaysian chloroquine-resistant parasites of *Plasmodium falciparum*. In fact, it was observed that the eurycomanone constituents of *E. longifolia* were relatively more potent than chloroquine against a multidrug-resistant *P. falciparum in vitro*. Extracts of *E. longifolia* have also shown anticarcinogenic activities (Jiwajinda et al., 2002).

Veterinary uses: Although mostly used in humans, recent work (unpublished) has been initiated to evaluate its effect on the fertility of production animals.

7.3.5 *Morinda citrifolia* (Mengkudu or Noni)

Phytochemistry: Major components identified in *Morinda citrifolia* include coumarins, terpenoids, alkaloids, anthraquinones, and flavonoids (Wang et al., 2002).

Biological studies: The fruit juice of *M. citrifolia* has been popularized for the treatment of different kinds of illnesses, such as arthritis, diabetes, and cardiovascular disease, among many others (Chan-Blanco et al., 2006). It also has been reported to be used in the treatment of colds and influenza, broken bones, bruises, sores, and wounds.

Veterinary uses: The alcoholic extracts of another Morinda Species, namely Morinda lucida, exhibited a dose-dependent inhibition of Haemonchus (Hounzangbe-Adote et al., 2005).

7.3.6 *Orthosiphon stamineus* (Misai kucing)

Phytochemistry: Orthosiphon stamineus has been reported to possess flavones, diterpenes, and triterpenes (Olah et al., 2003)

Biological studies: Orthosiphon stamineus has been used for the treatment of diabetes, rheumatism, and catarrh of the bladder. It has been shown to have diuretic, antiallergic, anti-inflammatory, antihypertensive, and antitumor properties (Matsubara et al., 1999; Nirnoy and Muangman, 1991).

Veterinary uses: Although not widely used in ethnoveterinary medicine, there were indications of interactions with conventional diuretics.

7.3.7 *Phyllanthus niruri* (Dukung anak)

Phytochemistry: The types of phytochemicals reported in *Phyllanthus niruri* include triterpenoids, steroids, alkaloids, flavonoids, and tannins (Khatoon et al., 2006).

Biological studies: Phyllanthus niruri has been traditionally used to treat jaundice, gonorrhea, frequent menstruation, dysentery, diabetes, skin ulcers, sores, swelling, and itchiness. Current research is focused on its potential antiviral effects, especially against hepatitis B virus. It has also been reported to have antitumor and antiparasitic effects, especially against malaria parasite.

Veterinary uses: It is used as a liver tonic addressing conditions of the liver such as jaundice and viral infections. It is also used for the treatment of chronic diarrhea and infections of the genitourinary tract.

7.3.8 *Zingiber officinale* (Halia)

Phytochemistry: The extracts of *Z. officinale* have been reported to contain monoterpenes and sesquiterpenes (Ali et al., 2008).

Biological studies: Zingiber officinale has been traditionally used as a digestive aid. It has also been reported to have other beneficial properties, such as cardiotonic, antilipidemic, antinausea/antiemetic, carminative, antiulcer, anti-inflammatory, hypoglycemic, antiplatelet, antiviral, antibacterial, antifungal, antineoplastic, and antioxidant properties. It has also been used in the treatment of baldness, toothache, snakebite, heart disease, hypercholesteremia, and arthritis.

Veterinary uses: It is used as an anti-inflammatory, antipyretic, antioxidant, nematodicidal, insect repellent, and antifilarial agent.

7.4 USE OF MEDICINAL AND AROMATIC PLANTS IN ANIMALS

Most of the application of ethnoveterinary medicine is in the rural areas where access to modern medicine is scarce or costly. A number of plants traditionally used for animal treatment, their preparation, and application are shown in Table 7.3.

TABLE 7.3
Common Plants Used for Ethnoveterinary Applications in SEA

Plant Species	Parts Used	Indications	Preparation	Dosages
Allium sativum	Cloves	Poisoning	Burn 3 heads of garlic, pulverize, and mix with 1 glass of water.	*Adults:* Drench with 1 cup of the mixture. *Young:* Drench with 0.5 cup of the mixture. Repeat the procedure if animal does not vomit.
Areca catechu	Fresh nut	Intestinal worms	Pound and mix with enough water to facilitate drenching.	*Chickens:* One nut as big as a peanut *Cattle and water buffalo:* 8–10 nuts *Goats and pigs*: 3 nuts Repeat the procedure after 2 weeks.
Blumea balsamifera	Dried seeds	Fluke infestation	Pound air-dried ripe seeds and add a little water.	Drench 1 time/day for 6 days. 1 kg seeds for large ruminants, 0.5 kg for small ruminants.
Carica papaya	Fresh leaves	Cold, cough, fever	Boil leaves for 15–20 minutes in 1 L of water.	Drench. 0.5 L twice a day for 1–3 days.
Chrysophyllum cainito	Fresh leaves	Diarrhea	Boil 0.5 kg of leaves in 3 glasses of water.	*Adult cattle and buffalo:* Drench: 1–2 cups three times/day for 1–3 days. *Calves, sheep, and goats:* Drench: Half the amount given above.
Gliricidia septum	Fresh leaves	External parasites	Pound the leaves (the number of leaves will depend on the severity of the infestation).	*Swine and ruminants:* Apply the juice of leaves on the affected area. Repeat the procedure two or three times/day until the parasites are eliminated.
		Warts	Pound the leaves (the number of leaves will depend on the severity of the infestation).	Apply the juice and massage the affected part two or three times/day until the warts disappear.
Cocos nucifera	Water	Dehydration	Mix water from 3–5 young coconuts with a cup of brown sugar and a little salt.	*Adult cattle and buffalo*: Drench with 2–3 L. *Calves, sheep, goats*: Drench with 1–1.5 L three times/day until animal recovers.

TABLE 7.3 (CONTINUED)
Common Plants Used for Ethnoveterinary Applications in SEA

Plant Species	Parts Used	Indications	Preparation	Dosages
	Charcoal (from shell)	Diarrhea	In feed: Pound and mix with feed. In water: Grind coconut shell charcoal to a powder and mix with water.	*In feed:* Divide the mixture into several rations and feed the animal one ration at a time until it recovers. As drench: three times/day. *Adult cattle and buffalo*: 1 cup of powdered coconut shell mixed in 500 mL of water. *Calves, goats, and sheep*: Half of the above amount.
	Powder (from the external part of the shell)	External wound or cuts	Scrape external part of the shell to collect powder (amount will depend on the size of the wound).	Apply on clean wound two or three times/day for 3 days.
Momordica charantia	Fresh leaves	Intestinal worms	Pound 0.5 kg of leaves. Extract the juice of the leaves and divide into small doses.	Drench: 1 dose of the juice one to three times/day. Repeat the procedure after 2 weeks.
		Anemia	Pound a handful of leaves.	*Piglets*: Give 5 drops (1 cc) of juice per piglet twice/day for 3–4 days. The juice extracted should be enough for 7–8 piglets
Moringa oleifera	Fresh leaves	Anemia	Pound 0.5 to 1 kg of leaves and extract the juice.	*Piglets*: Give 5 drops (1 cc) of juice per piglet as drench two to three times/day for 3–4 days.
Musa sapientum	Fresh leaves	Bloat	Chop 2 leaves for calves, sheep, and goats or 3–5 leaves for adult cattle and buffalo.	Feed to animal twice a day until it recovers. Only works at the early stage of bloat.
		Bleeding wound	Steam 1 leaf over boiling water.	*Ruminants and pigs*: Apply the leaf on the wound two to three times/day until the wound heals.
Premna odorata	Fresh leaves	Wound with maggots	Pound 3–5 leaves and extract the juice	Apply the juice on the wound two or three times/day for 3–4 days.

(*continued*)

TABLE 7.3 (CONTINUED)
Common Plants Used for Ethnoveterinary Applications in SEA

Plant Species	Parts Used	Indications	Preparation	Dosages
		Fever, colds, cough	Boil 8–15 leaves in 2–3 glasses of water.	Drench: 0.5–1 cup two or three times/day for 3 days.
		Mite infestation	A bunch of fresh leaves.	*Poultry*: Hang leaves in roosts/house until the leaves are fully dried.
Psidium guajava	Fresh leaves	Diarrhea	Boil 0.5 kg of leaves in 3 glasses of water.	*Adult cattle and buffalo*: Drench twice/day for 3–4 days. *Calves, sheep, and goats*: Half of the amount given above.
		Severe wounds	Pound 5–10 leaves.	*Ruminants, pigs, and poultry*: Apply juice on clean wound two or three times/day for 3–4 days.
Tamarindus indica	Fresh leaves	Fever, colds, cough	Boil 1 kg of leaves in 1 gallon of water. Divide into small doses.	*Adult ruminants*: Drench the animal with 1 dose two or three times/day until the animal recovers. *Young ruminants*: One-third to one-half the dosage given above.
Tinosphora rumpii	Fresh stems	Intestinal parasites	Soak 0.5 kg of pounded stems in 1 L of water.	*Adult cattle and buffalo*: Drench with 1 L. Repeat after 2 weeks. *Calves, sheep, and goats*: Half of the dosage given above.
Vitex negundo	Fresh leaves	Fever, colds, cough	Boil 0.5 kg of leaves for 15–20 min in 2 L of water.	*Adult ruminants*: Drench two or three times/day for 3–4 days. *Young ruminants*: One-third to one-half of the dosage given above.
		Mite infestation	A bunch of fresh leaves.	*Poultry*: Hang leaves in roosts/house until the leaves are fully dried.

Source: International Institute of Rural Reconstruction (1994).

Due to the widespread use of MAPs in traditional medicine, most countries of SEA have some regulatory measures to control their commercial trade and use. Informal use is not, and cannot be, regulated. For instance, the Philippines enacted the Traditional and Alternative Medicine Act in 1997, while in Indonesia (WHO, 2001) it is regulated by Health Law Act 23 (1992). This regulatory issue has become more important in view of a number of toxicities reported following the use of herbal products. For instance, Ang, Lee, and Kiyoshi (2004) observed that about 15% of the

TABLE 7.4
Agricultural Product Exports of Major SEA Countries (Million US$)

Country	1991	1998	1999	2000	2001	Growth Rate/Year
Cambodia	35.20	27.20	44.20	26.60	23.10	3.80%
Indonesia	3,122.50	5,054.30	5,135.70	4,946.40	4,368.40	4.10%
Lao PDR	26.80	28.00	33.80	34.30	29.30	–1.90%
Malaysia	4,421.80	7,756.40	7,117.50	5,821.00	5,515.50	2.80%
Myanmar	220.40	300.20	267.30	364.80	453.70	4.80%
Philippines	1,260.90	1,717.90	1,358.40	1,539.80	1,443.60	1.30%
Thailand	5,880.50	7,097.40	7,158.60	7,275.30	7,422.50	1.80%
Vietnam	617.40	2,373.10	2,428.40	2,191.80	2,036.40	14.20%

Source: Food and Agriculture Organization (2003).

products they evaluated contained high levels of lead. Despite these observations, herbal medicine can be regarded as generally safe.

In general, the use of pharmaceutical plant products (botanicals or phytomedicines) in veterinary health care in SEA is limited. However, recent years have seen an increase in the use of such products in veterinary practice, especially in production animals. An example is Utrifit®, an herbal product used by cattle farmers to tone the uterus and improve breeding efficiency. This product is a combination of a number of herbs, such as *Citrullus colocynthis, Piper longum, Piper nigrum,* and *Zingiber officinale.*

However, most such products are at the moment imported. There are intensified research efforts to evaluate indigenous medicinal plants for use especially in production animals in view of the magnitude of the agricultural and livestock industries in SEA (Tables 7.4 and 7.5, respectively) and the anticipated expansion.

TABLE 7.5
Livestock Production Indices of Major SEA Countries

Country	1992	1994	1996	1998	2000	2002
Cambodia	120.3	124.2	134.8	147.7	169.7	156.7
Indonesia	110.3	132.6	134.4	116.3	121	128.4
Lao PDR	111.6	133.3	140	151.4	174.8	196.1
Malaysia	125.9	145.4	149.1	152.9	136.7	150.7
Myanmar	103.6	107.7	121.3	137.7	157.6	181.7
Philippines	102.9	117.7	137	154	162.9	197.7
Thailand	121.7	126.7	132.1	132.3	130	139.9
Vietnam	111.9	125.2	137.8	159.1	183.4	210.6

Net livestock products increase, assume base year 1989–1991 = 100
Source: Food and Agriculture Organization (2003).

Furthermore, the total ban by the European Union on the use of antibiotic feed additives (AFAs) and growth promoters in January 2006 has resulted in the search for plant alternatives. This is basically due to the fact that a number of plants have potent antimicrobial components. Wiart et al. (2004) screened 72 methanol extracts from the leaves, bark, and roots of 50 plant species used in traditional medicine for their putative antibacterial and antifungal properties. They observed that 10 species—*Peristrophe tinctoria, Polyalthia lateriflora, Knema malayana, Solanum torvum, Celosia argentea, Eclipta prostrata, Ancistrocladus tectorius, Dillenia suffruticosa, Piper stylosum,* and *Rafflesia hasseltii*—displayed broad antimicrobial activities. At present, the detailed phytochemical constituents of most plants have not been fully ascertained, but these plants were selected following their ethnomedical uses (Wiart et al., 2004).

7.5 SCIENTIFIC STUDIES ON THE USE OF PLANT EXTRACTS IN ANIMALS

Besides their antimicrobial properties, MAPs are being evaluated for their anthelmintic effects in ruminants. Some of the studies conducted and the research approach adopted as well as associated issues concerning the use of MAPs in livestock production are discussed next.

As far as ruminants are concerned, parasitic infections remain a constant problem in ruminant production, especially in goats and sheep. A survey (Chandrawathani, 2004) of the prevalence of anthelmintic resistance in Peninsula Malaysia revealed that about 75% of the goat farms were infected with nematode parasites. The study also observed the presence of resistance of nematodes to benzimidazoles, levamisole, and ivermectin. The urgent need for new approaches to worm control led to the evaluation of MAPs for nematode control. For instance, the feeding of neem leaves (*Azadirachta indica*) to sheep was observed to reduce *Haemonchus contortus* fecal egg counts and larval recoveries after 5 weeks (Chandrawathani et al., 2002). However, other studies failed to show the effectiveness of neem against nematodes. Another study (Shanmugavelu, Marugaiyah and Rohana, 2007) was conducted to evaluate the effects of feeding *Melia excelsa* (same family as neem) leaves in controlling helminths in goats. The *in vitro* study revealed high efficacy, but the *in vivo* investigation failed to show significant reduction in fecal egg counts (Figure 7.2). Nevertheless, the coefficient of variation (CV) in fecal egg counts was reduced in the animals fed *Melia excelsa* leaves. Research efforts are ongoing to identify new MAPs for this application.

Other studies on the use of MAPs in nonruminant production animals, namely poultry and swine, have been targeted toward their use as feed additives in an attempt to substitute AFAs. Most studies investigated the effects of plant extracts on the gut microflora; however, in one study, Shanmugavelu, Acamovic, and Cowieson (2004) evaluated the effects of thyme oil (*Thymus vulgaris*) and garlic (*Allium sativum*) powder on the digestibility of nutrients in the chicken. This was done because any improvement to digestibility of feed would indirectly translate to increased utilization by the animal and less by the gut microbes (Shanmugavelu et al., 2006). The authors observed that thyme oil and garlic did not improve the nutritional value of

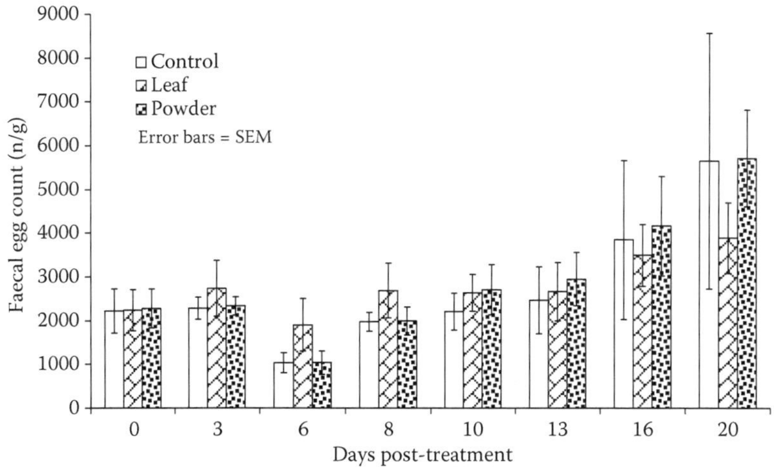

FIGURE 7.2 Effect of feeding *Melia excelsa* on the total fecal egg counts.

soybean meal (Table 7.6). The differences in digestibility parameters between thyme oil and garlic powder could have been due to their different antimicrobial effects.

The authors further evaluated thyme oil and garlic powder for *in vitro* antimicrobial effects on chicken gut microbes (Shanmugavelu, Brooker, et al., 2004). Total gut microbiota of the chicken rather than selected microbes were used in the assay. The interesting observation in the study was the potent antimicrobial effects of thyme oil, which virtually reduced the fermentative activity (an indication of microbial viability) of most microbes after 6 h of incubation *in vitro*. Nevertheless, fermentative

TABLE 7.6
Effect of Addition of Thyme Oil and Garlic Powder on the Nutritive Value of Soybean Meal

	Treatment Groups			
Parameters	Soybean Meal (SBM)	Soybean Meal Plus Thyme Oil	Soybean Meal Plus Garlic Powder	*p*
ADMD	0.43	0.47	0.43	.081
TDMD	0.58	0.62	0.57	.076
AME (MJ/kg)	10.89[a,b]	11.76[a]	10.79[b]	.028
TME (MJ/kg)	13.20[a,b]	14.08[a]	13.10[b]	.026
AMEn (MJ/kg)	10.92[a,b]	11.64[a]	10.82[b]	.039
TMEn (MJ/kg)	12.32[a,b]	13.04[a]	12.22[b]	.037

ADMD = apparent dry matter digestibility coefficient; TDME = true dry matter digestibility coeffcient; AME = apparent metabolizable energy; TME = true metabolizable energy; AMEn and TMEn are AME and TME corrected for nitrogen, respectively.

[a,b] Different superscripts between columns differ significantly ($p < .05$).

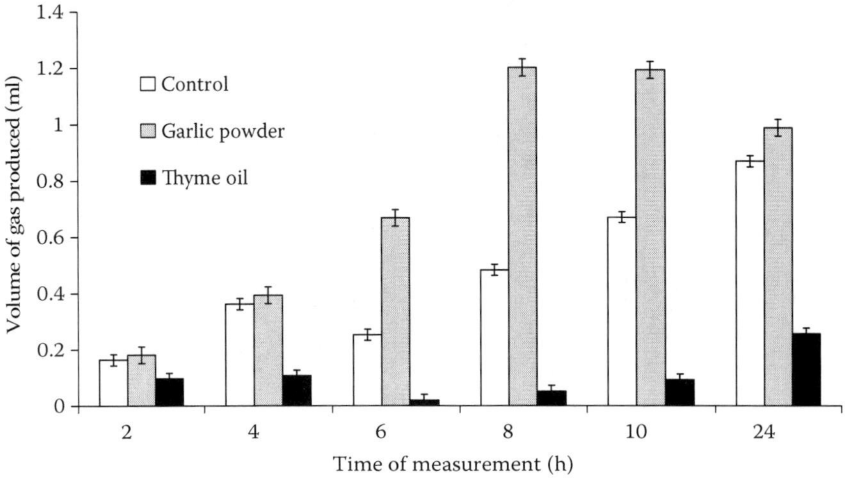

FIGURE 7.3 Effect of thyme oil and garlic powder on the fermentative characteristics of chicken duodenum and jejunum gut microbes.

activity resumed after 24 hours (Figure 7.3). This observation led to the evaluation of this herbal extract in an *in vivo* study in which thyme oil was fed at two concentrations (1 and 10 g/kg) compared with the control in a diet based on maize-soybean meal (Shanmugavelu et al., 2005). The potent antimicrobial activity demonstrated *in vitro* failed to improve performance similar to those fed AFAs (Table 7.7).

TABLE 7.7
Performance of Birds Fed Maize-/Soybean-Based Diets Supplemented with Enzyme and Thyme Oil

	Treatment Groups						
	Control		Thyme Oil				
			1 g/kg		10 g/kg		
Parameters	**Mean**	**SEM**	**Mean**	**SEM**	**Mean**	**SEM**	***p***
Body weight (g, 21 days)	731.60[a]	15.503	725.50[a]	15.899	585.82[b]	16.032	.0001
Average daily feed intake (g)	52.43[a]	0.982	52.88[a]	1.191	45.49[b]	0.788	.0001
Average daily gain (g)	36.71[a]	0.756	36.41[a]	0.774	28.64[b]	0.800	.0001
Feed conversion efficiency	0.70[a]	0.005	0.69[a]	0.004	0.63[b]	0.012	.0001
Coefficient of ADMD (19–21 days)	0.75	0.004	0.75	0.005	0.73	0.006	.0582
AME (MJ/kg, 5–7 days)	14.10[ab]	0.164	13.88[b]	0.129	14.40[a]	0.246	.0226

ADMD = apparent dry matter digestibility; AME = apparent metabolizable energy; SEM = standard error of mean.

[ab] Different superscripts between columns differ significantly ($p < .05$).

Further analysis of the microbial community was done using percent-guanine-plus cytosine (%G+C) analysis according to the methods of Apajalahti et al. (1998). It was observed that there was a shift toward an increase of microbes within the 34–37%G+C, which happens to be the range within which the bacterium *Clostridium* falls (data not shown here). Although it could not be concluded that the predominant bacterium was *Clostridium*, this study revealed the complications and various aspects that have to be considered when developing an MAP for use as a dietary supplement.

7.6 FUTURE POTENTIAL FOR THE DEVELOPMENT OF MAPs IN VETERINARY HEALTH CARE

The development of MAPs in veterinary health care is relatively new in SEA. However, the existing framework for its application in humans can be utilized for the research and development of MAPs for animals. Research in most SEA countries has been targeted toward the domestication of popular medicinal plants (generally regarded as safe) to ensure better yield and improved quality of raw materials. Similarly, chemical profiling and toxicology studies are also being carried out by a number of research agencies and universities. However, without extensive scientific research to characterize and evaluate the efficacy of these MAPs, they will remain traditional and be sidelined by the mainstream allopathic medicine.

7.7 CONCLUSIONS

Southeast Asia has the right environment for the development of MAPs for use for both human and animal application by virtue of its rich biological and cultural diversity, available infrastructure, and scientific expertise. The current interest and support by governments of the region and additional collaborative ventures with other organizations outside SEA may help realize these efforts to develop products for use in both humans and animals.

REFERENCES

Ali, B.H., Blunden, G., Tanira, M.O., and Nemmar, A. (2008). Some phytochemical, pharmacological and toxicological properties of ginger (*Zingiber officinale* Roscoe): a review of recent research. *Food and Chemical Toxicology* 46 (2): 409–420.

An, L.Q. (2000). Genetic resources and traditional knowledge use and commercialization in Viet Nam: the case of the crop plants and medicinal plants. In: *UNCTAD Expert Meeting on Systems and National Experiences for Protecting Traditional Knowledge, Innovations and Practices*, Geneva, October 30–November 2000.

Andersen, J., Nilsson, C., and De Richelieu, T. (2003). Local use of forest products in Kuyongon, Sabah, Malaysia. *ASEAN Review of Biodiversity and Environmental Conservation (ARBEC)*, January–March: 1–18.

Ang, H.H., Lee, K.L., and Kiyoshi, M. (2004). Determination of lead in *Smilax myosotiflora* herbal preparations obtained in Malaysia. *International Journal of Environmental Health Research*, 14: 261–272.

Ang, H.H., and Sim, M.K. (1998). *Eurycoma longifolia* increases sexual motivation in sexually naive male rats. *Archives of Pharmacal Research*, 21: 779–81.

Anon. (2004). *Known and threatened species*. United Nations Environment Programme World Conservation Monitoring Centre (UNEP-WCMC), International Union for Conservation of Nature and Natural Resources (IUCN), Stevenage, Hertforshire, UK.

Anon. (2006). *Third ASEAN state of the environment report*. UNEP, ASEAN Secretariat, Malaysia, November. ISBN 978-3496-39-8, printed in Malaysia.

Anon. (2008a). Herbal collections. Malaysian Agricultural Research and Development Centre (MARDI). http://www.mardi.my/bdc/herba/english/GotuKola.html (accessed September 19, 2008).

Apajalahti, J.H., Sarkilahti, L.K., Maki, B.R., Heikkinen, J.P., Nurminen, P.H., and Holben, W.E. (1998). Effective recovery of bacterial DNA and percent-guanine-plus-cytosine-based analysis of community structure in the gastrointestinal tract of broiler chickens. *Applied and Environmental Microbiology*, 64: 4084.

Apostol, V.J.M. (2003). The healing arts of the Philippines. http://asiapacificuniverse.com/asia_pacific_features3.htm (accessed September 19, 2008).

Aziz, Z. (2004). Herbal medicines: predictors of recommendation by physicians. *Journal of Clinical Pharmacy and Therapeutics*, 29: 241–246.

Chan-Blanco, Y., Vaillant, F., Mercedes Perez, A., Reynes, M., Brillouet, J.M., and Brat, P. (2006). The noni fruit (*Morinda citrifolia* L.): a review of agricultural research, nutritional and therapeutic properties. *Journal of Food Composition and Analysis*, 19: 645–654.

Chandrawathani, P. (2004). Problems in the control of nematode parasites of small ruminants in Malaysia: resistance to anthelmintics and the biological control alternative. Doctoral thesis, Swedish University of Agricultural Sciences, Uppsala, Sweden.

Chandrawathani, P., Brelin, D., Nor Fasihah, A.S., Adnan, M., Jamnah, O., Sani, R.A., Hoglund, J., and Waller, P.J. (2002). Evaluation of the Neem tree (*Azadirachta indica*) as an herbal anthelmintic for nematode parasite control in small ruminants in Malaysia. *Tropical Biomedicine 19*, 41–48.

Department of Veterinary Services (DVS). (2004). *Livestock statistics*. Department of Veterinary Services, Putrajaya, Malaysia.

Erdelen, W.R., Adimihardja, K., Moesdarsono, H., and Sidik. (1999). Biodiversity, traditional medicine and the sustainable use of indigenous medicinal plants in Indonesia. *Indigenous Knowledge and Development Monitor*, 7: 3–6.

Factbook. (2008). *Central Intelligence Agency world factbook*. Intelligence Agency Office of Public Affairs, Washington, DC. https://www.cia.gov/library /publications/the-world-factbook/reference_maps/southeast_asia.html (accessed September 19, 2008).

Food and Agricultural Organization (FAO). (2002). In: Vantomme, P., Markkula, A., and Leslie, R. N. (Eds.), *Non-wood forest products in 15 countries of tropical Asia: an overview*. GCP/RAS/173/EC. European Commission and FAO, Bangkok, Thailand.

Food and Agricultural Organization (FAO). (2003). *Selected indicators of food and agriculture development in Asia-Pacific region 1992–2002*. Rap Publication: 2003/10. Food and Agriculture Organization of the United Nations regional office for Asia and the Pacific, Bangkok, October.

Handa, S.S., Rakesh, D.D. and Vasisht, K. (2006). *Compendium of medicinal and aromatic plants Asia*. United Nations Industrial Development Organization and the International Centre for Science and High Technology (ICS-UNIDO), Trieste, Italy.

Hounzangbe-Adote, M.S., Paolini, V., Fouraste, I., Moutairou, K., and Hoste, H. (2005). In vitro effects of four tropical plants on three life-cycle stages of the parasitic nematode, *Haemonchus contortus*. *Research in Veterinary Science,* 78(2): 155–160.

International Institute of Rural Reconstruction (IIRR). (1994). *Ethnoveterinary medicine in Asia: An information kit on traditional animal health care practices*. 4 vols. International Institute of Rural Reconstruction, Silang, Cavite, Philippines.

Khatoon, S., Rai, V., Rawat, A.K.S., and Mehrotra, S. (2006). Comparative pharmacognostic studies of three *Phyllanthus* species. *Journal of Ethnopharmacology,* 104: 79–86.

Kim, H.K., Jeon, W.K., and Ko, B.S. (2001). Flavone glycosids from *Citrus junos* and their anti-influenza activity. Planta Medica, 67: 548–549.

Kraisintu, K. (2003). The status of medicinal and aromatic plants in Cambodia, Laos, the Philippines, Thailand and Viet Nam. In: Vasisht, K., and Kumar, V. (Eds.), *Medicinal plants and their utilization*. ICS-UNIDO, Trieste, Italy, pp. 3–53.

Ladinsky, J.L., Volk, N.D., and Robinson, M. (1987). The influence of traditional medicine in shaping medical care practices in Vietnam today. *Social Science and Medicine*, 25: 1105–1110.

Lawrence, B.M., Hogg, J.W., Terhune, S.J., and Podimuang, V. (1971). Constituents of the leaf and peel oils of *Citrus hystrix*, D.C. *Phytochemistry*, 10: 1404–1405.

Ling, P.K. (2004). Triterpene production in *Centella asiatica* (L.) Urban (Pegaga) callus and cell suspension cultures. PhD dissertation, Universiti Putra Malaysia, Serdang, Selangor, Malaysia.

Matsubara, T., Bohgaki, T., Watarai, M., Suzuki, H., Ohashi, K., and Shibuya, H. (1999). Antihypertensive actions of methylripario chromene A from *Orthosiphon aristatus*, an Indonesian traditional medicinal plant. *Biological and Pharmaceutical Bulletin*, 22: 1083–1088.

Myers, N., Mittermeier, R.A., Mittermeier, C.G., Da Fonseca, G.A.B., and Kent, J. (2000). Biodiversity hotspots for conservation priorities. *Nature*, 403: 853–858.

Nirnoy, M., and Muangman, V. (1991). Effect of *Folia orthosiphonis* on urinary stone promoters and inhibitors. *Journal of the Medical Association of Thailand*, 74: 318–321.

Olah, N.K., Radu, L., Mogosan, C., Hanganu, D., and Gocan, S. (2003). Phytochemical and pharmacological studies on *Orthosiphon stamineus* Benth. (Lamiaceae) hydroalcoholic extracts. *Journal of Pharmaceutical and Biomedical Analysis*, 33: 117–123.

Plotkin, M. (1999). Linking biodiversity conservation, traditional knowledge and sustainable use to community health, INRIK NEWS in 1999. http://www.geocities.com/inrik/iwim.htm (accessed September 24, 2008).

Shanmugavelu, S., Acamovic, T., Brooker, J.D., and Cowieson, A.J. (2005). Performance of male broilers fed maize-soybean-based diets supplemented with exogenous enzymes and thyme oil. *British Poultry Abstracts*, 1: 44–46.

Shanmugavelu, S., Acamovic, T., and Cowieson, A.J. (2004). Effect of thyme oil and garlic powder on the nutritive value of soybean meal. *British Poultry Science*, 45: S54–S55.

Shanmugavelu, S., Brooker, J.D., Acamovic, T., and Cowieson, A.J. (2004). Effect of thyme oil and garlic powder on microbial fermentation in various sections of the gastrointestinal tract of broilers. *British Poultry Science*, 45: S9–S10.

Shanmugavelu, S., Murugaiyah, M., and Rehana, A. S. (2007). Effectiveness of *Melia excelsa* leaves in controlling helminth in goats. In *Proceedings of the 28th Conference of the Malaysian Society of Animal Production*. Kuching, Sarawak, Malaysia, pp. 123–124.

Shanmugavelu, S., Ruzickova, G., Zrustova, J., and Brooker, J.D. (2006). A fermentation assay to evaluate the effectiveness of antimicrobial agents on gut microflora. *Journal of Microbiological Methods*, 67: 93–101.

Shukla, A., Rasik, A.M., Jain, G.K., Shankar, R., Kulshrestha, D.K., and Dhawan, B.N. (1999). In vitro and in vivo wound healing activity of asiaticoside isolated from *Centella asiatica*. *Journal of Ethnopharmacology* 65(1): 1–11.

Soepadmo, E. (1999). Botanical study of Malaysian medicinal plants: an appraisal. In Manaf Ali, A., et al. (Eds.), *Phytochemicals and biopharmaceutins from the Malaysian rain forest*. FRIM, Kuala Lumpur, Malaysia, pp. 1–14.

Stevensen, C. (1999). JAMU: an Indonesian herbal tradition with a long past, a little known present and an uncertain future. *Complementary Therapies in Nursing and Midwifery*, 5: 1–3.

Sydara, K., Gneunphonsavath, S., Wahlstrom, R., Freudenthal, S., Houamboun, K., Tomson, G., and Falkenberg, T. (2005). Use of traditional medicine in Lao PDR. *Complementary Therapies in Medicine*, 13: 199–205.

Tan, J.Z.G. (2003). The need for a national colloquium on medicinal plants research and their business opportunities. http://cdnet.stpi.org.tw/techroom/report/med_plant_res_bus_opp_2003.pdf (accessed September 19, 2008).

Tuschinsky, C. (1995). Balancing hot and cold—balancing power and weakness: social and cultural aspects of Malay Jamu in Singapore. *Social Science and Medicine*, 41: 1587–1595.

Uawonggul, N., Chaveerach, A., Thammasirirak, S., Arkaravichien, T., Chuachan, C., and Daduang, S. (2006). Screening of plants acting against *Heterometrus laoticus* scorpion venom activity on fibroblast cell lysis. *Journal of Ethnopharmacology,* 103 (2): 201–207.

Viera, D.L.M., and Scariot, A. (2006). Principles of natural regeneration of tropical dry forests for restoration. *Restoration Ecology,* 14: 11–20.

Wang, M.Y., West, B.J., Jensen, C.J., Nowicki, D., Chen, S., Paul, A.K., and Anderson, G. (2002). *Morinda citrifolia* (Noni): a literature review and recent advances in Noni research. *Acta Pharmacolgica Sinica*, 23: 1127–1141

Wiart, C., Mogana, S., Khalifah, S., Mahan, M., Ismail, S., Buckle, M., Narayana, A.K., and Sulaiman, M. (2004). Antimicrobial screening of plants used for traditional medicine in the state of Perak, Peninsular Malaysia. *Fitoterapia,* 75: 68–73.

World Health Organization (WHO). (2001). *Legal status of traditional medicine and complementary/alternative medicine: a worldwide review.* WHO/EDM/TRM/2001.2, WHO, Geneva, p. 188.

World Health Organization (WHO). (2005). *National policy on traditional medicine and regulation of herbal medicines, report of a WHO global survey.* WHO, Geneva, May 2005, p. 136.

Yanfang, Z., Xingping, L., Zongde, Z., Liren, C., and Yongmin, L. (2006). Simultaneous determination of andrographolide and dehydroandrographolide in *Andrographis paniculata* and Chinese medicinal preparations by microemulsion electrokinetic chromatography. *Journal of Pharmaceutical and Biomedical Analysis*, 40: 157–161

APPENDIX 7.1

Medicinal and Aromatic Plants in Common Use in Countries of Southeast Asia

Botanical Name	Uses/Indications	Country
Abrus precatorius	Stomatitis	Indonesia
	Conjunctivitis, colic, flu, fever, cold	Philippines
	Expectorant, anti-inflammatory, antiallergenic	Vietnam
Acacia arabica	Astringent, in diarrhea	Myanmar
Acorus calamus	Sedative	Indonesia
	Analgesic, in toothache, headache	Vietnam
Acorus gramineus	Arrhythmia, asthma, as stomachic, cardiotonic	Lao PDR
Adina cordifolia	Tonic, febrifuge, antiseptic	Myanmar
Ageratum conyzoides	In bleeding disorders	Lao PDR
	As homeostatic, in scabies, colic	Vietnam
Allium cepa	In asthenia, fatigue	Cambodia
	As diuretic, emmenagogue	Philippines
Allium odorum	Digestive, carminative, in cough	Cambodia
	In wounds, bruises	Philippines
Allium sativum	As antivenomous agent, in hypertension	Cambodia
	In candidiasis, as hypolipidemic	Indonesia
	In blood pressure	Philippines
Alocasia macrorrhiza	In malaria, asthma	Lao PDR
Aloe barbadensis	For hair care, as facial cleanser	Malaysia
Aloe vera	In falling hair, baldness	Philippines
Alpinia galanga	As aromatic	Myanmar
Alpinia officinarum	In cholic, as stomachic	Lao PDR
Alstonia scholaris	In malaria, gastric disorders	Lao PDR
	In fever, diarrhea, diabetes; as anthelmintic	Philippines
Amaranthus spinosus	As amenorrheic	Vietnam
Amomum aromaticum	Spice, medicine	Vietnam
Amomum krevanh	As carminative, expectorant	Cambodia
Amomum sp.	For digestion	Lao PDR
Amomum xanthioides	In flatulence	Lao PDR
Amorphophallus riviera	In malaria	Lao PDR
Anacardium occidentale	As analgesic	Indonesia
Andrographis paniculata	As anti-inflammatory to sore throat, for relief of cold	Thailand
	In diabetes, high blood pressure, skin irritation, and insect bites	Brunei
	In infections, dysentery, as antiseptic, tonic	Lao PDR
	As antiseptic, antidiabetic	Indonesia
	As herbal tea	Malaysia
Angiopteris evecta	In piles, boils, difficulty in urination, headaches, skin eczema, irritation	Brunei

(*continued*)

Botanical Name	Uses/Indications	Country
Apium graveolens	As hypotensive	Indonesia
Aquilaria agallocha	As aromatic, stimulant, aphrodisiac, tonic, diuretic	Myanmar
Aquilaria crassna	In colitis, nausea, hiccup	Lao PDR
	As incense, in cosmetics	Vietnam
Areca catechu	In tapeworm, roundworm infestation	Malaysia
Aristolochia tagala	As tonic, carminative	Myanmar
Artabotrys odoratissimus	As diuretic, carminative	Cambodia
Artemisia annua	In malaria, fever	Lao PDR
	In skin diseases, jaundice	Vietnam
Artemisia vulgaris	In headache, menstrual disorder	Lao PDR
	As expectorant, antispasmodic, carminative, emmenagogue	Philippines
	As digestive tonic, antiseptic	Vietnam
Artocarpus lakoocha	For tapeworm infestation	Lao PDR
Asparagus cochinchinensis	As diuretic, in cough	Lao PDR
Atalantia monophylla	As digestive, in cough	Cambodia
Azadirachta indica	In malaria	Lao PDR
	As antibiotic, organic pesticide and fungicide	Vietnam
Baeckea frutescens	As fragrance	Malaysia
Bidens pilosa	In headache, caries, hiccough	Lao PDR
Blumea balamifera	As analgesic, antipyretic	Indonesia
	In cold, coryza	Lao PDR
	As health tonic, lotion	Malaysia
	As antispasmodic, astringent, expectorant	Philippines
	As laxative, febrifuge	Cambodia
Butea frondosa	As astringent	Myanmar
Butea monosperma	As febrifuge, aphrodisiac, purgative	Vietnam
Caesalpinia sappan	As tonic, in dysentery	Lao PDR
Cananga odorata	For hair care, in perfumery	Malaysia
	As fragrance for soap	Vietnam
Canangium odorata	In hypertension	Lao PDR
Carica papaya	As antimalarial, male contraceptive	Indonesia
Carmona retusa	In colic, as antidiarrheal	Philippines
Carthamus tinctorius	As carminative	Philippines
Cassia acutifolia	As laxative	Philippines
Cassia alata	For eczema	Lao PDR
	In skin diseases, athlete's foot	Philippines
	As antiseptic	Malaysia
Cassia angustifolia	As laxative	Myanmar
Cassia fistula	As laxative	Philippines
Cassia occidentalis	In eye disorders, constipation	Lao PDR
Catharanthus roseus	In leukemia	Lao PDR
	As antidiabetic, vermifuge, purgative	Philippines

Botanical Name	Uses/Indications	Country
Centella asiatica	In urinary tract infection, stones	Brunei
	As diuretic, antiseptic, hypotensive	Indonesia
	As diuretic, in wounds	Philippines
	As health tonic	Malaysia
	In cancer, diabetes	Lao PDR
Chenopodium ambrosioides	Anthelmintic	Lao PDR
Chrysanthemum indicum	In stomach pain, colic, cervix infection	Philippines
Cinchona ledgeriana	In malaria, as tonic	Lao PDR
Cinnamomum cambodianum	As carminative, digestive	Cambodia
Cinnamomum camphora	In colic, chest pain, as cardiostimulant	Lao PDR
	In various drugs	Vietnam
Cinnamomum cassia	As tonic, balm ingredient, in coryza	Lao PDR
	In bakery products, as flavoring	Vietnam
Cinnamomum cortechinii	In influenza	Lao PDR
Cinnamomum iners	In influenza	Lao PDR
Cinnamomum sintoc	In cosmetics	Malaysia
Cinnamomum zeylanicum	As spice, fragrance	Malaysia
	In urinary, kidney, and stomach troubles	Philippines
Citrus aurantiifolia	As antitussive	Indonesia
Citrus aurantium	Digestive, in cough	Cambodia
Citrus hystrix	In influenza	Cambodia
Citrus medica	Digestive, antispasmodic, in cough	Cambodia
	In scurvy	Myanmar
Citrus nobilis	Digestive, in cough	Cambodia
Citrus sinensis	Digestive, in cough	Cambodia
Citrus spp.	In perfumery, cleanser	Malaysia
Clausena anisum-olens	In headache, dizziness, cough, asthma, bronchitis	Philippines
Coleus amboinicus	In cough	Cambodia
Coleus blumei	In bruises and contusions	Philippines
Coptis chinensis	As antibacterial, emollient	Vietnam
Coriandrum sativum	As digestive	Cambodia
	In chicken pox	Philippines
Coscinium fenestratum	In colic, stomachache	Vietnam
Coscinium usitatum	In dysentery, as cholagogue	Lao PDR
Costus speciosus	In arthritis	Lao PDR
Crataegus pinnatifidus	As stomachic, cardiotonic	Lao PDR
Crinum asiaticum	A poultice for aches, in sores and chaps, wounds, abscesses, piles	Singapore
Cucurbita moschata	In taeniasis	Indonesia
Curcuma comosa	In gynecological diseases	Thailand

(*continued*)

Botanical Name	Uses/Indications	Country
Curcuma domestica	In chronic hepatitis, as antiarthritic, antiseptic	Indonesia
	As cosmetic, food additive	Malaysia
Curcuma longa	For hemostasis, in pneumonia, gastritis, flatulence	Cambodia
	As anti-inflammatory	Myanmar
	In wounds, bruises	Philippines
	As antiflatulent, in peptic ulcers	Thailand
Curcuma xanthorrhiza	In chronic hepatitis	Indonesia
Curcuma zedoaria	In abdominal cramps, pains, amenorrhea	Philippines
Cymbopogon citratus	As diuretic, diaphoretic	Philippines
	As digestive, tonic, carminative, febrifuge, antifungal	Vietnam
Cymbopogon nadrus	As febrifuge, carminative	Cambodia
	In cosmetics, insect repellent	Malaysia
	In fever	Philippines
	In flu, influenza, cold	Lao PDR
Cyperus rotundus	As emmenagogue, in colic, diuretic	Cambodia
	As liver tonic, digestive tonic	Myanmar
Datura metel	In asthma, as narcotic, anodyne, antispasmodic	Philippines
	As antiasthmatic, febrifuge	Vietnam
Dichroa febrifuga	In malaria	Lao PDR
Dioscorea deltoidea	For producing steroid drugs	Vietnam
Dracaena angustifolia	As adaptogen	Vietnam
Drosera rotundifolia	In cough, as antispasmodic	Vietnam
Drynaria fortunei	In rheumatism	Lao PDR
Eclipta alba	In bleeding disorders	Lao PDR
	As antihepatotoxic	Myanmar
Eclipta prostrata	As liver tonic, styptic, in gynecologic disorders	Vietnam
Elephantopus	In allergy, diarrhea, fever, as diuretic	Lao PDR
Elsholtzia cristata	In cold, headache	Lao PDR
Embelia ribes	For tapeworm infestation	Lao PDR
Emblica officinalis	As antioxidant, tonic	Myanmar
Eryngium foetidum	As febrifuge, digestive	Cambodia
Eucalyptus globulus	As febrifuge, antiseptic	Cambodia
Eucalyptus	In cough, as anesthetic, antiseptic	Philippines
	In influenza	Lao PDR
Eugenia aromatica	In toothache	Malaysia
Eugenia caryophyllus	As local anesthetic	Cambodia
Eugenia zeylanica	As analgesic, in sprain	Cambodia
Eupatorium odoratum	In bleeding disorders, appendicitis	Lao PDR
Euphorbia hirta	In cataract, dermatitis	Philippines
	As antispasmodic, sedative	Vietnam
Eurycoma longifolia	As aphrodisiac, tonic, in fever	Malaysia
	In high blood pressure, fever, diarrhea, jaundice, tuberculosis, dysentery, as aphrodisiac, tonic	Singapore
Evodia lepta	In asthma	Lao PDR
Feroniella lucida	In influenza, as digestive	Cambodia

Botanical Name	Uses/Indications	Country
Ficus carica	As laxative, emollient, expectorant	Vietnam
Ficus religiosa	As astringent, laxative, antifungal	Vietnam
Foeniculum vulgare	In gaseous distention	Philippines
Fokienia hodginsii	In cosmetics	Vietnam
Garcinia mangostana	In diarrhea	Philippines
Glycosmis citrifolia	In asthma	Lao PDR
Gnaphalium indicum	In cough, bronchitis	Lao PDR
Graptophyllum pictum	In hemorrhoids	Indonesia
Guazuma ulmitolia	As hypolipidemic	Indonesia
Heliotropium indicum	In back pain	Lao PDR
Hibiscus abelmoschus	As diuretic, in constipation	Lao PDR
Hibiscus rosa-sinensis	As poultice for boils, in cancerous swelling, mumps	Philippines
Holarrhena antidysenterica	In dysentery	Lao PDR
Homalomena aromatica	In perfume	Vietnam
Homalomena occulta	In rheumatism	Lao PDR
Illicium verum Hook	As carminative, antispasmodic	Cambodia
	In alcohol beverages, as carminative	Vietnam
Isotoma longiflora	In wound healing	Philippines
Jasminum sambac	As antispasmodic	Cambodia
	As fragrance	Malaysia
	As antipyretic, decongestant, in diarrhea	Philippines
	Flavoring tea	Vietnam
Jasminum subtriplinerve	For recovery after giving birth	Vietnam
Jatropha curcas	As poultice for sprains and dislocation	Philippines
Justicia adhatoda	In bone fractures	Lao PDR
Kaempferia galanga	In aching joints, flavoring, tonic	Brunei
	As spice	Malaysia
	In sore throat, rheumatism, swelling, as tonic	Philippines
	As expectorant, carminative	Vietnam
Kaempferia parviflora	As tonic, for men	Thailand
Labisia pumila	In postpartum preparation	Malaysia
Lactuca indica	In chest ache, sore eyes	Vietnam
Lagerstroemia speciosa	In fever, diabetes, diarrhea, as diuretic	Philippines
Lantana camara	As an antidote, as poultice for sprains	Philippines
Lasia spinosa	In angina, edema	Lao PDR
Lawsonia alba	As emmenagogue, anthelmintic	Philippines
Lawsonia inermis	In cosmetics	Malaysia
Leea sambusina	As tonic, in rheumatism	Lao PDR
Leptospermum flavescens	As health tonic	Malaysia
Leucosyke capitellata	In cough, headache, tuberculosis	Philippines
Litsea cubeba	As stomachic	Lao PDR
	As source of citral	Vietnam

(continued)

Botanical Name	Uses/Indications	Country
Litsea glutinosa	In gastrointestinal disorders	Myanmar
Lonicera japonica	As diuretic, liver tonic, astringent, hypotensive	Vietnam
Loranthus sp.	As anticancer agent	Indonesia
Lygodium flexuosum	In skin problems	Philippines
Mahonia bealei	In conjunctivitis	Lao PDR
Mangifera indica	As diuretic, in sore throat, cough, colds	Philippines
Melaleuca cajuputi	As analgesic	Malaysia
Melaleuca leucadendron	As febrifuge, in sprain, cough	Cambodia
Melastoma decemfidum	In intestinal measles	Malaysia
Melia azedarach	As anthelmintic	Lao PDR
Mentha arvensis	In toothpaste	Malaysia
	As antispasmodic, carminative	Philippines
Mentha cordifolia	In headache, toothache, stomachache	Philippines
Mentha spicata	In colds, gaseous distention	Philippines
Michelia champaca	As carminative, febrifuge	Cambodia
	In cosmetics	Malaysia
Micromelum falcatum	In asthma	Lao PDR
Momordica charantia	As antidiabetic	Indonesia
	In chronic colitis, bacillary dysentery, cough, as colic astringent	Philippines
Momordica spp.	For flavoring drugs	Vietnam
Morinda citrifolia	In arterial hypertension, constipation, pain, diabetes	Lao PDR
	As health tonic	Malaysia
	As antibacterial, antiviral, antifungal, analgesic, hypotensive, immune-enhancing effects, anti-inflammatory	Singapore
Morinda officinalis	As kidney tonic, impotence remedy	Vietnam
Murraya paniculata	As diuretic, carminative	Cambodia
Myristica fragrans	As sedative	Indonesia
	In diarrhea, vomiting, indigestion, abdominal pain	Malaysia
	As carminative, antispasmodic	Vietnam
Nigella sativa	As anthelmintic, carminative	Myanmar
Nopalea cochenillifera	In rheumatism, earache, toothache	Philippines
Ocimum basilicum	As antispasmodic, carminative, digestive	Cambodia
	In nausea, fever, kidney diseases	Lao PDR
	In cosmetics	Malaysia
Ocimum gratissimum	In influenza	Cambodia
	In cold, caries	Lao PDR
Ocimum sanctum	As antispasmodic, carminative	Cambodia
	In malaria	Lao PDR
	As aromatic, stimulant, in paralysis, cough, colds	Philippines
Oldenlandia diffusa	In dysentery, snakebite	Malaysia
Orthosiphon stamineus	As diuretic	Indonesia
Pandanus odoratissimus	As purgative, cardiotonic, in smallpox	Philippines
Passiflora foetida	As tranquilizer, in anxiety	Lao PDR

Botanical Name	Uses/Indications	Country
Phyllanthus amarus	In jaundice, gonorrhea. diabetes	Myanmar
Pinus khesya	In cough, as antiseptic	Lao PDR
Piper betel	As antiseptic, poultice for boils	Myanmar
	To relieve stomachache, as antibacterial	Brunei
	As carminative, in cough	Cambodia
	As antiseptic	Indonesia
Piper cubeba	In migraine, rheumatism, boils	Cambodia
Piper lolot	As carminative, in diarrhea	Cambodia
Piper nigrum	As digestive, carminative	Cambodia
	As flavor	Malaysia
Piper sarmentosum	In coughs, flu, rheumatism, pleurisy, toothache, lumbago, as antioxidant	Singapore
Plantago major	As expectorant, decongestant, irritable bowel relief	Vietnam
Polygonum kingianum	As tonic	Lao PDR
Polygonum multiflorum	In cholera	Vietnam
Polygonum odoratum	In alimentary infections	Cambodia
Portulaca oleracea	In wounds, ulcers, cough, colds	Philippines
Premna odorata	In fever, as expectorant	Philippines
Psidium guajava	As antidiarrheal	Indonesia
Pterocarpus santalinus	As astringent, in diarrhea	Myanmar
Punica granatum	As antiseptic	Indonesia
Quisqualis indica	In ascariasis	Indonesia
	As antihelmintic	Philippines
Rauvolfia canescens	In hypertension	Lao PDR
Rauvolfia serpentina	In hypertension	Lao PDR
	As hypotensive, sedative	Myanmar
Rauvolfia verticillata	In hypertension	Lao PDR
Ricinus communis	As purgative, antirheumatic, arthritis, in paralysis, epilepsy	Philippines
Ruta graveolens	As antiseptic, antipyretic	Indonesia
Santalum album	As antiseptic	Myanmar
Sauropus androgynus	As breast milk stimulator	Indonesia
Schefflera elliptica	As tonic, in rheumatism	Lao PDR
Smilax glabra	As tonic	Lao PDR
	In backache	Vietnam
Smilax myosotiflora	As health tonic	Malaysia
Solanum surattense	In boils	Philippines
Solanum trilobatum	As antitussive and expectorant	Thailand
Sonchus arvensis	In nephrolitiasis, as diuretic	Indonesia
Spilanthes acmella	In caries	Lao PDR
	As analgesic, febrifuge	Cambodia
Stephania glabra	As tranquilizer	Lao PDR
Sterculia lychnophora	In constipation	Lao PDR
Streptocaulon extensus	As anthelmintic	Lao PDR
Streptocaulon juventas	As tonic	Lao PDR

(continued)

Botanical Name	Uses/Indications	Country
Strychnos nux vomica	In neurasthenia, as tonic	Lao PDR
	As muscle relaxant, drug-flavoring agent	Vietnam
Styrax tonkinesis	In cough, bronchitis	Lao PDR
Tagetes erecta	Detoxicant, in cough	Cambodia
	As tonic, in dysmenorrhea	Philippines
Terminalia chebula	As laxative, astringent	Vietnam
Tetrapanax papyrifer	As diuretic	Lao PDR
Tinospora crispa	As antimalarial, antidiabetic	Indonesia
	As bitter tonic, in joint pain	Lao PDR
Triphasia trifolia	In colic, diarrhea, skin diseases	Philippines
Uncaria sp.	In hypertension, rheumatism, fever convulsion	Lao PDR
Verbena officinalis	As nerve tonic, tonic, sedative	Vietnam
Vitex negundo	In colds, cough, fever, muscular pain	Philippines
Vitex trifolia	As antiseptic	Indonesia
	As analgesic, diuretic, emmenagogue,	Philippines
Wedelia calendulacea	As antibiotic	Lao PDR
Woodflordia floribunda	As antiseptic, diuretic	Indonesia
Xanthium strumarium	In allergy, goiter	Lao PDR
Zanthoxylum rhetsa	As antibacterial, in gastrointestinal disorders	Vietnam
Zanthoxylum spp.	In flatulence	Lao PDR
Zingiber officinale	As hypocholesterolemiac, tonic, in rheumatism	Cambodia
	As analgesic, antipyretic, anti-inflammatory	Indonesia
	In influenza, as stimulant, carminative	Lao PDR
	In rheumatism, sore throat, cough	Philippines
Zingiber zerumbet	In Jamu preparations	Malaysia

Source: After Handa, S.S., Rakesh, D.D. and Vasisht, K. (2006). *Compendium of medicinal and aromatic plants Asia.* United Nations Industrial Development Organization and the International Centre for Science and High Technology (ICS-UNIDO), Trieste, Italy.

8 Evidence-Based Botanicals in North America

Wendy Pearson and Michael I. Lindinger

CONTENTS

8.1 INTRODUCTION

North America has rich historical resources from which to draw information on ethnoveterinary medicine. The relationship that Native Americans wrought between themselves and the plants and animals around them was unique and seamless and existed within the mantra that all—humans, animals, and plants—were part of the natural world without divisions of hierarchy or ownership. Native Americans embraced animals as their equals within the natural world and believed that animals, like humans, held their own spiritual power.[1] Similarly, plants were an important and integral coefficient in the equation of natural life. Plants offered food for families and livestock, medicines for the sick, smoking for social and spiritual enlightenment, and poisons for hunting wild game. This inherent, spiritual harmony within which Native Americans, plants, and animals existed provided for medicine systems for humans and animals that were virtually indistinct from each other. The knowledge that Native Americans had of the uses of plants around them was applied without prejudice to the animals in their care.

Native Americans did not usually regard physical problems in themselves or in their animals as separate from spiritual problems, and even when a physical remedy was applied, the cause of the ailment might still be ascribed to evil spirits. Treatment was generally applied by the shaman, who was most often also an herbalist. In excess of 220 natural remedies of the Native Americans have been listed in the U.S. Pharmacopeia or the National Formulary since 1820, a testament to the efficacy of their ancient art of healing.

Rather than providing a glossary of herbs used by North American Indians in treating animal ailments,* in this chapter we instead try to selectively associate popular Native American herbs with supporting evidence for their use catalogued in the PubMed and ToxNet databases under headings of important veterinary issues. Many of the herbs discussed are not in themselves native to North America but were introduced between the 17th and 19th centuries. These nonetheless were readily adopted by the Native Americans, who recognized and valued their medicinal qualities.

8.2 NATIVE AMERICAN HERBS FOR TREATING RESPIRATORY AILMENTS

Great mullein (*Verbascum thapsus*) was introduced to North America from Europe and was highly regarded as a demulcent and expectorant. Native tribes, including Menominees, Forest Potawatomis, Creeks, and Chochtaws, referred to the plant as "wild tobacco" and, ironically, smoked the leaves to relieve respiratory ailments[2] or boiled them to produce a decoction that treated cough in their children. Contemporary scientific literature on the respiratory effects of the plant is substantial and has been reviewed.[2] The plant contains saponins,[3] polysaccharide mucilage,[4] iridoid glycosides,[5] and flavonoids,[6] all of which jointly or severally contribute to antibacterial,[7] emollient,[8] and anti-inflammatory[9] properties that are of importance when managing cases of respiratory disease or dysfunction.

There is mounting experimental and clinical evidence for the application of *Verbascum* sp. to respiratory disease, and the plant is recognized as an effective respiratory tonic in contemporary ethnoveterinary medicine.[10] An aqueous extract of the plant strongly inhibited replication and virulence of the influenza virus[6,11,12] and herpesvirus 1[6,13,14] *in vitro*. Research on *Verbascum* sp. in animals is limited to those demonstrating effective antiulcerogenic properties in rats[15] and antispasmodic effects in rabbits.[16] The only adverse effect associated with the herb is occupational dermal irritation from the hairy leaves.[17] There are no reports of LD_{50} or minimum lethal dose (MLD) for *Verbascum* sp.

Based on available scientific evidence, veterinary applications of mullein appear to be warranted in cases of viral respiratory disease. Influenza virus and herpesvirus are major contributing pathogens to respiratory disease in horses,[18,19] cattle,[20] dogs,[21] and cats.[22] As the growing herb is unpalatable to grazing livestock due in large part to the irritating hairs on the leaves,[23] it is recommended that the herb be administered as an aqueous extract of the flower. A standardized dose of mullein that is efficacious in respiratory disease has not been reported. However, lyophilized aqueous extract of *Verbascum* sp. of 1,227 mg/kg significantly improved ulcer score in gastrically challenged rats[15] and can be considered a reasonable starting point for dosage.

Garlic (*Allium sativum*) is one of the oldest medications used in human culture.[24] The *Allium* species encompasses more than 600 varieties, including garlic, onions, shallots, leeks, and scallions. Its place of origin is considered to be Central Asia, and the earliest report of medicinal applications of garlic has been dated to circa

* A good discussion of general principles of North American Indian medicine can be found online at http://www.healing-arts.org/mehl-madrona/mmtraditionalpaper.htm

2600 BC.[25] It was cultivated by the Choctaw Indians in gardens prior to 1775 and was used for such varied applications as treating skin diseases, increasing fertility, stimulating appetite, suppressing coughs, and treating rheumatism, abdominal diseases, hemorrhoids, and spleen enlargement.[26] An ostensible panacea, garlic has also been one of the most highly studied plants in contemporary pharmacology and ethnobotanical science. It is now well accepted that fresh garlic, or garlic processed as an extract or dried powder, can be an effective antibiotic and anti-inflammatory agent, with application to the treatment and prevention of respiratory disease. Aqueous garlic extract strongly promotes clearance of pathogenic pulmonary bacteria in mice[27] and is among the most potent herbal inhibitors of a wide range of pathogenic organisms *in vitro*, including *Staphylococcus aureus*, *Streptococcus faecalis*, *Candida albicans*, *Klebsiella pneumoniae*, *Pseudomonas aeruginosa*, *Escherichia coli*, and *Salmonella* spp.[28,29] Furthermore, dietary garlic supplementation is associated with significant protection of children against acute respiratory tract infection,[30] perhaps in part by inhibiting expression of cell adhesion molecules on endothelial cells[31] and limiting migration of neutrophils into pulmonary tissues.[32,33] These data indicate the important prophylactic potential of garlic in animals. Indeed, oral garlic supplementation can prevent the onset of the common cold in people.[34]

Although the evidence is strong for prophylactic and treatment roles of garlic and its derivates in respiratory disease of animals, it is prudent to note the potential toxicity of garlic in excess. High dietary inclusion rates of garlic, and its various derivatives, has been associated with hemolytic anemia in rodents,[35] horses,[36] dogs,[37–39] and sheep.[40] When administering large amounts of garlic products to animals, it is important to monitor the oxidation status of red blood cells to preempt inadvertent anemia. Furthermore, garlic may alter foregut microflora in ruminants,[41] which may affect digestibility of other feed ingredients. The oral LD_{50} of aqueous garlic (*Allium sativum*) extract is 174 mL/kg in rabbits, and the oral LD_{50} of allicin (the primary bioactive precursor) is 204 µg/kg.[42] An alkaloidal extract of Eqyptian garlic (*Allium ampeloprasum*) has a subcutaneous injection LD_{50} of 128 mg/kg and MLD of 100 mg/kg.[43]

North American natives also commonly utilized bark from native trees to treat respiratory ailments.[44] But, despite occasional reports of efficacy,[45] there is little scientific evidence to support this practice. Bark of the choke cherry (*Prunus serotina*) was considered a particularly important cough suppressant by Native Americans, but its use is not advisable due to association of *Prunus* sp. ingestion with cyanide poisoning of goats[46] and cows[47] and congenital defects in pigs.[48]

8.3 NATIVE AMERICAN HERBS AS ANTI-INFLAMMATORY AGENTS

Musculoskeletal inflammation—particularly lameness—is among the most important veterinary conditions affecting domesticated animals. It is associated with poor racing performance in racehorses[49] and adversely affects the economics of commercial swine[50] and dairy[51] industries. Similarly, inflammatory diseases—especially arthritic diseases—played a particularly destructive role in indigenous populations from ancient times[52] to the present.[53,54] As such, Native Americans commonly used plants to counter the pain associated with arthritis, and this practice continues even in contemporary ethnoveterinary medicine.[10]

Cayenne (*Capsicum annum*) has a 9,000-year history of use by Central and South American Indians.[55] The plant was a specific treatment of pain and inflammation and is today the most widely consumed herb worldwide.[55] Cayenne is native to Central and South America and became naturalized in North America during the 16th century. *Capsicum* species were cultivated in the Tehucan Valley in Mexico as early as 4000 BC, during the Coxcatlan era.[56] *Capsicum* is currently approved by the U.S. Food and Drug Administration for the topical relief of pain due to rheumatoid arthritis, osteoarthritis, and various neuralgias.

The majority of published efficacy research relating to cayenne is on a class of its putative bioactive principles, the capsaicinoids, which are found in the fruit. The capsaicinoid content of whole powdered fruit varies from 0.1% to 1.5%.[57,58] Capsaicin, a phenolic vanillyl amide of isodecenoic acid, accounts for 40–50% of the total capsaicinoid content. Capsaicin can be administered orally[59,60] or injected subcutaneously,[61] resulting in stimulation of catecholamine (adrenaline) secretion from the adrenal medulla.[61] This causes stimulatory, sympathomimetic effects similar to those associated with ingestion of coffee and tea. The most common mode of administration of capsaicin, however, is percutaneously. Absorption across the skin has been demonstrated in humans, rats, mice, rabbits, and pigs,[59] supporting its use in topical tinctures and creams. Clinical reports on treatment of knee[62] and hand[63] osteoarthritis with topical capsaicin have revealed significant improvement in pain scores after 2 weeks of treatment. These results are confirmed by a meta-analysis of studies using topical creams containing capsaicin[64] and in a randomized, double-blind, placebo-controlled clinical trial.[65] A comprehensive review of the medical literature has concluded that the pain-relieving effect of topical capsaicin was comparable to acetaminophen, intra-articular steroids and hyaluronic acid, and nonsteroidal anti-inflammatory drugs (NSAIDs) in patients with knee osteoarthritis.[66] However, a subsequent review evaluating effectiveness of capsaicin on chronic musculoskeletal or neuropathic pain concluded its effectiveness was only moderate to poor.[67]

Much has been published on potential mechanisms by which capsaicin exerts its anti-inflammatory activity in arthritis. In general, two primary mechanisms have been proposed: (1) Capsaicin depletes stores of substance P from capsaicin-sensitive nerve endings within the synovial membrane,[68] which may be responsible for the inhibitory action of capsaicin on neuron-dependent progression of osteoarthritis,[70,71] and (2) capsaicin stimulates metabolic activity of synoviocytes, as evidenced by dose-dependent increase DNA and collagen synthesis by the cells when exposed to capsaicin.[72]

Capsicum is generally recognized as safe (GRAS) by the U.S. Food and Drug Administration, and capsaicin is considered to be safe and effective as an external analgesic counterirritant.[73] Capsaicinoids at moderate to high concentrations are irritant to mucosal surfaces lining the respiratory and gastrointestinal systems,[73] and contact dermatitis may result from the direct handling of chili peppers containing capsaicin.[74] The oral LD_{50} for capsaicin in rats is between 160 and 190 mg/kg and in mice as low as 118 mg/kg, with hemorrhage of the gastric fundus observed in some of the animals.[58,73]

White willow (*Salix alba*) is native to Europe and introduced to temperate areas of North America. In Native American culture, willow has a long history of use as an anti-inflammatory, and its bark, leaves and catkins have been used to treat a wide array of health conditions.[75] There are over 500 species of *Salix*, of which the main ones in North America are *S. nigra* (black willow), *S. hookeriana, S. lucida, S. scouleriana, S. sitchenis,* and *S. barclayi.* The glycoside salicilin was isolated from willow bark in 1838 by the Italian chemist Rafaele Piria and developed into acetylsalicyclic acid (ASA; aspirin) by Bayer in the late 1930s. The salicin content of *S. alba* bark is only about 0.5% compared to 1–10% in other species. Other phenolic glycosides also found in willow include salicortin, salireposide, picein, trandrin, the acetylated forms of salicin, and esters of salicylic acid and salicyl alcohol. Flavonol glycosides include narigenin, isolaipurpouroside, and isoquercitin in leaves and bark.[58,76–78]

In contrast to many ethnobotanical therapies, there have been a number of clinical trials assessing the safety and efficacy of willow bark extracts. Such extracts are recommended for treatment of fever, rheumatism, and general pain by the European Scientific Cooperative on Phytotherapy (ESCOP). Research assessing efficacy of willow bark extract in reducing pain has produced equivocal results, although largely supportive of its use. On the positive side, clinical trials have demonstrated that willow bark extract is significantly better than placebo for treating patients with chronic low back pain after about 1 month of treatment[79,80] and not significantly different from rofecoxib (Cox-2 inhibitor) after 6 months of treating arthritis patients.[81] Similarly, a standardized willow bark extract (240 mg salicin/day) was provided to osteoarthritis patients, who observed a significant improvement in pain after 2 weeks.[81] Extract of white willow containing less than one-sixth of the amount of salicin as an ASA product showed at least equivalent efficacy as ASA in inhibiting leukocytic infiltration, preventing elevation of the rise in cytokines, and suppressing prostaglandin production, and was even more effective than ASA in inhibiting leukotriene formation and Cox-2 production.[82] These results suggest that the anti-inflammatory activity arises from more than just salicylin and raises the importance of further research characterizing white willow with respect to other bioactive compounds.[83]

However, other authors have reported poor analgesic effect of white willow. Using white willow extract standardized to the same dose of salicin as reported by Chrubasik et al.,[79] Bierget et al.[84] reported no significant improvement in pain scores of patients with osteoarthritis and rheumatoid arthritis. The lack of improvement observed in these patients may result from other bioactive constituents present in white willow that were not standardized in the same way as salicin.

There is a dearth of safety information available on white willow extract. Most adverse effects are associated with the content of salicin, which are significantly less in incidence and magnitude than those of ASA.[85–87] Salicin is poorly absorbed by the gastrointestinal tract,[85] which probably contributes to its very high LD_{50} in rats, which is 1,890 mg/kg.[76] Generally, white willow extract is considered safe and well tolerated.[80,81]

Contemporary North American ethnoveterinary medicine continues to use *Salix* sp. as a remedy for pain, arthritis, and muscle soreness in horses[10] and ruminants.[88] There is currently no research reporting effects of *Salix* sp. in companion animals or livestock. However, research in laboratory animals and humans supported its use as a moderate analgesic and anti-inflammatory in veterinary medicine.

Stinging nettle (*Urtica dioica*), like cayenne and willow, has a long history of use as a native anti-inflammatory, particularly for the treatment of inflammatory joint diseases and urinary tract infections.[89] Nettles, despite their perfect adaptation to North America, are not native but were brought from England by John Josselyn in the mid-17th century. This hardy little plant thrives throughout temperate regions across the globe.

The phytochemistry of nettle has been reasonably well characterized[58,90] and consists predominantly of flavonoids (glucosides and rutinosides of quercetin, kaempferol, and isorhamnetin), acids, β-sitosterol (-3-*O*-glucoside), polysaccharides, fatty acids, and minerals.

A detailed review of clinical anti-inflammatory literature on nettle has been published.[90] Of eight *in vivo* studies performed on humans with osteoarthritis, rheumatoid arthritis, or other rheumatic diseases, all studies showed significant improvement in pain score over time, with some studies showing significant improvement compared to placebo. *In vitro*, treatment of various cell cultures with nettle extract has resulted in reduced release of the inflammatory mediators tumor necrosis factor α (TNF-α) and interleukin 1β.[91,92] An extract of nettle has also demonstrated a significant capacity for inhibiting expression of chondrocyte matrix metalloproteinases,[93] the enzymes responsible for the destruction of articular cartilage in arthritic conditions. The precise mechanisms are not known but may be associated with inhibition of nuclear factor kappa B (NFκB) activation.[94]

The toxicity literature for stinging nettle has also been summarized by Chrubasik et al.[90] The intraperitoneal LD_{50} of aqueous extract in mice is 3,625 mg/kg. Higher doses were associated with a decrease in spontaneous activity, loss of muscle tone, and hypothermia. Low toxicity was observed after oral and intraperitoneal administration of up to 2 g/kg. Intravenous doses of 4,500 mg/kg caused transient hypotension and cardiac arrhythmias. The LD_{50} for intravenous injection in mice is 1.9 g/kg, and for chronic oral application in rats is 1,310 mg/kg. Chronic subcutaneous administration is associated with diarrhea. There is a report[95] of three horses with an apparent transient (resolved within 4 h) neurological disorder resulting from nettle rash that showed signs of ataxia, distress, and muscle weakness, and two of them had urticaria. In general, nettle is considered safe and may be an effective treatment for inflammatory disorders.[90]

8.4 NATIVE AMERICAN HERBS FOR CONTROLLING ENDOPARASITES

As a people whose diet consisted of browsing their environment, the North American natives were highly susceptible to internal parasites, a problem that persists even today in some native American cultures.[96–98] Endoparasites in livestock were, and continue to be, treated by Native Americans with many different herbs,[10] including the ubiquitous garlic. In addition to garlic's inhibitory effect on respiratory pathogens (see Section 8.2), garlic is an effective antiparasitic agent in poultry,[99,100] and a hexane extract of the bulb is a reasonably effective anthelmintic for fish.[101] Others have found that a wide range of human and animal parasitic pathogens are strongly inhibited by a metabolite of garlic (diallyl trisulfide) *in vitro*.[102] One study, however, reported that garlic was totally ineffective in treating parasitic infections in donkeys.[103] This result was most likely due to the fact that the garlic used in the study was cooked and crushed, a process that probably inactivated the bioactive principles (allyl sulfides).[104,105] There are no reports on the efficacy of garlic to treat parasitic load in companion animals. Also, although use of garlic as an antiparasitic in animals may be warranted, it is important to monitor potential development of anemia in animals fed diets with high garlic inclusion rates (see respiratory health).

Elecampane (*Inula helenium*) is an herb that found initial popularity not as a medicine for people but as a veterinary medicine. It was used in the European Middle Ages as an ostensible panacea in horses, hence one of its common names "horseheal."[75] It came to the New World with the early American colonists and was adopted by Native Americans as a treatment for bronchitis and chronic lung diseases* and to dispel intestinal parasites. There is a small body of contemporary literature supporting the use of *I. helenium* as an anthelmintic. An aqueous extract of *I. helenium* effectively caused atrophy, degeneration, and necrosis of the viscera of liver flukes, while reducing their recovery to about 2% in experimentally infected rabbits.[106] This extract also reduced egg-laying capacity of liver flukes in rabbits.[107] Intestinal roundworms are also susceptible to elecampane. Infective larvae of roundworms exposed to an extract of *I. helenium in vitro* were killed in about 40 days, and the eggs of exposed adult worms were killed in about 20 days.[108] There are no scientific reports of elecampane being used as a general anthelmintic in livestock or companion animals. The main adverse effect associated with elecampane is hypersensitivity.[109,110] Development of parasitic resistance to conventional anthelmintics is an important problem across the full spectrum of domesticated animals.[108,112] It would be of value to determine the chronic effectiveness of elecampane in animals with respect to the development of parasite resistance and to identify any species-specific characteristics of the herb.

Other common herbs used by North American natives to combat endoparasites include juniper branches (*Juniperus communis*), parsley (*Petroselinum crispum*), stinging nettle (*Urtica dioica*), celery (*Apium graveolens*), carrots (*Daucus carota*), and sunflower (*Helianthus annuus*) seeds.[10] Evidence for efficacy of these plants for managing parasite load is sparse. One article described the ability of sunflower meal combined with lucerne to reduce the need for anthelmintic drenching of grazing sheep and to improve their productivity without directly affecting wormload.[113] It is unclear whether the effect resulted from the sunflower meal, the lucerne, or the combination of the two. In contrast, significant direct cidal effect of a purified extract of carrot on roundworms (*Caenorhabditis elegans*) and microworms (*Panagrellus redivivus*) has been reported,[114] and the essential oil of celery is toxic to cercariae of *Schistosoma mansoni*.[115]

Gastric ulceration is another critically important disease in livestock and veterinary practice, with clinically and economically important morbidity and mortality

* Significant relaxant effect of elecampane volatile oils has been demonstrated on tracheal smooth muscle in guinea pigs (Reiter and Brandt 1985), providing a scientific basis for its use as a respiratory tonic.

in most major livestock species, including as many as 93% of racehorses,[116] 75% of veal and beef calves fed milk replacers,[117,118] and in the swine industry, up to 90% of necropsied pigs showed lesions of esophago-gastric ulcer syndrome.[119,120]

Over the centuries, a number of North American plants have been used to treat gastrointestinal complaints. Some, including marshmallow, American licorice, yarrow, and meadowsweet, have shown antiulcer properties under scientific investigation. In general, the antiulcer properties of these plants are thought to be due to their high content of flavonoids. Flavonoids function within plants as antioxidants and enzyme inhibitors and are involved with photosensitization, photosynthesis, and defense against infection.[121] Extensive literature describes beneficial effects of plant flavonoids in the prevention or treatment of gastric ulceration in mammals. The mechanisms of gastroprotective action of flavonoids are not well understood and are likely acting through multiple pathways. Hypotheses include an ability of flavonoids to reduce leukocyte activation, which reduces their adherence to the endothelial surface.[122] Increased leukocyte adherence is associated with reduced perfusion of the mucosa, which predisposes to injury. Activation of these cells (by inflammatory mediators such as LTB_4) leads to release of oxygen-derived free radicals, which damage the endothelium and mucosa.[123] In pylorus-ligated pigs, the flavonoid 3-methoxy-5,7,3′,4′-tetrahydroxyflavan was as effective as omeprazole in reducing free and total acids in gastric secretions and appears to function as a specific histidine decarboxylase inhibitor, thus reducing gastric mucosal histamine.[124]

Marshmallow *(Althaea officinalis* L.) is native to southern and western Europe and western Asia and was long ago introduced to the northeastern region of North America, where it grows well in salt marshes, damp low-lying land, river banks, and coastal regions. It has been used for centuries as both a food and a medicine, and its medicinal uses date back to ancient Greeks, who used it for the treatment of bruises and blood loss. Medieval European herbalists used marshmallow for treating coughs in cats,[124] sore throats, indigestion, and diarrhea. Currently, herbalists use it for asthma, bronchitis, common cold with sore throat, cough, inflammatory bowel diseases, stomach ulcers, weight loss, and wound healing.

The primary class of bioactive phytochemical constituents within marshmallow root is mucilage (18–35% by weight), consisting of a number of polysaccharides, apectin, asparagine, and tannins. Marshmallow also contains large amounts of vitamin A, calcium, and zinc and significant amounts of iron, sodium, iodine, and B-complex vitamins.

A primary use of the marshmallow root is to relieve digestive problems, and it is recommended for gastric and duodenal ulcers, ulcerative colitis, and Crohn's disease.[58] The mucilage acts as an excellent demulcent that coats the gastrointestinal mucosa, including the mouth and pharynx, providing protection against local irritation.[125] It also counters excess stomach acid and is mildly laxative. Marshmallow root has flavonoids that possess antiulcer activity (see discussion in this section).[126]

American licorice (*Glycyrrhiza lepidota*) is native to central Canada south to California, Texas, and Virginia but is absent from the southeastern states. It is also known as "wild licorice" to distinguish it from the related European liquorice (*Glycyrrhiza glabra*). The plant grows 40–100 cm (16–40 inches) in height; has long, tough brown roots, which are sweet; and has long been used as food and for medicinal purposes by Native Americans. American licorice is grazed by cattle and is considered an important ethnoveterinary medicine for gastric ulceration in Canada.[10,88] The sweetness of American licorice is from glycyrrhizin, the acid ammonium salt of the triterpenoid glycyrrhizic acid. Licorice extracts and its principal component, glycyrrhizin, have extensive use in foods and tobacco and in both allopathic and herbal medicine.[58] As a result, licorice and glycyrrhizin are immensely popular in the United States, with an estimated consumption of 0.027 to 3.6 mg glycyrrhizin/kg/day.[127] Both products have been approved for use in foods by most national and international regulatory agencies. Glycyrrhizin is believed to contribute to the herb's healing properties. Laboratory studies have reported that glycyrrhizin reduces inflammation, promotes secretion of mucus (usually through coughing), soothes irritation, protects the stomach and gastrointestinal tract, and stimulates the activity of the adrenal glands (regulates cortisol, the stress hormone).[128] The roots also contain coumarins, flavonoids, volatile oils, and plant sterols.[128,129]

Licorice root is often used to prevent and treat stomach ulcers by health care practitioners in Europe, Japan, and North America, and a "deglycyrrhizinated" product is increasing in popularity due to putative hypertensive effects of glycyrrhizin.[130] Deglycyrrhizinated licorice (DGL) has been used to treat gastric ulcers with positive results.[131] DGL reduces inflammation and is as effective as some prescription drugs for gastric ulcers.[132–134] DGL may offer protection against ulcer formation when taken with aspirin and enhance the effectiveness of antiulcer medications such as cimetidine or ranitidine.[131,133] The observed effectiveness of DGL has led to the search for another bioactive constituent; a strong candidate for the antiulcer efficacy of DGL is coumarin.[135] One animal study showed that aspirin coated with licorice reduced the number of ulcers in rats by 50%.[131] Prior studies in humans have found that preparations containing glycyrrhizin (an active compound in licorice) may be as effective as leading antiulcer medications in relieving pain associated with stomach ulcers and preventing the ulcers from recurring.[136]

North American yarrow (*Achillea millefolium*) is a popular native American herb for treatment of gastrointestinal complaints as well as respiratory diseases, bleeding, and wound healing.[136,137] The main constituents of yarrow are volatile oils (sabinene, β-pinene, 1,8-cineole, artemisia ketone, linalool, α-thujone, β-thujone, camphor, borneol, fenchyl acetate, bornyl acetate, (E)-beta-caryophyllene, germacrene D, caryophyllene oxide, beta-bisabolol, delta-cadinol, chamazulene)[139]; flavonoids (apigenin- and luteolin-7-glycosides, and rutin); and alkaloids.[58]

There is a scarcity of scientific evaluation of yarrow as an antiulcer medicine. An isolated report described the safety and gastroprotective efficacy of an aqueous extract of the aerial portions of yarrow in gastrically challenged rats.[137] Ulcer formation induced by ethanol and indomethacin was prevented by the extract, and existing chronic ulcers induced by acetic acid healed significantly faster when rats were fed the extract. Over the 90-day experimental period, there were also no signs of toxicity. The mechanism of gastroprotection is not known but may be associated with the flavonoid content of yarrow (see discussion of flavonoids). It is possible also that the anti-inflammatory effect of yarrow[140,141] may contribute to the healing properties of the plant.

8.5 CONCLUSIONS

Ethnoveterinary medicine in North America persists as an important health care strategy for livestock and companion animals. A rich and colorful history surrounds the use of a wide range of native and naturalized plants across the continent, and contemporary scientific investigation is providing new information on how well these plants are altering pathophysiological processes of important veterinary diseases. There is substantial evidence for willow and cayenne to treat inflammation and pain in animals, and there is good scientific rationale for the use of mullein and garlic in respiratory disease. Garlic also holds potential as an anthelmintic, as does elecampane, but further research is needed to confirm the anthelmintic action of these plants in a number of veterinary species. The safety profile of the majority of the plants discussed appears to be acceptable, but species-specific research is greatly lacking, and it is possible that toxicities may appear in some species that are not identifiable in others. The safety of garlic is guarded, with significant risk of anemia existing at higher doses. Veterinarians are cautioned to conduct regular hematological assessments of those animals chronically consuming garlic or those consuming garlic in high acute quantities.

REFERENCES

1. Anderson, V.D. *Creatures of empire: how domestic animals transformed early America.* Oxford University Press, Oxford, UK, 2002.
2. Turker, A.U., and Gurel, E. Common mullein (*Verbascum thapsus* L.): recent advances in research. *Phytother Res* 19, 733, 2005.
3. Miyase, T., Horikoshi, C., Yabe, S., Miyasaka, S., Melek, F.R., and Kusano, G. Saikosaponin homologues from *Verbascum* spp. The structures of mulleinsaponins I–VII. *Chem Pharm Bull (Tokyo)* 45, 2029, 1997.
4. Warashina, T., Miyase, T., and Ueno, A. Phenylethanoid and lignan glycosides from *Verbascum thapsus. Phytochemistry* 31, 961, 1992.
5. Seifert, K., Lien, N.T., Schmidt, J., Johne, S., Popov, S.S., and Porzel, A. Iridoids from *Verbascum pulverulentum. Planta Med* 55, 470, 1989.
6. Zgorniak-Nowosielska, I., Grzybek, J., Manolova, N., Serkedjieva, J., and Zawiliáska, B. Antiviral activity of *Flos verbasci* infusion against influenza and herpes simplex viruses. *Arch Immunol Ther Exp (Warsz)* 39, 103, 1991.
7. Turker, A.U., and Camper, N.D. Biological activity of common mullein, a medicinal plant. *J Ethnopharmacol* 82, 117, 2002.
8. Vef-Georg, G. Mullein helps in hoarseness. *Pflege Aktuell* 59, 578, 2005.
9. Recio, M.C., Giner, R.M., Máñez, S., and Ríos, J.L. Structural considerations on the iridoids as anti-inflammatory agents. *Planta Med* 60, 232, 1994.
10. Lans, C., Turner, N., Brauer, G., Lourenco, G., and Georges, K. Ethnoveterinary medicines used for horses in Trinidad and in British Columbia, Canada. *J Ethnobiol Ethnomed* 2, 31, 2006.
11. Serkedjieva, J. Combined antiinfluenza virus activity of *Flos verbasci* infusion and amantadine derivatives. *Phytother Res* 14, 571, 2000.
12. Zanon, S.M., Ceriatti, F.S., Rovera, M., Sabini, L.J., and Ramos, B.A. Search for antiviral activity of certain medicinal plants from Cordoba, Argentina. *Rev Latinoam Microbiol* 41, 59, 1999.

13. McCutcheon, A.R., Roberts, T.E., Gibbons, E., Ellis, S.M., Babiuk, L.A., Hancock, R.E., and Towers, G.H. Antiviral screening of British Columbian medicinal plants. *J Ethnopharmacol* 49, 101, 1995.
14. Slagowska, A., Zgorniak-Nowosielska, I., and Grzybek, J. Inhibition of herpes simplex virus replication by *Flos verbasci* infusion. *Pol J Pharmacol Pharm* 39, 55, 1987.
15. Gurbuz, I., Ozkan, A.M., Yesilada, E., Kutsal, O. Anti-ulcerogenic activity of some plants used in folk medicine of Pinarbasi (Kayseri, Turkey). *J Ethnopharmacol* 101, 313, 2005.
16. Krushkov, I., Paskov, D., and Popivanov, D. Ganglion-blocking effect of some alkaloids of the plant *Verbascum nobile* vel. *Nauchni Tr Vissh Med Inst Sofiia* 49, 19, 1970.
17. Castro, A.I., Carmona, J.B., Gonzales, F.G., and Nestar, O.B. Occupational airborne dermatitis from gordolobo (*Verbascum densiflorum*). *Contact Dermatitis* 55, 301, 2006.
18. Daly, J.M. Newton, J.R., and Mumford, J.A. Current perspectives on control of equine influenza. *Vet Res* 35, 411, 2004.
19. Wood, J.L., Newton, J.R., Chanter, N., and Mumford, J.A. Association between respiratory disease and bacterial and viral infections in British racehorses. *J Clin Microbiol* 43, 120, 2005.
20. Gagea, M.I., Bateman, K.G., van Dreumel, T., McEwen, B.J., Carman, S., Archambault, M., Shanahan, R.A., and Caswell, J.L. Diseases and pathogens associated with mortality in Ontario beef feedlots. *J Vet Diagn Invest* 18, 18, 2006.
21. Erles, K., Dubovi, E.J., Brooks, H.W., and Brownlie, J. Longitudinal study of viruses associated with canine infectious respiratory disease. *J Clin Microbiol* 42, 4524, 2004.
22. Helps, C.R. Lait, P., Damhuis, A., Björnehammar, U., Bolta, D., Brovida, C., Chabanne, L., Egberink, H., Ferrand, G., Fontbonne, A., Pennisi, M.G., Gruffydd-Jones, T., Gunn-Moore, D., Hartmann, K., Lutz, H., Malandain, E., Möstl, K., Stengel, C., Harbour, D.A., and Graat, E.A. Factors associated with upper respiratory tract disease caused by feline herpesvirus, feline calicivirus, *Chlamydophila felis* and *Bordetella bronchiseptica* in cats: experience from 218 European catteries. *Vet Rec* 156, 669, 2005.
23. Whitson TD. et al., *Weeds of the west.* Pioneer of Jackson Hole Printing. Jackson, WY, 1992.
24. Rivlin, R.S. Historical perspective on the use of garlic. *J Nutr* 131, 951S, 2001.
25. Hyams, E. *Plants in the service of man: 10,000 years of domestication.* Dent, London, 1971.
26. Hahn, G. History, folk medicine, and legendary uses of garlic. In: *Garlic. The science and therapeutic application of Allium sativum L. and related species.* 2nd ed. Koch and Lawson (Eds.). Williams and Wilkins, Baltimore, MD, 1996, pp. 1–19.
27. Bjarnsholt, T., Jensen, P.Ø., Rasmussen, T.B., Christophersen, L., Calum, H., Hentzer, M., Hougen, H.P., Rygaard, J., Moser, C., Eberl, L., Høiby, N., and Givskov, M. Garlic blocks quorum sensing and promotes rapid clearing of pulmonary *Pseudomonas aeruginosa* infections. *Microbiology* 151, 3873, 2005.
28. Adeleye, I.A., and Opiah, L. Antimicrobial activity of extracts of local cough mixtures on upper respiratory tract bacterial pathogens. *West Indian Med J* 52, 188, 2003.
29. Dikasso, D., Lemma, H., Urga, K., Debella, A., Addis, G., Tadele, A., and Yirsaw, K. Investigation on the antibacterial properties of garlic (*Allium sativum*) on pneumonia causing bacteria. *Ethiop Med J* 40, 241, 2002.
30. Andrianova, I.V., Sobenin, I.A., Sereda, E.V., Borodina, L.I., and Studenikin, M.I. Effect of long-acting garlic tablets "allicor" on the incidence of acute respiratory viral infections in children. *Ter Arkh* 75, 53, 2003.
31. Rassoul, F., Salvetter, J., Reissig, D., Schneider, W., Thiery, J., and Richter, V. The influence of garlic (*Allium sativum*) extract on interleukin 1alpha-induced expression of endothelial intercellular adhesion molecule-1 and vascular cell adhesion molecule-1. *Phytomedicine* 13, 230, 2006.

32. Hofbauer, R., Frass, M., Gmeiner, B., Kaye, A.D., and Frost, E.A. Garlic extract (*Allium sativum*) reduces migration of neutrophils through endothelial cell monolayers. *Middle East J Anesthesiol* 15, 649, 2000.
33. Hofbauer, R., Frass, M., Gmeiner, B., Kaye, A.D., and Frost, E.A. Effects of garlic extract (*Allium sativum*) on neutrophil migration at the cellular level. *Heart Dis* 3, 14, 2001.
34. Josling, P. Preventing the common cold with a garlic supplement: a double-blind, placebo-controlled survey. *Adv Ther* 18, 189, 2001.
35. Munday, R., Munday, C.M., and Munday, J.S. Hemolytic anemia and induction of phase II detoxification enzymes by diprop-1-enyl sulfide in rats: dose-response study. *J Agric Food Chem* 53, 9695, 2005.
36. Pearson, W., Boermans, H.J., Bettger, W.J., McBride, B.W., and Lindinger, M.I. Association of maximum voluntary dietary intake of freeze-dried garlic with Heinz body anemia in horses. *Am J Vet Res* 66, 457, 2005.
37. Caldin, M., Carli, E., Furlanello, T., Solano-Gallego, L., Tasca, S., Patron, C., and Lubas, G. A retrospective study of 60 cases of eccentrocytosis in the dog. *Vet Clin Pathol* 34, 224, 2005.
38. Yamato, O., Kasai, E., Katsura, T., Takahashi, S., Shiota, T., Tajima, M., Yamasaki, M., and Maede, Y. Heinz body hemolytic anemia with eccentrocytosis from ingestion of Chinese chive (*Allium tuberosum*) and garlic (*Allium sativum*) in a dog. *J Am Anim Hosp Assoc* 41, 68, 2005.
39. Lee, K.W., Yamato, O., Tajima, M., Kuraoka, M., Omae, S., and Maede, Y. Hematologic changes associated with the appearance of eccentrocytes after intragastric administration of garlic extract to dogs. *Am J Vet Res* 61, 1446, 2000.
40. Stevens, H. Suspected wild garlic poisoning in sheep. *Vet Rec* 115, 363, 1984.
41. Busquet, M., Calsamiglia, S., Ferret, A., and Kamel, C. Plant extracts affect in vitro rumen microbial fermentation. *J Dairy Sci* 89, 761, 2006.
42. Chowdhury, A.K., Ahsan, M., Islam, S.N., and Ahmed, Z.U. Efficacy of aqueous extract of garlic and allicin in experimental Shigellosis in rabbits. *Indian J Med Res* 93, 33, 1991.
43. Kamel, S.H. Toxicity of the Egyptian garlic alkaloid injected into rats kept under various conditions of temperature and relative humidity. *Zentralbl Veterinarmed A* 13, 662, 1966.
44. Turner, N.J., and Hebda, R.J. Contemporary use of bark for medicine by two Salishan native elders of southeast Vancouver Island, Canada. *J Ethnopharmacol* 29, 59, 1990.
45. Baranețchi, C., Segal, B., and Segal, R. Administration of a water-alcohol extract of the fir tree (*Abies alba*) as an aerosol in respiratory tract diseases. *Rev Med Chir Soc Med Nat Iasi* 89, 123, 1985.
46. Radi, Z.A., Styer, E.L., and Thompson, L.J. *Prunus* spp. intoxication in ruminants: a case in a goat and diagnosis by identification of leaf fragments in rumen contents. *J Vet Diagn Invest* 16, 593, 2004.
47. Sargison, N.D., Williamson, D.S., Duncan, J.R., and McCance, R.W. *Prunus padus* (bird cherry) poisoning in cattle. *Vet Rec* 138, 188, 1996.
48. Selby, L.A., Menges, R.W., Houser, E.C., Flatt, R.E., and Case, A.A. Outbreak of swine malformations associated with the wild black cherry, *Prunus serotina*. *Arch Environ Health* 22, 496, 1971.
49. Wilsher, S., Allen, W.R., and Wood, J.L. Factors associated with failure of thoroughbred horses to train and race. *Equine Vet J* 38, 113, 2006.
50. Zoric, M., Stern, S., Lundeheim, N., and Wallgren, P. Four-year study of lameness in piglets at a research station. *Vet Rec* 153, 323, 2003.
51. Booth, C.J., Warnick, L.D., Gröhn, Y.T., Maizon, D.O., Guard, C.L., and Janssen, D. Effect of lameness on culling in dairy cows. *J Dairy Sci* 87, 4115, 2004.

52. Underwood, T.J. Though the years, an arthritis timeline. *Arthritis Today* January–February 2000. http://arthritisinsight.com/medical/disease/ra/history.html (accessed June 27, 2007).
53. Mauldin, J., Cameron, H.D., Jeanotte, D., Solomon, G., and Jarvis, J.N. Chronic arthritis in children and adolescents in two Indian health service user populations. *BMC Musculoskelet Disord* 5, 30, 2004.
54. Ferucci, E.D., Templin, D.W., and Lanier, A.P. Rheumatoid arthritis in American Indians and Alaska Natives: a review of the literature. *Semin Arthritis Rheum* 34, 662, 2005.
55. Foster, S. Cayenne: pepper power! *Better Nutrition* July 1996. http://findarticles.com/p/articles/mi_m0FKA/is_n7_v58/ai_18453786 (accessed June 27, 2007).
56. Heiser, C.B. *Nightshades*. Freeman, San Francisco, 1969.
57. Evans, W.C. *Trease and Evans pharmacognosy*. Bailliere Tindall, London, 1989.
58. Newall, C.A., Anderson, L.A., and Phillipson, J.D. *Herbal medicines: a guide for health-care professionals*. Pharmaceutical Press, London, 1996.
59. Kawada, T., and Iwai, K. In vivo and in vitro metabolism of dihydrocapsaicin, a pungent principle of hot pepper, in rats. *Agric Biol Chem* 49, 441, 1985.
60. Kawada, T., Suzuki, T., Takahashi, M., and Iwai, K. Gastrointestinal absorption and metabolism of capsaicin and dihydrocapsaicin in rats. *Toxicol Appl Pharmacol* 72, 449, 1984.
61. Watanabe, T., Kawada, T., Yamamoto, M., and Iwai, K. Capsaicin, a pungent principle of hot red pepper, evokes catecholamine secretion from the adrenal medulla of anesthetized rats. *Biochem Biophys Res Commun* 142, 259, 1987.
62. Deal, C.L., Schnitzer, T.J., Lipstein, E., Seibold, J.R., Stevens, R.M., Levy, M.D., Albert, D., and Renold, F. Treatment of arthritis with topical capsaicin: a double-blind trial. *Clin Ther* 13, 383, 1991.
63. McCarthy, G.M., and McCarty, D.J. Effect of topical capsaicin in the therapy of painful osteoarthritis of the hands. *J Rheumatol* 19, 604, 1992.
64. Zhang, W.Y., and Li Wan Po, A. The effectiveness of topically applied capsaicin. A meta-analysis. *Eur J Clin Pharmacol* 46, 517, 1994.
65. McKay, L., Gemmell, H., Jacobson, B., and Hayes, B. Effect of a topical herbal cream on the pain and stiffness of osteoarthritis: a randomized double-blind, placebo-controlled clinical trial. *J Clin Rheumatol* 9, 164, 2003.
66. Towheed, T.E., and Hochberg, M.C. A systematic review of randomized controlled trials of pharmacological therapy in osteoarthritis of the knee, with an emphasis on trial methodology. *Semin Arthritis Rheum* 26, 755, 1997.
67. Mason, L., Moore, R.A., Derry, S., Edwards, J.E., and McQuay, H.J. Systematic review of topical capsaicin for the treatment of chronic pain. *BMJ* 328, 991, 2004.
68. Cordell, G.A., and Araujo, O.E. Capsaicin: identification, nomenclature, and pharmacotherapy. *Ann Pharmacother* 27, 330, 1993.
69. Virus, R.M., and Gebhart, G.F. Pharmacologic actions of capsaicin: apparent involvement of substance P and serotonin. *Life Sci* 25, 1273, 1979.
70. He, X., Schmidt, R.F., and Schmittner, H. Effects of capsaicin on articular afferents of the cat's knee joint. *Agents Actions* 25, 222, 1988.
71. Anonymous. Final report on the safety assessment of *Capsicum annuum* extract, *Capsicum annuum* fruit extract, *Capsicum annuum* resin, *Capsicum annuum* fruit powder, *Capsicum frutescens* fruit, *Capsicum frutescens* fruit extract, *Capsicum frutescens* resin, and capsaicin. *Int J Toxicol* 26, 3, 2007.
72. Hood, V.C., Cruwys, S.C., Urban, L., and Kidd, B.L. The neurogenic contribution to synovial leucocyte infiltration and other outcome measures in a guinea pig model of arthritis. *Neurosci Lett* 299, 201, 2001.
73. Matucci-Cerinic, M., Marabini, S., Jantsch, S., Cagnoni, M., and Partsch, G. Effects of capsaicin on the metabolism of rheumatoid arthritis synoviocytes in vitro. *Ann Rheum Dis* 49, 598, 1990.

74. Williams, S.R., Clark, R.F., and Dunford, J.V. Contact dermatitis associated with capsaicin: Hunan hand syndrome. *Ann Emerg Med* 25, 713, 1995.
75. Castleman, M. *The healing herbs: the ultimate guide to the curative power of nature's medicines*. Rodale Press, Emmaus, PA, 1995.
76. Duke, J. *CRC handbook of medicinal herbs*. CRC Press, Boca Raton, FL, 1985.
77. Bradley, P.R. *British herbal compendium*. British Herbal Medicine Association, Bournemouth, UK, 1992.
78. Felter, H.W., and Lloyd, J.U. King's American dispensatory. 1898. http://www.henriettesherbal.com/eclectic/kings/index.html.
79. Chrubasik, S., Eisenberg, E., Balan, E., Weinberger, T., Luzzati, R., and Conradt, C. Treatment of low back pain exacerbations with willow bark extract: a randomized double-blind study. *Am J Med* 109, 9, 2000.
80. Schmid, B., Lüdtke, R., Selbmann, H.K., Kötter, I., Tschirdewahn, B., Schaffner, W., and Heide, L. Effectiveness and tolerance of standardized willow bark extract in arthrosis patients. Randomized, placebo controlled double-blind study. *Z. Rheumatol* 59, 314, 2000.
81. Schmid, B., Lüdtke, R., Selbmann, H.K., Kötter, I., Tschirdewahn, B., Schaffner, W., and Heide, L. Efficacy and tolerability of a standardized willow bark extract in patients with osteoarthritis: randomized placebo-controlled, double blind clinical trial. *Phytother Res* 15, 344, 2001.
82. Khayyal, M.T., El-Ghazaly, M.A., Abdallah, D.M., Okpanyi, S.N., Kelber, O., and Weiser, D. Mechanisms involved in the anti-inflammatory effect of a standardized willow bark extract. *Arzneimittelforschung* 55, 677, 2005.
83. Fiebich, B., and Chrubasik, S. Effects of an ethanolic *Salix* extract on the release of selected inflammatory mediators in vitro. *Phytomedicine* 11,135, 2004.
84. Biegert, C., Wagner, I., Lüdtke, R., Kötter, I., Lohmüller, C., Günaydin, I., Taxis, K., and Heide, L. Efficacy and safety of willow bark extract in the treatment of osteoarthritis and rheumatoid arthritis: results of two randomized double-blind controlled trials. *J Rheumatol* 31, 2121, 2004.
85. Akao, T., Yoshino, T., Kobashi, K., and Hattori, M. Evaluation of salicin as an antipyretic prodrug that does not cause gastric injury. *Planta Med* 68, 714, 2002.
86. März, R.W., and Kemper, F. Willow bark extract—effects and effectiveness. Status of current knowledge regarding pharmacology, toxicology and clinical aspects. *Wien Med Wochenschr* 152, 354, 2002.
87. Opelz, G., and Terasaki, P.I. Suppression of lymphocyte transformation by aspirin. *Lancet* 2, 478, 1973.
88. Lans, C. Turner, N., Khan, T., Brauer, G., and Boepple, W. Ethnoveterinary medicines used for ruminants in British Columbia, Canada. *J Ethnobiol Ethnomed* 3, 11, 2007.
89. Anonymous. *Urticae folium/herba*. In: *ESCOP Monographs*. Thieme, Stuttgart, 2003, pp. 521–527.
90. Chrubasik, J.E., et al. A comprehensive review on nettle effect and efficacy profiles, Part I: *Herba urticae*. *Phytomedicine* 14, 423, 2007.
91. Obertreis, B., Ruttkowski, T., Teucher, T., Behnke, B., and Schmitz, H. Ex-vivo in-vitro inhibition of lipopolysaccharide stimulated tumor necrosis factor-alpha and interleukin-1 beta secretion in human whole blood by extractum urticae dioicae foliorum. *Arzneimittelforschung* 46, 389, 1996.
92. Teucher, T., Obertreis, B., Ruttkowski, T., and Schmitz, H. Cytokine secretion in whole blood of healthy subjects following oral administration of *Urtica dioica* L. plant extract. *Arzneimittelforschung* 46, 906, 1996.
93. Schulze-Tanzil, G., de, S.P., Behnke, B., Klingelhoefer, S., Scheid, A., *and* Shakibaei, M. Effects of the antirheumatic remedy hox alpha—a new stinging nettle leaf extract on matrix metalloproteinases in human chondrocytes in vitro. *Histol Histopathol* 17, 477, 2002.

94. Riehemann, K., Behnke, B., and Schulze-Osthoff K. Plant extracts from stinging nettle (*Urtica dioica*), an antirheumatic remedy, inhibit the proinflammatory transcription factor NF-kappaB. *FEBS Lett* 442, 89, 1999.
95. Bathe, A.P. An unusual manifestation of nettle rash in three horses. *Vet Rec* 134, 11, 1994.
96. Stürchler, D., DiGiacomo, R.F., and Rausch, L. Parasitic infections in Yakima Indians. *Ann Trop Med Parasitol* 81, 291, 1987.
97. Taranto, N.J., Cajal, S.P., De Marzi, M.C., Fernández, M.M., Frank, F.M., Brú, A.M., Minvielle, M.C., Basualdo, J.A., and Malchiodi, E.L. Clinical status and parasitic infection in a Wichí Aboriginal community in Salta, Argentina. *Trans R Soc Trop Med Hyg* 97, 554, 2003.
98. Basombrio, M.A., Segovia, A., Peralta Ramos, M., Esteban, E., Stumpf, R., Jurgensen, P., Winkler, M.A., Sayre, K., and Ferrer, J.F. Endemic *Trypanosoma cruzi* infection in Indian populations of the Gran Chaco territory of South America: performance of diagnostic assays and epidemiological features. *Ann Trop Med Parasitol* 93, 41, 1999.
99. Hafez, H.M., and Hauck, R. Efficacy of a herbal product against *Histomonas meleagridis* after experimental infection of turkey poults. *Arch Anim Nutr* 60, 436, 2006.
100. Zenner, L., Callait, M.P., Granier, C., and Chauve, C. In vitro effect of essential oils from *Cinnamomum aromaticum*, *Citrus limon* and *Allium sativum* on two intestinal flagellates of poultry, *Tetratrichomonas gallinarum* and *Histomonas meleagridis*. *Parasite* 10, 153, 2003.
101. Peña, N., Auró, A., and Sumano, H. A comparative trial of garlic, its extract and ammonium-potassium tartrate as anthelmintics in carp. *J Ethnopharmacol* 24, 199, 1988.
102. Lun, Z.R., Burri, C., Menzinger, M., and Kaminsky, R. Antiparasitic activity of diallyl trisulfide (Dasuansu) on human and animal pathogenic protozoa (*Trypanosoma* sp., *Entamoeba histolytica* and *Giardia lamblia*) in vitro. *Ann Soc Belge Med Trop* 74, 51, 1994.
103. Abells Sutton, G., and Haik, R. Efficacy of garlic as an anthelmintic in donkeys. *Israel J Vet Med* 54,1, 1999.
104. Cavagnaro, P.F., Camargo, A., Galmarini, C.R., and Simon, P.W. Effect of cooking on garlic (*Allium sativum* L.) antiplatelet activity and thiosulfinates content. *J Agric Food Chem* 55, 1280, 2007.
105. Ali, M., Bordia, T., and Mustafa, T. Effect of raw versus boiled aqueous extract of garlic and onion on platelet aggregation. *Prostaglandins Leukot Essent Fatty Acids* 60, 43, 1999.
106. Rhee, J.K., Baek, B.K., and Ahn, B.Z. Structural investigation on the effects of the herbs on *Clonorchis sinensis* in rabbits. *Am J Chin Med* 13, 119, 1985.
107. Rhee, J.K., Baek. B.K., and Ahn, B.Z. Alternations of *Clonorchis sinensis* EPG by administration of herbs in rabbits. *Am J Chin Med* 13, 65, 1985.
108. El Garhy, M.F., and Mahmoud, L.H. Anthelminthic efficacy of traditional herbs on *Ascaris lumbricoides*. *J Egypt Soc Parasitol* 32, 893, 2002.
109. Paulsen, E. Contact sensitization from Compositae-containing herbal remedies and cosmetics. *Contact Dermatitis* 47, 189, 2002.
110. Lamminpää, A., Estlander, T., Jolanki, R., and Kanerva, L. Occupational allergic contact dermatitis caused by decorative plants. *Contact Dermatitis* 34, 330, 1996.
111. Stear, M.J., Doligalska, M., and Donskow-Schmelter, K. Alternatives to anthelmintics for the control of nematodes in livestock. *Parasitology* 134, 139, 2007.
112. Kopp, S.R., Kotze, A.C., McCarthy, J.S., and Coleman, G.T. High-level pyrantel resistance in the hookworm *Ancylostoma caninum*. *Vet Parasitol* 143, 299, 2007.
113. Van Houtert, M.F., Barger, I.A., and Steel, J.W. Dietary protein for young grazing sheep: interactions with gastrointestinal parasitism. *Vet Parasitol* 60, 283, 1995.
114. Momin, R.A., and Nair, M.G. Pest-managing efficacy of trans-asarone isolated from *Daucus carota* L. seeds. *J Agric Food Chem* 50(16), 4475–78, 2002.

115. Saleh, M.M., Zwaving, J.H., Malingré, T.M., and Bos, R. The essential oil of *Apium graveolens* var. *secalinum* and its cercaricidal activity. *Pharm Weekbl Sci* 7, 277, 1985.
116. Andrews, F.M., Frank, N., Sommardahl, C.S., Buchanan, B.R., Elliott, S.B., and Allen, V.A. Effects of intravenously administered omeprazole on gastric juice pH and gastric ulcer scores in adult horses. *J Vet Intern Med* 20, 1202, 2006.
117. Ahmed, A.F., Constable, P.D., and Misk, N.A. Effect of an orally administered antacid agent containing aluminum hydroxide and magnesium hydroxide on abomasal luminal pH in clinically normal milk-fed calves. *J Am Vet Med Assoc* 220, 74, 2002.
118. Mattiello, S., Canali, E., Ferrante, V., Caniatti, M., Gottardo, F., Cozzi, G., Andrighetto, I., and Verga, M. The provision of solid feeds to veal calves: II. Behavior, physiology, and abomasal damage. *J Anim Sci* 80, 367, 2002.
119. Friendship, R.M. Gastric ulcers: an under-recognized cause of mortality and morbidity. *Adv Pork Prod* 14, 159, 2003.
120. Melnichouk, S.I. Mortality associated with gastric ulceration in swine. *Can Vet J* 43, 223, 2002.
121. Middleton, E. Plant flavonoid effects on mammalian cell systems. *J Herbs Spices Med Plants* 3, 103, 1988.
122. Blank, M., Ems, B.L., O'Brien, L.M., Weisshaar, P.S., Ares, J.J., Abel, P.W., McCafferty, D.M., and Wallace, J.L. Flavonoid-induced gastroprotection in rats: role of blood flow and leukocyte adherence. *Digestion* 58, 147, 1997.
123. Martin Calero, M., La Casa, C., Motilva, V., Lopez, A., and Alarcon de la Lastra, C. Healing process induced by a flavonic fraction of *Bidens aurea* on chronic gastric lesions in rat. Role of angiogenesis and neutrophil inhibition. *Z Naturforsch* 51, 570, 1996.
124. Parmar, N.S., and Hennings, G. The gastric antisecretory activity of 3-methoxy-5,7,3′,4′-tetrahydroxyflavan (ME)—a specific histidine decarboxylase inhibitor in rats. *Agents Actions* 15, 143, 1984.
125. Kardosova, A., and Capek, P. Antitussive activity of a rhamnogalacturonan from the roots of *Althaea officinalis*, var. *robusta*. *J Carbohydrate Chem* 12, 589, 1993.
126. Gudej, J. Flavonoids, phenolic acids and coumarins from the roots of *Althaea officinalis*. *Planta Med* 57, 284, 1991.
127. Isbrucker, R.A., and Burdock, G.A. Risk and safety assessment on the consumption of licorice root (*Glycyrrhiza* sp.), its extract and powder as a food ingredient, with emphasis on the pharmacology and toxicology of glycyrrhizin. *Regul Toxicol Pharmacol* 46, 167, 2006.
128. Shibata, S. A drug over the millennia: pharmacognosy, chemistry, and pharmacology of licorice. *Yakugaku Zasshi* 120, 849, 2000.
129. Hsiang, C.Y., Lai, I.L., Chao, D.C., and Ho, T.Y. Differential regulation of activator protein 1 activity by glycyrrhizin. *Life Sci* 70, 1643, 2002.
130. Yasue, H., Itoh, T., Mizuno, Y., and Harada, E. Severe hypokalemia, rhabdomyolysis, muscle paralysis, and respiratory impairment in a hypertensive patient taking herbal medicines containing licorice. *Intern Med* 46, 575, 2007.
131. Dehpour, A.R., Zolfaghari, M.E., and Samadian, T. The protective effect of liquorice components and their derivatives against gastric ulcer induced by aspirin in rats. *J Pharm Pharmacol* 46, 148, 1994.
132. Russell, R.I., Morgan, R.J., and Nelson, L.M. Studies on the protective effect of deglycyrrhinised liquorice against aspirin (ASA) and ASA plus bile acid-induced gastric mucosal damage, and ASA absorption in rats. *Scand J Gastroenterol* 92, 97, 1984.
133. Bennett, A., Clark-Wibberley, T., Stamford, I.F., and Wright, J.E. Aspirin-induced gastric mucosal damage in rats: cimetidine and deglycyrrhizinated liquorice together give greater protection than low doses of either drug alone. *J Pharm Pharmacol* 32, 150, 1980.
134. Aly, A.M., Al-Alousi, L., and Salem, H.A. Licorice: a possible anti-inflammatory and anti-ulcer drug. *AAPS Pharm SciTech* 6, E74–82, 2005.

135. Khayyal, M.T., Seif-El-Nasr, M., El-Ghazaly, M.A., Okpanyi, S.N., Kelber, O., and Weiser, D. Mechanisms involved in the gastro-protective effect of STW 5 (Iberogast) and its components against ulcers and rebound acidity. *Phytomedicine* 5, 56, 2006.
136. Rees, W.D.W., Rhodes, J., Wright, J.E., Stamford, L.F., and Bennett, A. Effect of deglycyrrhizinated liquorice on gastric mucosal damage by aspirin. *Scand J Gastroenterol* 14, 605, 1979.
137. Cavalcanti, A. M., Baggio, C.H., Freitas, C.S., Rieck, L., de Sousa, R.S., Da Silva-Santos, J.E., Mesia-Vela, S., and Marques, M.C. Safety and antiulcer efficacy studies of *Achillea millefolium* L. after chronic treatment in Wistar rats. *J Ethnopharmacol* 107, 277, 2006.
138. Chandler, F. Maritime Indian herbal remedies. *J Ethnopharmacol* 9, 323, 1983.
139. Orav, A., Arak, E., and Raal, A. Phytochemical analysis of the essential oil of *Achillea millefolium* L. from various European countries. *Nat Prod Res* 20, 1082, 2006.
140. Shipochleiv, T., and Fournadjiev, G. Spectrum of the antiinflammatory effect of *Arctostaphylos uva ursi* and *Achillea millefolium. Probl Vutr Med* 12, 99, 1984.
141. Goldberg, A.S., Mueller, E.C., Eigen, E., and Desalva, S. Isolation of the anti-inflammatory principles from *Achillea millefolium* (Compositae). *J Pharm Sci* 58, 938, 1969.

9 The Medicinal Use of Native North American Plants in Domestic Animals

Ronette Gehring and Kelly Kindscher

CONTENTS

9.1 INTRODUCTION

More than 250 plant species have been documented as being used by native North Americans for the treatment of animals (UKCropNet, 2008; Moerman, 2003). Research investigating the safety and efficacy of these remedies is relatively scarce, however, especially when compared with the data available for ethnoveterinary systems from other parts of the world. Reasons for this may include the comparatively minor socioeconomic role of domestic livestock in most native North American cultures (with the exception of horses and sheep in some societies, especially in central and western regions, following contact with Europeans from the 17th century onward), the predominant use of European herbal remedies by immigrants to the North American continent, and the market leadership of commercial pharmaceutical companies focusing on synthetically produced medicinal products. Many ethnoveterinary treatments used by native North Americans are likely to have been learned relatively recently, since the European introduction of domesticated chickens, horses, and sheep to the North American continent. This knowledge may have been gleaned from experiences with the treatment of humans.

This chapter discusses indigenous North American plants with documented ethnoveterinary uses for which supporting research has been reported in the scientific literature. The focus is on *Echinacea* species, which is the only plant for which clinical veterinary studies have been conducted. Other plants for which *in vitro* data or studies in experimental animal models are available to support the documented veterinary uses are also discussed.

9.2 *ECHINACEA* SPECIES

Echinacea angustifolia has been the most widely used medicinal plant of native American cultures in the Great Plains region of North America, where it is commonly found in grassland habitats. (Although these peoples have often been grouped together and collectively referred to as Plains Indians, they include a variety of formerly independent nations with distinct languages and cultures.) *Echinacea* species were used to treat many different conditions, including colds, infections, bee and insect stings, snakebite, headache, toothache, and wounds (Kindscher, 1992). These uses may be attributable to immunomodulatory, anti-inflammatory, local anesthetic, or anti-infective properties of the plant. Scientific research in both humans and domestic animals has focused on the immunostimulatory characteristics of the plant and its chemical constituents.

9.2.1 Botany and Chemical Constituents

Echinacea is a genus of perennial herbs that is native to North America and a member of the daisy family (Compositae or Asteraceae) (Kindscher, 1992). The three most widely distributed species, which also have a long tradition of medicinal use, are *Echinacea angustifolia,* which occurs throughout much of the western Great Plains; *E. purpurea*, which grows sporadically throughout much of eastern North America; and *E. pallida*, which occurs in the Midwest (Foster, 2008). *Echinacea purpurea* is the species that is most easily cultivated and most commonly used commercially. The whole plant of this species is used in herbal remedies (root, leaf, flower, and seed), whereas only the root of *E. angustifolia* and *E. pallida* is used (Mills and Bone, 2000; Pierce, 1999). The importance of correctly identifying the species used for experimental and clinical trials is highlighted by the fact that some earlier studies have been invalidated due to misidentification and adulteration of the preparations that were used (Pierce, 1999; Kindscher, 1992).

Several potentially active constituents have been isolated from *Echinacea*. These include (1) caffeic acid derivatives (echinacoside, cichoric and chlorogenic acids, and cynarin), which may play a role in stimulating phagocytosis; (2) alkylamides (echinacein and several isobutylamides), which may be responsible for local anesthetic and anti-inflammatory activity; (3) inulin and other high-molecular-weight polysaccharides, which stimulate macrophages and may possess anti-inflammatory activity; and (4) essential oil components (humulene, echolone, vanillin, germacrene, and borneol), several of which have insecticidal properties. Echinacea also contains polyacetylenes, nontoxic pyrrolizidine alkaloids, and flavonoids, of which the biological effects, if any, are presently unknown (Combest and Nemecz, 2002; Mills and Bone, 2000). Cichoric acid and the isobutylamides have been used as quality markers for echinacea products, but there is as yet no consensus about which phytochemicals should serve as the definitive standardization markers (Letchamo et al., 1999; Osowski et al., 2000). Variability in active phytochemical composition due to growth conditions, time of harvest, milling, and storage conditions have been postulated as the cause of variable activity of echinacea-containing preparations.

9.2.2 Use in Human Medicine

Medicinal use of echinacea by the Plains Indians long pre-dates European contact. It is also one of the few native North American plants to be popularized by Western folk practitioners at the turn of the 20th century. Early European settlers and doctors used this plant as a blood purifier, wound healer, and anti-infective (Kindscher, 1992). Today, preparations containing echinacea are primarily used as immunostimulants for the prophylaxis and treatment of infections (Melchart et al., 1994). Preparations containing echinacea ranked sixth on the list of best-selling herbal medicines in the United States (Blumenthal et al., 2008; Ernst, 2005). These are also very popular medications in Europe, particularly Germany.

Authors of a comprehensive and rigorous review of clinical trials investigating the efficacy of echinacea preparations for the treatment and prevention of the common cold in humans concluded that there is some evidence that preparations based on the aerial parts of *Echinacea purpurea* might be effective for the early treatment of colds in adults (Linde et al., 2006).

Echinacea is generally regarded as safe for use in humans, based on its widespread use and the low incidence of reports of adverse reactions. Its toxicity has not been studied in long-term clinical trials. Allergic reactions have been documented, and it is suggested that individuals with known allergies to plants in the daisy family should exercise caution in using echinacea products (Mills and Bone, 2000; Pierce, 1999; Myers and Wohlmuth, 1998).

9.2.3 *In Vitro* Studies and Animal Models

The results of *in vitro* studies and animal models to test the pharmacological effects of echinacea have been extensively reviewed (Barrett, 2003). The most widely reported activity is activation of macrophage and polymorphonuclear immune cells, which has been shown for various different extracts (hydrophilic, lipophilic, and ethanol) of all three medicinal species. Experiments used to demonstrate these effects included yeast-particle ingestion and carbon clearance. There are also reports of simultaneous enhanced cytokine production (nitric oxide [NO], interleukins [IL] 1 and 6, and tumor necrosis factor [TNF]). Claims of anti-inflammatory effects and wound-healing activity were supported by studies showing inhibition of hyaluronidase, cyclo-oxygenase, and 5-lipoxygenase activity *in vitro* (Wagner et al., 1989; Busing, 1952). Topical application of an extract of *E. angustifolia* inhibited edema in the croton oil mouse ear test, and application of an ointment made from expressed juice of *E. purpurea* (aerial parts) significantly improved wound healing in an experimental model (Tragni et al., 1985; Kinkel, Plate, and Tullner, 1984). Antiviral and antimicrobial effects are most likely secondary to immunomodulation.

9.2.4 Documented Uses in Veterinary Medicine

The majority of documented uses of *Echinacea* species by Native Americans are for conditions in humans, but there is specific mention of the use of a smoke fumigant of *Echinacea angustifolia* by Great Plains tribes to treat distemper in horses (Gilmore, 1977). Also known as strangles, this condition is caused by a bacterial (*Streptococcus equi*) infection of the upper respiratory tract. Interestingly, the first use of *Echinacea* to be mentioned in the literature was for the treatment of sores in horses (Gronovius, 1762/1946). *Echinacea purpurea* is also mentioned in the *Materia Medica Americana* as being used to treat saddle sores on a horse's back (Schopf, 1787/1903). In his *Treatise on* Echinacea, the pharmacist J. U. Lloyd told of the following statement by Dr. J. S. Leachman of Sharon, Oklahoma, in the October 1914 issue of *The Gleaner*: "Old settlers all believe firmly in the virtues of Echinacea root, and use it as an aid in nearly every sickness. If a cow or horse does not eat well, the people administer Echinacea, cut up and put in feed. I have noticed that puny stock treated in this manner soon begin to thrive" (Lloyd, 1924a, p. 20).

More recently, there has been a growing interest in investigating the potential benefits of echinacea in food-producing animals as an alternative to production-enhancing antimicrobials. This has been stimulated by concerns about the selection for antibiotic-resistant organisms and their spread through the food chain as a result of chronic oral exposure to subtherapeutic doses of antimicrobial drugs (van den Bogaard and Stobberingh, 2000; Holden, McKean, and Franzenburg, 1999).

Although echinacea has been well studied in laboratory animals to support its use in humans, there are relatively few species-specific data supporting its use in livestock and other domestic animals.

9.2.4.1 Swine

The species for which the most information on echinacea is available is swine. The earliest research suggesting that echinacea may be effective in stimulating immune responses in this species is an *in vitro* study of the effect of a commercially available combination product (Influex™) on the function of porcine polymorphonuclear neutrophils (PMNs) (Stahl et al., 1990). Influex, which is an alcoholic extract of *Echinacea, Aconitum, Apis,* and *Lachesis* containing 2.4×10^{-2}% ethanol, was diluted with 0.9% NaCl to obtain a standardized echinacoside concentration of 0.03 μg/mL.

This was then added to PMN preparations and experiments conducted to measure adherence, chemotaxis, phagocytosis, and chemiluminescence. A positive control (2.4×10^{-2}% ethanol in 0.9% NaCl) and a negative control (0.9% NaCl solution) were included in each experiment. The Influex drops significantly increased PMN adherence, chemotaxis, and phagocytosis compared with both the positive and negative controls. Interestingly, the ethanol alone (positive control) inhibited all these activities compared to the 0.9% NaCl alone (negative control). Chemiluminescence was inhibited to a greater extent by ethanol alone than by Influex, but both these groups showed lower spontaneous and elicited chemiluminescence compared with the negative control. The authors suggested that the Influex drops may act independently and in a contrary manner on the oxidative metabolism (inhibitory) and contractile apparatus (stimulatory) of porcine PMNs. Another explanation offered for the observations was that the inhibitory effect on oxidative metabolism reduces exposure of the contractile apparatus to breakdown by reactive oxygen species, thereby indirectly stimulating the reactive oxygen species. The authors concluded that Influex is a potentially effective remedy to strengthen nonspecific immune defense mechanisms in pigs. A disadvantage of this study is that a combination product was used, making it impossible to distinguish between effects of echinacea and the other constituents.

The first *in vivo* experiments in pigs were conducted by researchers at Iowa State University, who investigated whether adding echinacea to pig rations could improve performance under intensive production conditions (Holden, McKean, and Franzenburg, 1999). Five groups of 20 pigs each were fed rations to which echinacea was added at different levels (0%, 0.1%, 0.5%, and 2.0%). For the fifth ration, 50 g/ton Mecadox® (carbadox) was added as a positive control. The echinacea used in this study contained 0.08% echinacoside on a weight/weight basis. The pigs were fed for 5 weeks, and production parameters, including average daily gain (ADG), average daily feed intake (ADFI), and weight gain-to-feed ratio (G:F) were measured. In the 0- to 2-week period, there was a significant decrease in feed conversion efficiency in the group fed 0% echinacea compared to all the other groups. In the periods 0–3 and 0–4 weeks, the groups that were fed rations with the two higher levels of echinacea (0.5% and 2.0%) were significantly more efficient at feed conversion than the groups fed the lower levels (0.0% and 0.1%). Total performance for the entire experiment, weeks 0–5, was not statistically different between any of the groups, including the positive control. This suggests that subclinical stress was minimal for the duration of the experiment, and further research would be needed to determine whether echinacea affords an advantage to pigs that are stressed by disease and population pressure. Another finding from this study was that the meat from a single animal that was fed the highest level of echinacea (2.0%) had a higher off-flavor score than the others that were slaughtered. The off flavor was, however, not objectionable.

Researchers in Germany also investigated the effect of supplementing feed with aerial parts of *Echinacea purpurea* on weight gain in piglets (Ahrens, Lang, and Erhard, 2003). At weaning, animals ($n = 52$) were divided into treatment and control groups, and the diet of the treatment groups was supplemented with either 1% or 5% echinacea. Standardization of the supplement by a specific chemical constituent was not mentioned in the publication. The weight of animals in the group fed echinacea at 5% was significantly lower compared to the other two groups on days 63 and 77 postweaning (the last 2 days measured). The authors attributed this to lower feed intake due to the bitter taste of the ration. Serum and salivary samples were also analyzed for immunoglobulins. Immunoglobulin G (IgG) levels were found to be significantly higher on day 35 postweaning in the group fed 5% echinacea, but salivary IgA levels were significantly lower in the group fed 1% echinacea on day 77 postweaning. The authors concluded that these results suggest that echinacea should be supplemented at a high level for a short period.

The potential for echinacea to provide resistance against swine viral diseases has also been investigated (Hermann et al., 2003). The authors looked at the effect of dietary *Echinacea purpurea* on growth performance, level of viremia, and ontogeny of antibody response in nursery-age pigs challenged with porcine reproductive and respiratory syndrome virus (PRRSV). Weaned pigs (25 ± 1 day of age; $n = 120$) from a PRRSV-naïve herd were randomly allotted to one of eight pens ($n = 5$) in two separate rooms (four pens/room). The pigs were fed one of four isocaloric and isolysinic diets: (1) basal diet; (2) basal diet plus carabdox (0.055 g/kg of diet on an as-fed basis); (3) basal diet plus echinacea (2% of total diet); and (4) basal diet plus echinacea (4% of total diet). The echinacea used in this study contained 1.35% cichoric acid. Seven days after starting the diets, all pigs in one room were inoculated intranasally with PRRSV. Body weight and serum samples were collected weekly from all the animals, and serum samples were analyzed for the presence of PRRSV and PRRSV-specific antibodies. All challenged pigs became infected with PRRSV, and all unchallenged pigs remained free of infection. Feeding echinacea at either 2% or 4% of the diet did not significantly increase ADG, ADFI, or G:F in PRRSV-challenged animals, and it did not affect the rate or level of antibody response detectable by enzyme-linked immunosorbent assay (ELISA) or the level and duration of PRRSV in serum. However, there were also no statistically significant differences in ADG, ADFI, or G:F ratio between PRRSV-challenged and unchallenged animals, regardless of the diet. It is therefore possible that, like in the study by Holden, McKean, and Franzenburg (1999), the conditions of this study were not sufficient to elicit measurable responses in the PRRSV-challenged pigs that could be significantly mitigated by the echinacea.

Most recently, researchers in Germany studied the influence of administering *Echinacea purpurea* to pregnant and nursing sows on the immune system, health status, growth performance, and carcass quality of their offspring (Kuhn et al., 2005). Expressed juice of the aerial plant parts (a standardized human preparation made from 4.2% dry matter diluted in 22.5% ethanol) was supplemented in the feed of pregnant and suckling sows at a dose of 0.125 mL/kg body weight divided into two daily doses in six intervals (5 days of treatment followed by 2 weeks of break, respectively). Treatment started 30 days postinsemination and continued until 21 days postpartum. Blood samples were taken from sows and piglets at various times during treatment and analyzed for immunoglobulin content (IgG and IgA) as well as C-reactive protein (CRP). In the piglets from echinacea-treated sows, the greatest increase in IgA, IgG, and CRP was observed on the first day postpartum ($p = .0004$, $p < .0001$ and $p = .05$, respectively). This difference became less as the piglets grew older. In contrast, the sows did not show significant differences in IgA, IgG, or CRP until 21 days postpartum, when IgA levels were significantly higher in the control (untreated) group ($p = .002$). There was a trend toward lower rates of therapeutic treatment of piglets from echinacea-treated sows up to day 70 of age ($p = .08$). There were no significant differences in growth performance and carcass quality of the offspring from echinacea-treated and untreated sows.

In summary, results of studies in swine suggested that echinacea may potentially be useful for stimulating nonspecific immunity and protecting animals from disease as well as enhancing production. However, the differences between treated and control groups or the sample sizes used in the studies were not large enough to demonstrate significant effects *in vivo.*

9.2.4.2 Cattle

The potential of echinacea to modulate immune function in cattle was supported by two studies, one *in vitro* and the other *in vivo.* In the *in vitro* study, flow cytometric methods were used to evaluate the effects of culturing bovine leukocytes for 44 h in the presence of echinacea, either alone or as a combination product together with *Thuja occidentalis* and elemental phosphorus (EquiMun®) (Schuberth, Riedel-Caspari, and Leibold, 2002). The authors found that both EquiMun and echinacea reduced the size and enhanced the viability of polymorphonuclear cells but had no effect on mononuclear cells.

The *in vivo* study evaluated the effects of echinacea on the immune function of transitional dairy calves using a single, small ($n = 10$), placebo-controlled design (Gill et al., 2002). The treatment group received a dose of 2.5 g echinacea supplemented daily for 17 days in the normal calf ration. Formulation and concentration of marker compounds in the supplement were not mentioned in the article. Findings of the study included significant decreases in the percentage of monocytes and B cells. $CD4^+$ T-cell levels were significantly increased. No significant effects on neutrophil function, packed cell volume, serum protein concentrations, or body weight were found in this study.

9.2.4.3 Poultry

Echinacea has been shown to significantly increase weight gain in broiler chickens following live vaccination and subsequent challenge with multiple coccidia species (Allen, 2003). Birds were supplemented for 2 weeks with a ground root preparation of *Echinacea purpurea* mixed into a broiler starter ration at 0% (control), 0.1%, and 0.5% (wt/wt). Half the birds in each treatment group were immunized orally with a half-dose of a live vaccine containing a proprietary mixture of a small number of live oocysts of four coccidia species (Immucox®, Vetech Laboratories, Ontario, Canada). After 2 weeks, the birds were switched to an unsupplemented ration and then challenged by oral administration of 1,000 times the dose of Immucox after another 2 weeks. The higher weight gain in the supplemented animals persisted through the 2-week supplement of withdrawal and subsequent challenge infection. Supplementation also significantly lowered total gut lesions scores at 6 days postchallenge.

The influence of echinacea on nonspecific immunity in turkey-hens was studied using a commercially available product containing echinacea and synthetic vitamin C (Echinovit C®, P. W. Mikita, Poland) (Truchlinski et al., 2006). The Echinovit C was administered continuously at increasing doses (0.5, 1.0, and 1.5 mL/L water) in the water provided to the birds from the 7th to the 16th week of age. Blood samples were taken three times during the study to determine the hematocrit, hemoglobin volume, number of red and white blood cells, as well as percentage of white blood cells. Leukocyte phagocytic activity, the ability of peripheral neutrophils to kill phagocytosed microorganisms, and lysozyme activity of peripheral blood serum were also determined using standard laboratory methods. Administration of Echinovit C significantly increased the number of white blood cells, leukocyte phagocytic activity, and ability of neutrophils to kill phagocytosed microorganisms at some, but not all, sampling times.

The effects of echinacea on the humoral immunity of chickens have also been studied (Schranner et al., 1989). These authors showed that administration of 0.1 mL of an ethanolic extract (unknown concentration) increased immunoglobulin concentrations (IgA, IgG, and IgM in decreasing amounts) compared to controls injected with ethanol alone.

9.2.4.4 Horses

In a double-blind, placebo-controlled, crossover trial with eight healthy horses, it was found that echinacea increased the phagocytic ability of isolated neutrophils and boosted peripheral lymphocyte counts but decreased the number of neutrophils at certain sample times over the course of a 42-day treatment regimen (O'Neill, McKee, and Clarke, 2002). The authors attributed the last observation to neutrophil migration from the peripheral circulation into the tissues. An increase in the size and concentration of peripheral red blood cells and the concentration of hemoglobin and packed cell volume were also observed.

This study (O'Neill, McKee, and Clarke, 2002) used a powdered extract of *Echinacea angustifolia* root, standardized to 4% echinacoside, administered twice daily for 42 days as a top-dress on the feed given to the horses. There was a 14-day washout period between treatments. The strength of this study is its rigorous experimental design. However, the statistical analysis (paired difference *t* tests comparing treated vs. untreated at each sample time) did not account for possible period effects; therefore, differences in conditions between the two periods cannot be excluded. Also, the clinical significance of the observations is unknown in horses.

9.2.4.5 Dogs

Researchers in Germany and Switzerland conducted an open multicenter clinical trial to assess the efficacy of echinacea powder for the treatment of chronic and seasonal upper respiratory tract infections in dogs (Reichling et al., 2003). A total of 41 dogs were enrolled in the study, and subjects varied widely with regard to age, breed, and weight. Criteria for inclusion in the study were nonspecific and included "kennel cough," "nonthriving young animals," and inflammation of the upper respiratory tract. *Chronic disease* was defined as a condition expected, in the opinion of the investigator, to continue for more than 8 weeks if left untreated. Animals receiving concurrent treatment were excluded from the study.

Echinacea powder (standardized to between 0.6% and 2.1% chicoric acid) was administered daily by mixing it in moist food at a dose of 0.3 g echinacea per 10 kg body weight. Overall efficacy of the treatment was indicated by comparing the severity and resolution of clinical symptoms that existed before and after 4 and 8 weeks of treatment in each individual. The scale for this parameter was very good, good, moderate, or insufficient. In addition, the severity of 12 different clinical symptoms was scored on a scale of 1 to 4 at examinations after 4 and 8 weeks of treatment. Body weight and temperature were also recorded.

Improvement in the disease condition was assessed as either "good" or "very good" in 92% of dogs after 4 weeks of treatment and in 95% of dogs after 8 weeks of treatment. Small, but statistically significant, decreases in body temperature were observed. There were also significant reductions in the severity of clear nasal secretions, enlarged lymph nodes, dry cough, dyspnea, and dry lung sounds after both 4 and 8 weeks of treatment.

The authors concluded that echinacea is an acceptable alternative for the treatment of chronic and seasonal respiratory tract infections in dogs. However, several weaknesses in the study design preclude a definitive conclusion regarding its efficacy. These weaknesses include lack of a control group and specific diagnosis prior to inclusion in the study, as well as the subjective nature of the outcome measures.

9.3 OTHER PLANTS

9.3.1 *Eupatorium* Species L. (Asteraceae)

Eupatorium perfoliatum, which occurs throughout the eastern half of the United States, is documented to have been used by the Iroquois for the treatment of fever in horses (Herrick, 1977). This is consistent with its use as a febrifuge in humans. American Indians and early settlers used this plant to treat a wide range of conditions, ranging from colds and influenza to dengue fever, malaria, and typhoid. It was also used, apparently successfully, for the treatment and prevention of influenza in the "Spanish flu" epidemic of 1918, as well as flu epidemics of the 19th century (Lloyd, 1 AD, 1924b). The remedy was taken as a hot tea to induce sweating.

The plant's common name, boneset, alludes to the term *breakbone fever*, or a fever that is so severe "it feels like one's bones have broken." Breakbone fever is also the old name for dengue fever (Foster and Tyler, 1999). Although boneset does not have documented antipyretic or anti-inflammatory properties, xyloglucurans from the polysaccharide fractions of aqueous extracts have been shown to increase phagocytosis (Wagner et al., 1985). This suggests that at least some of this remedy's efficacy in the treatment of a wide range of infectious conditions could be attributed to immunostimulatory activity. Eupatorium species are a source of sesquiterpene lactones, which have been shown to have antitumor activity (Kupchan et al., 1969; Herz, Kalyanaraman, and Ramakrishnan, 1977) as well as flavonoid glycosides (Ramachandran Nair et al., 1995). Arabino-(glucurono)xylans isolated from

E. perfoliatum have been shown to have antimicrobial (Christakopoulos et al., 2003) and immunomodulatory activity (Ebringerovb et al., 2002).

Herrick (1977) also documented the Iroquois as using *Eupatorium rugosum* (*Ageratina altissima,* white snakeroot) to "stop sweating" in horses. There are, however, serious safety concerns regarding this remedy. White snakeroot is known to cause fatal toxicity in both humans and animals. The toxic principle has been identified as tremetone, an unstable alcohol that requires microsomal activation to cause toxicity in the animal (Beier et al., 1993). Horses are considered to be particularly sensitive to poisoning, and clinical symptoms in this species include muscle tremors, ataxia, reluctance to walk, heavy sweating, and myoglobinuria (Olson et al., 1984). Poisoning in humans is usually secondary to the consumption of milk or tissues from exposed animals.

9.3.2 *Rumex patientia* L. (Polygonaceae)

Rumex patientia (patience dock) is an herbaceous weed that occurs widely throughout the United States. The beaten roots of this plant are documented to have been fed by the Cherokee to horses "with a sick stomach" (Hamel and Chiltoskey, 1975). There are no clinical studies in horses to support this use. However, experimental studies have demonstrated gastroprotective and antiulcerogenic activity following oral administration of aqueous extracts of *Rumex patientia* roots (500 mg/kg) and fruits (Gurbuz et al., 2005; Suleyman et al., 2002; Suleyman, Demirezer, and Kuruuzum-Uz, 2004). Indomethacin ethanol-induced gastric ulcer models in rats were used for these studies. The effects have been attributed to antioxidant effects of the plant extracts.

Anthraquinones (some naturally halogenated) and tannins with antioxidant and anti-inflammatory properties have been isolated from this species (Demirezer et al., 2001; Dembitsky and Tolstikov, 2003). Emodin, which is one of the identified anthraquinones, has been shown to occur in several other plant taxa, including *Aloe ferox*. It has a multitude of actions, including vasorelaxant (Huang et al., 1991), proapoptotic (Su et al., 2005), hepatoprotective (Lin et al., 1996), and anti-inflammatory (Chang et al., 1996).

9.3.3 *Rhus aromatica* and *Rhus trilobata*

Rhus species (sumac) are shrubs and small trees that grow in subtropical and warm temperate regions throughout the world. The Cheyenne are documented to have used the fruits of a species from this genus, *Rhus aromatica*, for the treatment of urinary disorders and "to prevent tiredness" in horses (Hart, 1981). This ethnoveterinary use is, at least partially, supported by results from a study by researchers in Germany, who investigated using *in vitro* studies of *Rhus aromatica* for the treatment of overactive bladder syndrome in humans (Borchert et al., 2004). However, recent changes in nomenclature make it uncertain whether the species used in the German studies was indeed the same as the species used by the Cheyenne. This is because *Rhus aromatica* was split into two species, *Rhus aromatica* and *Rhus trilobata*. The Cheyenne are more likely to have used *Rhus trilobata* since its range stretches from the Midwest through the western United States. Following the change in nomenclature, the range of *Rhus aromatica* is limited to the Eastern United States.

Borchert et al. (2004) showed that an aqueous extract of the root cortex of *Rhus aromatica* inhibited carbachol- and KCl-induced contraction of human bladder muscle strips and whole rat bladders in a dose-dependent manner (range 0.01–0.1%). Interactions with the muscarinic receptors were also demonstrated using radioligand and inositol phosphate accumulation studies. The authors concluded that the inhibition of muscarinic-mediated contraction of bladder muscle by extracts of *Rhus aromatica* could be beneficial to patients with bladder dysfunction, but its clinical efficacy would need to be demonstrated in controlled clinical trials. Likewise, the benefit in horses must still be investigated. *Rhus* spp. are a rich source of condensed tannins (Min et al., 2007).

9.3.4 *Lepidium virginicum* L. (Brassicaceae)

Cherokees used infusions of *Lepidium virginicum* (Virginia pepperweed) to treat sick chickens. The plant was also mixed into chicken feed to increase egg production (Hamel and Chiltoskey, 1975). The plant grows throughout the United States as an annual or biennial weed in disturbed soil. Antiprotozoal activity has been demonstrated for a crude extract from the roots. Specifically, a benzyl glucosinolate was identified, through bioassay-guided fractionation, that showed *in vitro* activity against *Entamoeba histolytica* (median inhibitory concentration = 20.4 μg/mL) (Calzada, Barbosa, and Cedillo-Rivera, 2003). Efficacy of this documented ethnoveterinary use could be attributed to control of protozoal diseases, such as coccidiosis, histomoniasis, or trichominiasis, in chickens. In humans, *L. virginicum* roots are used in the control of diarrhea and dysentery in the highlands of Chiapas, Mexico.

There is no literature that we could find on the chemistry of *L. virginicum,* although its location in the family Brassicaraceae may point to shared chemistry with the common brassicas.

9.3.5 *Clematis hirsutissima* Pursh (Ranunculaceae)

Native Americans from the western and central regions of North America used a variety of plants as "stimulants" for horses to improve their performance in warfare, hunting, and horse racing. One of these, for which some scientific evidence is available, is the hairy clematis (*Clematis hirsutissima).* This herbaceous plant occurs in the western United States and was used by the Nez Perces and Teton Sioux as a stimulant and restorative for horses (Morgan, 1981). Blankinship (1905) documented that the scraped root was held to the nostril of a fallen horse. Anemonin, a dilactone of cyclobutane-1,2-diol-1,2-diacrylic acid derived from the cyclodimerization of protoanemonin, has been isolated from this plant (Kern and Cardellina, 1983). Anemonin is a known blistering agent, and vapors of this substance are reported to cause irritation of the nasal passages, nausea, and headache (Kern and Cardellina, 1983).

Clematis hirsutissima is related to the European pulsatillas *Amemone pulsatilla* and *Anemone pratensis* (Kern and Cardellina, 1983). These herbs were traditionally used for the treatment of hysteria in women and "affected powerfully not only the nerve centers but the heart itself" (Remington et al., 1918, p. 79). The use of *Clematis hirsutissima* as a stimulant in horses may therefore be related to its activity on the central nervous system. However, Remington also expresses skepticism as to the efficacy of pulsatillas on the central nervous system, since the herb is administered at infinitesimally small doses to prevent adverse reactions.

9.4 CONTEMPORARY USES

Native North American medicinal plants are rarely mentioned in contemporary texts on veterinary herbal medicine. Echinacea is a notable exception. Topical formulations of *Echinacea* spp. are listed as treatments for wounds, skin infections, snakebites, and insect bites. Administered systemically, this herb is used in animals as an immune stimulant, specifically for the treatment and prevention of upper respiratory tract infections. It may also have benefit in the prevention of cancer (Wynn and Fougere, 2007).

Another native North American plant that is listed in modern veterinary herbal *Materia Medica* is *Eupatorium perfoliatum*. Potential veterinary indications for this herb are listed as fever, upper respiratory infection, bone or muscular pain, and fever in horses (Wynn and Fougere, 2007).

9.5 CONCLUSION

Native North American plants offer a rich resource of medicinally active compounds that remains largely untapped. Historically, herbal remedies were not considered a priority by funding agencies and the pharmaceutical industry. In addition, the use of indigenous plants was largely surpassed by European herbs as a Westernized culture came to predominate in North American societies. More recently, however, there has been resurgence in interest in natural remedies and products that are produced locally from indigenous resources. This offers considerable opportunities to contribute to the knowledge base on these plants through basic and applied veterinary research.

Healthy population of *Echinacea angustifolia* stand in native rangeland near Miles City, Montana.

Flowers of *Echinacea Pallida* from a wild stand in east-central Kansas.

Flowers of the cultivated *Echinacea purpurea*.

The flowering plants of white snakeroot, *Ageratina altissima* (*Eupatorium rugosum*).

REFERENCES

Ahrens F, Lang E, and Erhard M (2003). Effects of aerial plant parts of *Echinacea purpurea* on humoral immune system and body weight gain in piglets. *Proc Soc Nutr Physiol* 12, 73.

Allen PC (2003). Dietary supplementation with *Echinacea* and development of immunity to challenge infection with coccidia. *Parasitol Res* 91, 74–78.

Barrett B (2003). Medicinal properties of *Echinacea*: a critical review. *Phytomedicine* 10, 66–86.

Beier RC, Norman JO, Reagor JC, Rees MS, and Mundy BP (1993). Isolation of the major component in white snakeroot that is toxic after microsomal activation: possible explanation of sporadic toxicity of white snakeroot plants and extracts. *Nat Toxins* 1, 286–293.

Blankinship JW (1905). Native economic plants of Montana. *Montana Agricultural College Experimental Station Bulletin* 56, 10.

Blumenthal M, Cavaliere C, and Rea P (2008). Herbal supplement sales in United States show growth in all channels. *HerbalGram* 78, 60–63.

Borchert VE, Czyborra P, Fetscher C, Goepel M, and Michel MC (2004). Extracts from *Rhois aromatica* and *Solidaginis virgaurea* inhibit rat and human bladder contraction. *Naunyn Schmiedebergs Arch Pharmacol* 369, 281–286.

Busing KH (1952). [Inhibition of hyaluronidase by echiancine]. *Arzneimittelforschung* 2, 467–469.

Calzada F, Barbosa E, and Cedillo-Rivera R (2003). Antiamoebic activity of benzyl glucosinolate from *Lepidium virginicum*. *Phytother Res* 17, 618–619.

Chang CH, Lin CC, Yang JJ, Namba T, and Hattori M (1996). Anti-inflammatory effects of emodin from *Ventilago leiocarpa*. *Am J Chin Med* 24, 139–142.

Christakopoulos P, Katapodis P, Kalogeris E, Kekos D, Macris BJ, Stamatis H, and Skaltsa H (2003). Antimicrobial activity of acidic xylo-oligosaccharides produced by family 10 and 11 endoxylanases. *Int J Biol Macromol* 31, 171–175.

Combest W and Nemecz G (1997). Echinacea. *US Pharmacist*, October: 126–132.

Dembitsky VM and Tolstikov GA (2003). Natural halogenated complex phenols. *Chem Sustain Dev* 11, 811–822.

Demirezer LO, Kuruuzum-Uz A, Bergere I, Schiewe HJ, and Zeeck A (2001). The structures of antioxidant and cytotoxic agents from natural source: anthraquinones and tannins from roots of *Rumex patientia*. *Phytochemistry* 58, 1213–1217.

Ebringerovb A, Kardosovb A, Hrombdkovb Z, MalovÆkovb A, and Hríbalová V (2002). Immunomodulatory activity of acidic xylans in relation to their structural and molecular properties. *Int J Biol Macromol* 30, 1–6.

Ernst E (2005). The efficacy of herbal medicine—an overview. *Fundam Clin Pharmacol* 19, 405–409.

Foster S (2008). *Echinacea: nature's immune enhancer*, 3rd ed. Healing Arts Press, Rochester, VT.

Foster S and Tyler VE (1999). *Tyler's honest herbal*, 4th ed. MJF Books, New York.

Gill NR, Powell C, Franklin ST, and Meek KI (2002). The effects of *Echinacea* on immune function of transitional calves. *J Anim Sci* Supplement 1, 1633.

Gilmore MR (1977). *Use of plants by the Indians of the Missouri River region*. University of Nebraska Press, Lincoln.

Gronovius LT (1946). *Flora Virginica*, 2nd ed. Arnold Arboretum, Cambridge, MA. (Orig. pub. 1762.)

Gurbuz I, Ozkan AM, Yesilada E, and Kutsal O (2005). Anti-ulcerogenic activity of some plants used in folk medicine of Pinarbasi (Kayseri, Turkey). *J Ethnopharmacol* 101, 313–318.

Hamel PB and Chiltoskey MU (1975). *Cherokee plants*. Herald, Sylva, NC, p. 32.

Hart JA (1981). The ethnobotany of the northern Cheyenne Indians of Montana. *J Ethnopharmacol* 4, 1–55.

Hermann JR, Honeyman MS, Zimmerman JJ, Thacker BJ, Holden PJ, and Chang CC (2003). Effect of dietary *Echinacea purpurea* on viremia and performance in porcine reproductive and respiratory syndrome virus-infected nursery pigs. *J Anim Sci* 81, 2139–2144.

Herrick JW (1977). *Iroquois medical botany.* University Microfilms International, Ann Arbor, MI.

Herz W, Kalyanaraman PS, and Ramakrishnan G (1977). Sesquiterpene lactones of *Eupatorium perfoliatum. J Org Chem* 42, 2264–2271.

Holden PJ, McKean J, and Franzenburg E (1999). *Botanicals for pigs—echinacea.* Iowa State University Swine Research Report 27–30. (Accessed Jan 2010 from www.extension.iastate.edu/Pages/ansci/swinereports/asi-1560.pdf)

Huang HC, Lee CR, Chao PD, Chen CC, and Chu SH (1991). Vasorelaxant effect of emodin, an anthraquinone from a Chinese herb. *Eur J Pharmacol* 205, 289–294.

Kern JR and Cardellina JH (1983). Native American medicinal plants. Anemonin from the horse stimulant *Clematis hirsutissima. J Ethnopharmacol* 8, 121–123.

Kindscher K (1992). *Medicinal wild plants of the prairie.* University Press of Kansas, Lawrence.

Kinkel HJ, Plate M, and Tullner HU (1984). Effect of Echinacin ointment in healing of skin lesions. *Medizinische Klinik* 79, 580–583.

Kuhn G, Ender K, Thomann R, Tuchscherer M, Tuchscherer A, Stabenow B, Krueger M, and Schroedl W (2005). Einsatz von Echinacea-extrakt bei tragenden und saugenden Sauen. *Arch Tierz Dummerstorf* 48, 270–282.

Kupchan SM, Kelsey JE, Maruyama M, Cassady JM, Hemingway JC, and Knox JR (1969). Tumor inhibitors. XLI. Structural elucidation of tumor-inhibitory sesquiterpene lactones from *Eupatorium rotundifolium. J Org Chem* 34, 3876–3883.

Letchamo W, Livesey J, Arnason TJ, Bergeron C, and Krutilina VS (1999). Cichoric acid and isobutylamide content in *Echinacea purpurea* as influenced by flower developmental stages. In *Perspectives on new crops and new uses*, ed. Janick J, pp. 494–498. ASHS Press, Alexandria, VA.

Lin CC, Chang CH, Yang JJ, Namba T, and Hattori M (1996). Hepatoprotective effects of emodin from *Ventilago leiocarpa. J Ethnopharmacol* 52, 107–111.

Linde K, Barrett B, Wolkart K, Bauer R, and Melchart D (2006). Echinacea for preventing and treating the common cold. *Cochrane Database Syst Rev* CD000530.

Lloyd J (1 A.D.). *A treatise on* Eupatorium perfoliatum. Lloyd Brothers, Pharmacists, Cincinnati, OH.

Lloyd JU (1924a). *A treatise on* Echinacea. *Drug Treatise No. 30.* Lloyd Brothers, Pharmacists, Cincinnati, OH.

Lloyd J (1924b). *A treatise on* Eupatorium perfoliatum. Lloyd Brothers, Pharmacists, Cincinatti, OH.

Melchart D, Linde K, Worku F, Bauer R, and Wagner H (1994). Immunomodulation with *Echinacea*—a systematic review of controlled clinical trials. *Phytomedicine* 1, 245–254.

Mills S and Bone K (2000). *Principles and practice of phytotherapy: modern herbal medicine.* Churchill Livingstone, London.

Min BR, Pinchak WE, Merkel R, Walker S, Tomita G, and Anderson RC (2007). Comparative antimicrobial activity of tannin extracts from perennial plants on mastitis pathogens. *Scientific Research and Essay* 3, 66–73.

Moerman D. Native American ethnobotany database. 2003 (accessed August 21, 2008 from http://herb.umd.umich.edu).

Morgan GR (1981). "Sugar bowls" (*Clematis hirsutissima*): a horse restorative of the Nez Perces. *J Ethnopharmacol* 4, 117–120.

Myers SP and Wohlmuth H (1998). Echinacea-associated anaphylaxis. *Med J Aust* 168, 583–584.

Olson CT, Keller WC, Gerken DF, and Reed SM (1984). Suspected tremetol poisoning in horses. *J Am Vet Med Assoc* 185, 1001–1003.

O'Neill W, McKee S, and Clarke AF (2002). Immunological and haematinic consequences of feeding a standardised echinacea (*Echinacea angustifolia*) extract to healthy horses. *Equine Vet J* 34, 222–227.

Osowski S, Rostock M, Bartsch HH, and Massing U (2000). [Pharmaceutical comparability of different therapeutic Echinacea preparations]. *Forsch Komplementarmed Klass Naturheilkd* 7, 294–300.

Pierce A (1999). *The American Pharmaceutical Association practical guide to natural medicines.* Stonesong Press, New York.

Ramachandran Nair AG, Gunasegaran R, Krishnan S, Bayet C, and Voirin B (1995). Flavonol glycosides from leaves of *Eupatorium glandulosum*. *Phytochemistry* 40, 283–285.

Reichling J, Fitzi J, Furst-Jucker J, Bucher S, and Saller R (2003). Echinacea powder: treatment for canine chronic and seasonal upper respiratory tract infections. *Schweiz Arch Tierheilkd* 145, 223–231.

Remington JP, Woods HC, and others (1918). *The US dispensatory*, 20th ed.

Schopf JD (1903). *Materia Medica Americana, potissimum regni vegitabilis.* Bulletin of the Lloyd Library of Botany, Pharmacy and Materia Medica. Bulletin No. 6. Reproduction Series. No. 3. Cincinnati, OH. (Orig. pub. 1787.)

Schranner I, Wurdinger M, Klumpp N, Losch U, and Okpanyi SN (1989). [Modification of avian humoral immunoreactions by Influex and *Echinacea angustifolia* extract]. *Zentralbl Veterinarmed B* 36, 353–364.

Schuberth HJ, Riedel-Caspari G, and Leibold W (2002). Flow cytometric testing of immunological effects of a phytomedicinal combination (equimun) and its compounds on bovine leucocytes. *J Vet Med A Physiol Pathol Clin Med* 49, 291–298.

Stahl M, Reifenberg K, Okpanyi S, and Losch U (1990). Porcine granulocyte functions: evaluation and modulation. *Zentralbl Veterinarmed B* 37, 261–267.

Su YT, Chang HL, Shyue SK, and Hsu SL (2005). Emodin induces apoptosis in human lung adenocarcinoma cells through a reactive oxygen species-dependent mitochondrial signaling pathway. *Biochem Pharmacol* 70, 229–241.

Suleyman H, Demirezer LO, and Kuruuzum-Uz A (2004). Effects of *Rumex patientia* root extract on indomethacine and ethanol induced gastric damage in rats. *Pharmazie* 59, 147–149.

Suleyman H, Demirezer LO, Kuruuzum-Uz A, and Akcay F (2002). Gastroprotective and antiulcerogenic effects of *Rumex patientia* L. extract. *Pharmazie* 57, 204–205.

Tragni E, Tubaro A, Melis S, and Galli CL (1985). Evidence from two classic irritation tests for an anti-inflammatory action of a natural extract, Echinacina B. *Food Chem Toxicol* 23, 317–319.

Truchlinski J, Krauze M, Cendrowska-Pinkosz M, and Modzelewska-Banachiewicz B (2006). Influence of garlic, synthetic 1,2,4-triasole derivative and herbal preparation Echinovit C on selected indices of turkey-hen non-specific immunity. *Polish J Vet Sci* 9, 51–55.

UKCropNet. Medicinal plants of native America database. 2008. http://ukcrop.net/perl/ace/search/MPNADB

van den Bogaard AE and Stobberingh EE (2000). Epidemiology of resistance to antibiotics. Links between animals and humans. *Int J Antimicrob Agents* 14, 327–335.

Wagner H, Breu W, Willer F, Wierer M, Remiger P, and Schwenker G (1989). In vitro inhibition of arachidonate metabolism by some alkamides and prenylated phenols. *Planta Med* 55, 566–567.

Wagner H, Proksch A, Riess-Maurer I, Vollmar A, Odenthal S, Stuppner H, Jurcic K, Le TM, and Fang JN (1985). [Immunostimulating action of polysaccharides (heteroglycans) from higher plants]. *Arzneimittelforschung* 35, 1069–1075.

Wynn SG and Fougere BJ (2007). *Veterinary herbal medicine.* Mosby Elsevier, St. Louis, MO.

10 Plants Used in Animal Health Care in South and Latin America

An Overview

Rômulo Romeu da Nóbrega Alves,
Raynner Rilke Duarte Barboza, and
Wedson de Medeiros Silva Souto

CONTENTS

10.1 INTRODUCTION

Medicinal plants have a long history of use in the treatment of both human illnesses and animal diseases.[1–6] However, research into ethnoveterinary medicine has been largely neglected in favor of human ethnomedicine. Although numerous ethnobotanical inventories of medicinal plant use in humans have been created, ethnoveterinary medicine remains poorly described. This scarce description of ethnoveterinary resources belies the important role they play for small-scale farmers, whose lack of regular access to essential medicines for their livestock greatly hampers productivity. According to the United Nations Food and Agricultural Organization (FAO), the lack of drugs to treat diseases and infections causes losses of 30–35% in the breeding sector of many developing countries, where poor animal health remains the major constraint to increased production.[7]

In many of these countries, there has been a decline in funding for veterinary services and for animal health care in general.[8] Traditional medicine based on phytotherapy may complement and offer alternatives for animal disease control, in particular for resource-poor breeders. However, more studies are needed to investigate the efficacy of these ethnobotanicals. The first step in this direction is to record and describe the plants used and to identify them correctly on a taxonomic level.[9]

Farmers and pastoralists in several countries use medicinal plants in the maintenance and management of livestock wellness. It is an established fact that medicinal plants, for several centuries, have been widely used as a primary source of prevention and control of livestock diseases. In fact, interest in such uses in the veterinary sector has resulted primarily from the increasing cost of animal health care and the search for new veterinary medicines and vaccines.[10]

Latin America is a vast region spanning parts of North America, almost all of South America, and much of the West Indies. It encompasses 19 countries as well as Puerto Rico, a commonwealth territory of the United States, and, arguably, even parts of the southwestern United States.[11] In Latin America and the Caribbean, the population stood at 577 million in 2008 and is projected to increase to 778 million by 2050.[12] It is the most urbanized region in the developing world, with around three-quarters of the population living in urban areas.[13] The population of Latin America is a composite of ancestries, ethnic groups, and races, making the region one of the most—if not the most—racially and ethnically diverse in the world. The specific composition varies from country to country: Some countries have a predominance of a mixed population, in others people of Amerindian origin are a majority, some are dominated by inhabitants of European ancestry, while others are primarily of African descent. Most Latin American countries also have large Asian minorities. Europeans are the largest single group, and they and people of part-European ancestry combine to make up approximately 80% of the population of the subcontinent.[14]

Besides being one of the world's principal culture regions, Latin America is distinguished from other world regions by a set of common cultural traits that include language, religion, social values, and civic institutions deriving principally from the Iberian Peninsula. Spanish and Portuguese are predominant languages. Catholicism is practiced by the vast majority of the region's inhabitants, and social customs and civic institutions bear many similarities to those of Spain. Nevertheless, the region is not culturally monolithic. Indigenous cultures and peoples have influenced national and subnational cultures, affecting language, religion, music, food habits, social customs, and civic institutions. African immigrants originally brought as slaves also have influenced the region's culture, although their effects have been most pronounced in Brazil, the Caribbean, and coastal areas of Central and northern South America. The cultural impact of other immigrants, including those from Italy, Asia, the Middle East, and even a few from North America, has been minor.[11]

Agriculture and rural economic activities are major sources of employment in Latin America.[15] In particular, livestock raising is an economic activity linked to local consumption or export. In this region, there are areas in which livestock production is the mainstay of the economy, second only to mining and timber.[16]

The total territory of Latin America, supracontinental in scale, contains the greatest ecological and biological diversity of the planet, mainly concentrated in the vast intertropical regions.[17] Together, the Latin American countries are home to a large part of the world's biodiversity. Brazil alone possesses about 20–22% of all existing plants and microorganisms. However, it is estimated that no more than 25,000 plant species have been the subject of any sort of scientific investigation. Latin American countries have thousands of plant species, and they are also very rich in animals, microorganisms, and marine resources.[18]

The adaptation of the various human groups to the rich biological resources has generated invaluable local knowledge systems that include extensive information on plant uses in general and medicinally useful species in particular. It is not surprising, therefore, that a vast majority of the world's biological diversity, as well as the richest sources of traditional knowledge of plant uses, are found in the less-developed countries where people live close to nature.[19] In spite of the very rich natural resources, these countries have never properly used their great biodiversity to the benefit of their own economic development. Furthermore, due to the uncontrolled exploitation, the biological diversity in these countries is being lost year by year, and important animal and plant species are disappearing.[18]

The medicinal flora of Central and South America and the Caribbean Islands has been the focus of much ethnobotanical research and bioprospecting over the last two decades. The current and potential value of medicinal plants as traditional remedies and commercial pharmaceuticals is widely acknowledged. A search of the literature clearly shows that over the last 25 years, few, if any, of the fields in Latin American science have progressed as rapidly as the research on natural products.[18] However, studies on the therapeutic use of plants for veterinary purposes have been neglected when compared to plants used for human medicine; little attention has been paid to the cultural, medical, or ecological significance of ethnoveterinary practices.

Latin and South America's rich biological and cultural diversity make them exceptional locations for examining and increasing our knowledge of floristic resources for veterinary use and to draw attention to the need to protect traditional knowledge and biodiversity. In that context, the aim of this chapter is to provide an overview of the use of plants in ethnoveterinary medicine in Latin and South America, identify those species used as folk remedies, and discuss the implications of their use. This study represents the first review of ethnoveterinary traditions in this region, and we hope to stimulate further discussions about this use of plants and its implications for conservation.

10.2 METHODOLOGY

To examine the diversity of plants used in traditional medicine in Latin America, all available references or reports of relevant folk remedies were examined. Only taxa that could be identified to species level are included in this review. Scientific names provided in publications were updated according to Tropicos of the Missouri Botanical Garden (http://www.tropicos.org/). The sources analyzed were Agra et al.,[20] Almeida et al.,[21,22] Barboza et al.,[23] Bussmann and Sharon,[24] Carretero,[25] De-la-Cruz,[26] Faria et al.,[27] Lans and Brown,[28,29] Lans et al.,[30–32] López,[33] Lucena et al.,[34] Marinho et al.,[35] Scarpa,[36] and Sikarwar.[37]

10.3 MEDICINAL PLANT USAGE

Early hominids were hunters and gatherers who relied on naturally occurring vegetation, fruits, nuts, carrion, and game for subsistence. Therefore, the use of flora and fauna by humans arises from prehistorical times, and the process of domestication of useful species can be linked to the dawn of civilization.[38] Once humans started taming certain animals for domestic utilization, therapeutic practices were developed to treat diseases that usually affected the target animals. In Latin America, as elsewhere in the world, humans use the fauna in several ways, for instance, as food and for fur, leather, pets, ceremonial rituals, medicines, and finally the controlled use of animal traction in production techniques and work.[2,39–46] Throughout the world, and in all production systems, livestock diseases lead to mortality and reduced productivity. Prevention, control, and eradication of diseases among domesticated animals are major concerns as such diseases lead to economic losses and possible transmission of the causative agents to humans, with possible public health implications. The use of plants to re-treat animal disease and improve production therefore naturally arose as a response to the problems that domestic fauna faced.

The obvious observation that we made was that this area is underresearched, as can be seen from the small number of studies available in the scientific literature. It is self-evident that there is an urgent need for more studies into ethnoveterinary practices in the region. The available literature revealed that at least 203 plants belonging to 66 families have been used in folk veterinary medicine in Latin America (Table 10.1). The number is certainly underestimated since the number of studies on the theme are limited and have only been done in Argentina, Bolivia, Brazil, Ecuador, Mexico, Peru, and Trinidad and Tobago. The family with the most medicinal plants utilized was Fabaceae (14%), followed by Asteraceae (6%), Euphorbiaceae (6%), and Solanaceae (4%). Other species recorded are distributed in 62 different families. Species of those families are also used frequently in Africa, Europe, and Asia. In Italy, for example, Asteraceae, Fabaceae, and Euphorbiaceae appear among the popularly used families.[47] In Ethiopia, Yineger et al.[48] reported that the families Asteraceae, Solanaceae, Fabaceae, and Lamiaceae were the most popular, while in studies in the African and European Mediterranean, Pieroni et al.[49] found that Asteraceae, Fabaceae, and Euphorbiaceae were the most widely used taxa in ethnoveterinary medicine. It therefore appears that the botanical families used in Latin America are also utilized broadly in the other parts of the world.

Analysis of the published information showed that some ethnoveterinary practices were common to various countries; that is, the same plant species or different parts of it can be used for treating animal diseases, although not necessarily for the same disease. *Chenopodium ambrosioides* L., for instance, is used as an antiparasitic in Argentina and Brazil.[23,36] Also in these countries it is a useful cicatrizant of wounds.[35,36] In Brazil and Trinidad and Tobago, *Momordica charantia* L. is used to treat ectoparasites in livestock.[30,35] The use of different herbal remedies for the same ailment is popularly practiced in the region,[50] and this may be due to the flexibility it allows when there are seasonal changes in the availability of certain plant species. Therefore, substitution is possible and quite common in the use of plants.

It was also observed that certain medicinal plants were recorded in just one country. For instance, *Funastrum gracile* (Decne.) Schltdl. is used as a galactagogue in cows in Argentina,[36] *Ziziphus joazeiro* Mart. bark is used as a vermifuge and cicatrizant in Brazil,[35] while *Staelia scabra* (C. Presl) Standl is popular in Mexican veterinary medicine.[33] This may be due to the availability of, and access to, the species, floral endemicity, and the unique cultural knowledge specific to certain areas. Some ethnobotanic studies have demonstrated that the diversity of plants known and used by human populations could be influenced by the plant diversity in the environment.[50–54]

TABLE 10.1
Plants Used in Ethnoveterinary Medicine in Latin America

Family/Plant Name	Parts Used	Veterinary Uses	Country	Author
Acanthaceae				
Barleria lupulina Lindl.	L	Snakebites	Trinidad and Tobago	30
Justicia secunda Vahl	L	Rashes	Trinidad and Tobago	30
Ruellia tuberosa L.	R	Anthelmintic, estrus induction	Trinidad and Tobago	28
Amaranthaceae				
Achyranthes indica (L.) Mill.	L, R	To induce estrus	Trinidad and Tobago	28
Amaranthus quitensis Kunth	AP	Horse hematuria	Argentina	36
Amaranthus viridis L.	AP	Horse hematuria	Argentina	36
Anacardiaceae				
Anacardium occidentale L.	BK	Antidiarrheal, anti-inflammatory, antiparasitical (endoparasites), cicatrizant	Brazil, Trinidad and Tobago	21, 28, 35
Astronium urundeuva (Allemão) Engl.	L	Cicatrizant, for treating myiasis	Brazil	21
Myracrodruon urundeuva Allemão	BK, L	Cicatrizant	Brazil	35
Schinus fasciculata (Griseb.) I.M. Johnst.	AP	Uterine atony in goats and cow, retained placenta	Argentina	36
Spondias mombin L.	Not mentioned	Retained placenta	Trinidad and Tobago	28
Annonaceae				
Annona squamosa L.	L	Scabies treatment	Brazil	35
Apiaceae				
Cicuta virosa L.	AP	To treat animal wounds	Ecuador	24
Apocynaceae				
Aspidosperma pyrifolium Mart.	AP, BK	External parasiticide	Brazil	21, 35
Aspidosperma quebracho-blanco Schltdl.	ST	Injuries, vulnerary	Argentina	36

(continued)

TABLE 10.1 (CONTINUED)
Plants Used in Ethnoveterinary Medicine in Latin America

Family/Plant Name	Parts Used	Veterinary Uses	Country	Author
Hancornia speciosa B.A. Gomes	BK	Cicatrizant	Brazil	35
Vallesia glabra (Cav.) Link	L, ST	Vulnerary; after the previous operation, the myiasis is bathed	Argentina	36
Aquifoliaceae				
Ilex paraguariensis A. St.-Hil.	AP	Purgative; infusion with oil and salt for acute constipation in horses	Argentina	36
Araceae				
Caladium bicolor (Aiton) Vent.	L	External cattle festers caused by worms	Brazil	37
Monstera dubia (Kunth) Engl. & K. Krause	L	Combine with leaves of the plants below to bathe dogs; "steam" to catch quenks	Trinidad and Tobago	30
Synandrospadix vermitoxicus (Griseb.) Engl.	TU	Injuries, vulnerary, and for skin myasis	Argentina	36
Syngonium podophyllum Schott	L	Combine with leaves of the plants below to bathe dogs; "steam" to catch quenks	Trinidad and Tobago	30
Araliaceae				
Dendropanax arboreus (L.) Decne. & Planch.		Combine with leaves of the plants above to bathe dogs; "steam" to catch quenks	Trinidad and Tobago	30
Arecaceae				
Acrocomia aculeata (Jacq.) Lodd. ex Mart.	R, SE	Not mentioned (R), makes cocks' skin greasy (SE)	Trinidad and Tobago	30, 32
Cocos nucifera L.	FR	To prevent dehydration, for reduced appetite in chickens	Brazil, Trinidad and Tobago	21, 29
Aristolochiaceae				
Aristolochia rugosa Lam.	AP	Snakebites	Trinidad and Tobago	30
Aristolochia trilobata L.	L	Snakebites	Trinidad and Tobago	30
Asclepiadaceae				

Asclepias curassavica L.	FL, L	Wounds	Trinidad and Tobago	28
Funastrum gracile (Decne.) Schltr.	AP	Treatment of galactagogue in cows	Argentina	36
Asteraceae				
Acanthospermum hispidum DC.	R	Cough	Brazil	35
Ageratum conyzoides L.	BK	Inflamations	Brazil	21
Ambrosia tenuifolia Spreng.	AP	For horses' insolation, internal parasiticide for horses	Argentina	36
Arnica montana L.	L	Cicatrizant, infections, inflammation in general, traumatism	Brazil	21
Baccharis salicifolia (Ruiz & Pav.) Pers.	AP	Diuretic, expectorant, vulnerary	Argentina	36
Eclipta alba (L.) Hassk.	L	Used for scorpion stings	Trinidad and Tobago	30
Egletes viscosa (L.) Less.	SE	Intestinal disorders	Brazil	35
Neurolaena lobata (L.) Cass.	L	For reduced appetite in chickens	Trinidad and Tobago	29
Pectis odorata Griseb.	AP	Oxytocic, for uterine atony in goats and cows, retained placenta	Argentina	36
Pluchea microcephala R.K. Godfrey	AP	Antidiarrheal, for bathing wounds, vulnerary,	Argentina	36
Pluchea sagittalis (Lam.) Cabrera	AP	Antidiarrheal, for bathing wounds, vulnerary	Argentina	36
Tessaria integrifolia Ruiz & Pav.	ST	Stick used to stop up the inside of goats' horns after treatment against tick fever	Argentina	36
Vernonia scorpioides (Lam.) Pers.		To stop bleeding, relieve inflammation; treatment of stomach aches, asthma, intestinal parasites and protection against snakebites	Trinidad and Tobago	30
Bignoniaceae				
Crescentia cujete L.	FR	Abortifacient	Mexico	37
Bixaceae				
Bixa orellana L.	SE	The pulp of the seeds is given to bulls before fights to make them more active and ferocious	Brazil	37
Bombacaceae				
Ceiba pentandra (L.) Gaertn.	Not mentioned	Not mentioned	Not mentioned	37

(*continued*)

TABLE 10.1 (CONTINUED)
Plants Used in Ethnoveterinary Medicine in Latin America

Family/Plant Name	Parts Used	Veterinary Uses	Country	Author
Pseudobombax marginatum (A. St.-Hil., Juss. & Cambess.) A. Robyns	BS	Renal problems	Brazil	35
Boraginaceae				
Cordia curassavica (Jacq.) Roem. & Schult.	L	To repel ectoparasites from chickens and ducks, grooming	Trinidad and Tobago	29, 31
Heliotropium indicum L.	L	For weakness and retained placenta	Brazil	35
Brassicaceae				
Nasturtium officinale R. Br.	L	Increase blood count	Trinidad and Tobago	31
Cactaceae				
Nopalea cochenillifera (L.) Salm-Dyck	AP	Used as diaphoretic, for tendon problems, and for wounds	Trinidad and Tobago	28, 31
Opuntia quimilo K. Schum.	ST	To remove thorns from horses	Argentina	36
Capparaceae				
Capparis flexuosa (L.) L.	BK, L	Intoxication	Brazil	35
Capparis speciosa Griseb.	BK	Vulnerary and for skin myiasis	Argentina	36
Capparis tweediana Eichler	L	Antidiarrheal	Argentina	36
Caricaceae				
Carica papaya L.	FL, L	Vermifuge in pigs	Trinidad and Tobago	32
Cecropiaceae				
Cecropia peltata L.	L, R	Used for snakebites (R), anhydrosis, kidney problems (L), to ease labor pains in pigs or fed as a postpartum cleanser (L)	Trinidad and Tobago	30, 31, 32
Chenopodiaceae				
Chenopodium ambrosioides L.	AP	Antidiarrheal, for internal parasiticide, oxytocic—for uterine atony in goats and cows and retained placenta, fractures, infections in general, inflammations, traumatism, and pneumonia	Argentina, Brazil, Trinidad and Tobago	21, 23, 32, 35, 36

Clusiaceae				
Mammea americana L.	BK	Used against exoparasites	Mexico, Trinidad and Tobago	37, 62
Convolvulaceae				
Ipomoea asarifolia (Desr.) Roem. & Schult.	BK, FR, L	For scabies treatment	Brazil	21, 35
Ipomoea carnea Jacq.	L	Vulnerary: leaves are boiled and left to cool, then applied directly on the wounds	Argentina	36
Operculina alata Urb.	TU	Vermifuge	Brazil	35
Operculina hamiltonii (G. Don) D.F. Austin & Staples	TU	Vermifuge	Brazil	22, 27
Operculina macrocarpa (Linn) Urb.		Cicatrizant, vermifuge	Brazil	21
Crassulaceae				
Kalanchoe brasiliensis Cambess.	L	Inflammation	Brazil	35
Kalanchoe pinnata (Lam.) Pers.	L	Used as emollient on burns and inflamed tissues	Trinidad and Tobago	28, 29
Cucurbitaceae				
Cucumis anguria L.	FR	For kidney stone treatment	Brazil	35
Cucurbita pepo L.	SE	Vermifuge	Brazil	22, 27, 35
Luffa acutangula (L.) Roxb.	FR	Unripe fruit is used against bowel disease of domestic fowl	Brazil	37
Luffa aegyptiaca Mill.	FR	Unripe fruit is used against bowel disease of domestic fowl	Brazil	37
Luffa operculata (L.) Cogn.	FR	Fever, bovine tick fever, and colic	Brazil	35
Momordica charantia L.	FR, L, ST	For pruritus treatment, vermifuge, used for sick chickens, used as tonic, for blood purifier and skin rashes	Brazil, Trinidad and Tobago	22, 27, 29, 30, 31, 35
Erythroxylaceae				
Erythroxylum coca Lam.	L	Oxytocic, for uterine atony in goats and cows, retained placenta	Argentina	36

(*continued*)

TABLE 10.1 (CONTINUED)
Plants Used in Ethnoveterinary Medicine in Latin America

Family/Plant Name	Parts Used	Veterinary Uses	Country	Author
Euphorbiaceae				
Cnidoscolus phyllacanthus (Müll. Arg.) Pax & L. Hoffm.	L	Not mentioned	Brazil	35
Croton blanchetianus Baill.	BK	Not mentioned	Brazil	34
Croton gossypiifolius Vahl	L	Used in dogs for successful hunting	Trinidad and Tobago	30
Croton sonderianus Müll. Arg.	BK	For colics	Brazil	35
Jatropha curcas L.	L	Used in dogs for successful hunting	Trinidad and Tobago	30
Jatropha gossypiifolia L.	BK, L	For blindness treatment (BK), used in dogs for successful hunting (L)	Trinidad and Tobago, Brazil	30, 35
Jatropha hieronymi Kuntze	FR	Purgative, for acute constipation in horse	Argentina	36
Jatropha mollissima (Pohl) Baill.	LA	To treat snakebites	Brazil	20
Phyllanthus niruri L.	L, ST, R	Against renal problems, used in dogs for successful hunting	Trinidad and Tobago, Brazil	30, 35
Phyllanthus urinaria L.	L	Used in dogs for successful hunting	Trinidad and Tobago	30
Ricinus communis L.	L, SE	To eliminate catarrh in equines (SE), tendon problems (L)	Brazil, Trinidad and Tobago	31, 35
Sapium haematospermum Müll. Arg.	L, ST	Stick used to stop up the inside of goats' horns after treatment for tick fever, wounds	Argentina	36
Fabaceae				
Acacia albicorticata Burkart	L	Vulnerary	Argentina	36
Acacia aroma Gillies ex Hook. & Arn.	L, BK	Vulnerary, for bathing wounds	Argentina, Bolivia	25, 36
Amburana cearensis (Allemão) A.C. Sm.	L	Antidiarrheal, antiparasitical (endoparasites)	Brazil	21
Anadenanthera colubrina (Vell.) Brenan	BK	For inflammation in general	Bolivia, Brazil	25, 35
Bauhinia cumanensis Kunth	AP	Used for snakebites	Trinidad and Tobago	30
Bauhinia guianensis Aubl.	AP	Used for snakebites	Trinidad and Tobago	30

Caesalpinia ferrea Mart.	L	Cicatrizant, for inflammation in general, to treat renal problems	Brazil	21, 35
Caesalpinia pyramidalis Tul.	L	Antidiarrheal, to treat intestinal infections	Brazil	35
Cassia angustifolia Vahl	L	Inflammations in general	Brazil	21
Cassia occidentalis L.	BK	Retained placenta	Brazil, Trinidad and Tobago	21, 28
Cassia tora L.	L	Antiparasitical (for ticks)	Brazil	37
Erythrina micropteryx Poepp. ex Urb.	L	Used to remove dead piglets from the uterus	Trinidad and Tobago	32
Erythrina pallida Britton	L	Used to remove dead piglets from the uterus	Trinidad and Tobago	32
Myroxylon peruiferum L. f.	Not mentioned	Not mentioned	Bolivia	25
Geoffroea decorticans (Gillies ex Hook. & Arn.) Burkart	BK, NU	Oxytocic; for uterine atony in goats and cows; retained placenta; the nutgalls employed are those that have an echinulate surface	Argentina	36
Leucaena glauca Benth.	AP	There is a prevalent belief that if horses, mules, or pigs eat any part of the plant, their hairs will fall out	Mexico	37
Mimosa albida Humb. & Bonpl. ex Willd.	R	For vision problems	Mexico	33
Mimosa pudica L.	R	Used to induce estrus	Trinidad and Tobago	28
Mimosa tenuiflora (Willd.) Poir.	L	Antibiotic, cicatrizant, for general inflammation	Brazil	21, 35
Mucuna pruriens (L.) DC.	L	Used in for enhanced performance	Trinidad and Tobago	31
Pithecellobium unguis-cati (L.) Benth.	Not mentioned	Not mentioned	Trinidad and Tobago	30
Prosopis alba Griseb.	FR, L	Galactagogue for cows; purgative; against acute constipation in horses	Argentina	36
Prosopis juliflora (Sw.) DC.	FR	To treat heart problems	Brazil	35
Prosopis ruscifolia Griseb.	FR, L	Galactagogue for cows, vulnerary	Argentina	36
Pueraria phaseoloides (Roxb.) Benth.	L	High protein feed	Trinidad and Tobago	31
Senna alexandrina Mill.	BK, whole plant	To treat foot-and-mouth disease (FMD)	Brazil	35
Senna morongii (Britton) H.S. Irwin & Barneby	L	Vulnerary; for bathing all kinds of wounds	Argentina	36

(*continued*)

TABLE 10.1 (CONTINUED)
Plants Used in Ethnoveterinary Medicine in Latin America

Family/Plant Name	Parts Used	Veterinary Uses	Country	Author
Tamarindus indica L.	FR	Used as laxative and anti-inflammatory	Brazil	35
Tipuana tipu (Benth.) Kuntze	Not mentioned	Not mentioned	Bolivia	25
Haemodoraceae				
Xiphidium caeruleum Aubl.	L	Used to help hunting dogs in Trinidad run fast and "brighten them up"	Trinidad and Tobago	30
Lamiaceae				
Calamintha officinalis Moench	AP	For colic and indigestion	Brazil	35
Hyptis suaveolens (L.) Poit.	FL, L	Antidiarrhetic	Brazil	35
Mentha villosa Huds.	L	To treat amoeba and giardia infections	Brazil	35
Ocimum basilicum L.	SE	For keratitis in cows; the seeds are placed directly in the animal's eyes	Argentina	36
Plectranthus amboinicus (Lour.) Spreng.	L	Cough	Brazil	35
Rosmarinus officinalis L.	Not mentioned	For stomach problems	Mexico	33
Lauraceae				
Laurus nobilis L.	L	Preventive; used to fumigate the stock pens together with lemon peel when the cattle die for no known reason	Argentina	36
Liliaceae				
Allium cepa L.	BU	To treat cataract	Brazil	35
Allium sativum L.	BU	Expectorant; against acute adenitis in horses, snakebites and insect bites; against canine distemper; for hypothermia caused by spider bites or malaria; against endoparasites; to treat mastitis and pneumonia; for reduced appetite in chickens	Argentina, Brazil, Trinidad and Tobago	21, 29, 35, 36

Aloe vera (L.) Burm. f.	L	Febrifuge: for fever caused by spider bites in horses and dogs; antidiarrheal; cicatrizant; purgative; to treat rabies; used for retained placenta and traumatism; vermifuge	Argentina, Brazil, Trinidad and Tobago	21, 28, 29, 31, 32, 36
Loranthaceae				
Struthanthus angustifolius (Griseb.) Hauman	AP	Oxytocic: for uterine atony in goats and cows, for retained placenta	Argentina	36
Tripodanthus acutifolius (Ruiz & Pav.) Tiegh.	AP	Oxytocic: for uterine atony in goats and cows, for retained placenta	Argentina	36
Lythraceae				
Heimia salicifolia Link	R	Antidiarrheal	Argentina	36
Malvaceae				
Malva sylvestris L.	L	Anti-inflammatory	Brazil	35
Meliaceae				
Azadirachta indica A. Juss.			Trinidad and Tobago	28, 29
Cedrela odorata L.	BK, L	Used for alimentary intoxication (BK), to repel ectoparasites from chickens and ducks and their environments (L)	Brazil, Trinidad and Tobago	29, 35
Monimiaceae				
Peumus boldus Molina	L	Antidiarrheal, antiemetic, and for liver and stomach disturbances	Brazil	21, 35
Myrtaceae				
Eucalyptus tereticornis Sm.	L	Expectorant	Argentina	36
Myrcia uniflora Barb. Rodr.	L	Diabetes	Brazil	35
Pimenta racemosa (Mill.) J.W. Moore	L	Used for respiratory conditions and as diaphoretic	Trinidad and Tobago	29, 31
Psidium guajava L.	BK, SP	Antidiarrhetic	Brazil, Trinidad and Tobago	21, 28, 31, 35
Olacaceae				
Ximenia americana L.	FR	Galactagogue for cows	Argentina, Brazil	21, 35, 36

(continued)

TABLE 10.1 (CONTINUED)
Plants Used in Ethnoveterinary Medicine in Latin America

Family/Plant Name	Parts Used	Veterinary Uses	Country	Author
Orchidaceae				
Sarcoglottis acaulis (Sm.) Schltr.	AP	Used in dogs for successful hunting	Trinidad and Tobago	30
Oxalidaceae				
Oxalis corniculata L.	Not mentioned	Enhanced performance in horses	Trinidad and Tobago	31
Papaveraceae				
Argemone subfusiformis G.B. Ownbey	L, LA, ST	Cicatrizant and to control cattle ticks	Peru	26
Phytolaccaceae				
Petiveria alliacea L.	Not mentioned	Used to induce estrus and as anthelmintic, make dogs more alert	Argentina, Trinidad and Tobago	28, 29, 30, 36
Piperaceae				
Lepianthes peltata (L.) Raf. ex R.A. Howard	L	Make dogs run faster	Trinidad and Tobago	30
Ottonia ovata (Vahl) Trel.	L, ST, R	To put it in the dog's nose or wash the dog's nose with decoction	Trinidad and Tobago	30
Piper scabrum Sw.	L	Leaves are used to bathe dogs	Trinidad and Tobago	30
Plantaginaceae				
Plantago major L.	L	Ocular injuries in backyard chickens	Trinidad and Tobago	32
Poaceae				
Bambusa vulgaris Schrad. ex J.C. Wendl.	L	For retained placenta and to ease pig labor pains or fed as a postpartum cleanser	Trinidad and Tobago	22, 26
Cymbopogon citratus (DC.) Stapf	L	Used as sedative	Brazil	35
Cynodon dactylon (L.) Pers.	AP	Emetic: dogs spontaneously eat its leaves when they suffer from indigestion	Argentina	36
Oryza sativa L.	Not mentioned	For retained placenta	Trinidad and Tobago	28
Panicum maximum Jacq.	L	Used for grooming	Trinidad and Tobago	31

Paspalum conjugatum P.J. Bergius	Not mentioned	Used with other plants to prepare hunting dogs	Brazil	37
Zea mays L.	FL, MR, STG	Against canine distemper (MR), diuretic for horses (STG), disinfectant of urinary tract (STG), oxytocic for uterine atony in horses and for retained placenta (FL)	Argentina, Brazil	35, 36
Polygonaceae				
Polygonum punctatum Elliott	L	Vulnerary: to bathe wounds	Argentina	36
Ruprechtia triflora Griseb.	L	Antidiarrheal	Argentina	36
Portulacaceae				
Portulaca oleracea L.	AP	Purgative	Argentina	36
Punicaceae				
Punica granatum L.	FR	Anti-inflammatory; for gastritis, ulcers	Brazil	35
Ranunculaceae				
Clematis dioica L.	L	To treat glanders in horses, donkeys, and mules	Mexico	33
Clematis montevidensis Spreng.	FR	Expectorant	Argentina	36
Rhamnaceae				
Ziziphus joazeiro Mart.	BK	Cicatrizant and to treat rabies	Brazil	35
Rubiaceae				
Coffea arabica L.	GR	Given to scouring pigs	Trinidad and Tobago	32
Coffea robusta Linden	GR	Given to scouring pigs	Trinidad and Tobago	32
Coutarea hexandra (Jacq.) K. Schum.	BK	Bovine tick fever	Brazil	35
Pogonopus tubulosus (A. Rich.) K. Schum.	Not mentioned	Used to heal castration wounds of small domestic animals	Bolivia	25
Psychotria ipecacuanha (Brot.) Stokes	BK	General inflammation, against endoparasites, to treat bronchitis and catarrh	Brazil	21, 35
Staelia scabra (C. Presl) Standl.	AP	To treat rages	Mexico	33
Rutaceae				
Citrus aurantifolia (Christm.) Swingle	FR	For respiratory conditions	Trinidad and Tobago	29
Citrus aurantium L.	FR	For respiratory conditions; juice included in pig diet before slaughter to reduce fat	Trinidad and Tobago	29, 32

(continued)

TABLE 10.1 (CONTINUED)
Plants Used in Ethnoveterinary Medicine in Latin America

Family/Plant Name	Parts Used	Veterinary Uses	Country	Author
Citrus limetta Risso	FR	For respiratory conditions	Trinidad and Tobago	29
Citrus limon (L.) Burm. f.	FR	Preventive; the peel is used to fumigate the stock pens together with leaves of *Laurus nobilis* when the cattle die for no known reason	Argentina	36
Citrus medica L.	FR	Used for upper respiratory problems and as a purgative in backyard chickens	Trinidad and Tobago	32
Citrus x limon (L.) Osbeck	FR	Against endoparasites, to treat pneumonia	Brazil	21
Salicaceae				
Salix humboldtiana Willd.	ST	Oxytocic: for uterine atony in goats and cow and for retained placenta	Argentina	36
Sapotaceae				
Bumelia sartorum Mart.	BK	To treat genital tract infections	Brazil	35
Manilkara zapota (L.) P. Royen	Not mentioned	Not mentioned	Trinidad and Tobago	62
Pouteria sapota (Jacq.) H.E. Moore & Stearn	Not mentioned	Not mentioned	Trinidad and Tobago	62
Sideroxylon obtusifolium (Humb. ex Roem. & Schult.) T.D. Penn.	AP, BK, R	Oxytocic: for uterine atony in goats and cows (AP, BK, R); purgative (R)	Argentina	36
Scrophulariaceae				
Scoparia dulcis L.	Not mentioned	Used as expectorant	Brazil	35
Solanaceae				
Capsicum annuum L.	L	Anhydrosis	Trinidad and Tobago	31
Capsicum chacoense Hunz.	FR	For anuria in horses	Argentina	36
Capsicum frutescens L.	Not mentioned	Not mentioned	Brazil	37
Lycopersicon esculentum Mill.	FR	Used to treat tumors	Brazil	35

Nicotiana tabacum L.	L	Hypothermia in horses caused by spider bites or malaria; cleans dog's nose to improve its ability to follow a scent	Argentina, Trinidad and Tobago	30, 36
Physalis viscosa L.	AP	Emetic	Argentina	36
Solanum aridum Morong	AP	Purgative for fighting roosters	Argentina	36
Solanum melongena L.	BK	Inflammations in general	Brazil	21
Solanum paniculatum L.	BK	Inflammations in general	Brazil	35
Sterculiaceae				
Cola nitida (Vent.) Schott & Endl.	SE	Used for snakebites	Trinidad and Tobago	30
Theobroma cacao L.	FR	Used for wounds	Trinidad and Tobago	28
Ulmaceae				
Celtis chichape (Wedd.) Miq.	L	Antidiarrheal, emetic	Argentina	36
Urticaceae				
Laportea aestuans (L.) Chew			Trinidad and Tobago	28
Parietaria debilis G. Forst.	AP	Purgative for fighting roosters	Argentina	36
Verbenaceae				
Aloysia polystachya (Griseb.) Moldenke	AP	Digestive	Argentina	36
Lippia alba (Mill.) N.E. Br.	AP	Antidiarrheal; against vomit, colic; to treat tympany	Brazil	21, 35
Stachytarpheta jamaicensis (L.) Vahl	L	Used to increase milk production and as anthelmintic	Trinidad and Tobago	28, 31
Verbena hispida Ruiz & Pav.	Not mentioned	Not mentioned	Bolivia	25
Viscaceae				
Phoradendron hieronymi Trel.	AP	Oxytocic: for uterine atony in goats and cows and for retained placenta	Argentina	36
Phoradendron liga (Gillies ex Hook. & Arn.) Eichler		Oxytocic: for uterine atony in goats and cows and for retained placenta	Argentina	36
Zingiberaceae				
Aframomum melegueta K. Schum.	SE	Dry seeds are ground to a powder and sprinkled on the dog's food	Trinidad and Tobago	30
Alpinia speciosa (Blume) D. Dietr.	FL	Cough	Brazil	35

(*continued*)

TABLE 10.1 (CONTINUED)
Plants Used in Ethnoveterinary Medicine in Latin America

Family/Plant Name	Parts Used	Veterinary Uses	Country	Author
Costus scaber Ruiz & Pav.	ST	Used for snakebites	Trinidad and Tobago	30
Curcuma longa L.	RI	For retained placenta, wounds	Trinidad and Tobago	28, 31
Renealmia alpinia (Rottb.) Maas	L	To repel ectoparasites from chickens and ducks and their environments	Trinidad and Tobago	29, 30
Zingiber officinale Roscoe	R	Antiseptic, antiphlogistic	Brazil	35
Zygophyllaceae				
Bulnesia sarmientoi Lorentz ex Griseb.	DU	Expectorant	Argentina	36

AP = aerial parts; BK = bark; BU = bulb; DU = duramen; FL = flower; FR = fruit; GR = grounds; LA = latex; L = leaves; MR = cob; NU = nutgalls; R = root; RI = rhizome; SE = seed; SP = sprouts; ST = stalk; STG = stigma; TU = tuber.

Access also exerts a significant role in influencing the type of medicine used. Phillips and Gentry,[51] investigating the useful plants of Tambopata, Peru, reported that easily accessed plants offer more possibilities for local populations to experiment with their uses, thus having a greater probability of being introduced into the local culture. As remarked by Alves and Rosa,[55] the use of local, more easily accessible resources is possibly related to historic aspects (i.e., medicinal knowledge focusing on species familiar to the locals, reflecting transmission of knowledge through generations) and to financial constraints that limit access to alloctonous resources.

Common plant uses by distantly located and distinct ethnic groups point to a similar evolution of indigenous knowledge and may confirm the credibility of the medicinal value of the plants. The information on popular plant uses in one location could be useful to other places having the same plants but not using them the same way. The value of studies comparing the ethnobotany of distant and distinct regions within a country or of different countries is universally recognized now. Such comparative studies yield valuable information on the uses and properties of plants and the width and depth of indigenous knowledge.[37]

It is known that the use of medicinal animals is common in several countries in Latin America,[2,56,57] and that often there are overlaps in the medicinal use of both plants and animals in traditional medicine for humans.[2,58] Similarly, in ethnoveterinary medicine, these overlaps also occur. In semiarid areas of Brazil, medicinal plants are used in association with animal parts for ethnoveterinary purposes. This is not surprising given that the use of animal products is common in traditional medicine, although scientific studies into these practices are rare. Barboza et al.[23] recorded the utilization of animals (zootherapeutics) as sources of medicines in folk veterinary medicine (ethnoveterinary medicine) in semiarid northeast Brazil and verified that 15 different animal species are used in the prevention or treatment of veterinary diseases in that region. There therefore appears to be a close association between phytotherapeutic and zootherapeutic practices in traditional medicine, both human and veterinary.

Animal and human medicine have been closely linked throughout history. Some diseases affect both animals and humans and can be treated with similar remedies. This way, many plants used in ethnoveterinary medicine are also used for the treatment of human diseases. In Brazil, for instance, São Caetano's melon (*Mormodica charantia* L.), the purga potato (*Operculina hamiltonii*), and the pumpkin's seed (*Cucurbita pepo* L.) are used as vermifuge in veterinary medicine, probably similar to the same way they are prepared in human medicine.[21,27] A similar trend has been observed by Lans et al.[32] in Canada, where the utilization of a popular medicine composed of *Juglans nigra* and *Artemisia annua* is used for the treatment of dogs with roundworms based on the usage of this medicine for humans. Studies in the traditional management of diarrhea in humans[59] have shown that *Psidium guajava* is widely used in ethnoveterinary medicine.[60] In Trinidad and Tobago, the ethnoveterinary uses of *Psidium guajava* and *Anacardium occidentale* for diarrhea have transferred directly from similar use in humans.[61] The use of the seeds of *Pouteria sapota, Manilkara zapota,* and *Mammea americana* for ectoparasite control in dogs is possibly derived from the human folk use against the chiggoe flea (*Tunga penetrans* Linn.).[62] This practice is also reported for *M. americana* seeds in Suriname[63] and by the Chocó Indians.[64] In Calabria region, southern Italy, erva lupara (*Scrophularia canina* L.) is used in ethnoveterinary medicine and in human ethnomedicine.[65] These results support other studies that have revealed that in most traditional societies there is no clear division between veterinary and human medicine.[66]

Scarpa[36] reported a strong parallel among the plants used in traditional veterinary medicine by the Chaco nation in the Argentinean northwest, where 60% of the reports stated as therapeutic correlate to plants used by humans. Such similarities of uses of plants is evidence of the fact that ethnoveterinary practices and human ethnomedicine have probably followed two main evolutionary pathways: one based on the observations of self-medication in animals (zoopharmacognosy) and the other related to human folk medicine, as reported by Pieroni et al.[49] The use of folk remedies in animals to treat diseases or conditions based on similar or identical illnesses that attack humans has been termed the "human model for diseases in animals" by Barboza et al.[23] The relationships between ethnoveterinary and human ethnomedicine on this perspective can be easily explainable since mammals are the main stock animals (e.g., cattle, sheeps, goats, pigs, among others). Regarding this, animals usually face health problems that also affect humans, being noticed in identical symptomatology by several people or communities. However, this cannot be taken as a rule for all particular cases.

10.4 SCIENTIFIC STUDIES

Many diseases affect livestock and other animals, and causal organisms of disease include protozoa, viruses, bacteria, fungi, and helminth parasites. Ticks are important vectors of several economically important ailments, such as heartwater, anaplasmosis, sweating sickness, and babesiosis, and several plant remedies are used to rid animals of ticks. Ethnoveterinary medicine may play a significant role in the management of such diseases in a cost-effective and accessible manner, and there are an increasing number of publications in the scientific literature reporting on the ethnopharmacological activity of plants used in traditional veterinary medicine. One example is *Aloe* species, which have been in use as broad-spectrum therapeutics in ethnomedicine.[67] Their antibacterial, antifungal, and antiviral activities have been established, as extensively reviewed by Reynolds and Dwecks.[68] *Aloe vera* has been shown to have antibacterial activity against a wide range of both Gram-positive and -negative bacteria.[67] Aloe-emodin, an anthraquinone from many species of aloe, was shown to have effect against four strains of methicillin-resistant *Staphylococcus aureus*[69] and showed activity against *S. aureus*, *Streptococcus pyogenes*, *Proteus vulgaris*, and *Bacillus subtilis*.[70] Also, high levels of antibacterial efficacy in extracts of *Ziziphus mucronata* (Rhamnaceae) were observed.[71] Compounds were isolated from leaves of this species: 2,3-dihydroxyl-up-20-en-28-oic acid and zizyberanalic acid.[72] The first compound demonstrated excellent activity against *Staphylococcus aureus*, promoting claims of the efficacy of a *Ziziphus mucronata* leaf paste in treating bacterial infections in animals and humans. A promising area of research involves the investigation of antitick activity of plant extracts, entailing observation of tick repellent and toxic effects of plants. *Allium* species were investigated for repellency and toxicity against *Hyalomma marginatum rufipes* adults.[73] The acetone extract of *Allium porrum* revealed a high repellency index (65–79.48%), and *Allium sativum* dichloromethane extract was toxic to 100% of ticks within an hour of exposure. Essential oils from the aerial parts of *Lippia javanica* and *Tagetes minuta* were repellent toward the ticks in a concentration-dependent manner.[73] In a growth inhibition bioassay, it was shown that *Tagetes minuta* essential oil delayed molting to adult stage of 60% of engorged nymphs of *Hyalomma marginatum rufipes*.[73]

As apparent in the examples above, the chemical constituents and pharmacological actions of some plant products are already known to some extent, and ethnopharmacological studies focused on plant remedies could be important in clarifying the eventual therapeutic usefulness of this class of biological remedies. Unfortunately, relatively little research has been conducted on the bioactivity of plants used for ethnoveterinary purposes in particular. As pointed by McGaw and Eloff[74] (see also Chapter 2), it is essential to evaluate not only the bioactivity but also the safety of such plant treatments if their use is to be promoted and potentially developed for commercial purposes. It is, however, expensive and ethically complex to conduct *in vivo* experiments, and large quantities of test material and specialized facilities are required.

10.5 CONCLUSION

Our results have found that a substantial number of plant species ($n = 201$) are used in veterinary traditional medicine, and that the vast majority of these species are collected from the wild. This indicates the very rich ethnomedical knowledge of the local population in relation to plants used in ethnoveterinary medicine. However, considering the wealth of biological resources in the subcontinent, this finding is a gross underestimation of the length and breadth of plants used for ethnoveterinary purposes. This justifies the call for more and extensive anthropological and scientific studies on the subject.

Medicinal plants represent an alternative or a complement to conventional medicine in Latin America as they do elsewhere in resource-poor countries, particularly in Africa and Asia. As pointed by Alves and Rosa,[75] the linkages between biodiversity and human health are especially important for developing countries. Biodiversity is a source of invaluable raw materials and materia medica. Traditional medicine is widely available and affordable, even in remote areas, and generally is accessible to most people. In many developing countries, a large part of the population, especially in rural areas, depends mainly on traditional medicine for their primary health care because it is cheaper and more accessible than orthodox medicine.[76–78]

The documentation and conservation of medicinal plants (for humans and animals) is therefore highly recommended. Many parts of Latin America are rapidly being altered, both ecologically and culturally. According to Calixto,[18] despite the many individual efforts of the governments to preserve the biological resources for future generations in Latin America, traditional knowledge, especially that associated with traditional medicine, is fast disappearing. The protection and revival of traditional medical knowledge and practice in thousands of ethnic communities is an important means of providing affordable and sustainable health care. The knowledge that traditional health practitioners, women, and farmers can bring to identifying, implementing, and managing environmental conservation and cultivation programs is seldom sought or utilized. Consequently, local health traditions, many of which are oral in nature and therefore largely undocumented, are being lost.

The medicinal use of plants for treating various disorders in humans and in their animals is a centuries-old tradition in many cultures. The possible benefit of plant-derived medications constitutes a rewarding area of research, particularly in areas such as Latin America that have a rich biodiversity of resources. The literature on ethnoveterinary botanicals from Latin America is sparse, and there is a great necessity for more studies, principally considering the cultural, socioeconomic, and ecological importance associated with plant utilization in ethnoveterinary medicine in the region. Conservation programs of both biological and cultural diversity in Latin America should be a priority of governments, researchers, and donors in the region.

REFERENCES

1. Alakbarli, F., *Medical manuscripts of Azerbaijan*. Heydar Aliyev Foundation, Baku, Azerbaijan, 2006.
2. Alves, R.R.N., Rosa, I.L., and Santana, G.G., The role of animal-derived remedies as complementary medicine in Brazil. *BioScience*, 57, 949, 2007.
3. David, J.P., and Anderson, L.A., Ethnopharmacology and western medicine. *Journal of Ethnopharmacology*, 25, 61, 1969.
4. Good, C., Ethno-medical systems in Africa and the LDCs: key issues in medical geography, in *Conceptual and methodological issues in medical geography*, Meade, M.S., Ed., University of North Carolina, Chapel Hill, 1980.
5. Gesler, W.M., Therapeutic landscapes: medical issues in light of the new cultural geography. *Social Science and Medicine*, 34, 735, 1992.
6. World Health Organization (WHO), *Millennium ecosystem assessment. Ecosystems and human well-being: synthesis*. WHO, Geneva, Switzerland, 2005.
7. Food and Agriculture Organization (FAO), *Genetics and animal health—spotlight*. FAO, Rome, 2002.
8. Centre Technique de Cooperation Agricole et Rurale (CTA), Primary animal health care in Africa, in *Synopsis of the Seminar held at Blantyre*. CTA, Lilongwe, Malawi, 1987, 28 pp.
9. Koné, W.M., and Atindehou, K.K., Ethnobotanical inventory of medicinal plants used in traditional veterinary medicine in Northern Côte d'Ivoire (West Africa). *South African Journal of Botany*, 74, 76, 2008.
10. Hoareau, L., and Silva, E.J., Medicinal plants: a re-emerging health aid. *Journal of Biotechnology*, 2, 123, 2001.
11. Kent, R.B., *Latin America: regions and people*. Guilford, New York, 2006.
12. Population Reference Bureau, World population highlights. *Population Bulletin*, 63, 3, 2008.
13. World Bank, *Urban development in Latin America and the Caribbean—the World Bank and Urban Development in Latin America and Caribbean region*. http://web.worldbank.org/WBSITE/EXTERNAL/COUNTRIES/LACEXT/EXTLACREGTOPURBDEV/0,contentMDK:20797274~menuPK:841079~pagePK:34004173~piPK:34003707~theSitePK:841043,00.html.
14. Central Intelligence Agency, *Field listing—ethnic groups*. https://www.cia.gov/library/publications/the-world-factbook/fields/2075.html.
15. International Fund for Agricultural Development (IFAD), Latin American and the Caribbean Division. *Regional strategy paper: IFAD strategy for rural poverty reduction in Latin America and the Caribbean*. IFAD, Rome, 2002.

16. Astudillo, V.M., Zottele, A.C., and Dora, F., Livestock development and animal health in Latin America. *Boletín del Centro Panamericano de Fiebre Aftosa,* 57, 1991.
17. Hauser, G., Little, M., and Roberts, D.F., Eds., *Man, culture and biodiversity. Understanding interdependences*. IUBS, Biology International, Paris, 1934, 32.
18. Calixto, J.B., Twenty-five years of research on medicinal plants in Latin America. *Journal of Ethnopharmacology,* 100, 131, 2005.
19. Elisabetsky, E., and Wannmacher, L., The status of ethnopharmacology in Brazil. *Journal of Ethnopharmacology,* 38, 131, 1993.
20. Agra, M.F., et al., Medicinal and poisonous diversity of the flora of "Cariri Paraibano," Brazil. *Journal of Ethnopharmacology*, 111, 383, 2007.
21. Almeida, K.S., Freitas, F.L.C., and Pereira, T.F.C. Etnoveterinária: a Fitoterapia na visão do futuro profissional veterinário. *Revista Verde*, 1, 67, 2006.
22. Almeida, W.V.F., et al., Avaliação de Plantas medicinais em caprinos da região do semi-árido paraibano naturalmente infectados por nematóides gastrintestinais. *Revista Caatinga*, 20, 1, 2007.
23. Barboza, R.R.D., Souto, W.M.S., and Mourão, J.S., The use of zootherapeutics in folk veterinary medicine in the district of Cubati, Paraíba State, Brazil. *Journal of Ethnobiology and Ethnomedicine*, 3, 1, 2007.
24. Bussmann, R.W., and Sharon, D., Traditional medicinal plant use in Loja Province, southern Ecuador. *Journal of Ethnobiology and Ethnomedicine*, 2, 1, 2006.
25. Carretero, A.L., Useful plants and traditional knowledge in the *Tucumano-Boliviano* Forest. MSc thesis, Aarhus University, Denmark, 2005.
26. De-la-Cruz, H., Vilcapoma, G., and Zevallos, P.A., Ethnobotanical study of medicinal plants used by the Andean people of Canta, Lima, Peru. *Journal of Ethnopharmacology*, 111, 284, 2007.
27. Faria, E.B., et al., Difusão da etnoveterinária como alternativa para o controle das verminoses de caprinos e ovinos do Alto Piranhas, PB, in *VIII Congresso Ibero-Americano de Extensão Universitária*, Sousa, A.I., et al., Eds., UFRJ, Rio de Janeiro, Brazil, 2005, 610.
28. Lans, C., and Brown, G., Ethnoveterinary medicines used for ruminants in Trinidad and Tobago. *Preventive Veterinary Medicine*, 35, 149, 1998.
29. Lans, C., and Brown, G., Observations on ethnoveterinary medicines in Trinidad and Tobago. *Preventive Veterinary Medicine*, 35, 125, 1998.
30. Lans, C., Harper, T., Georges K., and Bridgewater E., Medicinal and ethnoveterinary remedies of hunters in Trinidad. *BMC Complementary and Alternative Medicine*, 1, 1, 2001.
31. Lans, C., et al., Ethnoveterinary medicines used for horses in Trinidad and in British Columbia, Canada. *Journal of Ethnobiology and Ethnomedicine*, 2, 1, 2006.
32. Lans, C., Georges K., and Brown, G., Non-experimental validation of ethnoveterinary plants and indigenous knowledge used for backyard pigs and chickens in Trinidad and Tobago. *Tropical Animal Health and Production*, 39, 375, 2007.
33. López, V.M.E., Contribución etnobotánica en plantas medicinales utilizadas por dos grupos étnicos de Mecapalapa, municipio de Pantepec, Puebla. Graduate thesis, Escuela Nacional de Estudios Profesionales Iztacala, Universidad Nacional Autónoma de México, Los Reyes Iztacala, Edo. de México, 1988.
34. Lucena, R.F.P., et al., Useful plants of the semi-arid northeastern region of Brazil—a look at their conservation and sustainable use. *Environmental Monitoring and Assessment*, 125, 281, 2007.
35. Marinho, M.L., et al., A utilização de plantas medicinais em medicina veterinária: um resgate do saber popular. *Revista Brasileira de Plantas Medicinais*, 9, 64, 2007.
36. Scarpa, G.F., Plants employed in traditional veterinary medicine by the criollos of the northwestern Argentine Chaco. *Darwiniana*, 38, 253, 2000.

37. Sikarwar, R.L.S., Less-known ethnoveterinary uses of plants in India, in *Proceedings of an International Conference Held in Pune, India*, Mathias, E., Rangnekar, D.V., and McCorkle, C.M., Eds., BAIF Development Research Foundation, Pune, India, 1997, p. 103–107.
38. Gupta, A.K., Origin of agriculture and domestication of plants and animals linked to early Holocene climate amelioration. *Current Science*, 87, 1, 2004.
39. Pierret, P.V., and Dourojeanni, M.J., Importancia de la Caza para la Alimentación Humana en el Curso Inferior del Río Ucayali. *Revista Forestal del Perú,* 1, 1967.
40. Bennett, E.L., and Robinson, J.G., Hunting for sustainability: the start of a synthesis, in *Hunting for sustainability in tropical forests (Biology and Resource Series)*, Robinson, J.G., and Bennett, E.L., Eds., Columbia University Press, New York, 1999, p. 499–519.
41. Alves, R.R.N., and Nishida, A.K., A ecdise do caranguejo-uçá, *Ucides cordatus* L. (Decapoda, Brachyura) na visão dos caranguejeiros. *Interciencia*, 27, 110, 2002.
42. Alves, R.R.N., and Nishida, A.K., Aspectos socioeconômicos e percepção ambiental dos catadores de caranguejo-uçá *Ucides cordatus cordatus* (L. 1763) (Decapoda, Brachyura) do estuário do Rio Mamanguape, Nordeste do Brasil. *Interciencia*, 28, 36, 2003.
43. Naranjo, E.J., et al., Subsistence hunting by three ethnic groups of the Lacondon Forest, Mexico. *Journal of Ethnobiology*, 24, 384, 2004.
44. Rocha, M.S.P., et al., Aspectos da comercialização ilegal de aves nas feiras livres de Campina Grande, Paraíba, Brasil. *Revista de Biologia e Ciências da Terra,* 6, 204, 2006.
45. Alves, R.R.N., and Pereira-Filho, G.A., Commercialization and use of snakes on north and northeastern Brazil: implications for conservation and management. *Biodiversity and Conservation,* 16, 969, 2007.
46. Gilmore, R.M., Fauna e etnozoologia da América do Sul, in *Suma Etnológica Brasileira*, Ribeiro, D., Ed., FINEP/Vozes, Petrópolis, Brazil, 1987, p. 189–233.
47. Viegi, L., et al., A review of plants used in folk veterinary medicine in Italy as basis for a databank. *Journal of Ethnopharmacology*, 89, 221, 2003.
48. Yineger H., et al., Ethnoveterinary medicinal plants at Bale Mountains National Park, Ethiopia. *Journal of Ethnopharmacology*, 112, 55, 2007.
49. Pieroni, A., et al., Circum-Mediterranean cultural heritage and medicinal plant uses in traditional animal healthcare: a field survey in eight selected areas within the RUBIA Project. *Journal of Ethnobiology and Ethnomedicine*, 2, 1, 2006.
50. Bennett, B.C., Plants and people of the Amazonian rainforests. The role of ethnobotany in sustainable development. *Bioscience,* 42, 599, 1992.
51. Phillips, O., and Gentry, A.H., The useful plants of Tambopata, Peru. I. Statistical hypothesis tests with a new quantitative technique. *Economic Botany,* 47, 15, 1993.
52. Phillips, O., and Gentry, A.H., The useful plants of Tambopata, Peru. II. Additional hypothesis testing in quantitative ethnobotany. *Economic Botany,* 47, 33, 1993.
53. Milliken, W., and Albert, B., The use of medicinal plants by the Yanomami Indians of Brazil. Part II. *Economic Botany,* 51, 1997.
54. Galeano, G., Forest use at the Pacific Coast of Choco, Colombia: a quantitative approach. *Economic Botany*, 54, 358, 2000.
55. Alves, R.R.N., and Rosa, I.L., From cnidarians to mammals: the use of animals as remedies in fishing communities in NE Brazil. *Journal of Ethnopharmacology*, 107, 259, 2006.
56. Vázquez, P.E., et al. Uso medicinal de la fauna silvestre en los Altos de Chiapas, México. *Interciencia,* 31, 491, 2006.
57. Apaza, L., et al., Markets and the use of wild animals for traditional medicine: a case study among the Tsimane' Amerindians of the Bolivian rain forest. *Journal of Ethnobiology,* 23, 47, 2003.
58. Almeida, C.F.C.B.R., and Albuquerque, U.P., Uso e conservação de plantas e animais medicinais no estado de Pernambuco (Nordeste do Brasil): um estudo de caso. *Interciencia,* 27, 276, 2002.

59. Njoroge, G.N., and Mengo, D.M., Traditional management of diarrhoea in Kitui district (Kenya). Safety and efficacy in herbal medicines, in *Abstracts to the Plant Science Conference Makerere University*. Makerere University, Kampala, Uganda, 2005, p. 24.
60. Njoroge, G.N., and Bussmann, R.W., Ethnotherapeautic management of skin diseases among the Kikuyus of Central Kenya. *Journal of Ethnopharmacology,* 111, 303, 2007.
61. Lans, C., et al., Medicinal plants used for dogs in Trinidad and Tobago. *Preventive Veterinary Medicine*, 45, 201, 2000.
62. Lans, C., Ethnoveterinary practices used by livestock keepers in Trinidad and Tobago. MSc thesis, Wageningen Agricultural University, Wageningen, The Netherlands, 1996.
63. Morton, J.F., *Atlas of medicinal plants of middle America: Bahamas to Yucatan*. Charles C Thomas, Springfield, MO, 1981.
64. Duke, J.A., Ethnobotanical observations on the Chocó Indians. *Economic Botany*, 24, 1970.
65. Passalacqua, N.G., De Fine, G., and Guarrera, P.M., Contribution to the knowledge of the veterinary science and of the ethnobotany in Calabria region (Southern Italy). *Journal of Ethnobiology and Ethnomedicine*, 2, 1, 2006.
66. Ghirotti, M., Recourse to traditional versus modern medicine for cattle and people in Sidama, Ethiopia, in *Ethnoveterinary research and development,* McCorkle, C.M., et al., Eds., Intermediate Technology, London, 1996, p. 46–53.
67. Coats, B.C., and Holland, R.E., The veterinary use of *Aloe vera*, in Creatures in our care, Ahola, R., Ed., Private Publication, USA, 1985, 299 pp.
68. Reynolds, T., and Dwecks, A.C., *Aloe vera* leaf gel: a review update. *Journal of Ethnopharmacology,* 68, 3, 1999.
69. Hatano, T., et al., Phenolic constituents of cassia seeds and antibacterial effect of some naphthalenes and anthraquinones on methicillin-resistant *Staphylococcus aureus. Chemical and Pharmaceutical Bulletin,* 47, 1121, 1999.
70. Hernandez-Medel, M.R., Lopez-Perez, J.L., Trigos, A., and del-Hernandez-Medel, M.R., Anthraquinone from *Picramnia hirsute* stem. *Revista-Latinoamericana-de-Quimica,* 26, 86, 1998.
71. McGaw, L.J., Van der Merwe, D., and Eloff, J.N., In vitro anthelmintic, antibacterial and cytotoxic effects of extracts from plants used in South African ethnoveterinary medicine. *The Veterinary Journal,* 173, 366, 2007.
72. Moloto, M.P., Isolation and characterization of antibacterial, anthelmintic, antioxidant and cytotoxic compounds present in *Ziziphus mucronata*. MSc thesis, Medical University of South Africa, Pretoria, South Africa, 2004.
73. Nchu, F., Developing methods for the screening of ethnoveterinary plants for tick control. MSc thesis, Medical University of Southern Africa, Pretoria, South Africa, 2004.
74. McGaw, L.J., and Eloff, J.N., Ethnoveterinary use of southern African plants and scientific evaluation of their medicinal properties. *Journal of Ethnopharmacology*, 119, 559–574, 2008.
75. Alves, R.R.N., and Rosa, I.L., Zootherapeutic practices among fishing communities in north and northeast Brazil: a comparison, *Journal of Ethnopharmacology*, 111, 82, 2007.
76. Sofowora, A., *Medicinal plants and traditional medicine in Africa*, 2nd ed. Spectrum Books, Ibadan, Nigeria, 1993.
77. Luoga, E.J., Witkowski, E.T.F., and Balkwill, K., Differential utilization and ethnobotany of trees in Kitulanghalo forest reserve and surrounding communal lands, eastern Tanzania. *Economic Botany,* 54, 328, 2000
78. World Health Organization (WHO), *WHO traditional medicine strategy 2002–2005*. WHO, Geneva, Switzerland, 2002.

11 Ethnoveterinary Medicine in Southern Africa

Viola Maphosa, Peter Tshisikhawe, Kaizer Thembo, and Patrick Masika

CONTENTS

11.1 INTRODUCTION

Ethnoveterinary medicines (EVMs) have been in use since time immemorial. The era of treating EVM and any other ethnic knowledge system with suspicion and labeling it as myth, superstition, and witchcraft is long gone (Wanzala et al., 2005). The role of EVM in livestock development in resource-poor settings is beyond dispute (Martin, Mathias, and McCorkle, 2001), and a great number of professionals have since the 1970s recognized, documented, and studied the potential effectiveness of the traditional animal health care practices embodied in native and local communities. Part of this work has also extended to attempts to validate traditional use in laboratory *in vitro* assays. It is, however, important to keep in mind that lack of activity in a laboratory—based on an *in vitro* screening system—does not automatically correspond to lack of *in vivo* or clinical efficacy (McGaw and Eloff, 2008).

Traditional knowledge is usually passed on from generation to generation orally, and this poses the danger of it being distorted or totally lost. Masika, van Averbeeke, and Sonandi (2000) reported the urgent need for documentation of rapidly disappearing local EVM knowledge and conservation of medicinal plant resources.

Diseases of livestock potentially have severe economic impacts, ranging from production losses to mortality and morbidity, particularly in cultures in which animals are equated with wealth and are central to livelihoods (McGaw and Eloff, 2008). The search for effective and affordable remedies to combat diseases in animals is therefore inevitable and is ongoing. In southern Africa, livestock keepers have used traditional veterinary methods to treat livestock diseases for generations. This is socially and culturally accepted because the treatment is easily accessible and reasonably cheap (Githiori, Athanasiadou, and Thamsborg, 2006). Plants comprise the largest component of the diverse therapeutic armamentarium of traditional livestock health management practices (Sori et al., 2004).

In southern Africa, EVM continues to play an important part in animal health. Southern Africa consists of 11 countries, stretches from South Africa to Tanzania, and has a population of just over 165 million people. Cattle, goats, sheep, and chickens are of economic and cultural importance to people of this region. This chapter discusses different methods of traditional medicine preparations by farmers and herbalists.

11.2 MEDICINAL PREPARATIONS

Many aspects of EVM need to be taken into account, for example, methods of preparation and administration of the remedy, as well as management practices to limit the impact of the disease (McGaw and Eloff, 2008). Medicinal plants are prepared and administered in several ways; they may be prepared and used either individually or in combinations for therapeutic purposes. These will be in different formulations, such as powder, paste, decoction, infusion, and so on, and are usually administered either orally or topically.

A *decoction* is prepared by boiling herbs or plant materials such as leaves, roots, or barks. The decoction is then cooled, after which the water extract is strained and administered to animals either orally or topically. This differs from an *infusion*, which is prepared by soaking the plant material (maceration) (e.g., leaves, roots, or barks) in cold water for some time.

A *paste* is prepared by pounding or crushing fresh plant material, which can then be topically applied to affected areas, especially when treating wounds in animals. A small quantity of water may be added, especially when the paste is being prepared from woody plant species, whereas water may not be required for paste prepared from succulent species. Water may again be added if plant material was dried first.

The medicinal materials are first dried and then pounded into powder. The powder may be either dissolved in water for animals to be given orally or placed on affected areas, especially wounds, thereby covering the wounds.

11.3 PLANTS USED IN SOUTHERN AFRICA

Several plant species are used for animal health care in southern Africa (Table 11.1 and Table 11.2). Some of them are discussed in greater depth here. What is immediately interesting is the fact that many of the plant species used are known to be toxic (e.g., those of the families Amaryllidaceae, Hyacinthaceae, and Solanaceae). It is difficult to document fatalities among animals due to administration of these products, but they are likely to be negligible because the users are knowledgeable about dose and methods of preparation, which limit toxicity while optimizing the therapeutic benefits. This is similar to the use of herbals in humans or pharmaceuticals, which can only be done safely when dispensed by trained professionals. In this case, these "professionals" would be experienced farmers, herbalists, or traditional healers.

11.3.1 Amaryllidaceae

The Amaryllidaceae occur largely in the Southern Hemisphere with the distribution centered mainly in South America and southern Africa (Arroyo and Cutler, 1984). Of the 64 genera, there are those that are endemic and unique to the two regions.

Amaryllidaceae species are used widely for treating cattle, with the bulbs being the most commonly used, despite fatalities due to the toxic alkaloids that are known to occur in this taxa (Hutchings et al., 1996). Toxicity symptoms include excessive salivation, nausea, dizziness, dysrhythmias, visual disturbances, and dermatitis. Apart from the alkaloids, flavonoids have also been reported (Louw, Regnier, and Korsten, 2002).

Boophane disticha is used to treat redwater and constipation in cattle and to facilitate the healing of broken limbs (Dold and Cocks, 2001) as well as to induce abortion (Van der Merwe, Swan, and Botha, 2001). It is also used to treat wounds and has some antibacterial activity (Shale, Stirk, and Van Staden, 1999). It contains toxic and apparently hallucinogenic alkaloids (De Smet, 1996). Of the 130 species of *Crinum* worldwide, 30 are endemic to southern Africa and widely used in traditional medicine (Nair et al., 2000; Hutchings et al., 1996). In cattle, it is used to treat scabies (Begum and Nath, 2000). *Crinum* species contain phenanthridine alkaloids (Fennell and van Staden, 2001), which are used as analgesics and possess antitumor, antiviral, and anticholinergic activity (Lewis, 1990; Pham et al., 1998). In Zimbabwe, the bulbs of *Crinum macowanii* are used as emetics and to improve lactation in both animals and humans (Mavi, 1994).

11.3.2 Anacardiaceae

Anacardiaceae Lindl., more widely known as the cashew family, is economically important because of its widely cultivated edible fruits and nuts (e.g., mangoes and pistachio and cashew nuts; Pell, 2004). It has over 700 species distributed in 82 genera in the tropics.

In southern Africa, *Ozoroa*, *Rhus*, and *Sclerocarya* are the important medicinal taxa and are widely used in treating both humans and animals. *Ozoroa paniculosa* is

TABLE 11.1
Some of the Plants Used in South Africa for Ethnoveterinary Purposes

Family	Species	Indication	Plant Part Used	Chemical Constituents	Activity Screened for
Amaryllidaceae	*Ammocharis coranica* (Ker-Gawl.) Herb.	Used medicinally for cattle (Gerstner, 1938)	Unspecified parts	Alkaloids, organic acid, and hemolytic saponin (Watt and Breyer-Brandwijk, 1962)	
Amaryllidaceae	*Crinum moorei* Hook. F.	Used medicinally for cattle (Gerstner, 1938)	Unspecified parts	Lycorine, cherylline, crinamidine, crinidine, dihycrinidine, powelline, and phenols (Watt and Breyer-Brandwijk, 1962)	
Anacardiaceae	*Ozoroa paniculosa* (Sond.) R. & A. Fernandes	Abdominal problems in animals (Hutchings et al., 1996); diarrhea, red water, sweating sickness (Van der Merwe, Swan, and Botha, 2001)	Bark, root bark	Volatile oil (Watt and Breyer-Brandwijk, 1962)	
Anacardiaceae	*Rhus lancea* L.f.	Diarrhea, gall sickness (Van der Merwe, Swan, and Botha, 2001)	Roots, bark		Antibacterial, anthelmintic, brine shrimp toxicity (McGaw, Van der Merwe, and Eloff, 2007)
Anacardiaceae	*Sclerocarya birrea* (A. Rich.) Hochst.	Diarrhea, fractures (Van der Merwe, Swan, and Botha, 2001)	Bark		Antibacterial, anthelmintic, brine shrimp toxicity (McGaw, Van der Merwe, and Eloff, 2007)
Apiaceae	*Heteromorpha trifoliate* (Wendl.) Eckl. & Zeyh.	Zulus use bark for colic, scrofula, and vermifuge for horses (Gerstner, 1938); Xhosa use roots for threadworm in horses (Watt and Breyer-Brandwijk, 1962); red	Bark, roots	Falcarindiola and sarisan (antifungal) (Villegas et al., 1988)	

		water, gall sickness (Masika, van Averbeeke, and Sonandi, 2000)			
Apocynaceae	*Secamone filiformis* (L.f.) J.H. Ross	Infectious diseases in cattle (L.J. McGaw, personal communication, 2009)	Aerial parts		Antibacterial, anthelmintic, brine shrimp toxicity (McGaw, Van der Merwe, and Eloff, 2007)
Apocynaceae	*Strophanthus speciosus* (Ward & Harv.) Reber	Given to cattle for snakebite (Hutchings et al., 1996)	Unspecified	Cardiac glycosides stropeside and christyoside (Watt and Breyer-Brandwijk, 1962)	
Araliaceae	*Cossonia spicata* Thunb.	Leaves applied in hot fomentations to goats paralyzed in their hind quarters (Palmer and Pitman, 1972); bark used for retained placenta in stock, leaves used to treat endometritis and vaginitis in cows, bark decoction for gall sickness in cattle (Dold and Cocks, 2001; Masika, van Averbeeke, and Sonandi, 2000)	Leaves, bark		Antibacterial, anthelmintic, brine shrimp toxicity (McGaw, Van der Merwe, and Eloff, 2007); antibacterial, anti-inflammatory, mutagenic (Luseba et al., 2007)
Asclepiadaceae	*Sacrostemma viminale* (L.) R. Br.	Stem used to encourage lactation in cows, galactogue in cows (Dold and Cocks, 2001); wounds and maggots (Luseba and van der Merwe, 2006)	Stems, aerial parts		Antibacterial, anti-inflammatory, mutagenic (Luseba et al., 2007)
Asphodelaceae	*Aloe arborescens* Mill.	Used to drench sick calves (Hutchings et al., 1996)	Leaf decoctions	Aloin, barbaloin, aloes emodin, aloenin, polysaccharides, lectins, and other compounds (Hutchings et al., 1996)	

(continued)

TABLE 11.1 (CONTINUED)
Some of the Plants Used in South Africa for Ethnoveterinary Purposes

Family	Species	Indication	Plant Part Used	Chemical Constituents	Activity Screened for
Asphodelaceae	*Aloe marlothii* Berger	Newcastle disease in chickens (Luseba and van der Merwe, 2006); gall sickness, parasites, diarrhea, constipation, retained placenta, dystocia, maggots (Van der Merwe, Swan, and Botha, 2001)	Leaves		Antibacterial, anti-inflammatory, mutagenic (Luseba et al., 2007); antirickettsial (Naidoo, Zweygarth, and Swan, 2006); antibabesial (Naidoo et al., 2005)
Asteraceae	*Bidens pilosa* L.	Equine anthelmintics (Hutchings et al., 1996)	Unspecified	The polyacetylene phenylheptatrine and chalcomenes (Graham, Graham, and Towers, 1980; Hoffman and Hoelzl, 1988)	
Asteraceae	*Brachylaena discolor* DC.	Anthelmintics for calves, sheep, and goats (Hutchings et al., 1996)	Dried leaf milk infusions	Onopordopicrin (Zdero and Bohlmann, 1987)	
Asteraceae	*Callilepis laureola* DC.	Used to kill maggots in cattle (Watt and Breyer-Brandwijk, 1962)	Root paste	Atractyloside and its aglycone atractyligenin (Candy et al., 1977)	
Asteraceae	*Dicoma anomala* Sond.	Gall sickness in stock animals; powdered plants used for sores and wounds on horses (Watt and Breyer-Brandwijk, 1962)	Root decoctions	Germacranolides (Hutchings et al., 1996)	
Asteraceae	*Microglossa mespilifolia* (Less.) B.L. Robinson	Tonics for stock animals (Hutchings et al., 1996)	Infusions from leaves and stems	Epi-friedelinol and C_{17} acetylenic compounds (Bohlmann and Fritz, 1979)	

Asteraceae	*Printzia pyrifolia* Less.	Used for treating calves (Gerstner, 1939)	Roots	Matricaria ester and *p*-coumarate (Bohlmann and Zdero, 1978a)	
Asteraceae	*Schkuhria pinnata* (Lam.) Thell.	Eye infections, pneumonia, diarrhea, heartwater (Van der Merwe, Swan, and Botha, 2001)	Aerial parts		Antibacterial, anthelmintic, brine shrimp toxicity (McGaw, Van der Merwe, and Eloff, 2007); antibacterial, anti-inflammatory, mutagenic (Luseba et al., 2007)
Asteraceae	*Senecio oxyriifolius* DC.	Swellings in animals (Hutchings et al., 1996)	Leaves	A tricyclic sesquiterpenoid, angelogyl, and bisabolenes (Bohlmann and Zdero, 1978b)	
Asteraceae	*Vernonia neocorymbosa* Hilliard	Used by Zulus to treat calves (Gerstner, 1939); root decoctions administered by Lobedu as anthelmintics to donkeys (Hutchings et al., 1996); pounded leaf and root infusions administered by VhaVenda as anthelmintics to domestic animals (Mabogo, 1990)	Root, leaves	Squalene, vernolide, and vernodalin from aerial parts and 13-hydroxybisabol-2, 10-dien-1-one and small amounts of onopordopicrin in roots (Bohlmann, Ates, Jakupovic, 1983)	
Capparaceae	*Capparis tomentosa* Lam.	Root ash paste applied to sore teats in cows; root infusions used for stomach ailments in animals, particularly diarrhea in cattle (Watt and Breyer-Brandwijk, 1962; Pujol, 1990); root decoction for gall sickness in stock (Dold and Cocks, 2001)	Paste made from root ashes, root infusions, and decoctions	The alkaloids stachydrine (Dictionary of Natural Products, 1996) and 3-hydroxy-4-methoxy-3-methyl-oxyndole (Dekker et al., 1987)	

(*continued*)

TABLE 11.1 (CONTINUED)
Some of the Plants Used in South Africa for Ethnoveterinary Purposes

Family	Species	Indication	Plant Part Used	Chemical Constituents	Activity Screened for
Celastraceae	*Maytenus heterophylla* (Eckl. & Zeyh.) N.K.B. Robson	Administered by Zulus to stock animals for diarrhea (Watt and Breyer-Brandwijk, 1962)	Bark and leaf infusions	Dulcitol, a spermidine alkaloid, celacinnine, triterpenoids, maytansine (Hutchings et al., 1996)	
Chenopodiaceae	*Chenopodium album* L.	Decoctions made from plants mixed with *Chenopodiun ambrosioides* administered to goats and sheep for anemia (Hutchings et al., 1996)	Unspecified	Hydrocyanic acid, potassium oxalate, ascorbic acid, sitosterol, oleanic acid (Watt and Breyer-Brandwijk, 1962; Hutchings et al., 1996)	
Chenopodiaceae	*Chenopodium ambrosioides* L.	Decoctions made from plants mixed with *Chenopodium album* administered to goats and sheep for anemia (Hutchings et al., 1996)	Unspecified	Saponins (Watt and Breyer-Brandwijk, 1962); flavonoids quercetin and oxalic, malic, and succinic acids; triterpenoid glycosides, chenopodoside A and B, amino acids, ascaridole (Hutchings et al., 1996)	
Colchicaceae	*Gloriosa superba* L.	Used to kill lice, for skin eruptions, tick infestations, and screw-worm on cattle (Gerstner, 1939; Roberts, 1990)	Corms	Colchicine, chelidinic acid, various alkaloids, and other constituents (Hutchings et al., 1996)	
Combretaceae	*Combretum caffrum* (Eckl. & Zeyh.) Kuntze	Conjunctivitis (Masika, van Averbeeke, and Sonandi, 2000)	Drops from squeezed leaves		Antibacterial, antifungal (Masika and Afolayan, 2002)
Cucurbitaceae	*Cucumis africanus* L.f.	Used as animal medicines by the Xhosa (Hutchings et al., 1996)	Unspecified	Toxic cucurbitacins (Hutchings et al., 1996)	

Dioscroreaceae	*Dioscorea dregeana* (Kunth) Dur. & Schinz	Sores and wounds in animals and humans by Xhosa (Watt and Breyer-Brandwijk, 1962)	Water heated in scooped out tuber	An alkaloid and organic acids (Watt and Breyer-Brandwijk, 1962)	
Dioscroreaceae	*Dioscorea sylvatica* (Kunth) Eckl.	Swollen udders and uterine problems in cows (Watt and Breyer-Brandwijk, 1962)	Lotions from boiled crushed inner parts of tubers	Diosgenin (Watt and Breyer-Brandwijk, 1962)	
Euphorbiaceae	*Jatropha zeyheri* Sond.	General ailments, diarrhea (Luseba and van der Merwe, 2006)	Roots		Antibacterial, anti-inflammatory, mutagenic (Luseba et al., 2007)
Euphorbiaceae	*Ricinus communis* L.	Administered as purgative to calves refusing to suckle (Hutchings et al., 1996); constipation, internal parasites (Van der Merwe, Swan, and Botha, 2001)	Powdered seed	Seeds contain a fixed oil, ricin, lipases, and ricinine (Trease and Evans, 1983)	Antibacterial, anthelmintic, brine shrimp toxicity (McGaw, Van der Merwe, and Eloff, 2007); antibacterial, anti-inflammatory, mutagenic (Luseba et al., 2007)
Euphorbiaceae	*Synadenium cupulare* (Boiss.) L.C. Wheeler	Eye infection, blackquarter (Luseba and van der Merwe, 2006)	Milky latex		Antibacterial, anthelmintic, brine shrimp toxicity (McGaw, Van der Merwe, and Eloff, 2007)
Fabaceae	*Adenopodia spicata* (E. Mey.) Presl	Powdered roots used by Zulus to fatten goats (Hutchings et al., 1996); Mfengu use bark for colds in horses (Watt and Breyer-Brandwijk, 1962)	Roots, bark	Saponins (Watt and Breyer-Brandwijk, 1962)	
Fabaceae	*Calpurnia aurea* (Ait.) Benth.	Zulus use plant to destroy maggots in sores (Bryant, 1966)	Unspecified parts	Alkaloids (Hutchings et al., 1996)	
Fabaceae	*Calpurnia villossa* Harv.		Unspecified parts	The alkaloid oroboidine (Hutchings et al., 1996)	

(*continued*)

TABLE 11.1 (CONTINUED)
Some of the Plants Used in South Africa for Ethnoveterinary Purposes

Family	Species	Indication	Plant Part Used	Chemical Constituents	Activity Screened for
Fabaceae	*Elephantorrhiza elephantina* (Burch) Skeels	The Xhosa use roots for diarrhea and dysentery in cattle, horses, and humans (Watt and Breyer-Brandwijk, 1962); root given to cows for mange (Dold and Cocks, 2001); heartwater, blackquarter, appetite stimulant, or tonic (Luseba and van der Merwe, 2006); diarrhea, heartwater, coughing, pneumonia (Van der Merwe, Swan, and Botha, 2001)	Roots, aerial parts, and bulb	Tannin (Watt and Breyer-Brandwijk, 1962)	Antirickettsial (Naidoo, Zweygarth, and Swan, 2006); antibabesial (Naidoo et al., 2005)
Fabaceae	*Erythrophleum lasianthum* Corbishley	Lung sickness in cattle and abortions in dogs (Hutchings et al., 1996)	Bark	Seed and bark contain erythrophleine (Watt and Breyer-Brandwijk, 1962)	
Fabaceae	*Peltophorum africanum* Sond.	Tonic, diarrhea (Van der Merwe, Swan, and Botha, 2001)	Bark, root bark		Antibacterial, antioxidant (Bizimenyera et al., 2005; Bizimenyera, Githiori, Eloff, et al., 2006; Bizimenyera, Githiori, Swann, et al., 2006)
Fabaceae	*Pterocarpus angolensis* DC.	General illness, gall sicknes, intestinal worms, blackquarter (Luseba and van der Merwe, 2006)	Bark		Antibacterial, anthelmintic, brine shrimp toxicity (McGaw, Van der Merwe, and Eloff, 2007); antibacterial, anti-inflammatory, mutagenic (Luseba et al., 2007)
Fabaceae	*Schotia brachypetala* Sond.	Infectious diseases in cattle (L. J. McGaw, personal communication, 2009)	Unspecified		Antibacterial, anthelmintic, brine shrimp toxicity (McGaw, Van der Merwe, and Eloff, 2007)

Fabaceae	*Schotia latifolia* Jacq.	Red water in cattle (Dold and Cocks, 2001)	Bark decoction		Antibacterial, antifungal (Masika and Afolayan, 2002)
Fabaceae	*Tephrosia kraussiana* Meisn.	Plants used by Zulus for protecting cattle against quarter-evil and other diseases (Doke and Vilakazi, 1972)	Unspecified parts	Roots may contain saponin (Watt and Breyer-Brandwijk, 1962)	
Fabaceae	*Tephrosia macropoda* (E. Mey.) Harv.	Roots and seeds used for killing vermin on animals and humans (Gerstner, 1941); leaf extracts used as anthelmintics for cattle (Bryant, 1966)	Root, seeds, leaves	Roots contain toxicarol and deguelin (Watt and Breyer-Brandwijk, 1962)	
Geraniaceae	*Monsonia emarginata* (L.f.) L' Hérit.	Stomach ailments in calves, lambs, and humans (Watt and Breyer-Brandwijk, 1962)	Unspecified parts	Phloroglucin tannin (Watt and Breyer-Brandwijk, 1962)	
Geraniaceae	*Pelargonium luridum* (Andr.) Sweet	Administered to sick calves (Hutchings et al., 1996)	Leaf infusions	Coumarins in roots (Wagner and Baldt, 1975)	
Geraniaceae	*Pelargonium sidoides* DC.	Used as anthelmintics with *Ziziphus zeyyheriana* Sond. (Watt and Breyer-Brandwijk, 1962)	Decoctions of unspecified parts	Coumarins in roots (Wagner and Baldt, 1975)	
Gunneraceae	*Gunnera perpensa* L.	Used to facilitate expulsion of afterbirth in animals and women (Gerstner, 1939)	Roots	Bitter principle, celastrin (Watt and Breyer-Brandwijk, 1962)	
Hyacinthaceae	*Urginea physodes* (Jacq.) Bak	"Itch" in goats (Gerstner, 1941)	Unspecified		Antibabesial (Naidoo et al., 2005)
Lamiaceae	*Leonotis leonurus* (L.) R. Br.	Pounded roots and leaves are added to drinking water to prevent	Roots, leaves, drops from	Volatile oil diterpenoids (labdane-type lactones), for	

(*continued*)

TABLE 11.1 (CONTINUED)
Some of the Plants Used in South Africa for Ethnoveterinary Purposes

Family	Species	Indication	Plant Part Used	Chemical Constituents	Activity Screened for
		sickness in poultry and are used for gall sickness in cattle (Hulme, 1954); eye inflammation (Masika, van Averbeeke, and Sonandi, 2000)	squeezed leaf used for eyes	example, marrubin (Dictionary of Natural Products, 1996)	
Lamiaceae	*Tetradenia ripara*	Used for gall sickness and fevers in cattle (Hutchings et al., 1996)	Leaves	A diterpene diol, ibozol, and related diterpenoids, large amounts of α-pyrones (Dictionary of Natural Products, 1996)	
Moraceae	*Ficus ingens* (Miq.) Miq	Administered by Zulus to cows to increase milk production (Watt and Breyer-Brandwijk, 1962) and VhaVenda (Mabogo, 1990)	Bark decoctions	Tannin animals (Watt and Breyer-Brandwijk, 1962)	
Moraceae	*Ficus sur* Forssk.	Zulus use leaf and bark infusions as bovine galactagogues (Hutchings et al., 1996); VhaVenda use root decoctions for retained placenta in cows (Watt and Breyer-Brandwijk, 1962)	Leaves, bark, roots	Bark may contain tannin (Hutchings et al., 1996)	
Myrtaceae	*Heteropyxis natalensis* Harv.	Drench for stock animals (Watt and Breyer-Brandwijk, 1962)	Powdered leaves	Essential oils from ground dried leaves contain many constituents (Hutchings et al., 1996)	
Pedaliaceae	*Dicerocaryum eriocarpum* (Dcne.) J. Abels	Dystocia, drench for retained placenta (Luseba and van der Merwe, 2006; Van der Merwe,	Aerial parts, roots, whole plant		Antibacterial, anthelmintic, brine shrimp toxicity (McGaw, Van der Merwe, and Eloff, 2007);

	[and *Dicerocaryum senecioides* (Kltzsch.) J. Abels]	Swan, and Botha, 2001)			antibacterial, anti-inflammatory, mutagenic (Luseba et al., 2007)
Phytolaccaceae	*Phytolacca octandra* L.	Lung sickness in cattle (Watt and Breyer-Brandwijk, 1962)	Root infusions	Triterpenoid saponins, known as yiamoloside B (Moreno and Rodriguez, 1981)	
Polygonaceae	*Emex australis* Steinh.	Threadworm in horses (Hutchings et al., 1996)	Leaf decoctions	Anthraquinones (Watt and Breyer-Brandwijk, 1962)	
Polygonaceae	*Rumex lanceolatus* Thunb.	Used with *Euclea coriacea* A. DC. to treat gall sickness in stock animals (Guillarmod, 1971)	Unspecified parts	Chrysophanic acid, emodin, and volatile (Watt and Breyer-Brandwijk, 1962)	
Ptaeroxylaceae	*Ptaeroxylon obliquum* (Thunb.) Radlk.	Anthrax remedy, for ticks in cattle (Hutchings et al., 1996)	Wood	Powdered wood is irritating and includes sneezing (Hutchings et al., 1996); timber has high oil and resin content (Watt and Breyer-Brandwijk, 1962)	
Ranuncluaceae	*Clematis brachiata* Thunb.	Vermifuge and for bots in horses (Hutchings et al., 1996)	Infusions from shoots and leaves	Contains anemonol (Watt and Breyer-Brandwijk, 1962)	
Rhamnaceae	*Berchemia zeyheri* (Sond.) Grubov	Infectious diseases in cattle (L. J. McGaw, personal communication, 2009)	Bark		Antibacterial, anthelmintic, brine shrimp toxicity (McGaw, Van der Merwe, and Eloff, 2007)
Rhamnaceae	*Ziziphus mucronata* Willd.	Fertility enhancement, sores, burns (Van der Merwe, Swan, and Botha, 2001)	Roots, leaves		Antibacterial, anti-inflammatory, mutagenic (Luseba et al., 2007)

(*continued*)

TABLE 11.1 (CONTINUED)
Some of the Plants Used in South Africa for Ethnoveterinary Purposes

Family	Species	Indication	Plant Part Used	Chemical Constituents	Activity Screened for
Rutaceae	*Clausena anisata* (Willd.) Hook. f. ex Benth.	Dysentery in cattle (Hutchings et al., 1996)	Bark infusions	Terpenoid hydrocarbons, alkaloids, coumarins, and many other compounds (Hutchings et al., 1996)	
Salicaceae	*Salix* L. spp	Retained placenta (Masika, van Averbeeke, and Sonandi, 2000)	Decoction or infusion of unspecified parts		Antibacterial, antifungal (Masika and Afolayan, 2002)
Sapindaceae	*Hippobromus pauciflorus* (L.f.) Radlk.	Leaf and root infusions used to clear mucus from noses of sheep and goats (Watt and Breyer-Brandwijk, 1962); root infusions given to stock animals with cough (Hutchings et al., 1996); leaf sap used for inflamed eyes in animals and humans (Watt and Breyer-Brandwijk, 1962; Masika, van Averbeeke, and Sonandi, 2000); bark used for heartwater and diarrhea in cattle (Dold and Cocks, 2001)		Resinous and oily substances in wood and leaves (Watt and Breyer-Brandwijk, 1962)	Antibacterial, anthelmintic, brine shrimp toxicity (McGaw, Van der Merwe, and Eloff, 2007)
Sapotaceae	*Sideroxylone inerme* L.	Preparations administered by Zulus as tonics to calves and goats (Hutchings et al., 1996); Xhosa use bark for gall sickness in stock (Watt and Breyer-Brandwijk, 1962); bark	Unspecified parts, bark	Cinnamic acid, kaempferol, and leucanthocyanidins from bark (Hutchings et al., 1996)	

		decoction for red water in cattle (Dold and Cocks, 2001)			
Solanaceae	*Datura stramonium* L.	Powdered leaves are applied by Zulus to animal and human bruise and wounds to draw out inflammation and pus (Watt and Breyer-Brandwijk, 1962)	Powdered leaves	Alkaloids, including hyoscyamine and hyoscine (Oliver-Bever, 1986)	
Solanaceae	*Solanun aculeastrum* Dun.	Ringworm in cattle and horses, also for anthrax (Hutchings et al., 1996)	Fruit	Solanine (Watt and Breyer-Brandwijk, 1962)	
Solanaceae	*Solanum hermannii* Dun.	Fruit sap and leaf paste used for sores on sheep and horses by Xhosa and Sotho (Watt and Breyer-Brandwijk, 1962)	Fruit, leaves	Solanin, solanidine, azosolanidin, quinhydrone solasonine, solasodine, solasodamine from fruit (Watt and Breyer-Brandwijk, 1962)	
Solanaceae	*Withania somnifera* (L.) Dun.	Used to stimulate milk production in cows (Gerstner, 1941); roots used for black gall sickness in cattle (Luseba and Van der Merwe 2006) diarrhea (L.J. McGaw, personal communication, 2009)	Unspecified parts, roots	Many compounds, including choline, tropanol, glycowithanolides, withanolides, withaferine, and withasomnine (Hutchings et al., 1996 and references therein)	
Sterculiaceae	*Dombeya rotundifolia* (Hochst.) Planch.	Newcastle disease in chicken (Luseba and Van der Merwe 2006); infectious diseases in cattle (L.J. McGaw, personal communication, 2009)	Leaves and flowers		Antibacterial, anthelmintic, brine shrimp toxicity (McGaw, Van der Merwe, and Eloff, 2007)
Thymelaeaceae	*Gnidia capitata* L.f.	Heartwater in cows, anthrax (Dold and Cocks, 2001)	Root decoction		Antibacterial, anthelmintic, brine shrimp toxicity (McGaw, Van der Merwe, and Eloff, 2007)

(continued)

TABLE 11.1 (CONTINUED)
Some of the Plants Used in South Africa for Ethnoveterinary Purposes

Family	Species	Indication	Plant Part Used	Chemical Constituents	Activity Screened for
Thymelaeaceae	*Gnidia kraussiana* Meisn.	Preparations injected near the site of the fractured limbs of stock animals by the Sotho (Guillarmod, 1971)	Unspecified parts	Flavone heteroside from roots, toxic diterpenoid fraction, polysaccharides, daphnane orthoesters (Hutchings et al., 1996)	
Typhaceae	*Typha capensis* (Rohrb.) N.E. Br.	Decoctions taken or applied externally to aid expulsion of afterbirth in animals and humans (Roberts, 1990)	Unspecified parts	Quercetin 3′-dimethyl ether 4′-glucoside from leaf (Hutchings et al., 1996)	
Urticaceae	*Poulzozia mixta* Solms	Retained placenta, bloat, vaginal discharge (Van der Merwe, Swan, and Botha, 2001)	Roots, leaves, stems		Antibacterial, anthelmintic, brine shrimp toxicity (McGaw, Van der Merwe, and Eloff, 2007)
Verbenaceae	*Lantana rugosa* Thunb.	Pastes or infusions used for animal and human eye complaints (Watt and Breyer-Brandwijk, 1962)	Leaves	Volatile oil and the alkaloid lantanin (Watt and Breyer-Brandwijk, 1962)	
Vitaceae	*Cissus quadrangularis* L.	Used by Zulus as a drench for sick horses (Watt and Breyer-Brandwijk, 1962), aerial parts used as poultice for wounds, lumpy skin disease, and as tick repellent (Luseba and Van der Merwe, 2006)	Aerial parts	A steroidal mixture and triterpenoids (Hutchings et al., 1996)	Antibacterial, anthelmintic, brine shrimp toxicity (McGaw, Van der Merwe, and Eloff, 2007); antibacterial, anti-inflammatory, mutagenic (Luseba et al., 2007)

Vitaceae	*Rhoicissus tridentata* (L.f.) (Lam.) Willd and Drum.	Cattle diseases (Pujol, 1990); diarrhea in goats and sheep (Dold and Cocks. 2001); heartwater, redewater, internal parasites, general ailments, abortion (Van der Merwe, Swan, and Botha, 2001)	Tubers		Antibabesial (Naidoo et al., 2005)
Zingiberaceae	*Siphonochilus aethiopicus* (Schweinf.) B. L. Burtt	Administered to horses as prophylactics against horse sickness (Watt and Breyer-Brandwijk, 1962)	Rhizome infusions	Volatile oil with a characteristic sesquiterpenoid, α-terpineol and other monoterpenoids (Van Wyk, Van Oudtshoorn, and Gericke, 1997)	

TABLE 11.2
Traditional Medicinal Plants Used in Treatment of Commonly Encountered Disease Conditions in Farm Animals in Zimbabwe

Animal Condition	Remedy	Method of Application
Septic wounds	*Canthium* spp. (Muvengahonye)	Fresh leaves are ground and applied to the wound.
	Kigelia africana (Muvheva)	The inner core of dried fruit is applied as a powder on the wound.
	Cissus quadrangularis (Murenja)	Fruit is crushed, and the fluid is applied to the wound.
	Aloe spp. (Gavakava)	Dry leaves are crushed, and the powder is applied to the wound.
Eye problem	*Solanum indicum* (Nhundurwa)	Fruit is crushed, and the fluid is applied to the eye.
Bloat	*Pauzolzia mixta* (Munhanzva)	Leaves are crushed and water added; animal is made to swallow mixture.
Coccidiosis	*Aloe* spp. (Gavakava)	Grind fresh leaves and add to drinking water.
Worms	*Vernonia amygdalina* (Muzhozho)	Add water to ground fresh leaves; animal made to swallow mixture.
	Musa paradisiacal (Banana)	Add water to ground fresh roots; animal made to swallow mixture.
	Gavakava (*Aloe* spp.)	Add water to ground fresh leaves; animal made to swallow mixture.
Newcastle disease	Gusha (*Sesanum angustifolius*)	Crush fresh fruit and add to drinking water in poultry.
Retained afterbirth/placenta	Munhanzva (*Pouzolzia mixta*)	Fresh leaves are crushed and the slippery paste inserted into vagina.
Fertility	Gomarara (*Loranthus* spp.)	Feeding fresh leaves to rabbits improves kidding rate.
Snakebite	Munyoka (*Amaranthus gneizaus*)	Add water to crushed fresh roots; animal made to swallow mixture.
	Banana (*Musa paradisiacal*)	Add water to crushed dried roots; animal made to swallow mixture.
Delayed parturition	Murenja (*Cissus quadrangularis*)	Crush fresh stem and leaves, place in the vagina to hasten parturition.
Poor milk flow	Boabab (*Adansonia digitata*)	Inner core of dried fruit is removed, added to water; animal made to swallow mixture.
Fracture	Batanai (*Bulbophylum* spp.)	Bark is tied around fracture as supporting pad.
Fleas	Rutapatsikidzi (*Aneilema hockii*)	Branches of plant are placed near sleeping animals. Fleas are attracted by the herb and leave the animal.
Diarrhea	Murumanyama (*Xeroiderris stuhlmannii*)	Fresh leaves are crushed, water added; animal made to swallow mixture.

Source: From McGaw and Eloff (2008). *Journal of Ethnopharmacology* 119 (3): 329–330. With permission of Elsevier.

used for abdominal problems, diarrhea and red water in livestock (Hutchings et al., 1996; Van der Merwe, Swan, and Botha, 2001), while *Rhus* species are popular for diarrhea and gall sickness (Van der Merwe, Swan, and Botha, 2001).

Little chemistry has been done on *O. paniculosa*. Anacardic acids and ginkgoic acid, both compounds that showed *in vitro* anticancer activity, were isolated from the related *O. insignis* (Rea et al., 2003). The anacardic acids have been isolated from other members of the Anacardiaceae family as well as several tirucallane triterpenes (Liu and Abreu, 2006).

Rhus species are widely recognized to possess a host of biological activities, among them antifibrogenic, antifungal, anti-inflammatory, antimalarial, antimicrobial, and antimutagenic ones (Rayne and Mazza, 2007). They possess flavonoids and tannins (Ahmed, Galal, and Ross, 2001). The Morula tree (*Sclerocarya birrea*) is held in high esteem in the cultures of many Bantu people, especially the VhaVenda in Zimbabwe and South Africa. It is rich in flavonol and epicatechin derivatives (Galvez et al., 1992; Braca et al., 2003). It is probably the high tannin content in *S. birrea* and *Rhus* spp. that explains their popular use in treating diarrhea in both humans and animals. Ethanol extracts of *S. birrea* showed good activity against *Haemonchus contortus*, a pathogenic nematode in small ruminants (Mamidou et al., 2005). An infusion of *Protorhus longifolia* is used to treat heartwater and paratyphoid in cattle (Cocks, 2006).

11.3.3 Apocynaceae

There are about 90 genera and 700 species of Apocynaceae in southern Africa, with remarkable endemism, particularly in South Africa's KwaZulu Natal Province (Ollerton et al., 2003).

Powdered root decoctions of *Acokanthera oppositifolia* are used to treat pain, snakebite, anthrax, and tapeworm (Adedapo et al., 2008). In addition, polyphenolics occur that naturally have antioxidant activity. Powdered roots are administered orally or as snuff to treat pain and snakebite, and root decoctions are used against anthrax and tapeworm (Adedapo et al., 2008). However due to the occurrence of cardiac glycosides, this genus is also considered to be toxic to animals (Kellerman, Coetzer, and Naude, 1988).

11.3.4 Asparagaceae

Asparagaceae is a monogeneric family previously included under Liliaceae, and it has 100 perennials arising from tubers and rhizomes (Dinan, Savchenko, and Whiting, 2001). It is widely distributed in Africa, Europe, western Asia, and Australia.

In Zambia, root decoction of *Asparagus* spp. has been used for blackwater fever (Watt and Breyer-Brandwijk, 1962) and to treat fevers in Malawi, Zambia, and Tanzania (Morris, 1996).

Norlignans, carotenoids, alkaloids, flavonoids, and steroidal saponins have been isolated from various *Asparagus* species (Sharma, Sati, and Chand, 1982; Shao et al., 1997; Debella et al., 1999; Yang et al., 2004; Saxena and Chourasia, 2001). The phytoecdysteroids, which are typical of this Asparagaceae, have been shown to disrupt steroidal hormone action in pest species (Dinan, Savchenko, and Whiting, 2001). Some of the ethnoveterinary uses of the Asparagaceae can be attributed to the biological activity of the saponins and alkaloids, which have been shown to be antimicrobial, anti-inflammatory (Sparg, Light, and van Staden, 2004), and antiprotozoal (Francis et al., 2002) and widely recognized to be antifertility and uterotonic.

11.3.5 Asphodelaceae

Worldwide, there are some 800 species in the Asphodelaceae family, mostly herbs with rhizomes or tubers. There are 360 *Aloe* species described so far, and these are known to occur mainly in Africa and Arabia (Dagne and Yenesew, 1994). The African-Arabian succulent genus *Aloe* L. (Aloaceae/Asphodelaceae) is represented by approximately 120 infrageneric taxa in southern Africa, including *A. ferox* Mill., a species long used in commercial natural products (Glen and Hardy, 2000). Aloe has over the years proved to be one of the most important sources of biologically active compounds. Over 130 compounds belonging to different classes, including anthrones, chromones, pyrones, coumarins, alkaloids, glycoproteins, naphthalenes, and flavonoids, have been isolated from various species (Dagne et al., 2000; Reynolds, 2004). Aloin A is the most important commercial product of aloe, and it is used in the culinary and pharmaceutical industries (Abegaz et al., 1999)

The most frequently used species in southern Africa are *Aloe ferox* and *A. marlothii* because of their wide distribution (Grace et al., 2008). In South Africa, *Aloe ferox* has been reported in the ethnoveterinary control of ticks (Moyo and Masika, 2008), whereas in Tanzania, the extract of *Aloe secundiflora* was useful in the control of fowl typhoid in chickens (Waihenya et al., 2002). Leaf exudates of *Aloe* species are used to treat the drinking water of chicken as a bird flu prophylactic and tonic in most of rural southern Africa (personal observation).

11.3.6 Asteraceae

The Asteraceae (= Compositae) is the largest family of vascular plants, with over 23,000 species and wide distribution on all continents except Antarctica (Panero and Funk, 2008).

Plants from this family have been reported to possess polyacetylene phenylheptatrine and chalcomenes (Graham, Graham, and Towers, 1980; Hoffman and Hoelzl, 1988). *Bidens pilosa* and *Brachylaemia discolour* are used to control helminths in equines and in calves, sheep, and goats (Hutchings et al., 1996). Some plants from this family have been used to treat heartwater in goats, diarrhea in lambs, and gall sickness in livestock (Dold and Cocks, 2001).

11.3.7 Combretaceae

The Combretaceae occur in both tropical and subtropical climes and consist of 20 genera in about 300 species with *Combretum, Terminalia,* and *Quisqualis* the largest and most important (Eloff, Katerere, and McGaw, 2008).

Combretum caffrum is used to treat conjunctivitis (Masika, van Averbeeke, and Sonandi, 2000), and *C. paniculatum* is used for fertility problems in animals (Luseba and Van der Merwe, 2006). *Terminalia sericea* is used to treat wounds in animals (Luseba and Van der Merwe, 2006) as well as diarrhea (Van der Merwe, Swan, and Botha, 2001). In Kenya, combretums are used as anthelmentics for small ruminants (Githiori, Athanasiadou, and Thamsborg, 2006). Aqueous and organic leaf extracts of *C. apiculatum*, *C. hereroense,* and *C. mossambicense* showed good activity in *in vitro* assays against the free-living nematode *Caenorhabditis elegans* (McGaw et al., 2001).

Members of the Combretaceae have been extensively studied and shown to have broad-spectrum antimicrobial activity (Katerere et al., 2003; Martini, Katerere, and Eloff, 2001; Eloff, Katerere, and McGaw, 2008). Several flavonoids, triterpenoids, and stilbenoids have been isolated and shown to be biologically active.

11.3.8 Euphorbiaceae

Euphorbiaceae occur in different habitats of the tropics, and as result of their cosmopolitan nature, they have developed various life-forms, including herbs, shrubs, stunted succulents, and tall canopy trees (Balakrishnan and Chakrabarty, 2007). Phytochemical studies have shown the presence of flavonoids, saponins, diterpenes, phorbol esters (Bagalkotkar et al., 2006), and triterpenoids (Uzabakiliho, Largeau, and Casadevall, 1987).

Different species are used for diseases such as blackquarter, constipation in cattle and goats, and diarrhea (Luseba and van der Merwe, 2006), as well as eye infections, pneumonia, as a tonic, and for fertility enhancement in livestock (Van der Merwe, Swan, and Botha, 2001).

Luseba and van der Merwe (2006) reported that *Synadenium* (= *Euphorbia*) *cupulare* is used for eye infection and blackquarter. East Coast fever (ECF) is treated by topical application of the sap of *Synadenium grantii* (*Synadenium umbellatum* [SU] Pax [synonym: *Euphorbia umbellate*]) (Tabuti, Dhillion, and Lye, 2003).

Jatropha is a multipurpose plant that has become of particular interest in the burgeoning biodiesel industry (Openshaw, 2000). Although originally a tropical American plant (Mujumdar, Upadhye, and Misar, 2000), *Jatropha curcas* has now spread to Africa and Asia. It has been cited for use in gastrointestinal problems in cattle for both constipation and diarrhea (Luseba and van der Merwe, 2006). The antidiarrheal activity is thought to be effected via inhibition of elevated prostaglandin biosynthesis and reduced gut motility (Mujumdar, Upadhye, and Misar, 2000). Petroleum ether extracts of *J. curcas* were found to be active against the dengue vector *Aedes aegypti* and the lymphatic filariasis vector *Culex quinquefasciatus* (Rahuman et al., 2008). The oil of *J. curcas* showed good effect in the treatment of sarcoptic mange in sheep in a field trial (Dimri and Sharma, 2004). A toxalbumin, curcin, which occurs in *J. curcas* and *J. multifida*, makes these species, particularly the seeds, like those of *Ricinus communis*, highly toxic (Botha and Penrith, 2008).

Ricinus communis is administered as a purgative to calves refusing to suckle (Hutchings et al., 1996) and for constipation and internal parasites (Van der Merwe, Swan, and Botha, 2001). Methanolic extracts of *R. communis* have been shown to inhibit adult *Setaria digitata*, the cattle filarial worm (Nisha et al., 2007). The seeds are the main source of ricin and ricinine, which are toxic but can also be therapeutic

(Aslani et al., 2007). In spite of the risk of poisoning, of which horses and sheep are most prone, there are few cases of accidental poisoning in livestock (Aslani et al., 2007). The leaves contain flavonoids, phenolic acids, and tannins (Chen, Zhang, and Chen, 2008).

11.3.9 Fabaceae (Leguminosae)

Fabaceae is a large family consisting of trees, shrubs, vines, and herbs bearing bean pods. It is made up of 730 genera and more than 19,000 species (Wojciechowski, Mahn, and Jones, 2006). Many of the trees are an important source of tannins. Species from this family are used to treat various animal diseases, ranging from bacterial diseases to diseases and conditions caused by internal and external parasites (see Table 11.1). For example, *Peltophorum africanum* has many ethnomedical and ethnoveterinary uses; rural farmers use the root and bark extracts to treat diarrhea, dysentery, and infertility and to promote well-being and resistance to diseases in cattle (Bizimenyera., Githiori, Swan, et al., 2006). Leaf, root, and bark extracts of *P. africanum* inhibited egg hatching and larval development of *Trichostrongylus colubriformis*, an important cause of parasitic gastroenteritis in ruminants (Bizimenyera, Githiori, Eloff, et al., 2006). It contains coumarins (Mebe and Makuhunga, 1992), flavonoids, and tannins (El Sherbeiny et al., 1977). *Calpurnia aurea* is used to kill maggots in wounds (Bryant, 1966) and for the treatment of dysentery and diarrhea in animals and ticks (Adedapo et al., 2008). Quinolizidine alkaloids have been isolated from this species (Asres et al., 1986).

A bark decoction of *Schotia latifolia* is used in the treatment of red water in cattle. No previously recorded uses were found in the literature (Dold and Cocks, 2001). *Schotia brachypetala*, on the other hand, is used to treat infectious diseases in cattle (McGaw, Van der Merwe, and Eloff, 2007).

Stilbenes and phenolics are reported to be present in *S. brachypetala* (Drewes, 1971), and antibacterial fatty acids have been isolated from the leaves (McGaw, Jäger, and van Staden, 2002).

11.3.10 Geraniaceae

The Geraniaceae family consists of 800 species in seven genera. Coumarins and tannins have been isolated from most plants from this family (Watt and Breyer-Brandwijk, 1962). Pelargonium, a genus of this family, is one of the most important traditional medicines in southern Africa for both humans and animals. *Pelargonium reniforme* Curtis is used to treat diarrhea in goats and cows, heartwater in cattle, and liver disorders in cattle and sheep (Dold and Cocks, 2001). Roots have been used to prevent purging in horses and to dress wounds (Brendler and van Wyk, 2008). *Pelargonium sidoides*, on the other hand, while also cited for use in horses for similar use, is used to treat dysentery in cattle (Lewu, Grierson, and Afolayan, 2007) and, when mixed with *Ziziphus mucronata*, is employed as a dewormer in calves (Smith, 1966).

Anthocyanins (pelargonidin, cyanidin, and delphinidin, among them), various coumarin derivatives, flavonoids, proanthocyanidins, phenylpropanoid derivatives, and a diterpene have been isolated from both *P. sidiodes* and *P. reniforme* (Mitchell, Markham, and Boase, 1998; Kolodziej, 2007).

11.3.11 Hyacinthaceae

The Hyacinthaceae family consists of 700 genera and 1,000 species of herbaceous perennials (Pfosser and Septa, 1999). Saponins, homoisoflavanones, steroids, cardiac glycosides, and alkaloids have been reported from the Hyacinthaceae (Louw, Regnier, and Korsten, 2002), some of which confer toxicity on members of the family (Van Wyk and Gericke, 2000). Homoisoflavanones have anti-inflammatory, antibacterial, antihistaminic, antimutagenic, and angioprotective properties as well as being potent phosphodiesterase inhibitors (Heller and Tamm, 1981; Della et al., 1989). Species from this family have been used to treat diseases such as gall sickness in animals (Watt and Breyer-Brandwijk, 1962), diarrhea in goats, and intestinal parasites and retained afterbirth in cattle (Dold and Cocks, 2001)

Scilla natalensis (= *Merwilla natalensis*) is reported for use in cattle against "lung sickness" (Hutchings et al., 1996), while *S. nervosa* has been used as a purgative for calves (Gerstner, 1941). Organic extracts of *S. natalensis* showed high lethality against nematodes in anthelmintic assays (Sparg, van Staden, and Jäger, 2002). Homoisoflavanones have been isolated from the bulbs of *S. kraussii*, *S. dracomontana*, and *S. natalensis* (Crouch, Bangani, and Mulholland, 1999). *Scilla nervosa* has yielded homoisoflavanones and stilbenoids, including resveratrol (Bangani, Crouch, and Mulholland, 1999).

Ledebouria cooperi was reported by Watt and Breyer-Brandwijk (1962) to be used in cows to ensure that calves were of the same gender, while *Ledebouria revolute* is used to treat gall sickness and diarrhea. Homoisoflavanones and chalcones have been isolated from these species (Pohl et al., 2001).

Bulb decoctions of *Urginea* (Drimia) species are used for intestinal diseases and endoparasites and for retained placenta (Dold and Cocks, 2001; Van der Merwe, Swan, and Botha, 2001). *In vitro* antibabesial activity of extracts of *U. sanguinea* was negative (Naidoo et al., 2005). They, like *Scilla* spp., are also well-known poisonous plants that are an important component of materia medica in southern Africa. Toxicity is due to the presence of cardiac glycosides (one of which has been called transvaalin), and ingestion by cattle leads to respiratory problems, muscular tremors, and finally death (Kellerman, Coetzer, and Naude, 1988; Mulholland and Drewes, 2004). Fresh bulbs of *U. sanguinea* have yielded phloroglucinol and phenolic acids, in particular salicylic acid (Mulholland and Drewes, 2004). Several bufadienolides have been isolated from the bulbs of *Urginea maritime* (Kopp et al., 1996).

11.3.12 Lamiaceae (Labiatae)

Lamiaceae (commonly known as the mint family) contains over 200 genera and 7,000 species of most aromatic plants that are important culinary herbs (e.g., basil, mint, rosemary, and sage) (Heywood et al., 2007).

Many plants from this family are used to treat gall sickness (Dold and Cocks, 2001; Hutchings et al., 1996), heartwater and anthrax (Dold and Cocks, 2001), and eye inflammation (Masika, van Averbeeke, and Sonandi, 2000). A diterpene diol, ibozol and related diterpenoids, and large amounts of α-pyrones have been isolated from some plants of this family (Zelnik et al., 1978).

Leonotis is used in the drinking water of poultry to prevent disease and for gall sickness in cattle (Hulme, 1954). In Ethiopia, the Bale people use it to treat anthrax (Yineger et al., 2007). Diterpenes have been isolated from the Asian species *Leonotis nepetaefolia* (Govindasamy et al., 2002). Little work has been done on the southern Africa species despite their widespread occurrence and use.

The powdered aerial parts of *Plectranthus laxiflorus* Benth (= *Plectranthus albus* Gürke) are used as drenches for animals (Watt and Breyer-Brandwijk, 1962). *Plectranthus* is a large genus containing about 300 species of ornamental and medicinal value plants found in tropical Africa, Asia, and Australia. In East Africa, *Plectranthus* spp. are mostly used as dry season fodder (Lukhoba, Simmonds, and Paton, 2006). While the chemistry of *P. laxiflorus* remains unexplored, many related species have yielded abietane, phyllocladane, kaurane, clerodane, and labdane diterpenoids, triterpenes, and sesquiterpenes (Rijo et al., 2007), many of which have insect antifeedant and antibacterial activity (Wellsow et al., 2006).

11.3.13 Solanaceae

The Solanaceae contains both important agricultural food plants (e.g., potato and tomato) and famous toxic plants (e.g., Datura and Belladona [deadly nightshade]) (D'Arcy, 1986).

Many compounds, including glycowithanolides, withanolides, and hyoscyamine (Hutchings et al., 1996; Oliver-Bever, 1986), have been isolated in some taxa of this family. Species like *Solanum panduriforme* are used for the treatment of bacterial diseases, diarrhea, eye infections (Van der Merwe, Swan, and Botha, 2001), and respiratory problems (Luseba and Van der Merwe, 2006).

The Solanaceae are a source of steroidal saponins and glycoalkaloids that are pesticidal (Silva et al., 2005). Aqueous extracts of roots and berry of *S. aculeastrum* have shown molluscicidal activity due to steroidal alkaloid glycosides (Wanyonyi et al., 2002). Several species have potent molluscicidal activity (Silva et al., 2005) and larvicidal activity (Chowdhury, Ghosh, and Goutam, 2008).

Solanum nigrum, *S. lichtensteinii*, and *S. tettense* are recognized as poisonous plants in southern Africa (Botha and Penrith, 2008). The first two cause diarrhea, while *S. tettense* poisoning results in neurotoxicosis. Interestingly, of the three species, only *S. lichtensteinii* has been recorded for ethnoveterinary use. However, *S. nigrum* is an important plant in traditional medicine worldwide, and sapogenins, flavonoids, and alkaloids have been isolated from this species and tested in many different assays (Mallika and Devi, 2006; Ji et al., 2008; Kuo et al., 2000).

11.3.14 Vitaceae

Vitaceae are a family of dicotyledonous flowering plants, including *Cissus quadrangularis*, which is used as a poultice for wounds, for lumpy skin disease, and as a tick repellent (Luseba and Van der Merwe, 2006); *Rhoicissus tomentosa*, which is used as an anthelmintics for calves (Watt and Breyer-Brandwijk, 1962); and *Rhoicissus tridentata*, used to treat diarrhea in goats and sheep (Dold and Cocks, 2001), heartwater, red water, and internal parasites and for abortion (Van der Merwe, Swan, and Botha, 2001). Extracts of *Cissus quadrangularis* have been shown to possess *in vitro* activity in antibacterial, anthelmintic, and brine shrimp toxicity assays (McGaw, Van der Merwe, and Eloff, 2007). Iridioids, stilbenes, flavonoids, and triterpenoids have been isolated from the species (Singh, Rawat, and Maurya, 2007).

Rhoicissus tridentata has uterotonic activity that has been demonstrated on isolated rat uterus (Katsoulis Veale, and Havlik, 2000), which may validate its use in abortion and for retained placenta (Fennell and van Staden, 2001; Van der Merwe, Swan, and Botha, 2001). Activity against *Babesia caballi* in culture was not significant (Naidoo et al., 2005).

11.4 THE WAY FORWARD

EVMs do have advantages and disadvantages, as is the case with other practices (Gueye, 1999; Fielding, 2005; Mathias, 2001). The advantages are that they are culturally acceptable to livestock keepers due to familiarity; access is easy, and availability is free or affordable; and in general, they appear to be effective. However, EVM has limitations; for example, most medicinal plants have not been validated under controlled conditions for their safety and efficacy (Sori et al., 2004), the seasonality of some plants means that they may be unavailable at certain times of the year, the herbs used are not likely to treat acute viral diseases in animals, and commercialization may result in overharvesting and extinction of popular species.

In many developing countries, many veterinarians, development workers, and decision makers have not yet examined the potential of EVMs in treatment of livestock diseases. A change of attitude is still needed in this regard. Complex practices require systematic research to capture synergistic effects of the different aspects of EVM (Kudi, 2003).

Feasibility of promoting safe and effective herbal medicines should be investigated. Some of the aspects that should be addressed before promotion of herbal medicine include quality control, sustainability or cultivation, local self-sufficiency, and distribution. Viable alternatives to allopathic medicines that will reduce total expenditure on animal health should be promoted (Kudi, 2003)

11.5 CONCLUSION

Although the concept of local knowledge is global in its importance, its practical application is very much at the local level, where further investments should be concentrated on improving, if possible, a range of practices that are appropriate and sustainable. According to Mathias (2007), animal health care services in rural areas can be improved by urgently tapping from all available resources, including ethnoveterinary practices. It therefore requires people to get more information on the efficacy of EVM and more documentation of validated information. With the escalating costs of medicines for animal health care, as well as resistance to conventional drugs by some parasites, livestock owners will continue to do what they can to protect the health of their animals; hence, use of EVMs cannot be ignored.

REFERENCES

Abegaz, B.M.; Ngadjui, B.T.; Bezabih, M.; and Mdee, L.K. 1999. Novel natural products from marketed plants of eastern and southern Africa. *Pure and Applied Chemistry* 71 (6): 919–926.

Adedapo, A.A.; Jimoh, F.O.; Afolayan, A.J.; and Masika, P.J. 2008. Antioxidant activities and phenolic contents of the methanol extracts of the stems of *Acokanthera oppositifolia* and *Adenia gummifera*. *BMC Complementary and Alternative Medicine* 8: 54.

Ahmed, M.S.; Galal, A.M.; and Ross, S.A. 2001. A weakly antimalarial biflavanone from *Rhus retinorrhoea*. *Phytochemistry* 58 (4): 599–602.

Arroyo, S.C.; and Cutler, D.F. 1984. Evolutionary and taxonomic aspects of the internal morphology in Amaryllidaceae from South America and southern Africa. *Kew Bulletin* 39 (3): 467–498

Aslani, M.R.; Maleki, M.; Mohri, M.; Sharifi, K.; Najjar-Nezhad, V.; and Afshari, E. 2007. Castor bean (*Ricinus communis*) toxicosis in a sheep flock. *Toxicon* 49 (3): 400–406.

Asres, K.; Gibbons, W.A.; Phillipson, J.D.; and Mascagni, P. 1986. Alkaloids of Ethiopian *Calpurnia aurea* subsp. *aurea*. *Phytochemistry* 25 (6): 1443–1447.

Bagalkotkar, G.; Sagineedu, S.R.; Saad, M.S.; and Stanslas, J. 2006. Phytochemicals from *Phyllanthus niruri* Linn. and their pharmacological properties—a review. *The Journal of Pharmacy and Pharmacology* 8: 1559–1570.

Balakrishnan, N.P.; and Chakrabarty, T. 2007. *The family* Euphorbiacea *in India: A synopsis of its profile, taxonomy and bibliography*. Bishen Singh Mahendra Pal Singh, Dehradun, India, 500 pages.

Bangani, V.; Crouch, N.R.; and Mulholland, D.A. 1999. Homoisoflavanones and stilbenoids from *Scilla nervosa*. *Phytochemistry* 51(7): 947–951.

Begum, D.; and Nath, S.C. 2000. Ethnobotanical review of medicinal plants used for skin diseases and related problems in northeastern India. *Journal of Herbs, Spices and Medicinal Plants* 7 (3): 55–93.

Bizimenyera, S.E.; Swan, G.E.; Chikoto, H.; and Eloff, J.N. 2005. Rationale for using *Peltophorum africanum* (Fabaceae) extracts in veterinary medicine. *Journal of South African Veterinary Association* 76 (2): 54–58.

Bizimenyera, E.S.; Githiori, J.B.; Eloff, J.N.; and Swan, G.E. 2006. In vitro activity of *Peltophorum africanum* Sond. (Fabaceae) extracts on the egg hatching and larval development of the parasitic nematode *Trichostrongylus colubriformis*. *Veterinary Parasitology* 142: 336–343.

Bizimenyera, E.S.; Githiori, J.B.; Swan, G.E.; and Eloff, J.N. 2006. In vitro ovicidal and larvicidal activity of the leaf, bark and root extracts of *Peltophorum africanum* Sond. (Fabaceae) on *Haemonchus contortus*. *Journal of Animal and Veterinary Advances* 5: 608–614.

Bohlmann, F.; Ates, N.; and Jakupovic, J. 1983. Naturally occurring terpene derivatives. Part 464. Hirsutinolides from South African *Vernonia* species. *Phytochemistry* 22: 1159–1162.

Bohlmann, F.; and Fritz, U. 1979. New flavones from *Heteromorpha simplicifolium*. *Phytochemistry* 18: 1081.

Bohlmann, F.; and Zdero, C. 1978a. Naturally occurring terpene derivatives. Part 132. Diterpenes with a new carbon skeleton from *Printzia laxa*. *Phytochemistry* 17: 487–489.

Bohlmann, F., and Zdero, C., 1978b. Naturally occurring terpene derivatives. Part 150. A new type of sesquiterpene from *Senecio oxyriifolius*. *Phytochemistry* 17: 1669–1671.

Botha, C.J.; and Penrith, M.L. 2008. Poisonous plants of veterinary and human importance in southern Africa. *Journal of Ethnopharmacology* 119 (3): 549–558.

Braca, A.; Politi, M.; Sanogo, R.; Sanou, H.; Morelli, I.; Pizza, C.; and De Tommasi, N. 2003. Chemical composition and antioxidant activity of phenolic compounds from wild and cultivated *Sclerocarya birrea* (Anacardiaceae) leaves. *Journal of Agricultural and Food Chemistry* 51 (23): 6689–6695.

Brendler, T.; and van Wyk, B.E. 2008. A historical, scientific and commercial perspective on the medicinal use of *Pelargonium sidoides* (Geraniaceae). *Journal of Ethnopharmacology* 119 (3): 420–433.

Bryant, A.T. 1966. *Zulu medicine and medicine-men*. Struik, Cape Town.

Candy, H.A.; Pegel, K.H.; Brookes, B.; and Rodwell, M. 1977. The occurrence of atractyloside in *Callilepis laureola*. *Phytochemistry* 16: 1308–1309.

Chen, Z.; Zhang, J.; and Chen, G. 2008. Simultaneous determination of flavones and phenolic acids in the leaves of *Ricinus communis* Linn. by capillary electrophoresis with amperometric detection. *Journal of Chromatography* 863 (1): 101–106.

Chowdhury, N.; Ghosh, A.; and Goutam, C. 2008. Mosquito larvicidal activities of *Solanum villosum* berry extract against the dengue vector *Stegomyia aegypti*. *BMC Complementary and Alternative Medicine* 8: 10.

Cocks, M. 2006. Wild plant resources and cultural practices in rural and urban households in South Africa. PhD thesis, Wageningen University, Wageningen, The Netherlands.

Crouch, N.R.; Bangani, V.; and Mulholland, D.A. 1999. Homoisoflavanones from three South African: *Scilla* species. *Phytochemistry* 51 (7): 943–946.

Dagne, E.; Bisrat, D.; Viljoen, A.; and Van Wyk, B.E. 2000. Chemistry of *Aloe* species. *Current Organic Chemistry* 4 (10): 1055–1078.

Dagne, E.; and Yenesew, A. 1994. Anthraquinones and chemotaxonomy of the asphodelaceae. *Pure & Appl. Chem.* 66 (10/11): 2395–2398.

D'Arcy, W.G. 1986. *Solanacea*. Columbia University Press, New York.

Debella, A.; Haslinger, E.; Kunert, O.; Michl, G.; and Abebe, D. 1999. Steroidal saponins from *Asparagus africanus*. *Phytochemistry* 51 (8): 1069–1075.

Dekker, T.G.; Fourie, T.D.; Matthee, E.; and Snyckers, F.O., 1987. An oxindole from the roots of *Capparis tomentosa*. *Phytochemistry* 26: 1845–1846.

Della, R.; Loggia, P.; Del Negro, A.; Tubaro Barone, G.; and Parrilli, M. 1989. Homoisoflavanones as anti-inflammatory principles. *Planta Medica* 55: 587.

De Smet, P.A. 1996. Some ethnopharmacological notes on African hallucinogens. *Journal of Ethnopharmacology* 50 (3): 141–146.

Dictionary of Natural Products. 1996. CD-ROM, release 4:2. Chapman & Hall, London.

Dimri, U.; and Sharma, M.C. 2004. Effects of sarcoptic mange and its control with oil of *Cedrus deodara*, *Pongamia glabra*, *Jatropha curcas* and benzyl benzoate, both with and without ascorbic acid on growing sheep: epidemiology; assessment of clinical, haematological, cell-mediated humoral immune responses and pathology. *Journal of Veterinary Medicine. A, Physiology, Pathology, Clinical Medicine* 51 (2): 71–78.

Dinan, L.; Savchenko, T.; and Whiting, P. 2001. Phytoecdysteroids in the genus *Asparagus* (Asparagaceae). *Phytochemistry* 56 (6): 569–576.

Doke, C.M.; and Vilakazi, B.W. 1972. *Zulu-English dictionary*, 2nd ed. Witwatersrand University Press, Johannesburg, South Africa.

Dold, A.P.; and Cocks, M.L. 2001. Traditional veterinary medicine in the Alice district of the eastern Cape Province, South Africa. *South African Journal of Science* 97: 375–379.

Drewes, S.E. 1971. Isolation of 3,3′,4′,5,5′-pentahydroxystilbene from *Schotia* species. *Phytochemistry* 10 (11): 2837–2838.

Eloff, J.N.; Katerere, D.R.; and McGaw, L.J. 2008. The biological activity and chemistry of the southern African Combretaceae. *Journal of Ethnopharmacology* 119 (3): 686–699.

El Sherbeiny, A.E.; El Ansari, M.A.; Nawwar, M.A.; and El Sayed, N.H. 1977. Flavonol glycosides and flavonol glucoside gallates from *Peltophorum africanum*. *Planta Medica* 32 (2):165–170.

Fennell, C.W.; and van Staden, J. 2001. *Crinum* species in traditional and modern medicine. *Journal of Ethnopharmacology* 78 (1): 15–26.

Fielding, D. 2005. Ethnoveterinary medicine in the tropics—key issues and the way forward? *Farmaco* 50 (4): 245–256.

Francis, G.; Kerem, Z.; Harinder, P.; Makkar, S.; and Becker, K. 2002. The biological action of saponins in animal systems: a review. *British Journal of Nutrition* 88: 587–605.

Galvez, P.J.; Zarzuelo, A.; Busson, R.; Cobbaert, C.; and de Witte, P. 1992. (-)-Epicatechin-3-galloyl ester: a secretagogue compound from the bark of *Sclerocarya birrea. Planta Medica* 58(2): 174–175.

Gerstner, J. 1938. A preliminary checklist of Zulu names of plants with short notes. *Bantu Studies* 12: 215–236, 321–342.

Gerstner, J. 1939. A preliminary checklist of Zulu names of plants with short notes. *Bantu Studies* 13: 49–64, 131–149, 307–326.

Gerstner, J. 1941. A preliminary checklist of Zulu names of plants with short notes. *Bantu Studies* 15: 277–301, 369–383.

Githiori, J.B.; Athanasiadou, S.; and Thamsborg, S.M. 2006. Use of plants in novel approaches for control of gastrointestinal helminths in livestock with emphasis on small ruminants. *Veterinary Parasitology* 139 (4): 308–320.

Glen, H.F.; and Hardy D.S. 2000. *Aloaceae*. In *Flora of southern Africa*. Edited by G. Germishuizen, vol 5 (1) Fascicle 1, Part 1. pp 1–167. Pretoria: National Botanical Institute.

Govindasamy, L.; Rajakannan, V.; Velmurugan, D.; Banumathi, S.; and Vasanth, S. 2002. Structural studies on three plant diterpenoids from *Leonotis nepetaefolia. Crystal Research and Technology* 37 (8): 896–909.

Grace, O.M.; Simmonds, M.S.J.; Smith, G.F.; and van Wyk, A.E. 2008. Therapeutic uses of *Aloe* L. (Asphodelaceae) in southern Africa. *Journal of Ethnopharmacology* 119 (3): 604–614.

Graham, K.; Graham, E.A.; and Towers, G.H.N. 1980. Cercaricidal activity of phenyl hepta-triyne and -terthienyl naturally occurring compounds in species of Asteraceae Compositae. *Canadian Journal of Zoology* 58: 1955–1958.

Gueye, E.F. 1999. Disease control using ethno-veterinary medicine. *World Poultry Elsevier* 15 (7): 48–51.

Guillarmod, A.J. 1971. *Flora of Lesotho.* Cramer, (Lehre), 474 pages.

Heller, W.; and Tamm, C.H. 1981. Homoisoflavanones and biogenetically related compounds. *Forschritte der Chemie Organischer Naturstoffe* 40: 106.

Heywood, V.H.; Brummitt, R.K.; Culham, A.; and Seberg, O. 2007. *Flowering plant families of the world.* Royal Botanic Gardens, Kew, UK.

Hoffman, B.; and Hoelzl, J. 1988. New chalcones from *Bidens pilosa. Planta Medica* 54: 52–54.

Hulme, M.M. 1954. *Wild flowers of Natal.* Shuter and Shooter, Pietermaritzburg, South Africa.

Hutchings, A.; Scott, A.H.; Lewis, G.; and Cunningham, A.B. 1996. *Zulu medicinal plants: an inventory.* University of Natal Press, Pietermaritzburg, South Africa.

Ji, Y.B.; Gao, S.Y.; Ji, C.F.; and Zou, X. 2008. Induction of apoptosis in $HepG_2$ cells by solanine and Bcl-2 protein. *Journal of Ethnopharmacology* 115 (2): 194–202.

Katerere, D.R.; Gray, A.I.; Nash, R.J.; and Waigh, R.D. 2003. Antimicrobial activity of pentacyclic triterpenes isolated from African Combretaceae. *Phytochemistry* 63 (1): 81–88.

Katsoulis, L.C.; Veale, D.J.; and Havlik, I. 2000. The pharmacological action of *Rhoicissus tridentata* on isolated rat uterus and ileum. *Phytotherapy Research* 14 (6): 460–462.

Kellerman, T.S.; Coetzer, J.A.W.; and Naude, T.W. 1988. *Plant poisonings and mycotoxicoses of livestock in Southern Africa.* Oxford University Press, Cape Town, South Africa.

Kolodziej, H. 2007. Fascinating metabolic pools of *Pelargonium sidoides* and *Pelargonium reniforme*, traditional and phytomedicinal sources of the herbal medicine Umckaloabo. *Phytomedicine* 14: 19–17.

Kopp, B.; Krenn, L.; Draxler, M.; Hoyer, A.; Terkola, R.; Vollaster, P. and Robien, W. 1996. Bufadiernolides from *Urginea maritima* from Egypt. *Phytochemistry.* 42 (2): 513–522.

Kudi C. 2003. Ethno-veterinary, complementary and low cost treatment and management of working animals, p. 8. TAWS Workshop held April 24, Silsoe Research Institute, UK. http://www.taws.org.

Kuo, K.W.; Hsu, S.H.; Li, Y.P.; Lin, W.L.; Liu, L.F.; Chang, L.C.; Lin, C.C.; and Sheu, H.M. 2000. Anticancer activity evaluation of the Solanum glycoalkaloid solamargine—triggering apoptosis in human hepatoma cells. *Biochemical Pharmacology* 60 (12): 1865–1873.

Lewis, R. 1990. Amaryllidaceae alkaloids. *Natural Product Reports* 7: 549–556.

Lewu, F.B.; Grierson, D.S.; and Afolayan, A.J. 2007. Morphological diversity among accessions of *Pelargonium sidoides* DC. in the Eastern Cape, South Africa. *Genetic Resources and Crop Evolution* 54: 1–6.

Liu, Y.; and Abreu, P. 2006. Tirucallane triterpenes from the roots of *Ozoroa insignis*. *Phytochemistry* 67 (13): 1309–1315.

Louw, C.A.M.; Regnier, T.J.C.; and Korsten, L. 2002. Medicinal bulbous plants of South Africa and their traditional relevance in the control of infectious diseases. *Journal of Ethnopharmacology* 82 (2–3): 147–154.

Lukhoba, C.W.; Simmonds, MS.J.; and Paton, A.J. 2006. *Plectranthus*: a review of ethnobotanical uses. *Journal of Ethnopharmacology* 103 (1): 1–24.

Luseba, D.; Elgorashi, E.E.; Ntloedibe, N.T.; and Van Staden, J. 2007. Antibacterial, anti-inflammatory and mutagenic effects of some medicinal plants used in South Africa for the treatment of wounds and retained placenta in livestock. *South African Journal of Botany* 73: 378–383.

Luseba, D.; and Van der Merwe, D. 2006. Ethnoveterinary medicine practices among Tsonga speaking people of South Africa. *Onderstepoort Journal of Veterinary Research* 73: 115–122.

Mabogo, D.E.N. 1990. The ethnobotany of the Vhavenda. MSc thesis, University of Pretoria, South Africa.

Mallika, J.; and Devi, C.S.S. 2006. Antiulcerogenic and ulcer healing effects of *Solanum nigrum* (L.) on experimental ulcer models: possible mechanism for the inhibition of acid formation. *Journal of Ethnopharmacology* 104 (1–2): 156–163.

Mamidou, K.W.; Kamanzi, A.K.; Traoré, D.; and Bruno, B. 2005. Anthelmintic activity of medicinal plants used in northern Côte d'Ivoire against intestinal helminthiasis. *Pharmaceutical Biology* 43 (1): 72–78.

Martin, M.; Mathias, E.; and McCorkle, C.M. 2001. *Ethnoveterinary medicine: an annotated bibliography of community animal healthcare.* ITDG, London.

Martini, N. D.; Katerere, D.R.; and Eloff, J.N. 2001. Biological activity of five antibacterial flavonoids from *Combretum erythrophyllum* (Combretaceae). *Journal of Ethnopharmacology* 93 (2–3): 207–212.

Masika, P.J.; and Afolayan, A.J. 2002. Antimicrobial activity of some plants used for the treatment of livestock disease in the Eastern Cape, South Africa. *Journal of Ethnopharmacology* 83: 129–134.

Masika, P.J.; van Averbeeke, W.; and Sonandi, A. 2000. Use of herbal remedies by small-scale farmers to treat livestock diseases in central eastern Cape Province, South Africa. *Journal of the South African Veterinary Association* 71: 87–91.

Mathias, E. 2001. Introducing ethnoveterinary medicine. Retrieved May 23, 2009 from http://www.ethnovetweb.com/whatisvm.pdf

Mathias, E., 2007. Ethnoveterinary medicine in the era of evidence-based medicine: Mumbo-jumbo, or a valuable resource? *The Veterinary Journal*, 173 (2):241–242.

Mavi, S. 1994. Medicinal plants and their uses in Zimbabwe. In: B.J. Huntley, Editor. *Botanical diversity in southern Africa, Strelitzia 1.* vol. 1, pp. 169–174. National Botanical Institute, Pretoria.

McGaw, L.J.; Jäger, A.K.; and van Staden, J. 2002. Isolation of antibacterial fatty acids from *Schotia brachypetala. Fitoterapia* 73: 431–433.

McGaw, L.J.; Rabe, T.; Sparg, S.G.; Jäger, A.K.; Eloff, J. N.; and van Staden, J. 2001. An investigation on the biological activity of *Combretum* species. *Journal of Ethnopharmacology* 75(1): 45–50.

McGaw, L.J.; Van der Merwe, D.; and Eloff, J.N., 2007. In vitro anthelmintic, antibacterial and cytotoxic effects of extracts from plants used in South African ethnoveterinary medicine. *The Veterinary Journal* 173: 366–372.

McGaw, L.J.; and Eloff, J.N. 2008. Ethnoveterinary use of southern African plants and scientific evaluation of their medicinal properties. *Journal of Ethnopharmacology* 119 (3): 329–330.

Mebe, P.P.; and Makuhunga, P. 1992. 11-(*E*)-*p*-Coumaric acid ester of bergenin from *Peltophorum africanum*. *Phytochemistry* 31 (9): 3286–3287.

Mitchell, K.A.; Markham, K.R.; and Boase, M.R. 1998. Pigment chemistry and colour of *Pelargonium* flowers. *Phytochemistry* 47 (3): 355–361.

Moreno, M.; and Rodriguez, V.M. 1981. Medicinal pre-Hispanic plants. Part 9. Yiamoloside B, a fungistatic saponin of *Phytolacca octandra*. *Phytochemistry* 20: 1446–1447.

Morris, B. 1996. A study of herbalism in southern Malawi: monographs from the International African Institute. In: *Chewa Medical Botany 2*. Lit-Verlag, Hamburg, Germany, 188 pages.

Moyo, B.; and Masika, P.J. 2008. Tick control methods used by resource-limited farmers and the effect of ticks on cattle in rural areas of the eastern Cape Province, South Africa. *Tropical Animal Health and Production*. doi:10.1007/s11250-008-9216-4.

Mujumdar, A.M.; Upadhye, A.S.; and Misar, A.V. 2000. Studies on antidiarrhoeal activity of *Jatropha curcus* root extract in albino mice. *Journal of Ethnopharmacology* 70 (2): 183–187.

Mulholland, D.A.; and Drewes, S.E. 2004. Global phytochemistry: indigenous medicinal chemistry on track in southern Africa. *Phytochemistry* 65(7): 769–782.

Naidoo, V.; Zweygarth, E.; Eloff, J.N.; and Swan, G.E. 2005. Identification of anti-babesial activity for four ethnoveterinary plants in vitro. *Veterinary Parasitology* 130 (1–2): 9–13.

Naidoo, V.; Zweygarth, E.; and Swan, G. E. 2006. Determination and quantification of the in vitro activity of *Aloe marlothii* (A. Berger) subsp. *marlothii* and *Elephantorrhiza elephantina* (Burch.) skeels acetone extracts against *Ehrlichia ruminantium*. *Onderstepoort Journal of Veterinary Research* 73 (3): 175–178.

Nair, J.J.; Machocho, A.K.; Campbell, W.E.; Brun, R.; Viladomat, F.; Codina, C.; and Bastida, J. 2000. Alkaloids from *Crinum macowanii*. *Phytochemistry* 54 (8): 945–950.

Nisha, M.; Kalyanasundaram, M.; Paily, K.P.; Abidha, V.P.; and Balaraman, K. 2007. In vitro screening of medicinal plant extracts for macrofilaricidal activity. *Parasitology Research* 100 (3): 575–579.

Oliver-Bever, B. 1986. *Medicinal plants in tropical west Africa*. Cambridge University Press, London.

Ollerton, J.; Johnson, S.D.; Cranmer, L.; and Kellie, S. 2003. The pollination ecology of an assemblage of grassland asclepiads in South Africa. *Annals of Botany* 92: 807–834.

Openshaw, K. 2000. A review of *Jatropha curcas*: an oil plant of unfulfilled promise. *Biomass and Bioenergy* 19 (1): 1–15.

Palmer, E.; and Pitman, N. 1972. *Trees of southern Africa*. Vol. 3. Balkema, Cape Town, South Africa.

Panero, J.L.; and Funk, V.A. 2008. The value of sampling anomalous taxa in phylogenetic studies: major clades of the Asteraceae revealed. *Molecular Phylogenetics and Evolution* 47 (2): 757–782.

Pell, S.K. 2004. Molecular systematics of the cashew family (Anacardiaceae). PhD thesis, Louisiana State University, Baton Rouge, LA.

Pfosser, M.; and Septa, F. 1999. Phylogenetics of hyacinthaceae based on plastid DNA sequences. *Annals of the Missouri Botanical Garden* 86: 852–875.

Pham, L.H.; Döpke, W.; Wagner, J.; and Mügge, C. 1998. Alkaloids from *Crinum amabile*. *Phytochemistry* 48 (2): 371–376.

Pohl, T.; Koorbanally, C.; Crouch, N.R.; and Mulholland, D.A. 2001. Secondary metabolites of *Scilla plumbea*, *Ledebouria cooperi* and *Ledebouria ovatifolia* (Hyacinthaceae). *Biochemical Systematics and Ecology* 29 (8): 857–860.

Pujol, J. 1990. *Nature Africa—the herbalist handbook*. Jean Pujol Natural Healers' Foundation, Durban, South Africa.

Rahuman, A.A.; Gopalakrishnan, G.; Venkatesan, P.; and Geetha, K. 2008. Larvicidal activity of some Euphorbiaceae plant extracts against *Aedes aegypti* and *Culex quinquefasciatus* (Diptera: Culicidae). *Parasitology Research* 102 (5): 867–873.

Rayne, S.; and Mazza, G. 2007. Biological activities of extracts from sumac (*Rhus* spp.): a review. *Plant Foods for Human Nutrition* [formerly *Qualitas Plantarum*] 62 (4): 165–175.

Rea, A.I.; Schmidt, J.M.; Setzer, W.N.; Sibanda, S.; Taylor, C.; and Gwebu, E.T. 2003. Cytotoxic activity of *Ozoroa insignis* from Zimbabwe. *Fitoterapia* 74 (7–8): 732–735.

Reynolds, T. 2004. *Aloes: the genus* Aloe. CRC Press, Boca Raton, FL.

Rijo, P.; Gaspar-Marques, C.; Simões, M.F.; Jimeno, M.L.; and Rodríguez, B. 2007. Further diterpenoids from *Plectranthus ornatus* and *P. grandidentatus*. *Biochemical Systematics and Ecology* 35 (4): 215–221.

Roberts, M. 1990. *Indigenous healing plants*. Southern Book, Halfway House, 56–57.

Saxena, V. K.; and Chourasia, S. 2001. A new isoflavone from the roots of *Asparagus racemosus*. *Fitoterapia* 72 (3): 307–309.

Shale, T.L.; Stirk, W.A.; and Van Staden, J. 1999. Screening of medicinal plants used in Lesotho for antibacterial and anti-inflammatory activity. *Journal of Ethnopharmacology* 67 (3): 347–354.

Shao, Y.; Poobrasert, O.; Kennelly, E.J.; Chin, C.K.; Ho, C.T.; Huang, M.T.; Garrison, S.A.; and Cordell, G.A. 1997. Steroidal saponins from *Asparagus officinalis* and their cytotoxic activity. *Planta Medica* 63 (3): 258–262.

Sharma, S.C.; Sati, O.P.; and Chand, R. 1982. Steroidal saponins of *Asparagus curillus*. *Phytochemistry* 21(7): 1711–1714.

Silva, T.M.; Batista, M.M.; Camara, C.A.; and Agra, M.F. 2005. Molluscicidal activity of some Brazilian *Solanum* spp. (Solanaceae) against *Biomphalaria glabrata*. *Annals of Tropical Medicine and Parasitology* 99 (4): 419–425.

Singh, G.; Rawat, P.; and Maurya, R. 2007. Constituents of *Cissus quadrangularis*. *Natural Products Research* 21 (6): 522–528.

Smith, C.A. 1966. *Common names of South African plants. Memoirs of the Botanical Survey of South Africa 35*. Department of Agricultural Technical Services, Pretoria, South Africa.

Sori, T.; Bekana, M.; Adugna, G.; and Kelbessa, E. 2004. Medicinal plants in the ethnoveterinary practices of Borana pastoralists, southern Ethiopia. *International Journal of Applied Research in Veterinary Medicine* 2 (3): 220–225.

Sparg, S.G.; Light, M.E.; and van Staden, J. 2004. Biological activities and distribution of plant saponins. *Journal of Ethnopharmacology* 94 (2–3): 219–243.

Sparg, S. G.; van Staden, J.; and Jäger, A. K. 2002. Pharmacological and phytochemical screening of two Hyacinthaceae species: *Scilla natalensis* and *Ledebouria ovatifolia*. *Journal of Ethnopharmacology* 80 (1): 95–101.

Tabuti, J.R.S.; Dhillion, S.S.; and Lye, K.A. 2003. Ethnoveterinary medicines for cattle (*Bos indicus*) in Bulamogi country, Uganda: plant species and mode of use. *Journal of Ethnopharmacology* 88 (2–3): 279–286.

Trease, G.E.; and Evans, W.C. 1983. *Pharmacognosy*, 12th ed. Bailliere Tindall, London.

Uzabakiliho, B.; Largeau, C.; and Casadevall, E. 1987. Latex constituents of *Euphorbia candelabrum*, *E. Grantii*, *E. tirucalli* and *Synadenium grantii*. *Phytochemistry* 26: 3041–3045.

Van der Merwe, D.; Swan, G.E.; and Botha, C.J. 2001. Use of ethnoveterinary medicinal plants in cattle by Setswana-speaking people in the Madikwe area of the North West Province of South Africa. *Journal of the South African Veterinary Association* 72: 189–196.

Van Wyk, B.E.; and Gericke, N. 2000. *People's plants*. Briza, Pretoria, South Africa.

Van Wyk, B.E.; Van Oudtshoorn, B.; and Gericke, N. 1997. *Medicinal plants of South Africa.* Briza, Pretoria, South Africa.

Villegas, M.; Vargas, D.; Msonthi, J.D.; Marston, A.; and Hostettman, K. 1988. Isolation of the antifungal compounds falcarindiol and sarisan from *Heteromorpha trifoliata. Planta Medica* 54: 36–39.

Wagner, H.; and Baldt, S. 1975. Coumarins from South African *Pelargonium* spp. *Phytochemistry* 14: 2061–2064.

Waihenya, R.K.; Mtambo, M.A.; Nkwengulila, G.; and Minga, U.M. 2002. Efficacy of crude extract of *Aloe secundiflora* against *Salmonella gallinarum* in experimentally infected free-range chickens in Tanzania. *Journal of Ethnopharmacology* 79 (3): 317–323.

Wanyonyi, A.W.; Chhabra, C.S.; Mkoji, G.; Eilert, U.; and Njue, W.M. 2002. Bioactive steroidal alkaloid glycosides from *Solanum aculeastrum. Phytochemistry* 59 (1): 79–84.

Wanzala, W.; Zessin, K.H.; Kyule, N.M.; Baumann, M.P.O.; Mathias, E.; and Hassanali, A. 2005. Ethnoveterinary medicine: a critical review of its evolution, perception, understanding and the way forward. *Livestock Research and Development for Rural Development.* vol 17, article #119. Retrieved July 9, 2009 from http://www.lrrd.org/lrrd17/11/wanz17119.htm

Watt, J.M.; and Breyer-Brandwijk, M.G. 1962. *The medicinal and poisonous plants of southern and eastern Africa*, 2nd ed. Livingstone, London.

Wellsow, J.; Grayer, R.J.; Veitch, N.C.; Kokubun, T.; Lelli, R.; Kite, G.C.; and Simmonds, M.S. 2006. Insect-antifeedant and antibacterial activity of diterpenoids from species of *Plectranthus. Phytochemistry* 67 (16): 1818–1825.

Wojciechowski, M.F.; Mahn, J.; and Jones, B. 2006. Fabaceae. The Tree of Life Web Project. http://tolweb.org/Fabaceae/21093/2006.06.14.

Yang, C.X.; Huang, S.S.; Yang, X.P.; and Jia, Z.J. 2004. Nor-lignans and steroidal saponins from *Asparagus gobicus. Planta Medica* 70 (5): 446–451.

Yineger, H.; Kelbessa, E.; Bekele, T.; and Lulekal, E. 2007. Ethnoveterinary medicinal plants at Bale Mountains National Park, Ethiopia. *Journal of Ethnopharmacology* 112 (1): 55–70.

Zdero, C.; and Bohlmann, F. 1987. Sesquiterpene lactones from the genus *Brachylaena. Phytochemistry* 26: 2597–2601.

Zelnik, R.; Rabenhorst, E.; Matida, A.K.; Gottlieb, H.E.; Lavie, D.; and Panizza, S. 1978. Ibozol, a new diterpenoid from *Iboza riparia Phytochemistry* 17 (10): 1795–1797.

12 Ethnoveterinary Plants Used in East Africa

John B. Githiori and Peter K. Gathumbi

CONTENTS

12.1 INTRODUCTION

Ethnoveterinary medicine has always been a part of traditional medical knowledge in many parts of Africa. East Africa covers Kenya, Uganda, Tanzania, Eritrea, Ethiopia, Somalia, and Djibouti. In this region, livestock farming is a central part of the farming activities. For instance, in Kenya, about 69% of the area can only be used for livestock farming rather than crop agriculture. The farming systems involve mainly pastoralisms but also include small holder farmers who keep a few animals. Due to limitation in resources, many of the farmers in these areas rely more or less on traditional medicine for treatment of their livestock.

For East Africa, several publications are available on anecdotal information on usage of medicinal plants to combat diseases of livestock.[9,49,53–55,95] In addition, several monographs and studies in selected communities have been carried out to document the use of ethnoveterinary medicine.[14,33,56,58–60,67,81,92,93] The publications in most cases highlight different conditions affecting livestock as well as symptoms associated with several disease conditions. The conditions that are covered range from abscesses to ectoparasites to treatment of conditions caused by protozoan parasites such as theileriosis (East Coast fever, ECF) and trypanosomosis. In some instances, symptoms such as diarrhea are treated but not the underlying condition that may cause the diarrhea. A number of databases and sources such as PRELUDE, *ScienceDirect*, and books were searched for information on ethnoveterinary plants used for different conditions in East Africa.

From the databases, 193 plants were documented. A total of 161 plants were recorded in Kenya, while 31 were from Uganda. This, however, does not mean that there was no information available from the other countries in East Africa. Nonetheless, we concentrated on these two countries where most of the information was readily available, and it is likely that the ethnoveterinary practices documented are similar throughout the subcontinent because of many shared traditions and customs.

Various conditions were noted with different treatment being offered. A minimum of 26 conditions have been reportedly treated with ethnoveterinary preparations. These conditions ranged from abscesses, wounds, and burns to specific treatments like that for brucellosis. The highest number of different plant treatments available for a single condition was 19 in Kenya for treatment of wounds, while only one different plant species was available for each of the following conditions: heartwater, bloat, ephemeral fever, anaplasmosis, mastitis, coccidiosis in chicken, burns, and brucellosis. Theileriosis and endoparasites had 12 and 13 different plant species used for their treatment, respectively.

The different treatments used a total of 64 families of plants. These treatments are addressed as endo- and ectoparasitic conditions, hemoparasitic conditions, and other diseases.

12.2 PARASITIC DISEASES

Parasitic diseases have been described to have the highest effect on productivity of livestock owned by the poor worldwide. These diseases are the major causes of low productivity of livestock in tropical and subtropical regions, including East Africa. Various plants are used to treat different hemoparasitic diseases in East Africa; some are discussed in depth next.

12.2.1 Babesiosis

Babesiosis is caused by intraerythrocytic protozoan parasites of the genus *Babesia*. The disease, which is transmitted by ticks, affects a wide range of domestic and wild animals and occasionally humans. While the major economic impact of babesiosis is on the cattle industry, infections in other domestic animals, including horses, sheep, goats, pigs, and dogs, assume varying degrees of importance throughout the world. Two important species in cattle, *B. bigemina* and *B. bovis*, are widespread in tropical and subtropical areas worldwide. The main vectors of *B. bigemina* and *B. bovis* are one-host *Boophilus* spp. ticks, in which transmission occurs transovarially. The acute disease generally runs a course of about a week. Clinical signs include fever, inappetence, increased respiratory rate, muscle tremors, anemia, jaundice, weight loss, hemoglobinemia, and hemoglobinuria. Late-term pregnant cows may abort, and bulls may undergo temporary infertility due to transient fever.

Three plant species are used for treatment of babesiosis in cattle: the stem bark of *Acacia exocephaleia*, Mimosaceae, roots of *Erythrina abyssinica* Lam. Fabaceae,[1] and the bark of *Prunus africana* [Hook. f.) Kalkman, Rosaceae. They have been used for treatment of babesiosis in cattle,[9] and the last species has also been used as a laxative in livestock.[2]

12.2.1.1 *Erythrina abyssinica*

Erythrina abyssinica is a tree, rarely a shrub, 3–10 m tall, with a rounded crown, and it is found throughout East Africa. This plant has been found to contain mainly flavonoids.[18,19,97,98] The species was found to have antiprotozoal action and showed *in vitro* activity against plasmodium as well as against trypanosomosis. This antiprotozoal activity could explain its use in traditional treatment against babesiosis.

12.2.1.2 *Prunus africana*

Prunus (or *Pygeum*) *africana* is an endangered medicinal plant endemic to Africa. It is mainly found on higher-altitude areas, generally from 1,000 m above sea level. The African cherry [*Prunus africana* (Hook. f.) Kalm.] has been used in the treatment of benign prostatic hyperplasia and other disorders in men. The bark, from which the treatment is derived, is entirely wild collected. The major exporters of bark include Cameroon, Madagascar, Equatorial Guinea, and Kenya.[84] The active constituents of *Pygeum* extract that is derived from this tree include phytosterols (e.g., beta-sitosterol), which exert anti-inflammatory effects by inhibiting production of proinflammatory prostaglandins in the prostate. *Pygeum* also contains pentacyclic triterpenes (ursolic and oleanolic acids)[26] that have antiedema properties and ferulic acid esters (n-docosanol and tetracosanol) that reduce prolactin levels and block the accumulation of cholesterol in the prostate.[8,40,41,63,80] In animals, chloroform and aqueous extracts were observed to manifest hepatotoxicity at dosages of above 3 g/kg body weight.[30,31]

12.2.2 Trypanosomosis

Trypanosomosis is caused by protozoa of the genus *Trypanosoma* and affects all domestic animals. The major species are *T. congolense, T. vivax, T. brucei brucei*, and *T. simiae*. Cattle, sheep, and goats are infected, in order of importance, by *T. congolense, T. vivax*, and *T. brucei brucei*. In pigs, *T. simiae* is the most important. In dogs and cats, *T. brucei* is probably the most important. *Trypanosoma vivax* may occur outside tsetse-infested areas of sub-Saharan Africa. The trypanosomes that cause tsetse-transmitted trypanosomiasis (sleeping sickness) in humans, *T. brucei rhodesiense* and *T. brucei gambiense*, closely resemble *T. brucei brucei* from animals, and domestic animals act as reservoirs of human infections. Most tsetse transmission is cyclic and begins when blood from a trypanosome-infected animal is ingested by the fly. Tsetse flies (genus *Glossina*) are restricted to Africa from about latitude 15°N to 29°S. Mechanical transmission may also occur through tsetse or other biting flies. In the case of *T. vivax*, *Tabanus* spp. and other biting flies seem to be the primary mechanical vectors outside the tsetse areas, as in Central and South America. Mechanical transmission requires only that blood containing infectious trypanosomes be transferred from one animal to another. The clinical signs include intermittent fever, anemia, and weight loss, and in cattle the disease may have a chronic course with high mortality.

Several plant species have been used to treat trypanosomosis in East Africa. These include *Acacia reficiens, Adenia volkensii, Fagara chalybea, Salvadora persica,* and *Terminalia brownie,* which are used in Kenya. In Uganda, *in vitro* studies have been carried out to evaluate the antitrypanosomal activity of *Albizia gummifera, Ehretia amoena, Entada abyssinica, Securinega virosa,* and *Vernonia subuligera.*[29] In Tanzania, of 37 extracts derived from 15 plant species tested for antitrypanosomal activity, the chloroform extract of the root bark of *Asteranthe asterias* and the ethanol extract of the root bark of *Annickia kummeriae* showed the highest activity, with IC_{50} values of 0.8 and 0.7 μg/mL, respectively. Ten extracts exhibited antitrypanosomal activity, with IC_{50} values between 1 and 5 μg/mL. Among them are the extracts of *Asteranthe asterias*, *Commiphora emenii, Diospyros verrucosa, Enantia kummeriae, Hymenocardia ulmoides,* and *Zanthoxylum chalybeum.* Six extracts showed IC_{50} values between 5.1 and 20 μg/mL. All other extracts had less or no activity.[27,28]

Some of these species are discussed briefly next.

12.2.2.1 *Adenia volkensii* Harms, Passifloraceae (Kiliambiti)

Adenia volkensii Harms, Passifloraceae (Kiliambiti) is a woody shrub or herb that grows up to 1.5 m and arises from a tuberous rootstock or a succulent stem and is without tendrils. Roots are used in Kenya to treat trypanosomosis, pneumonia, and bronchitis in livestock.[9] This plant is known for the extreme toxicity of the sap. It has cyanogenic glycosides and lectins (ribosome-inactivating proteins) that destroy ribosomes and hence prevent protein synthesis.[12,16,17,68,85]

12.2.2.2 *Fagara chalybea* [Engl.] Engl. Rutaceae (= *Zanthoxylum chalybeum*) Engl.

Fagara chalybea [Engl.] Engl. Rutaceae (= *Zanthoxylum chalybeum*) Engl. is a deciduous shrub or small tree up to 6 m tall found in eastern and southern Africa. Apart from the use of stem bark, fruit, and seeds in the treatment of trypanosomosis, it has ethnoveterinary use as an antifebrile agent and for other livestock conditions, such as conjunctivitis, diarrhea, pneumonia, respiratory disease, and lymph node disease, which is the usual layperson's term used to describe theileriosis.[9,15,42,71] Alkaloids isolated from this plant were found to have mitochondrial inhibitory effects.[24,25,45,89] The quinolone alkaloids isolated from this species have been shown to have biocidal activity and are described as general growth inhibitors; that is, they can act as antifeedants, phytotoxics, and antifungals.[89] The quaternary benzophenanthridine alkaloids (e.g., sanguinarine, chelerythrine, and nitidine) have potent antiplasmodial and antitopomerase activities.

12.2.2.3 *Salvadora persica* L., Salvadoraceae

The toothbrush tree *Salvadora persica* (synonym *Galenia asiatica*, *Salvadora indica*) is a small tree or shrub with a crooked trunk. Its roots and bark are used to treat livestock of trypanosomosis and abscesses, mange, and retained afterbirth. It is also used for the treatment of anthrax.[67]

Various constituents isolated from different parts of *Salvadora* species have been described. These include alkaloids, steroids, terpenoids, flavonoids, and lignins.[44] Some useful activities reported from various parts of the plant and from the isolated chemicals are antiulcer, hypolipidemic, hypoglycemic, and anticonvulsant activities.[96] Due to the presence of benzyl isothiocyanate (BIT), it is widely used for oral hygiene (so-called chewing sticks) in different parts of the world.[7,20,61,76,82,96] A phytochemical investigation of stems from *Salvadora persica* demonstrated the novel presence of natural benzylamides.[48] The addition of *Salvadora persica* to the growing and mature BB male rabbit diets was demonstrated to improve growth and reproductive capabilities by raising plasma testosterone levels.[22] In Kenya, the leaves of the plant are a preferred camel diet.[47]

12.2.3 East Coast Fever (Theileriosis)

East Coast fever is an acute disease of cattle usually characterized by high fever, swelling of the lymph nodes, dyspnea, and high mortality. It is caused by *Theileria parva* and is a serious problem in east and central Africa. The disease is transmitted by the brown ear tick *Rhipicephalus appendiculatus* during feeding. Clinical signs vary according to the level of challenge and range from inapparent or mild to severe and fatal. Typically, fever occurs 7–10 days after parasites are introduced by feeding ticks, continues throughout the course of infection, and may be as high as 42°C. Lymph node swelling becomes pronounced and generalized. Anorexia develops, and the animal's condition deteriorates rapidly; lacrimation and nasal discharge may occur. Terminally, dyspnea is common. Just before death, a sharp fall in body temperature is usual, and pulmonary exudate pours from the nostrils. Death usually occurs 18–24 days after infection. The plant species discussed next have been cited for use in ECF in East Africa.

12.2.3.1 *Adansonia digitata* L., Bombacaceae

Leaves and fruits are used to treat ECF in Kenya.[1] The main stem of *Adansonia digitata* (baobab tree) may reach enormous proportions, up to 28 m in girth, although baobab trees seldom exceed a height of 25 m. The species is also used for treatment of trypanosomosis in West Africa.[10,11] It has been identified as possessing gastroprotective properties, probably due to astringent flavonoids and antioxidant properties, in rats and mice that have been artificially induced to develop ulcers.[46] Phytochemical screening of the fruit pulp indicated the presence of sterols or triterpenes, saponins, tannins, carbohydrates, and glycosides.[74]

12.2.3.2 *Aerva javanica* [Burm. f.] Juss. ex J.A. Schultes, Amaranthaceae

Flowers of *Aerva javanica* [Burm. f.] Juss. ex J.A. Schultes, Amaranthaceae are used in Kenya to treat theileriosis in cattle.[32] This is an erect, multi-branched perennial herb, 0.4–1.6 m high, that grows on sandy soils along drainage lines. It was found to stimulate mice in a hypnotic study.[6] Carbohydrates or glycosides, tannins, saponins, alkaloids, unsaturated sterols, triterpenes, and flavonoids have been demonstrated to be present in this species. Aqueous extracts of the species exhibited dose-dependent smooth muscle relaxant effects and significant antispasmodic activity.[94]

12.2.3.3 Euphorbiaceae Species

Several Euphorbiaceae species have been cited for use in treating theileriosis in East Africa. These include the bark of *Croton megalocarpus*, the whole plant of *Euphorbia triaculeata*, and a decoction of the leaves and bark of *Synadenium compactum*.[1,2,13,32,67–69]

Euphorbiaceae consists of about 322 genera and 8,900 species around the world in both arid and humid tropics, and the plants are found as herbs, shrubs, stunted succulents, and tall canopy trees.[100] The Euphorbiaceae have been shown to possess flavonoids, saponins, diterpenes, phorbol esters,[101] lectins,[102,103] and triterpenoids.[104]

Clerodane-type diterpenes have been isolated from *Croton megalocarpus*.[4] Euphorbiaceae are well known for skin-irritating and tumor-promoting diterpenoids (e.g., daphnane diterpenoids).[23] However, numerous macrocyclic diterpenes (e.g., jatrophane, ingol, and myrsinane) from this taxon have also shown good biological activity (i.e., anticancer, analgesic and antifeedant, among others).[66,91] In addition to the alkaloids, the Euphorbiaceae also contain phenolics, triterpenoids, and steroids.[43] *Synadenium* spp. have shown immunoregulatory,[75] fibrinolytic,[73] and antitumoral activity.[72]

Euphorbia candelabrum Trémaux ex Kotschy, Euphorbiaceae, latex and stem are also used to treat lymph node enlargement as well as wounds, abscesses, and conjunctivitis in livestock.[38] In Uganda, it is used to treat ECF.[21,65]

12.2.3.4 Other Species

Leaves and juice extracted from leaves of *Agave americana* L., Agavaceae, are also used in treatment of theileriosis wounds and act as a coagulant and an insect repellant.[1] Steroidal sapogenins with anti-inflammatory activity[70] and *in vitro* anti-leukemic activity[99] have been characterized from this species. Antibacterial and molluscidal constituents have also been isolated and characterized.[69,79,86] Agave is a source of fiber and produces steroidal sapogenins and saponins.[105]

Cissus quadrangularis, Vitaceae L., seeds, stem, and roots are used to treat theileriosis as well as diarrhea, external parasitism, foot rot, and pneumonia.[9,32,39,49,62,65,67,90] *Cissus quadrangularis* is a source of stilbenes, including resveratrol, flavonoids, and triterpenoids.[5,13,35,57] Extracts of the plant have shown a bone fracture healing property and antiosteoporotic effect[78] as well as antibacterial and antioxidant activities.[64]

Clerodendrum myricoides (Verbenaceae) root is also used in Kenya to treat theileriosis, diarrhea, and fever in livestock.[95] Spermidine alkaloids have been isolated from this species.[106] It is traditionally used to treat malaria in Kenya, and methanol extracts have shown significant suppression of plasmodial parasitemia.[107]

The leaves of the cucurbit, *Gerrardanthus lobatus* and the root of *Iboza multiflora* are used to treat theileriosis in cattle in Kenya,[9] while the leaves of the latter are used for ectoparasites.[49,67]

Plectranthus barbatus leaves are also used to treat ECF fever in Kenya.[67] Monoterpenoids, sesquiterpenoids, diterpenoids, and phenolics have been reported in species of *Plectranthus*.[107,108] *Plectranthus barbatus* is used in French Guyana for malaria.[109] *Albizia coriaria* and *A. zygia*, both Mimosaceae, root infusions are used in Uganda to treat ECF.[87,21] The related species *A. gummifera* is used traditionally to treat malaria and has spermine alkaloids that exhibited moderate activity against the malaria parasite *in vitro*.[110]

Other ethnoveterinary medicines reported for ECF in East Africa are leaf infusions of *Ananas comosus* (Bromeliaceae), *Aristolochia elegans* (Aristolochiaceae), *Asparagus racemosus* (Asparagaceae), *Boerhavia diffusa* (Nyctaginaceae), and *Clerodendrum myricoides* (Verbenaceae) and root infusions of *Harrisonia abyssinica* (Simaroubaceae), *Maytenus senegalensis* (Celastraceae), and *Milicia excelsa* (Moraceae).[88]

There appears to be a striking correlation between plants that are used to treat malaria in humans and the use of the plants in ECF. This may be a useful bioprospecting angle indicating that plant metabolites may have generalized antihemoparastic activity.

12.2.4 Heartwater

Heartwater is an infectious, noncontagious, rickettsial disease of ruminants in areas infested by ticks of the genus *Amblyomma*. These include regions of Africa south of the Sahara and the islands of the Comores, Zanzibar, Madagascar, São Tomé, Réunion, and Mauritius. Heartwater and its vector are also endemic on the islands of Guadeloupe and Antigua. Many ruminants, including some antelope species, are susceptible. Some animals may become subclinically infected and act as reservoirs. Indigenous African cattle breeds (*Bos indicus*) appear more resistant than *B. taurus* breeds. The causative organism is an obligate intracellular parasite, previously known as *Ehrlichia ruminantium*. Clinical signs are dramatic in the peracute and acute forms. In peracute cases, animals develop fever, followed rapidly by hyperesthesia, lacrimation, and convulsions. In the acute form, animals show anorexia and nervous signs such as depression, a high-stepping stiff gait, exaggerated blinking of eyes, and chewing movements. Both forms terminate in prostration and convulsions. Diarrhea is seen occasionally. In subacute cases, the signs are less marked, and central nervous system involvement is inconsistent.

Abrus precatorius (jequirity bean) is used in combination with *Carissa edulis* to treat heartwater.[9] Several toxic lectins have been isolated from the seeds of *A. precatorius*,[37] and isoflavanquinones isolated from the roots were found to have potent anti-inflammatory, antiplatelet, and antiallergic actions.[51]

Leaves and roots of *Carissa edulis* are used for heartwater, as a laxative and purgative, and for treatment of internal parasitism, colitis, gastritis, and peritonitis.[9] Lignans and sesquiterpenes have been isolated from this species.[2,3]

12.2.5 Anaplasmosis

Anaplasmosis is a vector-borne, infectious, hemolytic, rickettsial disease of cattle, sheep, goats, and other wild ruminants. In cattle, the most common etiological agent is *Anaplasma marginale*; cattle also are affected with *A. centrale,* generally resulting in mild disease. Anemia results from extravascular hemolysis when parasitized red blood cell (RBC) membranes are altered and recognized by the reticuloendothelial system. These RBCs are removed and finally destroyed. Accordingly, antibodies that develop against the altered cell membrane can cause destruction of uninfected erythrocytes. The severity of anaplasmosis depends on the species involved and age of the animal. Young calves seem to have an innate resistance to the disease, while the acute form generally occurs in cattle from 1 to 3 years. In cattle over 3 years, the peracute or most severe form, with rapid onset and death, predominates. Animals that survive anaplasmosis can become carriers for life and act as a reservoir of infection for susceptible animals. Economic losses from anaplasmosis include abortions, death, weight gain and loss, decreased milk production, bull infertility, and treatment expenses.

Roots, leaves, and bark of *Ficus sycomorus* L., Moraceae, are used to treat anaplasmosis and other conditions, including colitis, gastritis, and peritonitis and flushing of uterus after abortion in animals.[9,38] *Ficus sycomorus* is native to Africa south of the Sahel and north of the Tropic of Capricorn, also excluding the central-west rain forest areas. In Burkina Faso, it has been cited for use in malaria, but *in vitro* studies have not shown good activity against *Plasmodium falciparum*.[77]

Phytochemical analysis carried out on *F. sycomorus* demonstrated the presence of steroids, condensed tannins, flavone aglycones, and saponins.[37,38] Extracts of *F. sycomorus* have not shown good antibacterial or antifungal activity.[50,83] However, the *Ficus* genus is a popular herbal remedy that has been used to treat tumors (internal and external) as well as mastitis, bronchitis, tuberculosis, and diarrhea.[52] Phenanthroindolizidine alkaloids, coumarins, stilbenes, flavonoids, and triterpenods have been isolated from this taxon.[34,52]

12.3 ECTOPARASITES

A total of 25 plants have been documented to have activity against ectoparasites. Leaves are commonly used as insect repellant and for external parasites. Of the 25 species reported to be used for treatment of ectoparasites, 12 utilized leaves. These included *Acalypha fruticosa* (Euphorbiaceae), which is also used to treat wounds, abscesses, foot rot, burns, and pox as well,[1] This species was also found to attract ticks due to its odor.[36] It is also an important folk medicine in Saudi Arabia and India.[1] *Adenia veneta* Forssk. Passifloraceae leaves are used for treatment of external parasites and mange and as an antifungal.[65] Leaves of *Adenia multiflorum* Klotzsch Apocynaceae have been used as an insecticide and acaricide.[1,62] *Adenia* spp. possess entomotoxic lectins, which may explain their activity.[111]

12.4 CONCLUSION

Many plants have been identified for use in treatment of parasitic diseases in East Africa. For many of them, phytochemical analysis is yet to be done. However, there is a sufficient body of literature for many plants whose phytochemistry has been done. Most of the plants identified were for treatment of diseases in ruminants; in a few exceptions, plants were used to treat conditions in camels and poultry. Leaf and bark were the most commonly used plant parts. In some instances, whole plants were used, and a mixture of two or more plants has also been widely reported.

REFERENCES

1. Abulafatih HA: Medicinal plants in southwestern Saudi Arabia. *Economic Botany* 41: 354–360, 1987.
2. Achenbach H, Waibel R, Addae-Mensah I: Lignans and other constituents from *Carissa edulis*. *Phytochemistry* 22: 749–753, 1983.
3. Achenbach H, Waibel R, Addae-Mensah I: Sesquiterpenes from *Carissa edulis*. *Phytochemistry* 24: 2325–2328, 1985.
4. Addae-Mensah I, Achenbach H, Thoithi GN, Waibel R, Mwangi JW: Epoxychromodine and other constituents of *Croton megalocarpus*. *Phytochemistry* 31: 2055–2058, 1992.
5. Adesanya SA, Nia R, Martin M-T, Boukamcha N, Montagnac A, Païs M: Stilbene derivatives from *Cissus quadrangularis*. *Journal of Natural Products* 62: 1694–1695, 1999.
6. Alasbahi RHAS: Hypnotic activity of *Aerva javanica*. *University of Aden Journal of Natural and Applied Sciences* 8: 399–403, 2004.
7. Almas K, Al-Zeid Z: The immediate antimicrobial effect of a toothbrush and miswak on cariogenic bacteria: a clinical study. *Journal of Contemporary Dental Practice* 5: 105–114, 2004.
8. Andro MC, Riffaud JP: Pygeum africanum extract for the treatment of patients with benign prostatic hyperplasia: a review of 25 years of published experience. *Current Therapeutic Research* 56: 796–817, 1995.
9. Anonymous: *Ethnoveterinary medicine in Kenya: a field manual of traditional animal health practices*. Intermediate Technology Development Group and International Institute of Rural Reconstruction, Nairobi, Kenya, 1996.
10. Atawodi SE, Ameh DA, Ibrahim S, Andrew JN, Nzelibe HC, Onyike EO, Anigo KM, Abu EA, James DB, Njoku GC, Sallau AB: Indigenous knowledge system for treatment of trypanosomiasis in Kaduna State of Nigeria. *Journal of Ethnopharmacology* 79: 279–282, 2002.
11. Atawodi SE, Bulus T, Ibrahim S, Ameh DA, Nok AJ, Mamman M, Galadima M: In vitro trypanocidal effect of methanolic extract of some Nigerian savannah plants. *African Journal of Biotechnology* 2: 317–321, 2003.
12. Barbieri L, Falasca AI, Stirpe F: Volkensin, the toxin of *Adenia volkensii* (kilyambiti plant). *FEBS Letters* 171: 277–279, 1984.
13. Bhutani KK, Kapoor R, Atal CK: Two unsymmetric tetracyclic triterpenoids from *Cissus quadrangularis*. *Phytochemistry* 23: 407–410, 1984.
14. Bizimana N: Traditional veterinary practice in Africa. *Deutsche Gesellschaft für Technische Zusammenarbeit (GTZ)*, Rossdorf, Germany, 1994.
15. Bollig M: East Pokot camel husbandry. Institut für Völkerkunde, Köln University, Federal Republic of Germany, Cologne, 1992, 19 p.
16. Chambery A, Di Maro A, Monti MM, Stirpe F, Parente A: Volkensin from *Adenia volkensii* Harms (kilyambiti plant), a type 2 ribosome-inactivating protein. *European Journal of Biochemistry* 271: 108–117, 2004.

17. Chambery A, Severino V, Stirpe F, Parente A: Cloning and expression of the B chain of volkensin, type 2 ribosome inactivating protein from *Adenia volkensii* Harms: co-folding with the A chain for heterodimer reconstitution. *Protein Expression and Purification* 51: 209–215, 2007.
18. Cui L, Ndinteh DT, Na M, Thuong PT, Silike-Muruumu J, Njamen D, Mbafor JT, Fomum ZT, Ahn JS, Oh WK: Isoprenylated flavonoids from the stem bark of *Erythrina abyssinica*. *Journal of Natural Products* 70: 1039–1042, 2007.
19. Cui L, Thuong PT, Lee HS, Njamen D, Mbafor JT, Fomum ZT, Lee J, Kim YH, Oh WK: Four new chalcones from *Erythrina abyssinica*. *Planta Med* 74: 422–426, 2008.
20. Darmani H, Al-Hiyasat AS, Elbetieha AM, Alkofahi A: The effect of an extract of *Salvadora persica* (Meswak, chewing stick) on fertility of male and female mice. *Phytomedicine* 10: 63–65, 2003.
21. Ejobi F, Olila D, Arianga S, Eyudu P, Oluanah C, Komakech FA, Okubal P: Validation of selected ethno-veterinary medical practices in the Teso farming systems. In: *Final technical report of NARO/DIFID COARD*. National Agricultural Research Organisation, Uganda, Kampala, Uganda, 2004.
22. El-Kholy KH, Seleem TST, El-Aasar TA, Abdelharith AH: Effect of dietary addition of Arak (*Salvadora persica*) on growth and reproductive performance in Black Baladi rabbit males. *World Rabbit Science* 16: 21–27, 2008.
23. Evans FJ, Taylor SE: Pro-inflammatory, tumor promoting and antitumor diterpene of the plant families Euphorbiaceae and Thymelaeaceae. In: *Progress in the chemistry of organic natural products*, eds. Herz W, Grisebach H, Kirby GW, pp. 1–99. Springer-Verlag, New York, 1983.
24. Fish F, Waterman PG: Chloroform-soluble alkaloids from the root bark of *Fagara chalybea*. *Phytochemistry* 11: 1866–1867, 1972.
25. Fish F, Waterman PG: Methanol-soluble quaternary alkaloids from African *Fagara* species. (Chemosystematics in the Rutaceae. Part I.) *Phytochemistry* 11: 3007–3014, 1972.
26. Fourneau C, Hocquemiller R, Cavé A: Triterpenes from *Prunus africana* bark. *Phytochemistry* 42: 1387–1389, 1996.
27. Freiburghaus F, Jonker SA, Nkunya MH, Mwasumbi LB, Brun R: In vitro trypanocidal activity of some rare Tanzanian medicinal plants. *Acta Tropica* 67: 181–185, 1997.
28. Freiburghaus F, Jonker SA, Nkunya MH, Mwasumbi LB, Brun R: In vitro trypanocidal activity of some rare Tanzanian medicinal plants. *Acta Tropica* 66: 79–83, 1997.
29. Freiburghaus F, Ogwal EN, Nkunya MHH, Kaminsky R, Brun R: In vitro antitrypanosomal activity of African plants used in traditional medicine in Uganda to treat sleeping sickness. *Tropical Medicine and International Health* 1: 765–771, 1996.
30. Gathumbi PK, Mwangi JW, Mugera GM, Njiro SM: Toxicity of chloroform extract of *Prunus africana* stem bark in rats: gross and histological lesions. *Phytotherapy Research* 16: 244–247, 2002.
31. Gathumbi PK, Mwangi JW, Njiro SM, Mugera GM: Biochemical and haematological changes in rats administered an aqueous extract of *Prunus africana* stem-bark at various dosage levels. *The Onderstepoort Journal of Veterinary Research* 67: 123–128, 2000.
32. Glover PE, Stewart J, Gwynne MD: Masai and Kipsigis notes on East African plants, Part III—medicinal uses of plants. *East African Agricultural and Forestry* Journal 32: 200–207, 1966.
33. Grade JT, Longok A. Karamajong scientists: participatory field trial of a local dewormer. Uganda Veterinary Association Scientific Conference: The Veterinary Profession and Poverty Alleviation, 2000.
34. Greenham JR, Grayer RJ, Harborne JB, Reynolds V: Intra- and interspecific variations in vacuolar flavonoids among *Ficus* species from the Budongo Forest, Uganda. *Biochemical Systematics and Ecology* 35: 81–90, 2007.

35. Gupta MM, Verma RK: Unsymmetric tetracyclic triterpenoids from *Cissus quadrangularis*. *Phytochemistry* 29: 336–337, 1990.
36. Hassan SM, Dipeolu OO, Malonza MM: Natural attraction of livestock ticks by the leaves of a shrub. *Tropical Animal Health and Production* 26: 87–91, 1994.
37. Hegde R, Maiti TK, Podder SK: Purification and characterization of three toxins and two agglutinins from *Abrus precatorius* seed by using lactamyl-sepharose affinity chromatography. *Analytical Biochemistry* 194: 101–109, 1991.
38. Heine B, Brenzinger M: *Plant of the Borana (Ethiopia and Kenya)*, Part 4. Verlag Breitenbach, Saarbrücken, Germany, 1988.
39. Ichkawa M: A preliminary report on the ethnobotany of the Suiei Dorobo in Northern Kenya. *African Study Monographs Supplemental Issue* 7: 1–52, 1987.
40. Isaacs JT: Importance of the natural history of benign prostatic hyperplasia in the evaluation of pharmacologic intervention. *Prostate* 3: 1–7, 1990.
41. Ishani A, MacDonald R, Nelson D, Rutks I, Wilt TJ: *Pygeum africanum* for the treatment of patients with benign prostatic hyperplasia: a systematic review and quantitative meta-analysis. *American Journal of Medicine* 109: 654–664, 2000.
42. Itaru O: Turkana's view of livestock diseases. *African Study Monographs Supplemental issue* 3: 71–93, 1984.
43. Jassbi AR: Chemistry and biological activity of secondary metabolites in *Euphorbia* from Iran. *Phytochemistry* 67: 1977–1984, 2006.
44. Kamel MS, Ohtani K, Assaf MH, Kasai R, El-Shanawani MA, Yamasaki K, Ali AA, Tanaka O: Lignan glycosides from stems of *Salvadora persica. Phytochemistry* 31: 2469–2471, 1992.
45. Kamikawa T, Hanaoka Y, Fujie S, Saito K, Yamagiwa Y, Fukuhara K, Kubo I: SRS-A antagonist pyranoquinolone alkaloids from East African Fagara plants and their synthesis. *Bioorganic and Medicinal Chemistry* 4: 1317–1320, 1996.
46. Karumi Y, Augustine AI, Umar IA: Gastroprotective effects of aqueous extract of *Adansonia digitata* leaf on ethanol-induced ulceration in rats. *Journal of Biological Sciences* 8: 225–228, 2008.
47. Kassilly FN, Abate AL: Composition and quality of camel diets in Central Baringo, Kenya. *Bulletin of Animal Health and Production in Africa* 48: 139–147, 2000.
48. Khalil AT: Benzylamides from *Salvadora persica. Archives of Pharmacal Research* 29: 952–956, 2006.
49. Kokwaro JO: *Medicinal plants of East Africa*, 2nd ed. Kenya Literature Bureau, Nairobi, 1993.
50. Kubmarawa D, Ajoku GA, Enwerem NM, Okorie. DA: Preliminary phytochemical and antimicrobial screening of 50 medicinal plants from Nigeria. *African Journal of Biotechnology* 6: 1690–1696, 2007.
51. Kuo S, Chen S, Chen L, Wu J, Wang J, Teng C: Potent antiplatelet, anti-inflammatory and antiallergic isoflavanquinones from the roots of *Abrus precatorius. Planta Medica* 61: 307–312, 1995.
52. Lansky EP, Paavilainen HM, Pawlus AD, Newman RA: *Ficus* spp. (fig): Ethnobotany and potential as anticancer and anti-inflammatory agents. *Journal of Ethnopharmaco*logy 119: 195–213, 2008.
53. Leakey LSB: *The southern Kikuyu before 1903*, vol. 1. Academic Press, London, 1977.
54. Leakey LSB: *The southern Kikuyu before 1903*, vol. 2. Academic Press, London, 1977.
55. Leakey LSB: *The southern Kikuyu before 1903*, vol. 3, Academic Press, London, 1977.
56. Madibela OR, Jansen K: The use of indigenous parasitic plant (*Viscum verrocosum*) in reducing faecal egg counts in female Tswana goats. *Livestock Research for Rural Development* 15 (9), 2003. Retrieved Feb. 10, 2010 from http://www.lrrd.org/lrrd15/9/madi159.htm
57. Mehta M, Kaur N, Bhutani KK: Determination of marker constituents from *Cissus quadiangularis* Linn. and their quantitation by HPTLC and HPLC. *Phytochemical Analysis* 12: 91–95, 2001.

58. Minja MM: Medicinal plants used in the promotion of animal health in Tanzania. *Review of Science and Technology* 13: 905–925, 1994.
59. Minja MMJ. Collection of Tanzanian medicinal plants for biological activity studies. *Proceedings of the Seventh Tanzania Veterinary Association Scientific Conference*, Arusha, Tanzania, 1989, pp. 67–78.
60. Minja MMJ. The Maasai wonder plants. Paper presented at the People and Plants Training Workshop held at the Tropical Pesticides Research Institute, Arusha, Tanzania, March 15–18, 1999.
61. Monforte MT, Trovato A, Rossitto A, Forestieri AM, D'Aquino A, Miceli N, Galati EM: Anticonvulsant and sedative effects of *Salvadora persica* L. stem extracts. *Phytotherapy Research* 16: 395–397, 2002.
62. Morgan WTW: Ethnobotany of the Turkana: Use of plants by a pastoral people and their livestock in Kenya. *Economic Botany* 35: 96–130, 1981.
63. Murray MT: *The healing power of herbs*. Prima, Rocklin, CA, 1995.
64. Murthy KNC, Vnitha A, Swamy M, Ravishankar GA: Antioxidant and antimicrobial activity of *Cissus quadrangularis*. *Journal of Medical Food* 6: 99–105, 2003.
65. Ohta I: Symptoms are classified into diagnostic categories. Turkana's view of livestock diseases. *African Study Monograph Supplementary issue* 3: 71–90, 1984.
66. Öksüz S, Gürek F, Gil RR, Pengsuparp T, Pezzuto JM, Cordell GA: Four diterpene esters from *Euphorbia myrsinites*. *Phytochemistry* 38: 1457–1462, 1995.
67. Ole-Miaron JO: The Maasai ethnodiagnostic skill of livestock diseases: a lead to traditional bioprospecting. *Journal of Ethnopharmacology* 84: 79–83, 2003.
68. Pangalos MN, Francis PT, Pearson RC, Middlemiss DN, Bowen DM: Destruction of a sub-population of cortical neurones by suicide transport of volkensin, a lectin from *Adenia volkensii*. *Journal of Neuroscientific Methods* 40: 17–29, 1991.
69. Parmar VS, Jha HN, Gupta AK, Prasad AK, Gupta S, Boll PM, Tyagi OD: New antibacterial tetratriacontanol derivatives from *Agave americana* L. *Tetrahedron* 48: 1281–1284, 1992.
70. Peana AT, Moretti MDL, Manconi V, Desole G, Pippia P: Anti-inflammatory activity of aqueous extracts and steroidal sapogenins of *Agave americana*. *Planta Medica* 63: 199–202, 1997.
71. Politz J, Lekeleley J: The knowledge of the Samburu on animal diseases and their traditional methods of treatment. 1988.
72. Premaratna A, Shadaksharaswamy M, Nanjappa S: Isolation, purification and properties of a lectin from the latex of *Synadenium grantii* Hook f. *Indian Journal of Biochemistry and Biophysics* 18: 32–35, 1981.
73. Rajesh R, Nataraju A, Gowda CDR, Frey BM, Frey FJ, Vishwanath BS: Purification and characterization of a 34-kDa, heat stable glycoprotein from *Synadenium grantii* latex: action on human fibrinogen and fibrin clot. *Biochimie* 88: 1313–1322, 2006.
74. Ramadan A, Harraz FM, El-Mougy SA: Anti-inflammatory, analgesic and antipyretic effects of the fruit pulp of *Adansonia digitata*. *Fitoterapia* 65: 418–422, 1994.
75. Rogerio AP, Cardoso CR, Fontanari C, Souza MA, Afonso-Cardoso SR, Silva ÉV, Koyama NS, Basei FL, Soares EG, Calixto JB, Stowell SR, Dias-Baruffi M, Faccioli LH: Anti-asthmatic potential of a D-galactose-binding lectin from *Synadenium carinatum* latex. *Glycobiology* 17: 795–804, 2007.
76. Sanogo R, Monforte MT, Daquino A, Rossitto A, Maur DD, Galati EM: Antiulcer activity of *Salvadora persica* L.: structural modifications. *Phytomedicine* 6: 363–366, 1999.
77. Sanon S, Ollivier E, Azas N, Mahiou V, Gasquet M, Ouattaraa CT, Nebie I, Traore AS, Esposito F, Balansard G, Timon-David P, Fumoux F: Ethnobotanical survey and in vitro antiplasmodial activity of plants used in traditional medicine in Burkina Faso. *Journal of Ethnopharmacology* 86: 143–147, 2003.

78. Shirwaikar A, Khan S, Malini S: Antiosteoporotic effect of ethanol extract of *Cissus quadrangularis* Linn. on ovariectomized rat. *Journal of Ethnopharmacology* 89: 245–250, 2003.
79. Shoeb HA, Hassan AA, el-Sayed MM, Refahy L: The molluscicidal properties of *Agave decepiens* and *Agave americana* (var. *marginata*). *Journal of Egypt Society of Parasitology* 14: 265–273, 1984.
80. Simons AJ, Dawson IK, Dugumba B, Tchoundjeu Z: Passing problems: prostate and Prunus. *HerbalGram* 43: 49–53, 1998.
81. Sindiga I, Nyaigotti-Chacha C, Kanunah MP: *Traditional medicine in Africa*. East African Educational, Nairobi, 1995.
82. Sofrata A, Lingstrom P, Baljoon M, Gustafsson A: The effect of miswak extract on plaque pH. An in vivo study. *Caries Research* 41: 451–454, 2007.
83. Steenkamp V, Fernandes AC, van Rensburg CEJ: Antibacterial activity of Venda medicinal plants. *Filoterapia* 78: 561–564, 2007.
84. Stewart KM: The African cherry (*Prunus africana*): Can lessons be learned from an over-exploited medicinal tree? *Journal of Ethnopharmacology* 89: 3–13, 2003.
85. Stirpe F, Barbieri L, Abbondanza A, Falasca AI, Brown AN, Sandvig K, Olsnes S, Pihl A: Properties of volkensin, a toxic lectin from *Adenia volkensii*. *Journal of Biological Chemistry* 260: 14589–14595, 1985.
86. Sukumaran D, Parashar BD, Rao KM: Molluscicidal properties of *Agave americana* and *Balanites aegyptica*. *Pharmaceutical Biology* 32: 232–238, 1994.
87. Tabuti JR, Lye KA, Dhillion SS: Traditional herbal drugs of Bulamogi, Uganda: plants, use and administration. *Journal of Ethnopharmacology* 88: 19–44, 2003.
88. Tabuti JRS, Dhillion SS, Lye KA: Ethnoveterinary medicines for cattle (*Bos indicus*) in Bulamogi county, Uganda: plant species and mode of use. *Journal of Ethnopharmacology* 88: 279–286, 2003.
89. Taniguchi M, Haraguchi H, Higuchi F, Oi S, Chapya A, Kubo I: Biocidal alkaloids from African fagara plants: inhibition of mitochondrial function. *Agricultural and Biological Chemistry* 49: 3051–3052, 1985.
90. Timberlake JR: *Ethnobotany of the Pokot of Northern Kenya*. East African Herbarium, Nairobi. Center for Economic Botany, Royal Botanic Gardens, Kew, UK, 1987.
91. Wang L-Y, Wang N-L, Yao X-S, Miyata S, Kitanaka S: Diterpenes from the roots of *Euphorbia kansui* and their in vitro effects on the cell division of xenopus. *Journal of Natural Products* 65: 1246–1251, 2002.
92. Wanyama JB: Confidently used ethnoveterinary knowledge among pastoralists of Samburu, Kenya: Methodology and results, vol. 1: Methodology and Results. Intermediate Technology Kenya, Nairobi, Kenya, 1997.
93. Wanyama JB: Confidently used ethnoveterinary knowledge among pastoralists of Samburu, Kenya: preparation and administration, vol. 2: preparation and administration. Intermediate Technology Kenya, Nairobi, Kenya, 1997.
94. Wassel GM, Wahab SMA, Aboutabl EA, Ammar NM, Yassin N, Afifi M: Phytochemical and pharmacological investigation of *Aerva* species growing in Egypt. *Egyptian Journal of Pharmaceutical Sciences* 38: 43–52, 1997.
95. Watt JM, Breyer-Brandwijk MG: *Medicinal and poisonous plants of southern and eastern Africa*, 2nd ed. E & S Livingstone, Edinburgh, 1962.
96. Yadav JP, Saini S, Kalia AN: Botanical, cytological, phytochemical and pharmacognostical studies on *Salvadora* species. *Journal of Medicinal and Aromatic Plant Sciences* 28: 231–238, 2006.
97. Yenesew A, Derese S, Irungu B, Midiwo JO, Waters NC, Liyala P, Akala H, Heydenreich M, Peter MG: Flavonoids and isoflavonoids with antiplasmodial activities from the root bark of *Erythrina abyssinica*. *Planta Med* 69: 658–661, 2003.

98. Yenesew A, Induli M, Derese S, Midiwo JO, Heydenreich M, Peter MG, Akala H, Wangui J, Liyala P, Waters NC: Anti-plasmodial flavonoids from the stem bark of *Erythrina abyssinica. Phytochemistry* 65: 3029–3032, 2004.
99. Yokosuka A, Mimaki Y, Kuroda M, Sashida Y: A new steroidal saponin from the leaves of *Agave americana. Planta Medica* 66: 393–396, 2000.
100. Balakrishnan NP, and Chakrabarty T: *The family* Euphorbiacea *in India: A synopsis of its profile, taxonomy and bibliography*. Bishen Singh Mahendra Pal Singh, Dehradun, India, 2007, 500 pages.
101. Bagalkotkar G, Sagineedu SR, Saad MS, Stanslas J: Phytochemicals from *Phyllanthus niruri* Linn. and their pharmacological properties—a review. *Journal of Pharmacy and Pharmacology* 8: 1559–1570, 2006.
102. Premaratna A, Shadaksharaswamy M, Nanjappa S: Isolation, purification and properties of a lectin from the latex of *Synadenium grantii* Hook f. *Indian Journal of Biochemistry and Biophysics* 18: 32–35, 1981.
103. Rogerio AP et al.: Anti-asthmatic potential of a d-galactose-binding lectin from *Synadenium carinatum* latex. *Glycobiology* 17: 795–804, 2007.
104. Uzabakiliho B, Largeau C, Casadevall E: Latex constituents of *Euphorbia candelabrum, E. grantii, E. tirucalli* and *Synadenium grantii. Phytochemistry* 26: 3041–3045, 1987.
105. Tinto WF, Simmons-Boyce JL, McLean S, Reynolds WF: Constituents of *Agave americana* and *Agave barbadensis. Fitoterapia* 76: 594–597, 2005.
106. Bashwira S, Hootele C: Myricoidine and dihydromyricoidine, two new macrocyclic spermidine alkaloids from clerodendrum myricoides. *Tetrahedron* 44: 4521–4526, 1988.
107. Muregi FW, Ishih A, Miyase T, Suzuki T, Kino H, Amano T, Mkoji GM, Terada M: Antimalarial activity of methanolic extracts from plants used in Kenyan ethnomedicine and their interactions with chloroquine (CQ) against a CQ-tolerant rodent parasite, in mice. *Journal of Ethnopharmacology* 111: 190–195, 2007.
108. Lukhoba CW, Simmonds MSJ, Paton AJ: *Plectranthus*: A review of ethnobotanical uses. *Journal of Ethnopharmacology* 103: 1–24, 2006.
109. Vigneron M., Deparis X., Deharo E. and Bourdy G. Antimalarial remedies in French Guiana: a knowledge attitudes and practices study. *Journal of Ethnopharmacology* 98: 351–360, 2005.
110. Rukunga GM, Muregi FW, Tolo FM, et al: The antiplasmodial activity of spermine alkaloids isolated from *Albizia gummifera. Fitoterapia* 78: 455–459, 2007.
111. Carlini CR, Grossi-de-Sá MF: Plant toxic proteins with insecticidal properties. A review on their potentialities as bioinsecticides. *Toxicon* 40: 1515–1539, 2002.

13 Herbal Medicines for Animal Health in the Middle East and North Africa (MENA) Region

Laïd Boukraa, Hama Benbarek, and Mokhtar Benhanifia

CONTENTS

13.1 INTRODUCTION

Since time immemorial, medicinal plants have been used in virtually all cultures to maintain and restore health. The widespread use of herbal remedies and health care preparations, such as those described in ancient texts such as the Vedas and the Bible and obtained from commonly used traditional herbs and medicinal plants, has been traced to the occurrence of biologically active compounds.

The use of traditional medicine and medicinal plants in most developing countries for treating disease and maintaining good health has been widely observed (United

Nations Educational, Scientific, and Cultural Organization [UNESCO], 1996). Furthermore, an increasing reliance on the use of medicinal plants in the industrialized world has been traced to the extraction and development of several drugs and chemotherapeutics from these plants as well as from traditionally used remedies (UNESCO, 1998). Worldwide herbal remedies have become more popular in the treatment of minor ailments or for diseases that are refractory to allopathic medicine and because of the increasing costs of medical care. Indeed, the market and public demand has been so great that there is a great risk that many medicinal plants today face extinction due to overexploitation.

This chapter discusses traditional veterinary medicine in the Middle East and North Africa (MENA) region. Geographically, this area covers all the countries north of the Sahara in Africa and stretches to Egypt across to the Arabian peninsula, Iraq, and Turkey. The population is predominantly Arab and Moslem, but there are also ethnic minorities (e.g., Berbers, Druze, and Kurds). The area is semiarid to arid and prone to droughts, but there are also large expanses of forest, mountains, and fertile riverine valleys (e.g., the Nile and Euphrates valleys) (Intergovernmental Panel on Climate Change, 1997), which support irrigation and were historically the setting of great civilizations. Despite its aridity, the region contains 10% of the world's known species of higher plants of which half are endemic to the area (Intergovernmental Panel on Climate Change, 1997). The biological, cultural, and ethnic diversity of the region has arisen because it is the convergence of different climatic, phytogeographic, zoogeographic zones (i.e., the Mediterranean, the desert, and the Rift Valley) and an important crossroads of international trade and cultures (Lev, 2006). The passage of traders from East to West through the region has also led to exchange of cultural information and healing philosophies (e.g., doctrine of signatures) to and from both territories; for example, similar uses of medicaments have been witnessed elsewhere, such as the use of *Ammi visnaga* for kidney stones, *Malva nicaensis* for wounds, and so on (Azaizeh et al., 2006). Also, for example, the use of polyherbal preparations originating from India has been noted in faraway Morocco (Merzouki, Ed-derfoufi, and Molero Mesa, 2000).

Nomadic pastoralism, which was once dominant in the region, has been replaced since World War II by intensive agropastoral production on marginal grazing lands (Steinmann, 1998). In North Africa, rainfall diminishes from north to south, and the vegetation changes from perennial grass *(Stipa tenacissma)* and woody shrub *(Artemisia herba alba)* to a greater variety of woody shrubs and succulents (*Arthrophytum scoparium, Festuca algeriensis, Retama retam, Ziziphus lotus*) in the more arid Sahara (Steinmann, 1998). Annuals are the source of fodder and are also valued for medicinal purposes.

The Sahara region has vegetation that is generally homogeneous and confined to the wadi (beds) (Hammiche and Maiza, 2006). The Sahara—tropical and soudano—decannian flora of mainly *Acacia* and *Balanites* and *Aerva*, *Lepladenia,* or *Salsola* taxa, respectively, dominates in the south and changes to the saharo-mediterranean (e.g., *Myrtus*, *Lavandula,* and *Teucrium* species) along the northernmost coastal mountains (Hammiche and Maiza, 2006).

The eastern Mediterranean rim of the Middle East has plant species that are typical of the region (e.g., myrtle, sage, and wild majoram) and desert flora beyond that as it becomes more arid. For example, there are more than 2,600 plant species that have been inventoried from the hills of historical Palestine, of which about 700 have medicinal use (Azaizeh et al., 2006). However, this number and knowledge are under threat, and only a third of the medicinal plants are still in use in Arab traditional medicine; many of these are now rare or endangered, such as *Euphorbia hirsute* (used as an antiparasitic) and *Ophioglossum* spp. (for parasites and wounds) (Azaizeh et al., 2006). A similar decline in knowledge corresponding to loss of biodiversity has also been noted among informants in the West Bank (Palestine) (Said et al., 2002).

Important livestock in the region includes small ruminants (goats and sheep), bovines (cattle and buffalo), poultry (chicken and turkey), and camels and equines valued for draught power (Nordblom and Shomo, n.d.). For religious reasons, there are no pigs reared in the area.

13.2 VETERINARY MEDICINE IN MENA

With the demise of the Roman Empire, Arabia became the center of science and medicine, and it was there that the Hippocratic-Galenic theories were further developed and refined (Azaizeh et al., 2006; Lev and Amar, 2002).

Historical records reveal people's association with traditional veterinary treatment of animals in North Africa and the Middle East to be over 14,000 years old. The history reflects that by 9000 BC the dog was the first species to receive veterinary care, then sheep and goats in the Nile Valley civilization of ancient Egypt (Mellaart, 1967). Cattle became important livestock in Egypt from 4000 BC and in Anatolia from around 6000 BC. This evolution was followed by pig and horse cultures in view of their importance (Murray, 1968). There was a great deal of economic, cultural, social, and religious importance attached to each type of animal; naturally, veterinary medicine evolved specifically to take care of the health of animals, which were being domesticated (Thrusfield, 1986). Evidence of this is the recording of veterinary therapeutic techniques of Egyptian healers (priest healers) in the veterinary *Papyrus of Kahun* (ca. 1900 BC).

About 1,000 medicinal plants were documented in Sumerian records from over 4,000 years ago. The evolution and development of ethnoveterinary medicines (EVMs), however, have lagged behind that of humans (Kofi-Tsekpo and Kioy, 1998; Schillborn van Veen, 1997), but as history shows they should be viewed as complementary rather than separate fields (Lans, 2001). Schillborn van Veen (1997) mentioned that in addition to Sumerian records, veterinarians were providing services as early as 1800 BC in Babylon, and there are documents recording the creation of animal hospitals during the reign of King Ashoka between 269 and 232 BC in the Rock Edict 11. This kind of veterinary art must have facilitated the development of major veterinary centers, which appear in historical records in China and Egypt and later in Arabia (Schillborn van Veen, 1997).

Veterinary medicine as practiced today has its roots in herbal medicine as practiced in ancient China, India, and the Middle East (Schillhorn van Veen, 1996). The literature indicates that Arabia was the world center of veterinary and other medical knowledge in the early Middle Ages. With the spread of Islam, some of this knowledge made its way into Africa and was adopted by stock raisers

(Schillhorn van Veen, 1996). EVM was practiced as early as 1800 BC at the time of King Hamurabi of Babylon, who formulated laws on veterinary fees and charged for treating cattle and donkeys (Schillhorn van Veen, 1996).

13.3 CONTEMPORARY ETHNOVETERINARY MEDICINE

In contemporary rural NEMA, EVM often provides cheaper options to allopathic medicine, and the products are locally available and more easily accessible. In the face of these and other factors, there is increasing interest in the field of ethnoveterinary research and development.

Many pastoral societies have wide knowledge of what can only be termed *immunology*. For example, the North African Fulani (Ba, 1982; Wolfgang, 1983) and the Maasai (Schwabe, 1978) use similar immunization techniques for vaccinating against contagious bovine pleuropneumonia (CBPP) by inserting a piece of infected lung into the nostril of a healthy animal. Similarly, in Arabic pastoral cultures thorns from acacia plants are used for inoculation (Mathias-Mundy and McCorkle, 1995).

In North Africa, as elsewhere, more intensive herding and farming practices help sustain a growing human population. But overgrazing, soil erosion, expansion of cultivation, and irrigated agriculture threaten the long-term productivity of these rangelands and the livelihoods that depend on them (Swearingen, 1994). These problems persist in part because the perceptions and interests of land users are neglected in environmental change analysis and conservation planning (Blaikie, 1985).

In the Middle East and in Morocco specifically, researchers have generally addressed the causes of nomadic sedentarization and associated environmental consequences at the community or regional level. Their literature generally approaches the problem of landscape change through an analysis of demographic, historical, and political economic factors or transformations of regulatory institutions and land tenure systems (Abbaab, 1994; Blaikie, 1985; Bencherifa and Johnson, 1991; Mendes and Narjisse, 1992; Tozy, 1994; Trautmann, 1989). However, these analyses obscure the changing production decisions at the household level, where broader economic and environmental consequences are felt. The paucity of intrahousehold environmental research in the region also results from cultural values that limit contact between male researchers and local women. With few exceptions, male researchers have dominated fieldwork in pastoral and agropastoral communities in MENA. Furthermore, conservation and development projects are generally implemented and managed by men who work with male, but not female, land users. However, women are quite visible across the rural landscape: collecting water, fuel wood, and other plant resources; herding animals; and working in fields.

EVM has its advantages and limitations. The advantages, among others, are the ease of administration of the remedies with little or no danger to the pastoralists (in contrast to pharmaceuticals, which have complicated dose regimens and may require protective clothing), and the accessibility, availability, and affordability of the remedies and their familiarity to users (Mathias, 2004). The limitations are obvious and include the lack of scientific evidence to validate use; the inconvenience of collection times, which may be seasonal; and the time consumed in making the preparations. There are difficulties in standardizing herbal therapies as the concentration of active ingredient varies in different parts of the plants. It is also generally recognized by the users that EVM remedies may not be effective against infectious epidemic diseases.

To treat the diseases of their dromedaries, the Tuareg of North Africa use traditional medicines that are available in their immediate surroundings. Their treatments vary from bloodletting to the use of medicinal plants and mineral products. The most-used treatment is bloodletting, either alone or combined with the application of animal products or plant medicines. About 20 main pathological conditions have been described for which about 100 remedies are used. The collaboration between stockbreeders and veterinary breeding services—when the latter exist—is restricted due to dysfunctional social service systems in most of postcolonial Africa and Asia.

It is to worth mentioning that due to religious beliefs, pork breeding does not exist in North Africa and the Middle East, so no medicinal plants for pigs are reported. Instead, for climatic and geographic considerations, camel breeding is predominant in arid regions, so there is a great deal of knowledge of camel diseases and their treatments among the desert people (Tables 13.1 and 13.2). The camel is an important source of milk, meat, and transport in MENA and an important symbol of wealth (Abbas, Al-Qarawi, and Al-Hawas, 2002).

The camel plays an important role as a food animal in many hot and arid countries. The increase in the camel population was largely due to the growth of herds owned by former pastoralists who have settled as agropastoralists, as well as herds raised by wealthy merchants, who kept rather large herds of camels as a sign of family prestige (Abdel Rahim, Abdel Rahman, and El Nazier, 1994; Abbas, Al-Qarawi, and Al-Hawas, 1999). In spite of the general tendency toward settlement of pastoralists in Saudi Arabia (Fabietti, 1982), about 40% of camels in the country are kept by pastoralists who constitute less than 10% of the total population (Al Humaidi, 1994). Camel milk and meat are popular items in the Saudi menu, and the camel is a popular sport animal in all of the Arabian Gulf countries. Most of these plants were seasonal, and some were restricted to certain locations.

Several plants were used for the treatment of mycotic or allergic dermatitis, which included all skin conditions not diagnosed as mange. For this purpose, the leaves, seeds, or bark of *Tamarix aphylla*, *Blepharis ciliaris*, *Hamada elegans*, and *Euphorbia cuneata* were dried, well ground, and made into a fine paste using butter or petroleum jelly as base. The preparation was smeared over lesions once or twice daily for 3–7 days. Severe lesions, such as those due to contagious skin necrosis, ringworm in young calves, or teat lesions in adult females, were then covered or wrapped by several layers of *Calotropis procera* leaves for protection. *Rhazya stricta*, *Zygophylum album*, *Zygophylum coccineum*, and *Citrulus colocynthis* were the plants most commonly used by ethnoveterinarians for the treatment of helminthiosis in camels.

Calotropis procera promises to be a plant of multiple benefits in the arid zones (Fall, 1989). It contains cardenolides, which have broad antibacterial activity (Akhtar et al., 1992), and *C. procera* has both therapeutic (antitumor, anthelmintic, and anti-inflammatory) and toxic potential (Singh et al., 2000). Pastoralists in the Sahel used

TABLE 13.1
Medicinal Plants Used by Ethnoveterinarians in Saudi Arabia for Treatment of Camel Diseases

Latin Name (Arabic)	Uses
Tamarix aphylla (Athil)	Buds as paste for dermatitis; leaves and buds as eye wash
Rhazya stricta (Harmal)	Leaves as vermifuge, purgative; for mange
Zygophylum album (Um thraib)	Whole plant and seeds: vermifuge
Zygophylum coccineum (Humadd)	Whole plant: purgative, vermifuge
Artemesia abyssinica (Aazir)	Whole plant: laxative
Artemesia inculata (Sheeh)	Whole plant and seeds: laxative and vermifuge
Blepharis ciliaris (Shokadab)	Leaves and seeds: antiseptic; leaves: astringent
Zizyphus nummularia (Sid)	Leaves and seeds for old wounds; leave decoction as astringent
Hamada elegans (Rimth)	Leaves and bark: antiseptic, hasten wound healing; bark powder reduces scars
Citrullus colocynthis (Sheri)	Pulp: strong vermifuge, purgative; root and seeds for snakebite; tar for mange
Fagonia pruguieri (Shwaika)	Bark for mange
Lepidium sativum (Rashad)	Seeds and leaves for gastroenteritis and dermatitis; leaf for mange
Euphorbia cocuneata (Abu laban)	Leaves and latex: antiseptic; purgative; for dermatitis; latex for snakebite
Cymbopogon schoenanthus (Azkhar)	Root decoction for bloat and colic; whole plant laxative
Milolotus alba (Handagog)	Leaves and fruits for anorexia; leaves to induce milk let down
Lycium barbarum (Awsag)	Fruits and leaves for colic and anorexia; whole plant for myositis and arthritis; latex and leaves for snake bite
Heliotropium strigosum (Rimram)	Seed: vermifuge and wound antiseptic
Cleome arabica (Umzameel)	Bark: antiseptic, reduces scars
Haloxylon ammodendron (Ghadha)	Leaves decoction for cough
Calotropis procera (Ushar)	Fresh leaves and latex for arthritis, latex for scorpion sting; dry leaves for stomachache, tonic
Colocynthis vulgaris (Handhal)	Leaf for mange
Fagonia pruguieri (Shwaika)	Bark for mange
Rhazya stricta (Harmal)	Leaf for mange

Source: Abbas, Al-Qarawi, and Al-Hawas (2002). *Journal of Arid Environments*, 50: 367–379. With permission.

C. procera as a livestock famine food during the famous drought of the 1980s (Fall, 1989; Abbas, Eltayeb, and Sullieman, 1993), while in the Sudan it was fed to camels as a cure for night blindness, a condition precipitated by vitamin A deficiency (Abbas, 1997). Kumar and Basu (1994) ascribed anti-inflammatory effects to the latex of *C. procera,* while Al Qarawi et al. (2000) reported moderate anthelmintic activity of the latex against experimental hemonchosis in sheep.

TABLE 13.2
Some Camel Infections and Their Treatment with Local Herbs in Tuareg Land (North Africa)

Common Name	Local Name	Plant Used	Method
Dermatitis	Ajwid	*Balanites aegyptiaca*	Mixed to bone powder and boiled; put on the sick parts
Trypanosomiasis	Manchach	*Boscia senegalensis*	Macerate leaves with tobacco and give to animal as a drink
Piétin	Adyal	*Acacia senegalensis* *Calotropis procera*	Barks boiled in water and then put on wound Latex mixed to butter and put on wound
Beat wound	Tafaday	*Maerua crassifolia*	Macerated leaves cooked in butter and put on the wound
Sinusitis	Anafad	*Maerua crassifolia*	Macerated leaves cooked in butter and inserted into sinuses
Diarrhea	Touffite	*Acacia radiana* *Boscia senegalensis*	Macerated leaves mixed to argil, diluted in water; give to animal as drink
Viper bite	Assam	*Calotropis procera*	Powder of burned leaves put on swollen part

13.4 THE MOST COMMON PLANTS USED FOR ANIMAL HEALTH

Many plants are used for animal health in the MENA region (Tables 13.3 and 13.4). Some of the important plant species are discussed briefly next.

13.4.1 *Urtica dioica*

Stem and leaves of *Urtica dioica* are known for their anti-inflammatory effect, so they are used to fight rheumatism and inflammatory pain of the urinary tract. It is used in folk medicine as a diuretic and antianemic as it is rich in iron.

The leaves are known to be diuretic, anti-inflammatory, antiasthenia, antidiabetic, astringent, and wound healing. The roots are diuretic and decrease the urine quantity in the bladder.

The fruit is a tonic and galactogogue. Research has attributed anticancer, antioxidant, and antiviral effects to this plant (Celik and Tuluce, 2007; van der Meer et al., 2007).

13.4.2 *Artemisia herba-alba*

The *Artemisia herba-alba* plant species is rich in essential oils. It is known for its purgative effect, which plays a very important role in controlling intestinal worms. The leaves of this plant are used in folk medicine to treat diabetes, bronchitis, abscess, and diarrhea and as an anthelmintic (Le Floc'h, 1983).

Studies conducted in Morocco have shown the powerful effect of *Artemisia herba-alba* against leishmania. The strongest leishmanicidal activity was observed with the essential oil at 2 μg/mL versus the other two strains tested. The aqueous extract

TABLE 13.3
The Most Common Medicinal Plants Used

Scientific Name	Local Name	Plant Uses	Major Chemical Constituents
Cupressaceae *Juniperus phoenicea*	Arrar	Cough, wounds, and other skin problems	Diterpenes, phenylpropanoids, furanones
Urticaceae *Urtica dioica*	Tazia, Horigh	Urine retention, antianemic	Flavonoids, triterpenes, coumarin
Solanaceae *Hyoscyamus albus*	Tasakra	Retained placenta, fractures, and lameness	Sesquiterpenes, tropane and piperidone alkaloids, imino sugars
Lamiaceae *Rosmarinus officinalis*	Yazire, Hchiche, El-Arnab	Wounds and other skin problems, edema, bite, pneumonia, bronchitis, antiseptic, antispasmodic	Sesquiterpenes, diterpenes, essential oils
Thymus vulgaris	Zaatar	Indigestion, bloat	Essential oils, flavonoids
Marrubium vulgare	Timret, timriouet, merouket	Pneumonia, bronchitis, antiseptic	Diterpene, phenylethanoids, flavonoids
Origanum floribundum	Timrisitine, origan	Pneumonia, bronchitis, antiseptic, antispasmodic, appetite stimulant	Essential oils
Asteraceae *Artemisia absinthium*	Cheih, Chiba	Wounds and other problems of the skin	Essential oils, guaianolide sesquiterpenes, flavonoids
Artemisia herba-alba	Cheih	Intestinal worms	Sesquiterpenes, flavonoids
Globulariaceae *Globularia alypum*	Tasalka	Conjuctivis, keratitis, cataract	Phenylethanoids, iridoids, flavonoids
Lythraceae *Lawsonia alba*	Hana, henna	Fractures, diarrhea	Xanthones, triterpenoids, napthoquinones
Zygophyllaceae *Peganum harmala*	Harmal	Intestinal worms	Quinazoline, β-carboline alkaloids
Fabaceae			
Cassia acutifolia	Sna meki	Indigestion, constipation	Anthraquinones
Ceratonia siliqua	Kharoube, carob	Diarrhea	Tannins, flavonoids, complex sugars (mannans, galactans)
Geraniaceae *Pelargonium odoratissimum*	Atarchaa, tasekroute	Diarrhea, intestinal worms	Flavonoids, tannins, coumarins, phenolic acids
Alliaceae *Allium cepa*	Bsal	Indigestion, colic, intestinal worms	Flavonoids, thiosulfinates, triterpenes
Rutaceae *Ruta graveolens*	Fidjel	Anthelmintic, diuretic, antispasmodic	Coumarins, quinolone alkaloids
Rhamnaceae *Rhamnus alaternus*	Meliles-sfera	Diuretic	Anthocyanins, flavonoids

TABLE 13.4
Recorded Veterinary Uses of Mediterranean Plants in North Africa

Plant Name	Country	Part Used	Administration	Medical Use
Acacia nilotica	Egypt	Fruits and leaves	Decoction	Digestive
Acacia tortilis	Egypt	Leaves, fruits and gums	Decoction	Digestive
Achillea fragantissima	Egypt	Whole plant	Decoction, topic application	Digestive, skin disease
Allium cepa	Morocco	Bulb	Fodder	Placenta retention
Allium sativum	Algeria, Morocco	Bulb	Fodder	Digestive, vermifuge
Althaea officinalis	Morocco	Whole plant	Wrapped in linen and placed in the mouth of the horse	Oral inflammation
Anabasis articulate	Egypt	Aerial parts	Topic	Skin diseases
Angelica archangelica	Morocco		Macerate in vinegar	Digestive
Artemisia absinthium	Algeria, Morocco	Aerial parts	Fodder, decoction	Nervous disease, digestive
Artemisia herba-alba	Algeria	Aerial parts	Fodder, decoction, smoke	Vermifuge, diarrhea, acaricide
Artemisia judaica	Egypt	Aerial parts	Decoction	Urinary disorders

Source: From Pieroni et al. (2006). *Journal of Ethnobiology and Ethnomedicine*, 2: 16. With permission.

showed an antileshmanial activity at 4 μg/mL (Hatimi et al., 2001). Essential oil of *Artemisia herba-alba* was active against some gram-positive and gram-negative bacteria. The essential oil was fractionated by column chromatography, and these fractions were tested for antibacterial activity. The principal component of the most active fraction was santolina alcohol (Yashphe et al., 1979).

The antifungal activity of *Artemisia herba-alba* was found to be associated with two major volatile compounds (carvone and piperitone) isolated from the fresh leaves of the plant. Antifungal activity was measured against *Penicillium citrinum* (American Type Culture Collection [ATCC] 10499) and *Mucora rouxii* (ATCC 24905). The antifungal activity (IC_{50}) of the purified compounds was estimated to be 5 μg/mL, 2 μg/mL against *Penicillium citrinum* and 7 μg/mL, 1.5 μg/mL against *Mucora rouxii* for carvone and piperitone, respectively (Saleh, Belal, and el-Baroty, 2006).

Research in Tunisia has attributed an antioxidant role to *Artemisia herba-alba*. It has been shown that it had antioxidant effects equivalent to green and black tea decoctions on oxidative stress and related metabolic parameters in rats (Abid et al., 2007).

13.4.3 *Thymus capitatus*

Thyme (*Thymus capitatus*) is a well-known plant used in folk medicine. Its leaves are rich in essential oils largely used in phytotherapy (e.g., thymol, carvacrol, and β-caryophyllene). They are used as infusion to treat respiratory problems and wounds (Le Floc'h, 1983). Many studies have reported its anti-inflammatory properties (Vigo et al., 2004), antispasmodic effects (Meister et al., 1999), and antifungal properties (Pina-Vaz et al., 2004).

13.4.4 *Juniperus phoenicea*

Decoction of leaves of *Juniperus phoenicea* is used to treat diabetes, diarrhea, and locomotive problems. Dried fruits are used as powder for ulcer and abscess healing (Le Floc'h, 1983). It is also known for its use as an antiseptic, analeptic, and antiparasitic.

13.5 SCIENTIFIC EVIDENCE OF MEDICINAL PLANT PROPERTIES USED IN ANIMAL HEALTH

13.5.1 Gastrointestinal Complaints

Several plants are described for therapy of digestive problems, especially tympanism. This condition occurs after the ingestion of several fresh plants and can be solved with herbal remedies like *Thymus vulgaris*, *Artemisia absinthium*, *Allium cepa*, and yeast (*Saccharomyces cerevisiae*) (Renzetti and Taiani, 1988; Guarrera, 1981, 1987).

Decoctions of *Cassia accutifolia* and *Rhamnus alaternus* are used to treat reactive motility (motricity) of the gastrointestinal tract, like indigestion and constipation. *Artemisia herba-alba*, *Marrabium vulgare, Lawsonia alba,* and *Ceratonia siliqua* are used for intestinal anti-inflammatory and antidiarrheal purposes.

Several compounds isolated from *Rosmarinus officinalis* (especially the triterpenes ursolic acid and oleanolic acid) have been shown to have antimicrobial activity, such as growth inhibition of *Staphylococcus aureus*, gram-negative organisms, and *Microsporium lensoum*. Ursolic acid was identified as one of the active components in rosemary to inhibit the growth of some food-associated bacteria and yeast, while oleanolate saponins also exhibited a broad spectrum of antifungal activity, especially against the strain of *Candida glabrata.*

Essential oils of *Thymus vulgaris*, *Allium cepa,* and *Origanum floribundum* have been shown to stimulate the appetite of cattle, sheep, and horses (Pieroni et al., 2006).

13.5.2 Treatment and Prophylaxis of Helminthiasis

In traditional societies, there seem to be a number of plant remedies deemed suitable for parasitic disease (e.g., *Peganum harmala, Artemisia herba-alba, Artemisia absinthium, Juniperus phoenicia, Rosmarinus officinalis*) (Guarrera, 1999; Al-Qura'n, 2009). These plant species have compounds such as alkaloids, which have paralytic action on tapeworms, and sesquiterpenes (e.g., santonin and artimisin), which also have antiparasitic properties (Bennet-Jenkins and Bryant, 1996). The antiseptic and acaricidal properties of this oil are also well known by shepherds in the Pyrenees, who frequently use it (Agelet, 1999; Villar et al., 1992). Similar products derived from other *Juniperus* species (*Juniperus drupacea*, *Juniperus excelsa*, *Juniperus oxycedrus*) are used in Turkey as antiseptic and parasiticidal remedies in animals (Ertug, 1999). In Italy, *Juniperus oxycedrus* is employed for the same purpose (Viegi et al., 2003).

13.5.3 Respiratory Disorders

Many plants are used for respiratory disorders, such as *Juniperus phoenicia* (contains essential oils, pectin, camphen, resin, organic acid) and *Laurus nobilis* (lauric acid, phellandren, tannin, resin) with antiseptic and balsamic properties and *Marrubium vulgare* (marrubin, cholin, tannin, essential oils, glucosids, vitamin C) and *Marrubium deserti* (Hammiche, 2006). *Rosmarinus officinalis* (oleanolic acid, ursolic acid) has antiseptic and antimicrobial activity. The mechanisms of anti-inflammatory effects have been attributed to the inhibition of histamine release from mast cells and inhibition of cyclooxygenase and lipoxygenase activity. *Thymus vulgaris, Origanum floribundum*, and *Foeniculum vulgare* all have antiseptic, expectorant, and anti-inflammatory properties in cattle (Afifi and Abu-Irmaileh, 2000; Yesilada, 1995).

13.5.4 Urinary Disorders

Decoctions of *Urtica dioica* and *Laurus nobilis* are used as diuretics and depuratives in animals and humans (Vàzquez, 1997). Rosemary (*Rosmarinus officinalis*) is widely used as a folk medicine remedy in different countries for several diseases, including urinary ailments. Furthermore, several compounds isolated from *R. officinalis* L. have been characterized and shown to possess potent antioxidative activity (Liu, 1995). *Marrubium vulgare, Origanum floribundum*, and *Ruta graveolens* were also used in urinary disorders.

13.6 CONCLUSION

Herbal medicine is probably the oldest medicine for humans and, by extension, for animals. Unfortunately, the written tradition detailing historical uses of herbs in domestic animals is scant at best. Veterinary herbalists study the human literature for clues while using their knowledge of unique animal physiology and biochemistry to develop treatment recommendations. Herbal medicine can provide benefits that conventional medicine may not be able to offer. EVMs are often not as fast acting and potent as allopathic medicines. They may therefore be less suitable to control and treat epidemic and endemic infectious diseases (e.g., foot-and-mouth disease, rinderpest, hemorrhagic septicemia, anthrax, blackquarter, rabies) and acute life-threatening bacterial infections (e.g., generalized cases of coli- or pyogenes mastitis). For these problems, modern drugs might be the best choice. But, for common self-limiting diseases and conditions such as colds, skin diseases, worm infestation, wounds, reproductive disorders, nutritional deficiencies, and mild diarrhea, EVM has much to offer and can be a cheap and readily available alternative to costly imported drugs. For some diseases, a combination of modern and local remedies and management practices might be preferable.

Even with infectious diseases, ethnoveterinary treatments should not be dismissed out of hand. Many drugs used in chemotherapy are based on chemical substances of plant origin or on the semisynthetic derivatives of such substances. Some local preventive methods are effective and simple to apply; an example is the pox vaccination conducted by pastoralists.

The search for alternatives is especially important as now any unnecessary use of antibiotics and other chemical drugs is discouraged in the light of residue problems and the growing resistance of microorganisms to some drugs. Projects should therefore explore whether local treatments are available and should validate practices that are promising.

Far more research is needed in this domain to provide an understanding of the effects of specific plants on animal health in general and especially on the quality and quantity of meat and dairy products.

Many other in-depth ethnobotanical comparative studies will surely be necessary in the Mediterranean region, as in other parts of the world, before we can gain crucial clues about commonalities and differences in medicinal plant usages across different cultures. In particular, our review has demonstrated that there is an urgent need for the documentation of traditional and indigenous knowledge related to the intangible cultural heritage concerning traditional plant uses, and that such a heritage is much more complex than we may think. Ethnoveterinary data in the Mediterranean region could offer an extraordinary background for conducting serious studies aimed at implementing clinical phytotherapy in animal health care and the use of plant-derived nutraceuticals, with the aim of improving the quality of animal-derived food products.

ACKNOWLEDGMENTS

We are very thankful to all the contributors who have made this study possible: veterinarians, breeders, and Department of Veterinary Sciences staff.

Artemisia absinthium

Ruta graveolens

Thymus vulgaris

Urtica dioica

REFERENCES

Abaab, A. (1994). Dynamique des systemes fonciers et leurs impacts sur la gestion des ressources naturelles en zones steppiques maghrebines. Dakar, Senegal: Centre de Recherche Pour le Developpement.

Abbas, B. (1997). *The ethnoveterinary practices of camel pastoralists in Butana, Sudan: an illustrated monograph.* Addis Ababa, Ethiopia: Organization for Social Studies and Research in Eastern and Southern Africa (OSSREA), p. 97.

Abbas, B., Al Qarawi, A.A., and Al Hawas, A. (1999). Survey of camel raising in Qassim region, Saudi Arabia: herding strategies, productivity and mortality. *Revue D'Elevage et de Medicine Veterinaire des Pays Tropicaux*, 53: 236–247.

Abbas, B., Al-Qarawi, A.A., and Al-Hawas, A. (2002). The ethnoveterinary knowledge and practice of traditional healers in Qassim Region, Saudi Arabia. *Journal of Arid Environments*, 50: 367–379.

Abbas, B., Eltayeb, A.E., and Sullieman, S.R. (1993). *Calotropis procera* feed potential for arid zones. *The Veterinary Record*, 5: 132.

Abdel Rahim, S.A., Abdel Rahman, K., and El Nazier, A.E. (1994). Camel production and reproduction in Qassim, Saudi Arabia. *Journal of Arid Environments*, 26: 53–59.

Abid, Z.B., Feki, M., Hédhili, A., and Hamdaoui, M.H. (2007). *Artemisia herba-alba* Asso (Asteraceae) has equivalent effects to green and black tea decoctions on antioxidant processes and some metabolic parameters in rats. *Annals of Nutrition and Metabolism*, 51 (3): 216–22.

Afifi, F.U., and Abu-Irmaileh, B. (2000). Herbal medicine in Jordan with special emphasis on less commonly used medicinal herbs, *Journal of Ethnopharmacology*, 72: 101–110.

Agelet, A., (1999). Estudis d'etnobotànica farmacèutica al Pallars. PhD thesis, University of Barcelona, Spain.

Akhtar, N., Malik, A., Noor Ali, S., and Urooj Kazmit, S. (1992). Proceragenin, an antibacterial cardenolide from *Calotropis procera*. *Phytochemistry* 31(8): 2821–2824.

Al Humaidi, I.A. (1994). The role of rural offices in Saudi village development in Qassim region: an evaluation study. PhD thesis, University of Wales, Cardiff, UK.

Al Qarawi, A.A., Mahmoud, O.M., Sobaih, E.M., Haroun, E.M., and Adam, S.I.E. (2000). A preliminary study of the anthelmithic activity of *Calotropis procera* latex against *Haemonchus contortus* infection in Najdi sheep. *Veterinary Research Communications*, 25: 61–70.

Al-Qura'n, S. (2009). Ethnopharmacological survey of wild medicinal plants in Showbak, Jordan. *Journal of Ethnopharmacology*, 123: 45–50.

Azaizeh, H., Saad, B., Khalil, K., and Said, O. (2006). The state of the art of traditional Arab herbal medicine in the eastern region of the Mediterranean: A review. *eCAM* 3(2): 229–235.

Ba, A.S. (1982). L'art veterinaire des pasteurs saheliens. Dakar, Senegal: Fondation pour le Progrès de l'Homme et ENDA, pp. 73–82.

Bencherifa, A., and Johnson, D.L. (1991). Changing resource management strategies and their environmental impacts on the middle Atlas Mountains of Morocco. *Mountain Research and Development*, 11(4): 183–194.

Bennet-Jenkins, E., and Bryant, C. (1996). Novel anthelmintics. *International Journal of Parasitology*, 26: 937–947.

Blaikie, P. (1985). *The political economy of soil erosion in developing countries*. London: Longman.

Celik, I., and Tuluce, Y. (2007). Elevation protective role of *Camellia sinensis* and *Urtica dioica* infusion against trichloroacetic acid-exposed in rats. *Phytotherapy Research*, 21: 1039–1044.

Ertug, F. (1999). Plant, animal and human relationship in the folk medicine of Turkey. In: A. Pieroni (Ed.), *Herbs, humans and animals. Erbe, uomini e bestie*. Cologne, Germany: Experiences Verlag, pp. 45–63.

Fabietti, U. (1982). Sedenterization as a means of detribilization: some policies of the Saudi Arabian government towards the nomads. In: T. Niblock (Ed.), *State, society and the economy in Saudi Arabia*. London: Croom Helm, pp. 186–197.

Fall, S.T. (1989). *Utilization d'Acacia albide et de Calotropis procera pour amelioner les rations de petits ruminants au Senegal*. Bamenda, Cameroon ILCA: African Small Ruminant Network, pp. 16–24.

Guarrera, P.M. (1981). Ricerche etnobotaniche nelle Province di Macerata e di Ancona. EPPOS 2: 99–108; *Rivista Italiana EPPOS*, 4: 220–228.

Guarrera, P.M. (1987). Usi tradizionali delle piante nel territorio della Majella. In: *Fasc. Monograf, Erbe e piante medicinali nella storia e nelle tradizioni popolari abruzzesi, Reg. Abruzzo, Chieti, Tip*. Lanciano, Italy: Anxanum, pp. 17–45.

Guarrera, P.M. (1999). Traditional antihelmintic, antiparasitic and repellent uses of plants in central Italy. *Journal of Ethnopharmacology*, 68: 183–192.

Hammiche, V., and Maiza, K. (2006). Traditional medicine in Central Sahara: pharmacopoeia of Tassili N'ajjer. *Journal of Ethnopharmacology*, 105(3): 358–367.

Hatimi, S., Boudouma, M., Bichichi, M., Chaib, N., and Idrissi, N.G. (2001). In vitro evaluation of antileishmania activity of *Artemisia herba alba* Asso. *Bulletin de la Société de Pathologie Exotique*, 94(1): 29–31.

Huxley, A. (1977). Foreword. In: R. Le Strange (Ed.), *A history of herbal plants*. London: Angus and Robertson, 304 pp.

Intergovernmental Panel on Climate Change (IPCC). (1997). IPCC special report on the regional impacts of climate change: an assessment of vulnerability. Chapter 7: Middle East and Arid Asia. http://www.grida.no/climate/ipcc/regional/index.htm.

Liu, J. (1995). Pharmacology of oleanolic acid and ursolic acid. *Journal of Ethnopharmacology*, 49: 57–68.

Kofi-Tsekpo, W.M., and Kioy, W.D. (1998). Newer direction of plant drug research: The need to examine and investigate traditional remedies for pharmacotheraphy of animal and zoonotic diseases. *African Journal of Health Sciences*, 5: 12–14.

Kumar, V.L., and Basu, N.O. (1994). Anti-inflammatory activity of the latex of *Calotropis procera*. *Journal of Ethnopharmacology*, 44: 123–125.

Lans, C. (2001). Creole remedies. Case studies of ethnoveterinary medicine in Trinidad and Tobago. PhD thesis, University of Wageningen, The Netherlands.

Le Floc'h, E. (1983). *Contribution à une étude ethnobotanique de la flore tunisienne*. Editions Ministère de l'enseignement supérieur et de la recherche scientifique. Publications Scientifiques Tunisiennes, Algeria, Tunisia.

Lev, E. (2006) Ethno-diversity within current ethno-pharmacology as part of Israeli traditional medicine—a review. *Journal of Ethnobiology and Ethnomedicine*, 2: 4.

Lev, E., and Amar, Z. (2002). Ethnopharmacological survey of traditional drugs sold in the Kingdom of Jordan. *Journal of Ethnopharmacology*, 82, 131–145.

Martin, M., McCorkle, M.C., and Mathias, E. (2001). *Ethnoveterinary medicine. An annotated bibliography of community animal healthcare*. London: ITDG.

Mathias, E. (2004). Ethnoveterinary medicine: harnessing its potential. *Veterinary Bulletin*, 74(8): 27N–37N.

Mathias-Mundy, E., and McCorkle, C.M. (1989). Ethnoveterinary medicine: An annotated bibliography. Bibliographies in technology and social change, No. 6, Technology and Social Change Programme. Ames: Iowa State University.

Mathias-Mundy, E., and McCorkle, C.M. (1995). Ethnoveterinary medicine and development—a review of literature. In: D.M. Warren and L.J. Slikerveer (Eds.). *The cultural dimension of development. Indigenous knowledge systems*. New York: Intermediate Technology Publications Ltd. 583 pp.

Meister, A., Bernhardt, G., Christoffel, V., and Buschauer, A. (1999). Antispasmodic activity of *Thymus vulgaris* extract on the isolated guinea-pig trachea: discrimination between drug and ethanol effects. *Planta Medica*, 65(6): 512–516.

Mellaart, J.C. (1967). *Catal Huyuk*. New York: McGraw-Hill.

Mendes, L., and Narjisse, H. (1992). Range-animal ecology and agropastoralism in Morocco's western high Atlas mountains. In: C.M. McCorkle (Ed.), *Plants, animals, and people: agropastoral systems research*. Boulder, CO: Westview Press. 196 pp.

Merzouki, A., Ed-derfoufi, F., and Molero Mesa, J. (2000). Contribution to the knowledge of Rifian traditional medicini. II: Folk medcinine in Kasr Lakbir district (NW Morocco). *Fitoterapia*, 71: 278–307.

Muhammad, G., Khan, M.Z., Hussain, M.H., Iqbal, Z., and Athar, M. (2005). Ethnoveterinary practices of owners of pneumatic-cart pulling camels in Faisalabad City (Pakistan). *Journal of Ethnopharmacology*, 97: 241–246.

Murray, J. (1968). Some aspects of ovicaprid and pig breeding in neolithic Europe. In: J.M. Coles and D.A.A. Simpson (Eds.), *Studies in ancient Europe. Essays presented to Stuart Piggott*. Leicester, UK: Leicester University Press.

Nordblom, T.L., and Shomo, F. (n.d.). Livestock and feed trends in west Asia and North Africa: past, present and future. Aleppo, Syria: International Centre for Agricultural Research in the Dry Areas (ICARDA).

Pieroni, A., Giusti, M.E., de Pasquale, C., et al. (2006). Circum-Mediterranean cultural heritage and medicinal plant uses in traditional animal healthcare: a field survey in eight selected areas within the RUBIA project. *Journal of Ethnobiology and Ethnomedicine*, 2: 16.

Pina-Vaz, C., Gonçalves Rodrigues, A., Pinto, E., Costa-de-Oliveira, S., Tavares, C., Salgueiro, L., Cavaleiro, C., Gonçalves, M.J., and Martinez-de-Oliveira, J. (2004). Antifungal activity of Thymus oils and their major compounds. *Journal of European Academy of Dermatology and Venereology*, 18(1): 73–78.

Renzetti, E., and Taiani, R. (1988). Sulla pelle del villano. S. Michele all'Adige, Italy: Museo Usi e Costumi della Gente Trentina, pp. 147–164.

Said, O., Khalil, K., Fulder, S., and Azaizeh, H. (2002). Ethnopharmacological survey of medicinal herbs in Israel, the Golan Heights and the West Bank region. *Journal of Ethnopharmacology*, 83(3): 251–265.

Saleh, M.A., Belal, M.H., and el-Baroty, G. (2006). Fungicidal activity of *Artemisia herba alba* Asso (Asteraceae). *Journal of Environmental Science and Health*, 41(3): 237–244.

Schillhorn van Veen, T.W. (1996). Sense or nonsense? Traditional methods of animal disease prevention and control in African savannah. In: C.M. McCorkle, E. Mathias, and T.W. Schillhorn van Veen (Eds.), *Ethnoveterinary research and development*. London: Intermediate Technology Publications, p. 338.

Schillhorn van Veen, T.W. (1997). Sense or nonsense? Traditional methods of animal parasitic disease control. *Veterinary Parasitology*, 71: 177–194.

Schwabe, C.W. (1978). *Cattle priests and progress in medicine*. The Wesley W. Spink lectures on corporative medicine, Vol. 4. Minneapolis: University of Minnesota Press.

Sezik, E., Yeşilada, E., Honda, G., Takaishi, Y., Takeda, Y., and Tanaka, T. (2001). Traditional medicine in Turkey X. Folk medicine in Central Anatolia. *Journal of Ethnopharmacology*, 75: 2–3, 95–115.

Singh, H., Kumar, S., Dewan, S., and Kumar, V.L. 2000. Inflammation induced by latex of *Calotropis procera*—a new model to evaluate anti-inflammatory drugs. *Journal of Pharmacological and Toxicological Methods*, 43(3): 219–224.

Steinmann, S.H. (1998). Gender, animal management and environmental change in eastern Morocco. *The Arab World Geographer*, 1(2), 117–135.

Swearingen, W. (1994). Northwest Africa, case study. In: M.H. Glantz (Ed.), *Drought follows the plow*. Cambridge, UK: Cambridge University Press, pp. 117–133.

Thrusfield, M. (1986). *Veterinary epidemiology*. London: Butterworths, pp. 1–10.

Tozy, M. (1994). Les modes d'appropriation, gestion et conservation des ressources entre le droit positif et communautaire au Maghreb. Paper presented at the CRDI Conference on Land Degradation and Desertification, Dakar, Senegal, March 7–9.

Trautmann, W. (1989). The nomads of Algeria under French rule: a study of social and economic change. *Journal of Historical Geography*, 15: 126–138.

United Nations Educational, Scientific, and Cultural Organization (UNESCO). (1996). Culture and health, orientation texts—world decade for cultural development 1988–1997, Document CLT/DEC/PRO–1996. Paris.

United Nations Educational, Scientific, and Cultural Organization (UNESCO). (1998). FIT/504-RAF-48 terminal report: promotion of ethnobotany and the sustainable use of plant resources in Africa, Paris.

van der Meer, F.J., de Haan, C.A., Schuurman, N.M., Haijema, B.J., Peumans, W.J., Van Damme, E.J., Delputte, P.L., Balzarini, J., and Egberink, H.F. (2007). Antiviral activity of carbohydrate-binding agents against Nidovirales in cell culture. *Antiviral Research*, 76(1): 21–29.

Vàzquez, F.M., Suarez, M.A., and Pérez, A. (1997). Medicinal plants used in the Barros Area, Badajoz Province (Spain). *Journal of Ethnopharmacology*, 55: 81–85.

Viegi, L. Pieroni, A. Guarrera P.M and Vangelisti, R. (2003). A review of plants used in folk veterinary in Italy as basis for a databank. *Journal of Ethnopharmacology*, 89, 221–244.

Vigo, E., Cepeda, A., Gualillo, O., and Perez-Fernandez, R. (2004). In-vitro anti-inflammatory effect of *Eucalyptus globulus* and *Thymus vulgaris*: nitric oxide inhibition in J774A.1 murine macrophages. *Journal of Pharmacy and Pharmacology*, 56(2): 257–263.

Villar, L., Palacín, J.M., Gómez, D., and Montserrat, G. (1992). *Plantas medicinales del Pirineo Aragonés y demás tierras oscenses*, 2nd ed. Huesca, Spain: Diputación de Huesca, CSIC.

Wolfgang, K. (1983). An ethnoveterinary study of cattle health care by Fulbe herders of south central Upper Volta. Thesis, Hampshire College, Amherst, MA.

Yashphe, J., Segal, R., Breuer, A., and Erdreich-Naftalim G. (1979). Antibacterial activity of *Artemisia herba-alba*. *Journal of Pharmaceutical Sciences*, 68(7): 924–925.

Yesilada, E. (1995). Traditional medicine in Turkey. V. Folk medicine in the inner Taurus Mountains. *Journal of Ethnopharmacology*, 46: 133–152.

14 Ethnoveterinary Medicine and Sustainable Livestock Management in West Africa

Ifeanyi Charles Okoli, Hamidou Hamadou Tamboura, and Mawena Sylvie Hounzangbe-Adote

CONTENTS

14.1 INTRODUCTION

In-depth research into different indigenous animal production paradigms of nutrition, disease causes and treatment, and socioeconomic aspects of animal husbandry geared toward proper understanding of animal production systems of an area and formulation of appropriate interventions not only promotes the development of useful concepts in veterinary medicine but also encourages the maintenance of biological and cultural diversity (De Haans and Bekure, 1991). In West Africa, the highest proportion of livestock remains in the care of traditional herdsmen, among whom ethnoveterinary health care forms a major component of disease management (Lobi, 1984; Agbédé, Téguia, and Manyeli, 1995; Guèye, 1999; Okoli, Okoli, and Ebere, 2002; Okeudo, 2004; Fajimi and Taiwo, 2005; Hounzangbe-Adote, 2001, 2005a; Tamboura, Sawadogo, et al., 1998).

Traditional animal health care practice has remained popular because it is a readily available, low-cost alternative to the relatively costly inputs driven by modern veterinary services (McCorkle, 1995, 1996; Mathias-Mundy and McCorkle, 1991) in many rural communities of West African countries. In fact, animals such as the West African dwarf ruminants, local chickens, and ducks kept in the humid and subhumid parts of the region hardly receive any form of modern veterinary attention to face the wide range of diseases (Sonaiya, 1990).

Diseases of high incidence encountered in the area stretching from Senegal to Cameroon include helminthoses, trypanosomosis, babesiosis, ectoparasitoses, dermatophilosis, orf, brucellosis, tuberculosis, and heartwater. Major outbreaks of contagious bovine pleuropneumonia (CBPP), foot-and-mouth disease (FMD), peste des petits ruminants (PPR), and black quarter are also recorded.

Considering the popularity of ethnoveterinary practice among stock raisers, it would seem that the amount of published information available on its diversity and efficacy in the region is limited. It is, however, established that many of the traditional medications used in humans are also applied in ethnoveterinary practice (Wahua and Oji, 1987; Guissou, 1997; Adjanohoun, 1989; Houinato, 1999, 2002). Furthermore, a number of browse plants utilized in animal feeding are also used to treat the self-same animals when they fall ill (Wahua and Oji, 1987; Okoli, Okoli, and Ebere, 2002; Adewumi, 2004; Hounzangbe-Adote, 2004). Since plant substances continue to serve as important sources of drugs for the majority of the world's human and animal population and several plant-based drugs are in extensive clinical use, there is a need for the development of sustainable strategies to exploit Africa's rich biodiversity (Toyang and Wirmum, 1994; Bognounou, 1993). There is an urgent need to characterize the ethnoveterinary properties of increasingly endangered plant species to improve animal production on the continent (Ngeh et al., 1995).

According to Leeflang (1993), one of the most positive ways to help planners and development organizations to find workable solutions to problems of tropical animal health and production is to conduct research on indigenous veterinary medicine and to disseminate the findings. Better appreciation of indigenous veterinary skills and practices and greater sensitivity toward local knowledge systems will enable planners and development workers to view animal health and production through the eyes of the people whom their programs are supposed to benefit. Similarly, development programs could be better adapted to local problems and therefore become more likely to yield sustainable results. Insights into local knowledge systems would also improve communication with livestock owners and facilitate their participation in decision making about the health care of their animals (Padmakumar, 1998).

Etuk, Okoli, and Udedibie (2005) noted that before any of these animal health development strategies can be successfully employed in the tropics, issues of production systems need to be clearly outlined and integrated to the strategy to ensure that local problems and circumstances are properly accommodated. Furthermore, strategic issues that may be effective in animal health management in one region may at best be partially successful and at worst find no application in other regions, especially in the tropics. Most regions of the world have evolved relatively successful indigenous knowledge (IK)-based systems and ethnoveterinary medical practices in solving their problems almost successfully.

Formal veterinary services and structures in West Africa and its structures were originally created with the objective of implementing specific animal health policies developed by the colonial administrations. The development of the livestock industry was not the priority of the colonial administration in most of the countries. Rather, their major goals were to source raw materials for colonial factories and provide markets for their manufactured goods. Initial priorities were therefore research, delivery of limited veterinary services to a select few, and processing of products for export to the mother country (Maina, 2002). The legacy of colonial thinking has been a fragmented, dysfunctional, and underresourced animal health care service.

This chapter appraises the current facts about ethnoveterinary development in the West African subregion and highlights future research issues that could generate appropriate data for intervention strategies.

14.2 LIVESTOCK PRODUCTION IN WEST AFRICA

Livestock production provides a veritable means of livelihood to millions of people in West Africa. A large percentage of rural people in the region are involved in livestock production, which involves the rearing and marketing of cattle, sheep, goats, pigs, poultry, camels, and the like (Oyesola and Olujide, 2000; Okeudo, 2004). Over 80% of the West African livestock population, especially cattle, sheep, and goats, are in traditional herds. They are mostly owned by pastoral ethnic groups such as the Fulani, Hausa, and Tuareg, generally located in the northern grasslands and sahelian regions or dry areas (Ikeme, 1990; Bâ, 1994; Tamboura et al., 1998a). However, cattle remain the most prominent of all domesticated animals in West Africa (Tewe, 1995).

There are many breeds of cattle indigenous to West Africa. According to Pagot (1992), the popular breeds of cattle in West Africa include red Bororo, white Fulani, Sokoto Gudali, Muturu, Keteku, Ndama, Azawak, Kouri, Gobra, and Foulbe. Popular sheep include Macina woolsheep, Djallonke, Peul sheep, long-haired Moorish, Mossi sheep, and West African dwarf, while the goats are the dry zone goats called chèvre du Sahel, Savannah goats chèvre Mossi, West African dwarf goats, and the red Maradi.

In the region, the historical mobile pastoralism or transhumance is the dominant system of ruminant production. For a long time, it involved movement of the herdsmen, their families, and the herds from place to place, with the availability of fodder, water, and animal health as determining factors. But now, due to shortage of land space and occurrence of conflicts between herders and crop farmers, most of the transhumants have become permanent stakeholders and do keep in a fixed place part of the family and older or sick animals. Cattle are kept as a status symbol and cultural medium (they play a major role in marriages, weddings, sacrifices, and funerals), but also for beef and milk production, hides and skin, as well as for traction power and as a source of manure for cropping (Landais and Lhoste, 1993; Adekunle, Oladele, and Olukaiyeja, 2002).

In the subhumid savannah and humid south, small ruminant and poultry production predominates, with 85% of rural families keeping goats, sheep, chicken, and local guinea fowl primarily as an investment, a source of income and manure, or for meat at home or during festivals (Molokwu, 1982; Pagot, 1992; Okoli, 1993). These indigenous animals, although hardy and well adapted to the environment, grow slowly and are low in productivity. Most modern farms are commercial poultry and pig farms. They are growing in number, especially around major cities. In Nigeria specifically, cultural and religious prohibitions have restricted pig farming to the southern parts of the country, while small-scale producers dominate commercial poultry production (Okeudo, 2884; Etuk et al., 2005; Okoli et al., 2004). The greater proportion of poultry population in the West African region is thus made up of indigenous breeds kept under extensive scavenging and a free-range system (Guèye, 1999; Sonaiya, 2000; Okeudo, 2004; Ibrahim and Abdu, 1996).

Currently, livestock production in West Africa lends itself to small, medium, and large (industrial) production (Ikpi, 1992). According to Sonaiya (2000), families employ various but largely extensive management systems to take advantage of common village resources to produce milk, meat, and guinea fowl eggs. Industrial livestock production is also proving to be a veritable economic recovery tool in most West African countries (Okeudo, 2004; Etuk et al., 2005; Okoli et al., 2004; Okoli, Anyaegbunam, et al., 2005).

The Food and Agricultural Organization (FAO, 1998), Afolabi (2007) and Abowei and Hart (2008) estimated the animal protein intake of Nigeria at about 5 g per person per day, 7 g for the Côte d'Ivoire, Mali, and Burkina Faso zone and 8 g for Senegal, as against 60 g per person per day for North America. The need for increased animal food products in the daily food intake of West Africans can therefore not be overemphasized. Demand for animal products in the region, however, continues to rise, as shown by Delgado et al. (1998), and is driven by improvements in personal income, population growth, and increasing urbanization. Animal food product supply, especially production efficiency, however, remains low in West Africa despite the increasing number of yearly veterinary science graduates from various institutions of higher learning (Okoli, 2006). One way to improve livestock production systems in the region is through harnessing of ethnoveterinary practices.

14.3 IMPORTANCE OF LIVESTOCK

The potential of livestock in sustaining local economies ranging from the village family to national corporate entities has been acknowledged worldwide (Intermediate Technologies Development Group and International Institute of Rural Reconstruction [ITDG and IIRR], 1996; LEISA Editorial, 2002; Renard et al., 2004). In most sub-Saharan African countries where resource-poor farmers live, livestock production is almost always the first step in the difficult climb out of poverty. For many, animals are the prime or only source of livelihood. For many more, they provide additional sources of income, food, clothing, and labour or traction. Livestock act as a store of wealth and medium of exchange. They are also a vital part of culture and offer the only way to survive in many difficult environments (Pagot, 1992; Etuk, Okoli, and Udedibie, 2005). According to Oyesola and Olujide (2000), the livestock sector contributed on average 5.2% to the gross domestic product (GDP) of Nigeria between 1993 and 1997, while figures among the sahelian countries varied from 10% to 12% (Renard et al., 2004). In view of this acknowledged importance of the livestock production sector, factors that tend to impede its continued development and expansion should not only be identified but also adequate strategy should be developed and applied to ensure a sustainable growth.

14.4 CONSTRAINTS TO LIVESTOCK PRODUCTION

Several factors act as constraints to the optimal production of livestock in West Africa. These constraints are of varied importance depending on the geographical area and objective of production. Some of these constraints include, but not limited to, low breed productivity, scarcity and seasonal availability of feedstuff, as well as diseases and health problems (Pagot, 1992; Seifert, 1996).

Animal health problems are major constraints to livestock production in many countries, especially in the tropics (Tamboura, Bessin, et al., 1998; Etuk, Okoli, and Udedibie, 2005). Throughout recorded history, a major problem that has plagued herders is how to prevent contagious diseases (CBPP, Rift Valley fever, tick-borne diseases, among others) from spreading into new areas along travel and trade routes (Schwabe, 1969; Atang, 1974; Lindberg, 1995). Other major problems in tropical cattle breeding are parasites, mainly worms, ticks, and bots (Primavesi and Primavesi, 2002). LEISA Editorial (2002) reported that traditional systems of livestock production can be relatively productive while making optimal use of the available natural and human resources. Seifert (1996) added that irrespective of how livestock is valued, either as an economic entity or a social status symbol, the health of the herd is of fundamental importance to the achievement of the objective of the farmer. The health of animals (including humans) with regard to the spread of zoonosis is of extraordinary importance above all, especially in places where milk is consumed (FMANR, 1995; Edelsten, 1996; Atsanda, Agbede, and Ijagbona, 1999). This is of major concern especially among the nomads and pastoralists who control livestock production in many West African countries (Okoli, Aladi, et al., 2005). It is animal diseases, more than customs duties and quotas, that are now the major obstacles to international trade in animal products (Perry, 2003; Verstraete, 2005). Anyaegbunam (2003), Okoli, Aladi, et al. (2005), and Nwaodu (2005) have identified disease occurrence and inadequate veterinary services as major constraints to poultry production in southeastern Nigeria.

In view of the overriding importance of the health of livestock to the achievement of any objective of production, it becomes imperative for management strategies to be elucidated not only to solve particular problems (shotgun approach) but also in a holistic and sustainable manner incorporating all issues, including the objectives of production, knowledge, and welfare of the farmers, the livestock, and the environment (Etuk, Okoli, and Udedibie, 2005). As amply captured by Seifert (1996), the adaptation of animal health measures must make particular contribution to sustaining the production system.

Pests and diseases are the two greatest threats to the realization of the productive potential of livestock herds in tropical regions such as West Africa (Seifert, 1996; Etuk, Okoli, and Udedibie, 2005). The inadequacy of modern animal health care delivery systems calls for a look at alternative means of dealing with the menace of pests and diseases. Pest and disease control has always been carried out according to the traditional beliefs of the herdsmen prior to colonization, and these practices continue despite exposure to modern control methods.

14.5 ANIMAL HEALTH MANAGEMENT STRATEGIES

Animal health can be defined as the condition of the animal that enables it to attain acceptable levels of production within the farming system in which it is maintained (Etuk, Okoli, and Udedibie, 2005). Good health in animals will increase production efficiency, protection against epidemic diseases, and improvement of human health by safeguarding humans against zoonotic diseases (Tyler, 1990; Lindberg, 1995).

Modern poultry production, and indeed most modern livestock production that is normally termed industrial or factory production, involves the rearing of high-performing hybrid strains of livestock on a commercial scale. The venture presents certain challenges in terms of management, nutrition, and maintenance of the productive health of the flocks (Okoli, Okoli, and Ebere, 2002). The introduction and continuous promotion of this "modern" method of livestock production over other methods has given rise to the provision of veterinary support services as animal health management strategy. According to Etuk, Okoli, and Udedibie (2005), this operates either as a one-stop animal health management shop where all possible issues relating to the health of animals are tackled or as an avenue to consult only when all other strategies of disease control fail. This last is especially the case for the smallholder, often resource-poor farmers.

Provision of properly coordinated veterinary services has improved the productivity of livestock a great deal. However, the smallholder farmers who still dominate production in West Africa (and most of the developing world) have not found it easy to adapt to the mode of operation of these services, mostly because of the cost implication and accessibility problems (McCorkle, 1995; Okoli, Okoli, and Ebere, 2002). Thus, farmers and herders may only access veterinary services as a last option, especially when all others have failed. This was amply exemplified in Bolivia by van't Hooft (2003) when she described two basic livestock-keeping strategies: diversified livestock keeping and specialized keeping. In the former, health care is based on local practices and medicine embodied in the IK systems of the people, sometimes complemented with selected "modern practices" (Schillhorn van Veen, 1991).

It has been severally highlighted that in most West African countries, orthodox animal health care is plagued by many problems, which include inadequate manpower and logistics, scarce and erratic supply of veterinary drugs and supplies, high cost of said drugs and services, poor infrastructure, and counterproductive government policies that do not complement the development of alternative and indigenous medicine (Bâ, 1994; Tamboura, Sawadogo, et al., 1998; Adekunle, Oladele, and Olukaiyeja, 2002;

Hounzangbe-Adote, 2004; Hounzangbe-Adote et al., 2005b). The relationship between these problems and the current dependence on orthodox veterinary medicine has resulted in a failure to solve the majority of animal health problems in the subcontinent.

The IK of livestock owners, however, forms the foundation for and complements the success of all sustainable animal health care programs in developing countries. It is only recently that orthodox veterinarians and other scientists have begun to recognize the fact that livestock owners have holistic understanding and approach in dealing with diseases and other livestock production problems.

14.5.1 Ethnoveterinary Concepts in Livestock Management

Global awareness of the potential contribution of IK to sustainable development and poverty alleviation was recently heightened. Information originating from developing countries suggested that IK is playing important roles in many sectors of agriculture, such as intercropping techniques, animal production, pest control, crop diversity, animal health care, and seed varieties as well as other forms of natural resources management (Anon., 2002). As a result of this increased awareness and flow of information, academics, policy makers, and development practitioners have shown increasing interest in IK. For example, through the use of modern ethnobotanical research, studies of the diversity of ruminant browses in southeastern Nigeria (Okoli, Nwokeocha, et al., 2002; Okoli et al., 2003) and elsewhere in the subregion (Bognounou, 1993; Hounzangbe-Adote, 2004; Hounzangbe-Adote et al., 2005c; Tamboura, 2006) have yielded information of scientific promise on plants of ethnoveterinary importance.

Walter and Dietrich (1992) and Bâ (1994) reported that traditional medicine still plays an important role in the nomadic life. It has been practiced since time immemorial because it was the only medical system accessible to the majority of farmers living in remote areas. According to these authors, traditional healers and pastoralists know a lot about the transmission and spread of animal diseases. Therefore, disease prevention plays an important role; for instance, traditional tick control involves avoiding places with high infestation of ticks and feeding of certain plants that make ticks fall off an animal. Furthermore, before leaving the enclosure in the morning, women and children pick ticks from the animals and throw the ticks into a fire burning near the entrance to the enclosure. Shady trees are deliberately avoided in case of tick infestation, while tick eradication by burning the infested pasture is widely practiced (Bâ, 1994; Bary, 1998).

There is a silent revolution that is bringing back the previously neglected ethnoveterinary medicine and knowledge of indigenous people in addressing hitherto intractable animal health problems, especially localized ones. According to Provost (1974), disease control should be based on the geographical area since vegetation zone also influences the kind of disease prevalent in an area. The experience of ITDG and IIRR (1996) indicated that there is a large overlap between standard and indigenous veterinary practices, with many local practices having close equivalents in allopathic veterinary medicine. They therefore recommended that in the field, especially in the tropics, standard veterinary medicine and traditional practices should be complementary.

Pharmaceutical drugs are often more effective and convenient to use than traditional remedies, but the latter have the advantages that they are locally available and usually cheaper. In remote and often inaccessible locations of the tropics, ethnoveterinary interventions are often the first line of defense against potentially crippling health problems (Agaie et al., 2004). Thus, as stated, a strategic marriage of the two on a case-by-case basis could be a potentially successful tool in improving livestock productivity in the tropics (Etuk, Okoli, and Udedibie, 2005).

Furthermore, ethnoveterinary medicine lends itself easily to local adaptation and application. However, while efficacy of some traditional medicines has been validated (Bayala, 2005; Tamboura et al., 2004, 2005), standardization of extracts and dosage regimes needs to be done.

Based on this, health management needs to take cognizance not only of the production system and objectives but also of the prevalent IK when prescribing strategies in each locality. This is in agreement with the work of van't Hooft (2002), who stated that measures taken to reduce the mortality rate in diversified livestock keeping should be based on the strategies, the practices, and the knowledge of rural families, and that such measures should combine traditional and modern veterinary medicine.

14.5.2 Characteristics of Ethnoveterinary Practices in West Africa

Ethnoveterinary medicine differs not only from region to region but also among and within communities (Mathias, 2001; Tamboura, 2006). This variation is driven by the work division and professional specialization, which may make men know more about large animals, while women may be more familiar with small animals or with certain types of diseases, such as gynecological, mastitis, and neonatal care. Hunters may have a wealth of information on hunting dogs. In some parts of southeastern Nigeria where small ruminants are usually under permanent confinement, information on the diversity and utilization of indigenous browses has been found to reside mostly with women and children, who are traditionally saddled with the responsibility of collecting browses for the animals (Okoli et al., 2003). In others, like Burkina Faso, Mali, and Niger, nomadic herders are more specialized in cattle diseases, while poultry and small ruminants are well treated by sedentary agropastoralists (Aké-Assi, 1992; Tamboura, Sawadogo, et al., 1998). According to Mathias (2001), knowing about such differences can be crucial in the selection of respondents in research and partners for extension approaches, the design of training courses, and the selection of trainees for community-based animal health workers. Discussed next are some of the factors that need to be considered.

14.5.2.1 Clinical and Epizootiologic Knowledge

Stock raisers commonly know when their animals are sick. They can describe the disease, which season it commonly strikes, and what types of animals are affected. For example, the association between *Amblyomma* ticks and heartwater with streptothricosis has long been appreciated by West African stock raisers (Schillhorn van Veen, 1997). They also know where to find the best pasture, how to avoid tsetse-infested areas, where to find salt licks, and so on, and they use tactical and seasonal movements as management tools (Tamboura, Sawadogo, et al., 1998).

14.5.2.2 Practices

Practices are much more varied than just the use of herbal medicines. They also cover bone setting, vaccination against pox and other infectious diseases, branding, and integrated management practices.

One of the most common nonherbal treatments is scarification and bloodletting to free the animal from spoiled blood and use of red-hot plates on the body of the animal to burn diseased spots. Indications of this type of therapy are lameness, the rheumatic complex, skin diseases, and infectious diseases of the digestive and respiratory tracts. There are obvious dangers to these practices, and some of them should be discouraged if there is no apparent empirical basis for their use.

Also, Rajan and Sethuraman (1997) reported that indigenous disease control measures are carried out through herd management, such as using herd dispersion to reduce the risk of infecting all animals belonging to one household, choosing an animal for breeding based on the health of the animal, and preventing contact between healthy and ill animals. Disease prevention can cause a herdsman and his herds to move, while pastoralists avoid regions where vector insects abound or cattle rearers use medication without proper care.

In Nigeria, the Fulani nomads have been known to possess a reservoir of IK that has helped them survive even under extremely unfriendly conditions. Leeflang (1993) reported that the Fulani's response to FMD illustrates how IK sometimes outstrips contemporary Western science. The Fulani sometimes move their cattle upwind of infected herds to prevent the disease from spreading, and sometimes they move them downwind to expose the animals to FMD, knowing that a mild case of the disease will not be fatal and will confer immunity. Western scientists learned that the FMD virus could be aerially transmitted over long distances only after recent outbreaks of the disease in Europe.

Leeflang (1993) also reiterated that Nigerian cattle owners also know that the fluid in the tongue blisters of animals infected with FMD is infectious to other animals. To control the spread of the disease, they collect this fluid, dip a tree thorn in it, and scratch the tongue epithelium of apparently healthy animals to vaccinate them. Vaccination against CBPP is a standard procedure for the Fulani. They slice lung tissue from a diseased animal and implant it under the skin on the foreheads of their cattle, sealing the incisions with mud.

The Fulani recognize the role of insects in the spread of disease, that is, that trypanosomiasis is linked to tsetse fly bites, whereas ticks are known to transmit other blood diseases (Leeflang, 1993). Common preventive measures among the Fulani include applying effective homemade fly repellents, lighting smudge fires to drive off insects, and avoiding infested grazing areas and shade trees. To control ticks, they feed host animals salty plants so that ticks fall off, they pick off ticks and burn them, and they burn off infested rangeland. They also practice annual migration, which was historically aimed at avoiding trypanosome infection. In the rainy season, cattle would be brought to the Fulani's ancestral homeland in the Sudan zone, where tsetse flies are found only among river vegetation. As the dry season advances, however, the grass in this zone quickly dries up and disappears, and water supplies shrink or disappear as well. The Fulani would then move their cattle southward, following disease-free tracks that had been scouted in advance by one of their members.

14.5.2.3 Tools and Technologies

Tools and technologies range from simple tools such as thorns to vaccinate animals to complex animal housing adapted to local conditions. Farmers are familiar with the various materials available in their environment and skillfully take advantage of them.

14.5.2.4 Beliefs

Beliefs are commonly thought of as superstitious, something negative that has to be suppressed. Still, some beliefs can be useful because they improve the condition of the animals or prevent them from getting sick. Examples are the feeding of salt that has been blessed, protecting animals against evil winds, and not letting animals on pastures where other animals have died from diseases such as anthrax. Therefore, it is advisable to have a close look at beliefs and encourage these if they promote animal health.

14.5.2.5 Breeds

Local breeds, such as dairy buffalo, are the outcome of centuries of selection. At first sight, they may produce less than introduced breeds, but they may not score as poorly if both input costs and outputs are considered instead of only the outputs. Local breeds are presently receiving increased attention in connection with attempts to conserve their dwindling genetic diversity.

14.5.2.6 Human Resources

Knowledgeable farmers, herders, and local healers are repositories of knowledge and can be valuable partners in development projects.

14.5.2.7 Herbal Plants Used

Ethnoveterinary research has focused on understanding both the general practice and screening of commonly used medicinal plants. Many indigenous plants have been found to possess medicinal properties. The ethnoveterinary characteristics and active principles of some of the plants from West Africa against common infections have been studied (Gefu, Abdu, and Alawa, 2000; Adewumi, 2004; Fajimi and Taiwo, 2005), while the majority have not received much scientific attention.

Specifically in the last few decades, herbal remedies employed in animal production in Nigeria have attracted the attention of researchers. A review of the literature by Fajimi and Taiwo (2005) discussed the previous and current status of herbal remedies in animal parasitic diseases in Nigeria. This study presented in quantifiable terms the degree of efficacy of whole or plant parts and their extracts in percentages of efficacy. Dosages and concentrations were reported in certain instances, especially in confirmatory research trials for which scientific validation is necessary, unlike in "surveys" in which dosages were not recorded but presented only as "traditional practices among herdsmen." Similarly, comparative inferences were drawn between the efficacy of the tested herb and its counterpart in modern medicine.

Adekunle, Oladele, and Olukaiyeja (2002) examined the use of indigenous control methods for pests and diseases of cattle among herdsmen in northern Nigeria and showed that the majority of herdsmen used indigenous methods to control pests and diseases in their herd; this included hygiene (93%), herbs (87%), herd sharing (22%), bush burning (48%), holy books (35%), incantations (28.6%), and local concoctions (35%). They found significant relationships between the use of IK and age, marital status, and years of experience of the herdsmen.

Fieldwork at Tahara in northeastern Nigeria reported by Akingboye (1995) revealed that the majority of the Fulani herdsmen (Bororo) have knowledge of traditional plant preparations through which common herd diseases are cured. Seeds, roots, leaves, barks, tubers, and fruits are gathered for processing by grinding, boiling, or soaking in water and are used to tackle skin diseases, wounds, cold, and reduced appetite. The Fulani rely on the knowledge passed on by their forefathers to observe signs and symptoms of sickness in animals and to decide on the type of treatment. Commonly used species include baobab (*Adansonia digitata*) against diarrhea and skin disorders; ginger (*Zingiber officinale*) as a laxative, appetizer, and antibloat agent; garlic (*Allium sativum*) as an antidote; African locust beans (*Parkia filicoides*) for skin infections, wounds, and worms; tobacco (*Nicotiana tabacum*) against myiasis, hoof infections, and ectoparasites; and neem (*Azadirachta indica*) as an insect repellent (Hounzangbe et al., 2002, 2004; Ibrahim, 1996).

Similarly, Okoli et al. (2002a) surveyed the diversity of plants of ethnoveterinary importance in southeastern Nigeria. They found that indigenous farmers and healers utilized 24 plant species in the treatment of common livestock ailments, such as diarrhea, ecto- and endoparasitic infections, retained placenta, and dehydration, among others. Methods of preparation and administration were found to include direct feeding of the plant parts, drenching with aqueous decoction, or direct external application of plant juice on affected parts.

Adewumi (2004) reviewed the potential role of herbal plants in livestock production in Nigeria and highlighted the results of surveys carried out by different scientists in Nigeria that showed that herdsmen and indigenous livestock keepers were competent in the diagnosis of animal diseases and have various methods of preparation of medicinal plants for the treatment of their animals. These reports showed that herbs and plant extracts, seeds, leaves, and barks of certain trees, tubers, and roots were the most commonly used sources of drugs. The review, however, concluded that the toxicity of plants containing toxic components such as aristolochic acid and pyrrolizidine alkaloids should be made known to livestock keepers.

Ethnoveterinary medical practice is also widespread among herdsmen and village livestock producers in northern Nigeria, which harbors most of the livestock in the country (Gefu, Abdu, and Alawa, 2000). Modern veterinary inputs and services are usually not readily available to most of the villagers in the region. Therefore, they rely on local plants for livestock health management (Adewumi, 2004).

According to Alawa, Jokthan, and Atuk (2002), herdsmen and livestock owners readily identify signs of disease. They also reported various traditional drugs and methods of treating some common animal health and production problems among Fulani herdsmen and village producers. They also noted the use of less-conventional treatments such as kerosene and spent engine oil. The authors concluded that considering the combination of ingredients used by the traditional animal health practitioners, it is likely that additive, synergistic, and nutritional effects might be involved in alleviating the problem of ill health in animals. The descriptive signs for a specific disease were fairly common from one producer to another. The diseases and problems identified included ticks and flea infestations, swollen joints or feet, sprain, ringworm, retained placenta, pneumonia and other respiratory diseases, poisoning, and mineral deficiencies and how to improve milk yield. Other diseases and problems included mastitis, kid navel disease or protection, infertility, helminth infections, gastric or emetic problems, foot rot, diarrhea and dysentery, bloat, appetite promotion, anthrax, and acetonemia.

According to Kaikabo, Mustapha, and Dagona (2004), the ancient Bade pastoralists of Nigeria possessed a sophisticated body of ethnoveterinary knowledge about animal diseases, treatments, and management practices that spaned from generation to generation to date. Ethnoveterinary practices were adopted by 85.7% of the pastoralists. The major reasons for their adoption were low cost, effectiveness, accessibility, and practicability.

Sonaiya (2000) reported that the use of natural products in various aspects of family poultry is widespread in Nigeria. Owners of family poultry everywhere reported that they have preparations for treating Newcastle disease (NCD), a viral disease (Chavunduka, 1976; Nwude and Ibrahim, 1980). Also, the efficacy of natural products in fighting parasites has been established. The use of *Ficus exasperata* leaves to remove ectoparasites, especially mites and lice from the body of birds in coops or overnight housing, is widely reported (Sonaiya, 2000; Okoli et al., 2002b). Most poultry producers feed hot peppers (*Capsicum* spp.) to their birds as a prophylaxis or treatment for common colds (Guèye, 1999; Sonaiya, 2000). Other reports indicated that feeding peppers to laying hens, apart from enhancing yolk color, also increases egg size (Brown, 1996).

14.6 INVENTORY OF PLANTS AND REMEDIES USED IN ETHNOVETERINARY PRACTICES

Major plants and remedies utilized in indigenous veterinary health care in West Africa are highlighted in Table 14.1. Seeds, leaves, bark, tubers, and roots of plants are commonly used, but wood ash kaolin and potassium, local soaps, and spent engine oil have also been reported (Adewumi, 2004). Some of the important species are discussed briefly here. It should be noted that the biological activity has largely been tested in *in vitro* assays, and there is little phytochemical work conducted on many of the species in the table.

The use of a single remedy or plant extract in treating more than one disease as well as combinations of various plant extracts for broad-spectrum therapy are common. For example, extracts of or ingredients from mahogany (*Khaya senegalensis*) are used to treat anthrax, diarrhea, dysentery, foot rot, helminth infections. and ringworm (Maas, 1991; Tamboura, Sawadogo, et al., 1998; Fagnissè, 2006). Extracts from the mahogany tree are also used to improve appetite and fertility as well as to relieve animals in cases of gastric or emetic problems or poisoning and as a laxative (Alawa et al., 1996; Adewumi, 2004). The methods of processing vary from grinding or soaking in water to obtain solutions given by mouth to inclusion in feedstuff fed to the animal (Alawa et al., 1996; Jagun et al., 1998; Okoli, Okoli, and Ebere, 2002).

TABLE 14.1
Plants and Remedies Used by Indigenous Veterinary Healers in West Africa

Species	Family	Parts Used	Preparation	Diseases Treated	References
Sanseveria liberica	Agavaceae	Fresh leaves	—	Antivenom (naja)	Aké-Assi, 1992
Spondia mombin	Anacardiaceae	Leaves	Water and ethanolic extract	Diarrhea, galactogogue	Maas (1991), Gbego and Hounzangbe-Adote (2001)
Sclerocarya birrea	Anacardiaceae	Bark	Powdered	Anthrax	Tamboura, Sawadogo, et al., 1998
Lannea acida	Anacardiaceae	Bark	—	Galactogogue, bovine newborn nutrition	Aké-Assi, 1992
Lannea microcarpa	Anacardiaceae	Bark	—	Astringent	Aké-Assi, 1992
Annona senegalensis	Annonaceae	Bark, tillage, leaves, roots	—	Insecticid, worms, piroplasmicid, antiallergy	Aké-Assi, 1992; Nacoulma-Ouédraogo, 1996
Saba florida	Apocynaceae	Latex	—	Galactogogue	Aké-Assi, 1992
Saba senegalensis	Apocynaceae	Leaves	—	Galactogogue	Sawadogo, 1987
Leptadenia hastata	Asclepiadaceae	Roots + leaves + tillage + salt	Decoction or maceration	Placental retention	Tamboura, Sawadogo, et al., 1998
Calotropis procera	Asclepiadaceae	Leaves, tillage, latex	—	Insecticide, latex	Aké-Assi, 1992
Lesptadenia pirotechnica	Asclepiadaceae	Roots	—	Flatulence	Aké-Assi, 1992
Balanites aegyptiaca	Balanitaceae	Bark	Decoction on hygromas	Brucellosis	Tamboura, Sawadogo, et al., 1998
		Roots from young plants	Maceration	Anthrax	Fagnissè, 2006
		Bark	Decoction	Galactogogue	
Adansonia digitata	Bombacaceae	Fruits, roots, fibers	—	Artificial delivery, surgical sealing, astringent, dermatophilosis	Aké-Assi, 1992

(continued)

TABLE 14.1 (CONTINUED)
Plants and Remedies Used by Indigenous Veterinary Healers in West Africa

Species	Family	Parts Used	Preparation	Diseases Treated	References
		Bark or leaves	Decoction	Placental retention	Tamboura, Sawadogo, et al., 1998
Ceiba pentendra	Bombacaceae	Leaves, bark + salt	Maceration	Placental retention	Tamboura, Sawadogo, et al., 1998
Bombax costatum	Bombacaeae	Leaves	Decoction	Placental retention	Tamboura, Sawadogo, et al., 1998
				Hemorrhoid	
					Aké-Assi, 1992
Capparis corymbosa	Capparidaceae	Roots	—	Ticks	Aké-Assi, 1992
Boscia senegalensis	Capparidaceae	Leaves + fruits	—	Insecticide, anti-inflammatory, purgative	Aké-Assi, 1992
Boscia salicifolia	Capparidaceae	Leaves	—	Galactogogue, fever	Aké-Assi, 1992
Cadaba farinosa	Capparidaceae	Leaves	—	Galactogogue	Sawadogo, 1987
Maytenus senegalensis	Celastraceae	Roots	Decoction	Worms	Fagnissè, 2006
Afzelia africana	Cesalpinaceae	Bark	Maceration	Pasteurellosis	Tamboura, Sawadogo, et al., 1998
Bauhinia rufescens	Cesalpiniaceae	Leaves	—	Hemostatic	Aké-Assi, 1992
Cassia alata	Cesalpiniaceae	Whole plant	Water and ethanolic extract	Dermatophilosis	Emmanuel-Ali, 2002
Cassia sieberiana	Cesalpiniaceae	Leaves, root bark	Infusion, decoction	Diuretic	Aubreville,1950
				Blennorragy	
				Sex impotence	
				Teniasis fever, malaria	
				Dysmenorrhea	Adjanohoun,1986
				Female sterility, diuretic yellow fever, icteria	Von-Maydell, 1983
				Intestinal colic	
				Avian pestis	
				Worms	

Cassia tora	Cesalpiniaceae	Roots	Decoction	Intestine worm	Maas (1991), Gbego and Hounzangbe-Adote (2001)
Cesalipinia sp.	Cesalpiniaceae	Leaves	Maceration applied on teats	Galactogogue	Gbego and Hounzangbe-Adote, 2002
Cissus sp.	Cesalpiniaceae	Leaves	Maceration applied on teats	Galactogogue	Gbego and Hounzangbe-Adote, 2002
Daniellia olieri	Ceasalpiniaceae	Bark	Powdered	Galactogogue	Fagnissè, 2006
Manihot esculata	Cesalpiniaceae	Leaves	Maceration applied on teats	Galactogogue	Gbego and Hounzangbe-Adote, 2002
Mucuna sp.	Cesalpiniaceae	Leaves, fruits, seeds	Applied on teats	Mother refusing to suckle kids	Gbego and Hounzangbe-Adote, 2002
Spondia mombin	Cesalpiniaceae	Leaves	Grilled and eaten at the moment of lambing	Galactogogue	Gbego and Hounzangbe-Adote, 2002
Tamarindus indica	Cesalpiniaceae	Fruit	Maceration	Bloat	Tamboura, Sawadogo, et al., 1998
Vigna unguilata	Cesalpiniaceae	Seeds ground + salt	Maceration	Brucellosis	Tamboura, Sawadogo, et al., 1998
Anogeissus leiocarpus	Combretaceae	Bark	Decoction	Abdominal pain	Fagnissè, 2006
		Fruits	Decoction	Vermifuge	Tamboura, Sawadogo, et al., 1998
		Roots + seeds	—	Bovine pestis	Aké-Assi, 1992
Guiera senegalensis	Combretaceae	Whole plant	—	Heals, enhance physical performance	Aké-Assi, 1992
		Nodes on tillage	Maceration	Brucellosis	Tamboura, Sawadogo, et al., 1998
		Roots	Maceration (per os) + some drops in nose	Anthrax	
Combretum glutinosum	Combretaceae	Bark of tillage and roots	—	Cicatrisant, poison antidote, anticough	Aké-Assi, 1992
Combretum nigricans	Combretaceae	Leaves	—	Youngster tonic	Aké-Assi, 1992
Combretum micranthum	Combretaceae	Roots, seeds	Cataplasm of powdered	Eyes diseases, cough	Fagnissè, 2006
		Leaves	Decoction	Vermifuge	Tamboura, Sawadogo, et al., 1998

(*continued*)

TABLE 14.1 (CONTINUED)
Plants and Remedies Used by Indigenous Veterinary Healers in West Africa

Species	Family	Parts Used	Preparation	Diseases Treated	References
Terminala avicenoides	Combretaceae	Bark, roots	Powdered	Wounds, bovine pasteurellosis	Fagnissè, 2006
		Leaves and roots		Worm, eyes filariosis, African animal trypanosomias (AAT)	Aké-Assi, 1992
Ricinus communis	Euphorbiaceae	Whole plant	Water and ethanolic extract	Dermatophilosis, plague	Emmanuel-Ali, 2002
Euphorbia balsamifera	Euphorbiaceae	Latex and young leaves	—	Galactogogue, epizootic lymphangite	Aké-Assi, 1992
Euphorbia hirta	Euphorbiaceae	Young leaves	—	Galactogogue	Aké-Assi, 1992
Alysicarpus ovalifolius	Fabaceae	—	—	Antiallergy	Nacoulma-Ouédraogo, 1996
Lonchocarpus laxiflorus	Fabaceae	Bark	Decoction	Diarrhea	Fagnissè, 2006
Pterocarpus erinaceus	Fabaceae	Leaves, roots, bark	Decoction	Diarrhea, bloat, lack of appetite, anemia, bovine pasteurellosis, tick-borne disease	Fagnissè, 2006
		Bark	Maceration	Avian typhosis	Tamboura, Sawadogo, et al., 1998
Xeroderris stulhmaii	Fabaceae	Bark	Decoction	Abdominal pain	Fagnissè, 2006
Hyptis suaveolens	Labiaceae	Whole plant	Water extract	Small ruminants dermatophylosis	Wabi, 2000
Tapinanthus bangwensis	Loranthaceae	Leaves, seeds, fruits	—	Sterility	Aké-Assi, 1992
Gossypium arboreum	Malvaceae	Seeds	Powdered	Eye diseases	Fagnissè, 2006
Hibicus sabdariffa	Malvaceae	Seeds	Maceration	Placental retention; antiallergy	Tamboura, Sawadogo, et al., 1998 Nacoulma-Ouédraogo, 1996

Daniellia oliveri	Meliaceae	Bark	Decoction	Bloat	Tamboura, Sawadogo, et al., 1998
Khaya senegalensis	Meliaceae	Bark	Maceration, decoction	Abdominal pain, diarrhea, bloat, helminthosis, lack of appetite, anemia, wounds, bovine tuberculosis	Maas, 1991; Fagnissè, 2006.
Acacia ataxacantha	Mimosaceae	Bark	Decoction	Diarrhea, fasciolosis, snakebites, bovine tuberculosis	Fagnissè, 2006
Acacia albida	Mimosaceae	Bark	Maceration	Avian typhosis	Tamboura, Sawadogo, et al., 1998
Acacia hockii	Mimosaceae	Rubber	—	Bovine tuberculosis	Fagnissè, 2006
Acacia macrostachya	Mimosaceae	Leaves	Macerated leaves to be poured on	Snake bites	Tamboura, Sawadogo, et al., 1998
Acacia nilotica	Mimosaceae	Fruit, seeds	Powdered	Wounds, galactogogue	Fagnissè, 2006
		Leaves			Aké-Assi, 1992
Acacia pennata	Mimosaceae	Bark	Maceration	Snakebites, genital inflammation	Fagnissè, 2006
Acacia raddiana	Mimosaceae	Leaves	—	Youngster tonic	Aké-Assi, 1992
Acacia senegal	Mimosaceae	Rubber, latex	—	Intestinal pain	Aké-Assi, 1992
Acacia seyal	Mimosaceae	Bark of tillage	—	Keratite	Aké-Assi, 1992
Dichrostachys cinerea	Mimosaceae	Bark	Decoction in milk	Brucellosis	Tamboura, Sawadogo, et al., 1998
Parkia biglobosa	Mimosaceae	Seeds, fruits	—	Anorexy, abdominal	Aké-Assi, 1992
		Bark, fruit	Decoction	pain, diarrhea, bloat,	Fagnissè, 2006
		Roots		laxative	Tamboura, Sawadogo, et al., 1998
				Black leg disease	
Prosopis africana	Mimosaceae	Young leaves	—	Hemorroids	Aké-Assi, 1992
		Roots, fruits	Decoction	Snakebites, wounds, bovine pasteurellosis	Fagnissè, 2006

(*continued*)

TABLE 14.1 (CONTINUED)
Plants and Remedies Used by Indigenous Veterinary Healers in West Africa

Species	Family	Parts Used	Preparation	Diseases Treated	References
Ficus capensis	Moraceae	Leaves, fruits	—	Sterility, galactogogue	Sawadogo, 1987
Ficus ingens	Moraceae	Bark of tillage	—	Galactogogue	Sawadogo, 1987
Moringa oleifera	Moringaceae	—	—	Antiallergy	Nacoulma-Ouédraogo, 1996
Arachis hypogea	Papillonaceae	Oil	—	Constipation	Aké-Assi, 1992
Pterocarpus erinaceus	Papillonaceae	Bark of tillage	—	Diarrhea	Aké-Assi, 1992
Ceratotheca sesamoides	Pedaliaceae	Whole plant	—	Placental delivery, ocytocic Placental retention	Aké-Assi, 1992 Tamboura, Sawadogo, et al., 1998
Eragrostis tremula	Poaceae	Whole plant	—	Constipation	Aké-Assi, 1992
Sorghum bicolor	Poaceae	Germed seeds	—	Dysentery	Aké-Assi, 1992
		Young plant, seeds	Grilled	Bloat, snakebites	Fagnissè, 2006
Zea mays	Poaceae	Seeds	Powdered and given as porridge	Galactogogue	Gbego and Hounzangbe-Adote, 2002
Securidaca longepedunculata	Polygalaceae	Bark	Decoction	Worm, snakebites	Fagnissè, 2006
Ziziphus mauritiana	Rhamnaceae	Roots	Maceration	Vermifuge	Tamboura, Sawadogo, et al., 1998
Mitracarpus scaber	Rubiaceae	Whole plant	Water and ethanolic extract	Dermatophilosis, plague	Wabi, 2000
Mitragyna inermis	Rubiaceae	Bark of tillage	—	Sterility	Aké-Assi, 1992
Crossopteryx febrifuga	Rubiaceae	Bark	Maceration	Antivenom	Fagnissè, 2006
Sarcocephallus latifolius	Rubiaceae	Roots	Decoction	Diarrhea	Fagnissè, 2006
Vitellaria paradoxa	Sapotaceae	Fruit	Powdered	Babesiosis	Fagnissè, 2006
		Fresh leaves	Powdered	Galactagogue	Chabi Toko, 2005

Striga hermonthica	Scrophulariaceae	Whole plant Leaves	Decoction	Diarrhea, bloat Dermatosis, FMD	Fagnissè, 2006 Tamboura, Sawadogo, et al., 1998 Aké-Assi, 1992
Capsicus frutescens	Solanaceae	Fruits	—	Laxative, oxytocic	Aké-Assi, 1992
Sterculia setigera	Sterculiaceae	Bark of tillage Tillage	— Decoction	Anivenom Constipation in donkeys	Aké-Assi, 1992 Tamboura, Sawadogo, et al., 1998
Lantana camara	Verbenaceae	Whole plant	Water and ethanolic extract	Dermatophilosis, plague	Emmanuel-Ali, 2002
Cissus populnea	Vitaceae	Leaves and roots	Decoction	Snakebites	Fagnissè, 2006
Cissu quadragularis	Vitaceae	Whole plant	Powdered and then decoction	Wounds, bovine tuberculosis Dermatophilosis	Fagnissè, 2006 Aké-Assi, 1992
Khaya senegalensis + *Ximenia americano*	Melioceae Olacaceae	Bark	Maceration	Black leg disease	Tamboura, Sawadogo, et al., 1998
Tapinantus sur Balanites aegyptiaca	Loranthaceae Balanitaceae	Leaves	Powdered	Pasteurellosis	Tamboura, Sawadogo, et al., 1998
Piliostigma reticulatum + *Khaya senegalensis* + *Lonchocarpus cyanescens*	Caesalpiniaceae + Meliaceae + Fabaceae	Leaves + bark + roots	Decoction	FMD	Tamboura, Sawadogo, et al., 1998
Butyrospermum parkii + *Diospyros mespiliformis*	Sapotaceae + Ebenaceae	Bark + fruits + roots	Maceration	Avian typhosis	Tamboura, Sawadogo, et al., 1998
"K" solution from sorghum or millet tillage	—	—	Aqueous solution given per os	Bloat	Tamboura, Sawadogo, et al., 1998

Yohimbine, widely used for sexual virility in men, is also employed for use in veterinary medicine as an aphrodisiac (Nwude and Ibrahim, 1980). The anti-inflammatory as well as the antipyretic activities of *Azadirachta indica* (neem) have been evaluated and documented (Okpanyi and Ezeukwu, 1981). Neem is universally recognized as an effective insecticidal repellant (Birmah, 2000). The anti-inflammatory and analgesic properties of the methanolic extract of *Ramalina farinacea* at a dose of 1,600 mg/kg have been shown to suppress signs associated with inflammation and gave results comparable to those of indomethacin (Udem et al., 2001). The hot water leaf extract of *Ocimum grattisimum,* when given to dogs, produced significant reduction in the duration of emesis comparable to that of metaclopamide, a standard antiemetic drug (Udem and Opara, 2001).

Leaves of *Cassia occidentalis* are used as an anticonvulsant and as a purgative, *Adansonia digitala* (baobab) is used as an antidiarrheic in cattle, while *Erythrina senegalensis* has a potent diuretic property (Gefu, Abdu, and Alawa, 2000). *Guiera senegalensis, Anogeissus leocarpus,* and *Sclerocarya birrea* have been found useful in overcoming dystocia in domestic animals (Hassan and Zalla, 2005). The root extract of *Nauclea latifolia* was observed by Madubuinyi (1995) to possess an antihepatotoxic effect and inhibited the multiplication of *Trypanosoma brucei.* Alcoholic extract of a combination of *Sorghum bicolor* and *Telfaria occidentalis* reconstituted at a concentration of 4 mg/100 mL of distilled water proved a potent hematinic in the treatment of anemic rabbits and was better compared to commercial hematinic (Adedapo et al., 2002). The activities of *Carica papaya* extracts against *Salmonella typhi, Staphylococcus aureus* and *Escherichia coli* have been documented (Osato et al., 1993). Similar reports on the antimicrobial activities of certain Nigerian plants have been reported (Olukoya et al., 1993; Sofowora, 1982; Irobi, 1992; Iwu, 1994; Etkin, Ross, and Muazzami, 1990).

The antimicrobial properties of the crude extract of *Ageratum conyzoides* have been validated (Durodola, 1977). The aqueous extract of *Combretum padiculatum* was found to be bacteriocidal at a concentration of 10% and bacteriostatic at 5% against *Salmonella pullorum* (Atawodi, 2000). Extracts of *A. egypti* showed activity against *Staphylococcus aureus*, and methanol leaves extract of *Balanites aegyptiaca* inhibited *Staphylococcus albus* and *Shigella* using the zone inhibition method (Agaie and Usman, 2001).

The anti-infective activities of *Vernonia amygdalina* and *Anona senegalensis* leaves were effective in the treatment of helminthes in cattle (Alawa et al., 2000; Chiezey et al., 2000) and in chicken (Jisaka et al., 1992; Igile et al., 1994; Abdu and Faya, 2000). Abdu and Faya (2000) also reported the use of *Solanum incanum* fruit to treat coccidiosis in poultry, *M. balsamia* to treat fowl pox, and *Capsicum frutescens* for the treatment of NCD.

The kern oil of *B. aegyptica* proved highly efficacious against ectoparasite infestation of camels in northern Nigeria (Oliver, 1960). Skin parasite infections of goats in southern Nigeria have been successfully treated with leaves of *Sida carpinifolia* squeezed and rubbed on lesions (Dalziel, 1937). Burned leaves and twigs of *Guiera senegalensis* as well as burned whole plants of *Hyptis spicifigera* are used individually as insecticides in Nigeria (Nwude, 1997).

Shea butter (*Vitellaria paradoxa*) plus salt at a ratio of 100:1 has been efficacious in the treatment of localized psoroptic mange infection in rabbits, with about 90%

efficacy by the 14th day (Fajimi, Taiwo, et al., 2002). *Aloe variegata*, a green dagger-shaped plant with a clear viscous gel applied as a gel and spread over mange lesions, produced an efficacy of over 50% (Fajimi, Ayodeji, et al., 2002). Okolo and Unaigwe (1984) reported that mange is treated by scrubbing the skin lesions with the fibrous palm kernel fruit waste with the addition of the mineral lime, kitchen salt, lime juice (*Citrus aurantium*), and palm oil for a couple of weeks. Shittu and Bwala (1988) reported satisfactory results of the seed oil of mahogany (*Khaya ivoriensis*) against dermatitis associated with mange and dermatophilosis. Leaf and stem extracts of tobacco showed 100% efficacy against lice by the second day of application and maintained this efficacy for about 56 days postchallenge in West African dwarf goats (Okoli et al., 2002b; Fajimi et al., 2003). *Annona squamosa* and *Tephrosia vogelli,* through their powdered seeds, effectively controlled lice in ruminants.

14.7 CONCLUSION

The literature so far reviewed reveals the wealth of traditional knowledge embodied in ethnoveterinary medical practices of the peoples of West Africa. These clearly constitute untapped resources for possible deployment in sustainable animal health management systems in the region. Even though a sizable body of published literature now exists on the subject in the region, it deals mostly with pharmacological screening of common plants. Useful information on traditional animal health care practices, however, remains uninvestigated in a sizable proportion of livestock-raising communities in the region. There is therefore the need to investigate and integrate these results into primary animal health care delivery systems of the subcontinent.

Animal production curricula in the educational institutions in the region still reflect the ignorance of the special needs of traditional production systems in most countries, which still constitute the major segment of the industry (Sonaiya, 1990; Ikeme, 1990; Okeudo, 2004). It is therefore necessary to consider the development of other alternative pathways for harnessing livestock resources by developing appropriate curricula that will empower a new generation of properly equipped animal health practitioners. Etuk, Okoli, and Udedibie (2005) reiterated the need to predicate animal health development strategies in the tropics on issues of production objectives. There is also the need to clearly outline and integrate these strategies to ensure that local problems and circumstances are properly accommodated.

This review has also highlighted the fact that traditional animal health caregivers remain important components of livestock health care systems in West Africa. However, as observed by Mathias (2004) and Wanzala et al. (2005), traditional healers and their role in animal health care have been largely ignored by the modern veterinary community. Veterinary colleges and public veterinary services pay little or no attention to these local-level experts except perhaps to label them as quacks or witch doctors, whose practice should even be criminalized. Very little is currently known about ethnoveterinarians, especially their sociocultural practices and pharmacopoiea (McCorkle and Mathias-Mundy, 1992). There is an urgent need to study this important group and design steps for integrating them into the conventional animal health care systems.

Ethnoveterinary practices need to be validated before they can be widely promoted. This, according to Mathias (2001), constitutes an important component of research and could be pursued across several levels, which may include tapping the experience of local people, searching the literature for available information on the botany, phytochemistry, and *in vitro*, in silico, and *in vivo* tests, and other relevant aspects; conducting laboratory and clinical tests on station or in experimental herds; monitoring the use of remedies in the field; and studying a remedy's influence on production and agronomic parameters. The exact method or combination of methods will depend on the intended use of the practice to be tested and the purpose of the validation.

The impact of ethnoveterinary promotion on the environment has been shown by McCorkle (1986, 1998) to vary between positive and negative effects. According to these authors, local practices could be environmentally friendlier than imported ones. Local tick control methods, for example, are commonly less harmful to the environment than dipping with commercial chemicals, while projects that stimulate conservation measures and the establishment of herb gardens help maintain biodiversity. On the negative side, large-scale promotion and commercial production bring with them the danger that heavily used plant species may become scarce or even extinct. Therefore, projects promoting plant medicines on a large scale should be required to monitor their environmental impact and explore how far endangered species can be cultivated. There is the need to continuously monitor the impact of ethnoveterinary policies and programs in the region.

According to Mathias (2001), understanding local approaches to animal health care and production and being familiar with the information people have can facilitate the planning and implementation of appropriate development projects and training efforts. There are currently limited data on the impact of appropriate ethnoveterinary projects and trainings executed in the region. Similarly, literature offers little data on the economic impact of promoting ethnoveterinary projects in the region, while other issues like intellectual property rights (IPRs) have hardly received any serious regional attention. These issues will require future research attention by workers in the region (Wanzala et al., 2005).

Lannea microcarpa (Anacardiaceae). Photo courtesy of H.H. Tamboura, 2006.

Tamarindus indica (Caesalpiniaceae). Photo courtesy of H.H. Tamboura, 2006.

Calotropis procera. Photo courtesy of M.S. Hounzangbe, 2006.

Heliotropicum indicum. Photo courtesy of M.S. Hounzangbe, 2006.

REFERENCES

Abowei, J.F.N. and Hart, A. (2008). Artisanal Fisheries Characteristics of the Fresh Water Reaches of Lower Nun River, Niger Delta, Nigeria. *Journal of Applied Sciences and Environmental Management*, 12(1): 5–11.

Abdu, P. A., and Faya, J. N. (2000). The efficacy of some Nigerian plants on helminthes found in local chicken. In *Proceedings of an international workshop on ethnovet practices held between 14–18 August, 2000, Kaduna*. National Animal Production Research Institute, Ahmadu Bello University, Zaria, Nigeria, pp. 65–71.

Adedapo, A. A., Dina, O. A., Saba, A. B., Oladipo, O. D. (2002). Evaluation of *Telfaria occidentalis* and *Sorghum bicolor* extracts as potent haematinics in domestic rabbits. *Nig. J. Anim. Prod.,* 29(1): 88–93.

Adekunle, O. A., Oladele, O. I., and Olukaiyeja, T. D. (2002). Indigenous control methods for pests and diseases of cattle in northern Nigeria. *Livestock Research for Rural Development*, 14(2). http://www.cipav.org.co/lrrd/lrrd14/2/adek142.htm.

Adewumi, C. O. (2004). The potential role of ethno-veterinary pharmacy in livestock production in Nigeria. *Vom J. Vet. Sci.*, 1(1): 125–137.

Adjanohoun, E. J., Cusset, G., Waechter. P., Le, I., Keita, A., Le Bras M. and Lejoly, J. (1989). Banque de données de médecine traditionnelle et pharmacopée (PHARMEL). ACCT (Agence de Coopération Culturelle et echnique).Volume III. Numéro 1.

Afolabi, J.A. (2007). Evaluation of poultry egg marketing in south-western Nigeria. *International Journal of Poultry Science* 6 (5): 362–366.

Agaie, B. M., Onyeyili, P. A., Muhammad, B. Y., and Osunkwo, U. A. (2004). Phytomedicines used in the treatment of helminthosis of ruminants in Sokoto state, Nigeria. In *Proceedings of the 29th annual conference of the Nigerian Society of Animal Production*, March 21–25, 2004, Sokoto, Nigeria, pp. 64–65. Nigerian Society of Animal Production, Abuja, Nigeria.

Agaie, B. M., and Usman, I. D. (2001). Antibacterial activity of the leaf extract of *Balanites aegypti*. *Nig. Vet. J.*, 22(2): 42–46.

Agbédé, G. B., Téguia, A., and Manyeli, Y. (1995). Enquête sur l'élevage traditionnel des volailles au Cameroun. *Tropicultura,* 13(1): 22–24.

Aké-Assi, Y. (1992). Contribution au recensement des espèces végétales utilisées traditionnellement sur le plan zootechnique et vétérinaire en afrique de l'ouest. PhD thesis. ENV Université Claude Bernard, Lyon, France.

Akingboye, K. A. (1995). *The significance of ethnobotany in animal care***.** University of Ibadan, Ibadan, Nigeria.

Alawa, C.B.I., Adamu, A.M., Gefu, J.O., Ajanusi, O.J., Abdu, P.A., Chiezey, N.P., Alawa, J.N. and Bowmam, D.D. (2003). *In vitro* screening of two Nigerian medicinal plants (*Veronica amygdalina* and *Annona senegalensis*) for anthelmintic activity. *Veterinary Parasitology*, 113 : 73–81.

Alawa, J. P., Jokthan, G. E., and Atuk, K. (2002). Ethnoveterinary medicinal practice for ruminants in the subhumid zone of northern Nigeria. *Preventive Veterinary Medicine*, 54: 79–80.

Alawa, J. P., Lemu, N. S. A., Sackey, A. K., and Alawa, C. B. I. (1996). Studies of small ruminant production system in Zaria urban area, Kaduna State: Observations on aspects of health management of small ruminants. *Nigerian Journal of Agricultural Extension*, 9(1): 1–10.

Anon. (2002). Introduction. Opinion. http://www.scidev.net/dossiers/index.cfm?Fuseaction=dossierfulltext&Dossier=7.

Anyaegbunam, C. N. (2003). *A study of the characteristics of agro-livestock businesses in Imo state, Nigeria.* B. Agric. Tech. Project Report. Federal University of Technology, Owerri, Nigeria.

Atang, P. (1974). International approach to animal disease control. In *Animal production in the tropics.* G. M. Babatunde, V. A. Oyenuga, and J. K. Loosli (Eds.). pp. 379–387. Heinemann Educational Books, Ibadan.

Atawodi, S. E. (2000). Antibacterial effects of *Combretum glutinosum* and *Trapinasthus dodoneifolius* on *Salmonella gallinarium* and *Salmonella pullorum.* In: *Proceedings of an international workshop on ethnoveterinary practices held between 14 and 18 August, 2000, Kaduna.* National Animal Production Research Institute, Ahmadu Bello University, Zaria, Nigeria, pp. 72–77.

Atsanda, N.N., and Agbede, S.A., (1999). A survey for Brucellosis, Tuberculosis and Cysticercosis in cattle slaughtered in Ibadan and Maiduguri abattoirs. *Nigerian Veterinary Journal* 20 (1): 61–66.

Bâ, A. S. (1994). L'art vétérinaire et la pharmacopée traditionnelle en Afrique sahélienne. *Review of the Scientific and Technical Office of International Epizootiology.* 13(2): 373–396.

Bary, H. (1998). Le savoir-faire traditionnel des pasteurs: exemple du Burkina Faso. Doc PRASET/GTZ – VSF. Ouagadougou, Burkina, Faso.

Bayala, B. (2005). Activité de type oestrogénique et anti progestéronique des extraits aqueux de *Holarrhena floribunda* chez le rat. Doctoral thesis, University of Ouagadougou, Burkina Faso.

Birmah, A. K. (2000). Efficacy and commercialization of Neem products. African Books Collective. Lagos, Nigeria.

Bizimana, N. (1994). Traditional veterinary practice in Africa. *Schriftreihe der GTZ, No. 243*, Eschborn, Germany.

Bognounou, O. (1993). Réflexions sur les thérapeutiques traditionnelles en soins de santé animale et état des connaissances ethnobotaniques au Burkina Faso. In *Métissages en santé animale de Madagascar à Haïti.* M. Ansay and K. Kasonia (Eds.). Presses University, Namur, Belgium. pp. 181–201.

Brown, A. A. (1996). Evaluation of peppers (*Capsicum* spp.) as enhancers of egg production and quality in layers. PhD thesis, Obafemi Awolowo University, Ile-Ife, Nigeria.

Chabi Toko, R. (2005). Aspects zootechniques et économiques de l'utilisation des feuilles de *Vitalleria paradoxa* et de tourteau de coton en supplémentation sur la production laitière de la vache Borgou en saison hivernale. PhD thesis. University of Abomey-Calavi, Cotonou, Benin.

Chavunduka, D. M. (1976). Plants regarded by Africans as being of medicinal value to animals. *Rhodesian Veterinary Journal*, 7(1): 6–12.

Chiezey, N. P., Gefu, J. O., Jagun, A. G., Abdu, P. A., Alawa, C. B. I., Magaji, S. O., Adeyinka, I. A., and Eduvie, L. O. (2000). Evaluation of some Nigerian plants for anthelmintic activity in young cattle. In Gefu, J. O., Abdu, P. A., and Alawa, C. B. (Eds.). Ethnoveterinary Practices. *Proceedings of the International Workshop on Ethnoveterinary Practices* held 14–18 August, 2000, Kaduna, Nigeria. National Animal Production Research Institute, Ahmadu Bello University, Zaria, Nigeria.

Dalziel, M. (1937). *The useful plants of West Tropical Africa.* Crown Agents, London.

De Haans, S., and Bekure, A. (1991). *Animal health services in sub-Saharan Africa: Initial experience with alternative approaches.* World Bank, Washington, DC.

Delgado, C. L., Courbois, C. B., and Rosegrant, M. W. (1998). Global food demand and contribution of livestock as we enter the new millennium. In *Food, lands and livelihoods—setting research agendas for animal science.* M. Gill, T. Smith, G. E. Pollott, E. Owen, and T. L. J. Lawrence (Eds.). BSAS Occasional Publication, pp. 21, 27–42. IFPRI, Washington, D.C.

Durodola, J. I. (1977). Antimicrobial property of crude extracts from herbal wound remedy—*Ageratum conyzoides. Planta Medica*, 32: 388.

Edelsten, R. M. (1996). Tuberculosis in cattle in Africa—control measures and implications for human health. In *Veterinary medicine—impacts on human health and nutrition in Africa.* R. Lindberg (Ed.). SIPATH, Uppsala, pp. 23–31.

Emmanuel-Ali, N. (2002). Dermatophilose bovine en République du Bénin: epidémiologie, phytothérapie comparées. Doctoral thesis, University d'Abomey-Calavi, Benin.

Etkin, N. L., Ross, P. J., and Muazzami, I. (1990). The indigenization of pharmaceutical therapeutic transitions in natural Hausa land. *Social Science and Medicine*, 30(8): 919–928.

Etuk, E. B., Okoli, I. C., and Udedibie, A. B. I. (2005). Priority issues in tropical animal health management. *Animal Production Research Advances*, 1(2): 83–91.

Fagnissè, F. B. (2006). *Valorisation des plantes médicinales dans le traitement des maladies des ruminants (cas des bovins) aux alentours du parc W: inventaire ethnobotanique et perspectives.* Unpublished dissertation. University of Abomey-Calavi, Cotonou, Benin.

Fajimi, A. K., Ayodeji, I. O., Ajayi, F. T., Ogundola, F. I. (2002). Efficacy studies of topically administered aloe-vera in rabbits naturally infested with psoroptic mange mites. *Moor Journal of Agriculture Research*, 3(2): 199–202.

Fajimi, A. K., and Taiwo, A. A. (2005). Herbal remedies in animal parasitic diseases in Nigeria: a review. *African Journal of Biotechnology*, 4(4): 303–307.

Fajimi, A. K., Taiwo, A. A., Adebowale, E. A., and Ogundola, F. I. (2002). Evaluation of sheabutter plus salt against naturally acquired psoroptic mange in rabbits. *Tropical Veterinarian*, 20(1): 11–16.

Federal Ministry of Agriculture and Natural Resources (FMANR). (1995). Summary of the status of animal diseases in Nigeria. Unpublished report of FMANR, Abuja, Nigeria.

Food and Agricultural Organization (FAO). 1998. World production of animal proteins and need for new approach. *AGA: AAP 98.* FAO, Rome, Italy.

Gbego, I., and Hounzangbe-Adote, M. S. (2001). Efficacité de *Spondia mombin* dans l'amélioration de la croissance présevrage des chevreaux et des agneaux de type Djallonké.; 4e Atelier Scientifique du sud du Bénin, Abomey-Calavi du 11 au 13 décembre 2001. University of Abomey-Calavi, Cotonou, Benin.

Gefu, J. O., Abdu, P. A., and Alawa, C. B. (Eds.). (2000). Ethnoveterinary Practices. In *Proceedings of the workshop on ethnoveterinary practices.* 14–18 August, 2000, Kaduna, Nigeria. National Animal Production Research Institute, Ahmadu Bello University, Zaria, Nigeria.

Guèye, E. F. (1999). Newcastle disease in family poultry: prospects for its control through ethnoveterinary medicine. *Livestock Research for Rural Development*, 14(5). http://www.cipav.org.co/lrrd/lrrd14/5/guey145.htm.

Hassan, W. A., and Zalla, I. L. (2005). Overcoming parturition difficulties in domestic animals through ethno-veterinary practices in Zamfara state, Nigeria. In *Proceedings of the eighth annual conference on animal science of Nigeria* held on 16–18 September 2003, Minna, Nigeria.

Houinato, M. (1999). *L'étude de la pharmacopée populaire dans la région des mont Kouffé (Bénin).* Document Regipharm s.a Bruxelles/LEA-FSA-UNB. National University of Bénin.

Houinato, M. (2002). *Rapport d'étude de mission de la végétation (aspects ethnobotaniques) du complexe écologique du parc W.* FSA, Abomey-Calavi, Bénin.

Hounzangbe-Adote, M. S. (2000). *La pharmacopée en médecine vétérinaire au Sud Bénin (cas des ovins et caprins): Colloque Européen d'Ethnopharmacologie.* Société Française d'ethnopharmacologie, Metz, France, du 11 au 13 mai 2000. *Proceedings of the Colloquium of European Ethnopharmacology*, held in Metz, France, 11-13 May 2000. SFE, Paris, France.

Hounzangbe-Adote M. S. (2004). Propriétés anthelminthiques de 4 plantes tropicales testées in vitro et in vivo sur les nématodes gastro-intestinaux chez les petits ruminants Djallonké. PhD thesis, Université d'Abomey-Calavi, Bénin.

Hounzangbe-Adote, M. S., Moutairou, K., and Hoste, H. (2005a). In vitro effects of four tropical plants on three stages of the parasitic nematodes, *Trichostrongylus colubriformis. Journal of Helminthology*, 79: 29–33.

Hounzangbe-Adote, M. S., Paolini, V., Fouraste, I., Moutaïrou, K., and Hoste, H. (2005b). In vitro effects of four tropical plants on three stages of the parasitic nematode *Haemonchus contortus. Research in Veterinary Science*, 78: 155–60.

Hounzangbe-Adote, M. S., Zinsou, F. E., Affognon, K. J., Koutinhouin, B., Adamou N'diaye, M., and Moutairou, K. (2001). Efficacité antiparasitaire de la poudre de graines de papaye (*Carica papaya*) sur les strongles gastro-intestinaux des moutons Djallonké. *Revue d'Elevage et de Médecine Vétérinaire des pays Tropicaux*, 54(3–4): 225–229.

Hounzangbe-Adote, M. S., Zinsou, F. E., Hounkpe, V., Moutairou, K., and Hoste, H. (2005c). In vivo effects of fagara leaves on sheep infected with gastrointestinal nematodes. *Tropical Animal Health and Production*, 37: 205–214.

Hounzangbe-Adote, M. S., Zinsou, E., Ibrahim, S., Moutaïrou, K., and Hoste, H. (2002). Propriétés anthelminthiques de plantes locales sur les strongles du tube digestif de petits ruminants en station. *Atelier scientifique Sud* (3rd ed.), 11 and 12 December 2002 au CRA-Niaouli.

Ibrahim, M. A. (1996). Ethno-toxicology among Nigerian agropostoralists. In *Ethnoveterinary research and development.* C. M. McCorkle, E. Mathias, and T. W. Schillhorn van Veen (Eds.). IT Publications, London, pp. 54–59.

Ibrahim, M. A., and Abdu, P. U. (1996). Ethno-agroveterinary perspectives on poultry production in rural Nigeria. In *Ethnoveterinairy research and development.* C. M. McCorkle, E. Mathias, and T. W. Schillhorn van Veen (Eds.). IT Publications, London, pp. 54–59.

Igile, G. O., Oleszek, W., Jurzysta, M., Burda, S., Fafunso, M., and Fasanmade, A. A. (1994). Flavonoids from *Vernonia amygdalina* and their antioxidant activities. *Journal of Agricultural and Food Chemistry*, 42: 2445–2448.

Ikeme, A. I. (1990). *Meat science and technology. A comprehensive approach.* African Feb Publishers, Onitsha, Nigeria.

Ikpi, A. E. (1992). *The structure of livestock sector.* Federal Department of Livestock, Lagos, Nigeria.

Irobi, O. N. (1992). Activities of *Chromolaena odorata* (composite) leaf extract against *Sendomonas aeruginosa* and *Streptococcus faecalis. Journal of Ethnopharmacology*, 37(1): 51–53.

Intermediate Technologies Development Group and International Institute of Rural Reconstruction (ITDG and IIRR). (1996). *Ethno-veterinary medicine in Kenya: A field manual of traditional animal health care practices.* Intermediate Technologies Development Group and International Institute of Rural Reconstruction, Nairobi, Kenya.

Iwu, M. M. (1994). African medicinal plants in the search for new drugs based on ethnobotanical. *Ciba Foundation Symposium,* 185: 116–126.

Jagun, A. G., Abdu, P. A., Chiezey, N. P. , Magaji, S. O., Alawa, C. B. I., and Mohammed, A. K. (1998). *Evaluation of Nigerian herbal plants for anthelmintic activities.* Technical progress report. IDRC, Ottawa, Canada.

Jisaka, M., Ohigashi, H., Takagaki, T., Nozaki H., Tada, T., Hirota, M., Irie, M., Huffman, M. A., Nishida, T., Kaji, M., and Koshimizu, K. (1992). Bitter steroid glucosides, Vernosoisides A1, A2 and A3 and related B1 from a possible medicinal plant *Vernonia amygdalina* used by wild chimpamzee. *Tetrahedron*, 48: 625–630.

Kaikabo, A. A., Mustapha, A., and Dagona, A. M. (2004). Ethnoveterinary medicine among bade pastoralists in semi arid zone of northern Nigeria: studies on indigenous treatments and management practices. *African Journal of Livestock Extension*, 3: 59–64.

Landais, E., and Lhoste, P. (1993). Livestock systems and fertility transfer in the African savanna zone 2. *Farmyard manure management methods in crop-livestock interactions* 2 (1): 9–25. Cahiers Agricultures. France.

Leeflang, P. (1993). Some observations on ethnoveterinary medicine in northern Nigeria. *Veterinary Quarterly*, 15(2): 72–73.

LEISA Editorial. (2002). Livestock: Industrial or Integrated. *LEISA Magazine*, 18(1): 5.

Lindberg, R. (1995). *Veterinary medicine—impacts on human health and nutrition in Africa* (Preface). SIPATH, Uppsala, Sweden.

Lobi, B. B. (1984). Incidence de la vision et des pratiques traditionelles sur le développement de l'aviculture au Togo. MSc Thesis. Ecole Inter-Etats des Sciences et Médecine Vétérinaires de Dakar (Inter-State School for Veterinary Sciences and Medicine in Dakar) Senegal.

Maas, A. (1991). *Inventaire des produits traditionnels contre la diarrhée des petits ruminants. Enquête dans le département du Mono. Rapport de stage. Direction de la recherche agronomique.* Projet de recherche appliquée en milieu réel. Ministère du développement rural, Bénin.

Madubuinyi, I. I. (1995). Antihepatotoxic and trypanocidal activities of the ethanolic extract of *Nauclea latifolia* root bark. *Journal of Herbs, Spices, and Medicinal Plants*, 3: 23–35.

Maina, J. A. (2002). Legal framework for the advancement and protection of the veterinary profession in Nigeria. In *The proceedings of the 39th NVMA Annual Congress*, 27–31, October, 2002, Sokoto, Nigeria, pp. 10–14. NVMA, Abuja, Nigeria.

Mathias, E. (2004). Ethnoveterinary medicine: harnessing its potential. *Veterinary Bulletin*, 74(8): 27–37.

Mathias, E. (2001). Introducing ethnoveterinary medicine. http://www.google.com/search?q=cache:vLLQhlwfBcMJ:www.ethnovetweb.com/whatisevm.pdf+ethnov.

Mathias-Mundy, E., and McCorkle, C. (1991). *Ethnoveterinary medicine: an annotated bibliography.* Bibliographies in Technology and Social Change, No. 6. Technology and Social Change Program, Iowa State University, Ames.

McCorkle, C. M. (1986). An introduction to ethnoveterinary research and development. *Journal of Ethnobiology*, 6(1): 129–149.

McCorkle, C. M. (1995). Back to the future: lessons from ethnoveterinary RD&E for studying and applying local knowledge. *Agriculture and Human Values*, 12(2): 52–80.

McCorkle, C. M. (1998). Ethnoveterinary medicine: ethnoscience or just anti-science? In *Complementary and alternative veterinary medicine: principles and practice.* A. M. Schoen and S. G. Wynn (Eds.). Mosby, St. Louis.

McCorkle, C., and Mathias-Mundy, E. (1992). Ethnoveterinary medicine in Africa. *Africa*, 62(1): 59–93.

Molokwu, E. C. (1982). Goat production in Nigeria—prospects and problems. Paper presented at national workshop on small ruminant diseases and production in Nigeria, University of Nigeria, Nsukka, Nigeria.

Ngeh, J. T., Nuwanyakpa M., Ndi, C., Django, S., and Kinyuy, W. C. (1995). Ethnoveterinary medicine practices in the Northwest Province of Cameroon. http://www.nuffic.nl/ciran/ikdm/3–3/articles/toyang.html.

Nwaodu, C. H. (2005). *Studies on the management practices of turkey producers in Imo state, Nigeria*. B. Agic Tech. Project Report. Federal University of Technology, Owerri, Nigeria.

Nwude, N. (1997). Ethnoveterinary pharmacology practices in Nigeria. An overview. Paper presented at the inaugural review and planning workshop on Nationally Coordinated Research Programme on Livestock Diseases NVRI, March 24–28, Vom, Nigeria.

Nwude, N., and Ibrahim, M. A. (1980). Plants used in traditional veterinary medical practice in Nigeria. *Journal of Veterinary Pharmacology and Therapeutics*, 3: 261–273.

Okoli, I. C. (2006). Safeguarding human and animal health: the challenge of veterinary medicine in the early twenty first century. In *The Nigerian Project*. I. C. Okoye (Ed.). Tabanei Publishers, Nigeria. 215 pp.

Okoli, I. C., Aladi, N. O., Etuk, E. B., Opara, M. N., Anyanwu, G. C., and Okeudo, N. J. (2005). Current facts about the animal food products safety situation in Nigeria. *Ecology of Food and Nutrition*, 44: 359–373.

Okoli, I. C., Anyaegbunam, C. N., Etuk, E. B., Opara, M. N., and Udedibie, A. B. I. (2005). Entrepreneurial characteristics and constraints of poultry enterprises in Imo State, Nigeria. *Journal of Agriculture and Social Research*, 5(1): 25–32.

Okoli, I. C., Anyaegbunam, C. M., Etuk, E. B., Uchegbu, M. C., and Udedibie, A. B. I. (2004). Socio economic characteristics of poultry business entrepreneurs in Imo state, Nigeria. *Journal of Agriculture and Social Research*, 4(2): 100–111.

Okoli, I. C., Ebere, C. S., Uchegbu, M. C., Udah, C. A., and Ibeawuchi, I. I. (2003). A survey of the diversity of plants utilized for small ruminant feeding in South-eastern Nigeria. *Agriculture Ecosystem and Environment*, 96: 147–154.

Okoli, I. C., Nwokeocha, J. R., Anyanwu, G. A., and Okeudo, N. J. (2002a). Analysis of Imo State Abattoir records (1995–1999) II: Assessment of estimated volumes and values of slaughter animals. *International Journal of Environmental Health and Human Development*, 3(1): 16–20.

Okoli, I. C., Okoli, C. G., and Ebere, C. S. (2002b). Indigenous livestock production paradigms revisited: survey of plants of ethno-veterinary importance in southeastern Nigeria. *Tropical Ecology*, 43(2): 257–263.

Okolo, M. I. O., and Unaigwe, J. E. (1984). Studies on traditional veterinary practice in Anambra State of Nigeria. Diseases and treatment. *Nigerian Veterinary Journal*, 13(2): 14–22.

Okpanyi, S. N., and Ezeukwu, G. C. (1981). Anti-inflationary and antipyretic activities of *Azadirachta indica*. *Planta Medica*, 41: 34.

Oliver, B. (1960). *Medicinal plants in Nigeria*. Private ed. Nigerian College of Arts, Science and Technology, Ibadan, Nigeria.

Olukoya, D. K., Idika, N., and Odugbemi, T. (1993). Antibacterial activity of some medicinal plants from Nigeria. *Journal of Ethnopharmacology*, 39(1): 69–72.

Osato, J.A., Santiago, L.A., Remo, G.M., Cuadra, M.S. and Mori, A. (1993). Antimicrobial and antioxidant activities of unripe papaya. *Life Sciences*, 53: 1383–1389.

Oyesola, O. B., and Olujide, M. G. (2000). Participation of livestock farmers in Nigerian agricultural insurance scheme. A case study of Iseyin L. G. A. In *Proceedings of Animal Science Association of Nigeria Conference*, September 19–22, Port Harcourt.

Padmakumar, V. (1998). Farmers' reliance on ethnoveterinary practices to cope with common cattle ailments. *Indigenous Knowledge and Development Monitor*, 6(2): 20.

Pagot, J. (1992). Animal production in the tropics and subtropics. Macmillan Press Ltd. London and Basingstoke. 526 pp.

Perry, B. D. (2003). The cost of no measures in the impact and poverty reduction implications of foot-and-mouth disease control in southeastern Africa, with special reference to Zimbabwe. ILRI Nairobi, *Spore*, 105: 9.

Primavesi, A., and Primavesi, O. (2002). Optimizing climate-soil-pasture-cattle interactions in Brazil. *LEISA Magazine*, 18(1): 12–13.

Provost, A. (1974). Problems of disease control in livestock production. In *Animal production in the Tropics*. G. M. Babatunde, V. A. Oyenuga, and J. K. Loosli (Eds.). *Proceedings of the International Symposium of Animal Production in the Tropics*, March 26–29, Ibadan, Nigeria, pp. 307–312.

Rajan, S., and Sethuraman, M. (1997). Traditional veterinary practices in rural areas of Dindigul district, Tamilnadu, India. *Indigenous Knowledge and Development Monitor*, 5(3): 15.

Sawadogo L. (1987). Contribution à l'étude des plantes médicinales et de la pharmacopée traditionnelle africaine: cas des plantes lactogènes. PhD thesis. François Rabelais University, Tours, France.

Schillhorn van Veen, T. W. (1991). The present and future veterinary practitioners in the tropics. *The Veterinary Quarterly*, 15: 41–47.

Schillhorn van Veen, T. W. (1997). Sense or nonsense? Traditional methods of animal parasitic disease control. *Veterinary Parasitology*, 71: 177–194.

Schwabe, C. W. (1969). *Veterinary medicine and human health*, 2nd ed. Williams and Wilkins, Baltimore, MD.

Seifert, H. S. H. (1996). *Tropical animal health*. Kluwer Academic, London. 548 pp.

Shittu, M., and Bwala, H. (1988). Traditional veterinary care among the normadic herdsmen of southern Borno State. *Nigerian Livestock Farmer*, 8(2–4): 27–34.

Sofowora, A. (1982). *Medicinal plants and traditional medicine in Africa*. Wiley, New York.

Sonaiya, E. B. (1990). The context and prospects for development of smallholder rural poultry production in Africa. In *Proceedings CTA seminar on small holder rural poultry production*, October 9–13, 1990, Thessaloniki, Vol. 1. Results and Technical Papers, pp. 35–52. FAO, Rome, Italy.

Sonaiya, E. B. (2000) Family poultry and food security: research requirements in science, technology and socio-economics. In *Proceedings of the 21st World's Poultry Congress, Montreal, Canada* (CD-ROM), pp. 1–13. FAO, Rome, Italy.

Tamboura, H. H. (2006). Activité biologique des extraits aqueux de *Holarrhena floribunda* (G. Don. Durand & Schinz) (Apocynaceae): étude des effets de type hormone mâle chez le rat. PhD thesis, University of Ouagadougou, Burkina Faso.

Tamboura, H. H., Bayala, B., Lompo, M., Guissou, I. P., and Sawadogo, L. (2005). Acute toxicity of aqueous extract, ecological distribution and morphological characteristics of *Holarrhena floribunda* (G. Don) Durand & Schinz, *Leptadenia hastata* (Pers.) Decne and *Cassia sieberiana (D C)* used by veterinary healers in Burkina Faso. *Afr. J. Trad. C. Alt. Med.*, 2 (1): 13 – 24.

Tamboura, H. H., Bayala, B., Lompo, M., Somé, N. P., Ouédraogo, S., Guissou, P. I., and Sawadogo, L. (2004). Effet des extraits aqueux de *Holarrhena floribunda* (G. Don) Durand & Schinz sur l'activité androgénique chez le rat . *Rev. Afr. Santé Prod. Anim. Afr.*, 2(1): 75–78.

Tamboura, H. H., Bessin, R., Sidibé, M., Ouédraogo, L., Kaboré, J., and Kyelem, B. (1998a). Dominantes pathologiques de la reproduction dans les élevages de petits ruminants du centre-est, de l'est et du centre du Burkina Faso. *Bulletin of Animal Health and Production in Africa*, (46): 187–191.

Tamboura, H. H., Sawadogo, L., Kaboré, H., and Yaméogo, S. M. (1998b). Ethnomédecine vétérinaire et pharmacopée traditionnelle dans le plateau central du Burkina Faso: cas de la province du Passoré. *Revue Sciences and Techniques*, Série: *Science Naturelle*, 22(2): 100–118.

Tewe, O. O. (1995). Sustainability and development, paradigms for Nigeria's livestock industry. Inaugural lecture, University of Ibadan. University of Ibadan Press, Ibadan, Nigeria.

Toigbe, G. E. (1978). Contribution à l'étude de la médecine africaine traditionnelle: la pharmacopée des peul du Bénin et du Sénégal. Veterinary medicine thesis. Ecole Inter Etats des Sciences et Médécine Vétérinaires de Dakar (EISMV). Senegal.

Toyang, N. J., Nuwanyakpa, M., Ndi, C., Django, S., and Kinyuy, W. C. (1995). Ethnoveterinary medicine practices in the Northwest Province of Cameroon. *Indigenous Knowledge and Development Monitor*, 3(3), 24.

Tyler, L. (1990). Maintenance of health. In *An introduction to animal husbandry in the tropics*, 4th edition. W. J. A. W. Payne (Ed.). Longman Scientific and Technical, London.

Udem, S. C., Esimone, C. O., Obinwa, O. C., and Akah, P. A. (2001). In *Proceedings of the 38th Annual Congress of the Nigerian Veterinary Medical Association,* October 9–13, 2001, Topo, Badagry, Lagos State, Nigeria. NVMA.

Udem, S. C., and Opara, I. C. (2001). In *Proceedings of the 38th annual congress of the Nigerian Veterinary Medical Association,* October 9–13, 2001, Topo, Badagry, Lagos State, Nigeria. NVMA.

Van't Hooft, K. (2002). Optimizing livestock strategies in Bolivian mixed farming systems. *LEISA Magazine*, 18(1): 10–11.

Verstraete, F. (2005). EU legislation on mycotoxin in food and feed: Overview of the decision-making process and recent and future developments. In *Reducing impact of mycotoxins in tropical agriculture, with emphasis on health and trade in Africa.* September 13–16, 2005, Accra, Ghana. Abstract booklet, pp. 7–8. http://www.iita.org/mycotoxinconf/ABSTRACT_Final.pdf (accessed September 2008).

Wabi, A. (2000). Etude comparative de l'efficacité de trois plantes *Hyptis suaveolens*, *Mitracarpus scaber*, et *Ricinus communis* dans le traitement de la gale des petits ruminants. Mémoire de fin de Cycle pour l'obtention du Diplôme d'Ingénieur des Travaux (DIT). UNB/CPU/DPA. 51 p.

Wahua, T. A. T., and Oji, U. I. (1987). Survey of browse plants in upland area of Rivers state, Nigeria. In *Browse and small ruminant production in southeastern Nigeria, Proceedings of Symposium. International Livestock Center for Africa (ILCA).* L. Reynolds and Attah-Krah (Eds.). Humid Zone Program, Ibadan, Nigeria.

Walter, A., and Dietrich, F. (1992). Role of traditional medicine among nomads of Somalia. *Traditional Veterinary Practice in Africa.* GTZ No. 243, Eschborn, Germany.

Wanzala, W., Zessin, K. H., Kyule, N. M., Baumann, M. P. O., Mathias, E., and Hassanali, A. (2005). Ethnoveterinary medicine: a critical review of its evolution, perception, understanding and the way forward. *Livestock Research for Rural Development,* 17(11). http://www.cipav.org.co/lrrd/lrrd17/11/cont1711.htm.

15 Traditional Chinese Veterinary Medicine

Lida Teng, Debbie Shaw, and Joanne Barnes

CONTENTS

15.1 INTRODUCTION

Traditional Chinese veterinary medicine (TCVM) is the application of traditional Chinese medical theory to treating disease in animals and maintaining animal health, including for domestic animals, poultry, pets, marine life, and other wildlife. TCVM consists of traditional Chinese herbal veterinary medicine and physiotherapies such as acupuncture and cupping; it is based on the same traditional Chinese medical theory but clearly varies in therapeutic forms.

15.2 HISTORY AND DEVELOPMENT OF TCVM

TCVM has over 3,000 years of history in China. The Chinese words *veterinary* and *veterinarian* were first introduced in the Zhou Dynasty (11th century BC to 256 BC), according to an ancient book of etiquette, *The Rites of Zhou-Offices of Heaven* (written in the 7th to 4th centuries BC). The names of well-known Chinese veterinarians have been documented in China since the 10th century BC (Niu, 1991). Although herbs were used for animal health thousands of years ago, the systematic practice and theory of TCVM was only fully developed in the Han Dynasty (206 BC to 220 AD), when the traditional Chinese philosophy of yin-yang and the five elements was widely applied to medicine. From then up to 1840 BC, over 80 reference books and monographs on veterinary medicine and animal husbandry were written and published in ancient China (He, 2007), of which the *Yuanheng*

Liaoma Collections (published in the 1600s) were considered the most comprehensive ancient theoretical books for TCVM practice in China (Niu, 1991). The modern corrected edition of the *Yuanheng Liaoma Collections* includes about 175 herbal formulas and covers animal disease caused by internal and external factors as well as some miscellaneous conditions (Chinese Academy of Agricultural Sciences, 1963).

Compared with some other forms of indigenous ethnoveterinary medicine, the properties and traditional therapeutic effects of Chinese materia medica for veterinary use are well established in China. The earliest pharmacognosy monographs, *Shen Nong Ben Cao* (also known as *The Divine Farmer's Herb-Root Classic*; 206 BC to 8 AD), recorded more than 300 herbs for human and animal use. The first comprehensive herbal pharmacopoeia, *Ben Cao Gang Mu* (also known as the *Compendium of Materia Medica*), was completed in the late 16th century and included 1,892 herbs and over 11,000 formulas for medicinal uses. Excluding herbs for both human and animal use, *Ben Cao Gang Mu* recorded 219 herbal materials specifically for veterinary medicine, animal feeds, and animal growth (Feng, 2000). For example, Fu Zi (*Radix Aconiti Lateralis Preparata;* the processed root of *Aconitum carmichaeli*) was recorded as being used for treating coma and anal incontinence caused by physical overexertion in apes or monkeys. The herbal materia medica has continued to evolve and is now a key component of the veterinary pharmacopoeia. The most recent *Chinese Veterinary Pharmacopoeia* written in Chinese (*CVP*; Commission of Chinese Veterinary Pharmacopeia, 2005) recorded 685 different TCVM, 491 of which are single crude herbal materials. Each monograph in the *CVP* generally includes the Chinese name, scientific name, herbal species, microscopic and chemical identification, quantitative determination, processing methods, properties in traditional description, indications, and dosages. In addition, the dosages are documented for different animals. For example, the dosage of a common herb Dang Gui (*Radix Angelicae Sinensis*) is recorded as "*Equus* and *Bos* 15–60 g; *Camelus* 35–75 g; *Ovis* and *Suidae* 5–15 g; *Canis* and *Felis* 2–5 g; *Leporidae* and *Aves* 1–2 g." At present, most Chinese veterinary reference books are only available in the Chinese language. However, the first English version of *CVP* has been published (Commission of Chinese Veterinary Pharmacopeia, 2008).

15.3 BASIC THEORY OF TCVM

Although traditional Chinese medical theory is well known worldwide for its application in human health, it has rarely been referred to in veterinary science outside China. The concepts of Chinese medicine have been applied to the practice of both veterinary and human medicine for thousands of years in China and other Eastern countries. In practice, this includes holistic approaches to differential diagnosis. The concept of holism concerns the animal body as a whole, with the functions of each organ interdependent (Wang, 2005). Moreover, holism also considers animals and the living environment as a whole, especially focusing on the influences of environmental changes (e.g., weather and season) on animal health (Wang, 2005). TCVM uses the word *pattern* to differentiate the combination of symptoms and signs observed in different animals. The pattern or syndrome directs the choice of treatment, that is, acupuncture points or herbal combinations.

The general therapeutic principles of TCVM are based on the Chinese philosophy of yin-yang and the five elements (Wang, 2005). Yin-yang (see Figure 15.1) focuses on the opposite characteristics of different substances and phenomena.

FIGURE 15.1 Yin-yang.

Yin (the black area in Figure 15.1) represents dark, passive, downward, and internal features, while yang (the white area in Figure 15.1) represents bright, active, upward, and external features. In TCVM, all phenomena in nature are considered to have yin and yang features.

Yin and yang are opposite, but interdependent; they are transformable at a certain stage. The yin-yang concept is applied to interpreting TCVM anatomy and physiology; for example, the external part of the body (e.g., skin) is yang, and the internal part (e.g., organs) is yin. The stomach and intestines are considered to be yang organs as they are in contact with the exterior through different tracts and as they have yang functions (e.g., absorption and transforming foods to produce body energy); organs such as the heart, liver, kidneys, and lungs are yin organs as they are less dynamic and used to store essential substances (e.g., "energy" and blood). The yin-yang concept is also used to describe different diseases, diagnoses, and treatments. For example, animals suffering from toxins in the blood could appear to have high fever and other "hot" symptoms (yang); however, the extreme heat in the body can cause damage to body fluids and blood, so "cold" symptoms (yin) would appear, such as weakness and coldness in limbs. In TCVM, hot symptoms (yang) are usually treated with herbs with cold properties (yin) and vice versa.

The five-element theory gives further explanation of the body medical system. The theory uses the five elements wood, fire, earth, metal, and water to describe different substances and phenomena of similar nature. Animal body organs are classified according to the properties of each element; for example, the animal spleen relates to Earth, as in TCVM the function of the spleen is to transform food energy and produce the body's vital substances, which are considered similar to the nature of Earth (i.e., growing and generating). The relationship of the five elements is complex and includes intergeneration, interrestriction, overacting, and reverse restriction. As each organ is assigned a relevant element (Table 15.1), the relationship of

TABLE 15.1
Examples of Body Organs Classified by the Five Elements

Element	Yang Organ	Yin Organ	Sense Organ	Yin-Yang Properties
Wood	Gall bladder	Liver	Eyes	Less yang
Fire	Small intestine	Heart	Tongue	Utmost yang
Earth	Stomach	Spleen	Mouth	Neutral
Metal	Large intestine	Lung	Nose	Less yin
Water	Bladder	Kidney	Ears	Utmost yin

the body organs to each other is interpreted by the relationship of the five elements. However, it is important to clarify here that the medical terms used in TCVM are different from those of conventional or orthodox Western medicine, particularly in describing the name of the organ and the organ function. In TCVM, each organ is considered to be a holistic energy system rather than the actual anatomical organ as identified by Western physiology.

The yin-yang concept and the five-element theory fundamentally support the TCVM theory philosophically, and the basic treatment principle of TCVM concerns maintaining the balance of yin and yang and the balance of the five elements. However, these two theories are not sufficient to interpret more complicated cases in actual practice. Other important principles of TCVM include the theory of viscera (internal organs), the theory of meridian (the longitudinal pathways on the body where acupuncture points are distributed), and the identification of patterns according to states of vital substances such as Qi energy, blood, body fluids, and mind (Wang, 2005).

15.4 DIAGNOSIS AND THERAPEUTICS

TCVM diagnostic methods vary depending on the different animal species to which they are applied. Multiple factors have to be considered during diagnosis, including external factors (e.g., heat and damp climate) and internal factors (e.g., over-, or lack of, exercise). Four basic diagnostic techniques are applied to determine the correct therapeutic pattern for a certain animal and a certain disease; these techniques are observation (e.g., observing physical expression and symptoms of disease); listening and smelling (e.g., listening to the animal's sounds and breathing, smelling secretions and excretions); interrogation (e.g., asking the animal carer about the appearance or history of the disease); and palpation (e.g., touching or pressing the animal's body, feeling the pulse and other internal organs). After diagnosis, diseases are classified into various patterns (also known as *syndromes*), such as cold-hot patterns and exterior-interior patterns. These patterns or syndromes determine the forms of treatment that will be used for any particular animal, whether TCVM or physiotherapies such as acupuncture, cupping, or scraping.

15.5 HERBAL VETERINARY MEDICINES

Unlike veterinary acupuncture, a therapeutic model of TCVM that has been widely accepted in Western countries since the 1970s (Haas, 2004; Jaggar, 1992), traditional Chinese herbal medicines have not been widely accepted in veterinary use outside the Eastern subcontinent. Before conventional medicine was introduced to China (in the mid-1800s), herbal medicine was the only pharmacotherapy available for animal use in China. In modern Chinese society, medicinal herbs have gradually gained popularity for use in animal health, partly because herbal medicines or herbal additives in animal feeds are believed to result in fewer residues than from conventional drugs (Li, 2008).

Herbal treatments may involve nonherbal materials, as traditional Chinese materia medica includes not only plant species but also animal parts (e.g., toad venom, cicada slough) and minerals (e.g., talcum, calomel). However, as over 90% of TCVM crude materials are of plant origin, they are commonly referred to collectively as "herbal" medicines in most ancient and modern textbooks.

In traditional Chinese medicine, each crude herbal material has to be processed before clinical use. The simplest method of processing is to cut or slice and dry the medicinal parts of the plants. Many herbs need more complicated processing methods to reduce their toxicity or change their medicinal properties. For example, the toxic herb *Aconitum carmichaeli* is processed by soaking in frequently changed water for a prolonged period and then steaming or boiling. The aconitine alkaloids, responsible for cardiac toxicity, are reduced by up to 90% in the resulting processed herb, Fu Zi (*Radix Aconiti Lateralis Preparata*) (Chan et al., 1994).

TCVM herbs are mostly prescribed as compound formulas, and each formula includes 1–15 or more herbal ingredients. After diagnosis, appropriate herbs or standard formulas defined for a specific therapeutic pattern are prescribed. The order of the herbs in each formula reflects the functions of the herbs. The first herbs in a formula are known as the *principal herbs* and target the main symptoms; these are followed by *associate herbs*, which are thought to assist the function of the principal herbs and target other symptoms. The next part of the formula comprises the *adjuvant herbs* for strengthening the function of the principal herbs and reducing toxicity; finally the *messenger herbs* guide the function of all herbs into the right pattern and harmonize the function of the formula.

In TCVM, a course of treatment is highly individualized depending on the particular animal's characteristics (e.g., genera, age, gender) and the TCVM therapeutic patterns after diagnosis. Herbs are rarely listed by medical condition according to Western classification; rather, it is usual to describe which herbs could be considered as priority herbs for a specific diagnosis or disease. In general, approximately 500 herbs have been documented as commonly used in veterinary practice. In practice, the principal herbs in a customized remedy may be varied for the same condition, and some principal herbs in a certain remedy are used as associate herbs or adjuvant herbs in other customized remedies for similar conditions. For example, common constipation occurring in swine can be divided into a few different patterns on diagnosis: constipation due to "excess heat syndrome"; constipation due to "deficiency of yin and blood syndrome"; constipation due to "asthenia cold deficiency" (Wang, 2005). The common principal herb used for constipation of the excess heat syndrome type is Da Huang (*Radix et Rhizoma Rhei*); however, in the remedy for deficiency of yin and blood syndrome, Da Huang is only considered as an associate herb or is not used at all due to its strong purgative effects, which can cause "overdeficiency" or loss of the body's vital substances. For constipation of the asthenia cold deficiency type, different formulations, such as suppositories, are recommended since the oral administration of herbal decoctions may cause strong bowel reactions in swine with this type of constipation.

The latest *CVP* (Committee of China Veterinary Pharmacopeia, 2005) records around 200 standard herbal formulas for various veterinary conditions based on different TCVM therapeutic patterns. In most cases, the standard or documented formula for a certain pattern or syndrome is modified by the addition or removal of herbal ingredients depending on differences in the condition of the particular animal.

Methods for preparing herbal remedies include making a decoction of the mixture of herbs, making porridges of herbs and grains, or grinding herbs into granules (to be fed to the animal after mixing with water). Herbal decoctions are commonly prepared with water, but in some cases, herbs are boiled in rice water or with liquor to enhance their therapeutic functions. The time of administration (e.g., feeding with warm or cold medicine, before or after foraging) varies depending on the disease being treated. For example, animals suffering from spleen-stomach weakness, such as diarrhea, or suffering from painful conditions (or inflammation) are normally fed with decoctions before foraging; the associated herbs for this particular method include Hou Pu (*Cortex Magnoliae Officinalis)* and Sha Ren (*Frutus Amomi*). Decoctions are administered before foraging to improve absorption of the medicine and reduce digestive burdens on the stomach (Wang, 1991).

15.6 SCIENTIFIC EVIDENCE AND MODERN VETERINARY USES OF TCVM

TCVM uses a large number of natural materials and is applied to the majority of veterinary conditions in China. However, the therapeutic effects of Chinese herbs are mostly described in the literature by traditional classifications using Chinese medical terms, which may not be the same as the actual pharmacological effects of the treatment, particularly as described for conventional pharmaceutical drugs. Obtaining scientific evidence for TCVM theory itself is problematic since the theory is based on a highly abstract philosophy.

In China, many herbs have been used to replace pharmaceuticals because herbs are perceived to be less toxic and reduce the presence of animal drug residues; in particular, some herbs are used as additives in animal feeds alone or in conjunction with conventional drugs. However, the evidence that herbs could reduce therapeutic toxicity and produce fewer residues has not yet been well established.

Modern phytochemistry and pharmacological sciences have been widely applied to the study of traditional veterinary medicine. Some TCVMs have been found to have several pharmacological effects (described using Western medical terminology), including antimicrobial and immunomodulatory properties and ability to improve reproductivity of animals (Yang, 2005). As many plant species are used traditionally for both animal and human health, the results from some scientific studies initially designed to assess treatments for improving human health have been used in improving or modifying TCVM treatments. For example, the roots of Dang Shen (*Radix Codonopsis,* referred species *Codonopsis pilosula*) are traditionally used for tonic purposes. However, the other plant parts of *Codonopsis pilosula*, such as the stems and leaves, have been found to contain high concentrations of saponins, which can be used to feed poultry to increase their egg-laying rate (Liang, Zhang, and Li, 2005). Table 15.2 summarizes phytochemical and pharmacological information for several TCVM herbs mentioned in this chapter, as well as their traditional descriptions (herbs recorded in *CVP* [Committee of China Veterinary Pharmacopeia, 2005] are selected in the table as they are popular in TCVM practice).

Studies assessing the efficacy of Chinese herbs specifically for animal uses are published mostly in Chinese scientific journals, and few are available in the English

TABLE 15.2
Traditional Descriptions, Chemical Constituents, and Pharmacological Data of TCVM Herbs Described in This Chapter

Pharmacopoeial Name[a]	Chinese Name	Botanical Name[a]	TCVM Indications[a,b]	Common Dosage per Formula[a]	Main Chemical Constituents[b–d]	Pharmacological Effects[b,d]
Cortex Magnoliae Officinalis	Hou Pu	*Magnolia officinalis*	Indigestion, food accumulation due to Qi stagnation, abdominal distension, constipation, vomiting, nausea, coughing and wheezing due to Qi-stagnation	*Equus & Bos* 15–45 g; *Camelus* 30–60 g; *Ovis & Suidae* 5–15 g; *Leporidae & Aves* 1.5–3 g	Lignans (magnolol, tetrahydromagnolol, isomagnolol, honokiol, magnaldehyde), alkaloids (magnocurarine, magnoflorine, salicifoline), β-eudesmol	Antibacterial effects (gram-positive bacteria); antifungal effects; muscle relaxant activity; preventing water immersion stress ulceration and gastric hemorrhage induced by stress; antiallergic effects; lowering blood pressure; inhibiting blood platelet concentration
Flos Lonicerae	Jin Yin Hua	*Lonicera japonica*	Fever due to wind-heat, cough due to excessive lung heat, sore throat, dysentery due to heat toxins, abscesses	*Equus & Bos* 15–60 g; *Ovis & Suidae* 5–10 g; *Canis & Felis* 3–5 g; *Leporidae & Aves* 1–3 g; fishes 3–5 g/kg	Chlorogenic acids, triterpenoid saponins, flavonoids (luteolin, corymbosin), linalool, tannins, luteolin, lonicerin, inositol, loganin	Anti-inflammatory effects; antibacterial effects (*Staphylococcus aureus, S. albus, Salmonella typhi*); controlling influenza viruses in mice; inducing abortion in the early stages of pregnancy in animal tests
Fructus Crataegi	Shan Zha	*Crataegus pinnatifida*	Indigestion, stomach distension, food accumulation	*Equus & Bos* 20–60 g; *Ovis & Suidae* 10–15 g; *Canis & Felis* 3–6 g; *Leporidae & Aves* 1–2 g	Crataegolic acid, oleanolic acid, ursolic acid, chlorogenic acid, malic acid, citric acid, quercetin, vitexin, hyperoside, epicatechin, sitosterol, vanillic aldehyde, carotene, flavan polymers, glycosides	Improving digestion and increasing secretion of digestive juices; dilating blood vessels and lowering blood pressure; increasing blood flow in the coronary arteries; decreasing cholesterol concentrations

(*continued*)

TABLE 15.2 (CONTINUED)
Traditional Descriptions, Chemical Constituents, and Pharmacological Data of TCVM Herbs Described in This Chapter

Pharmacopoeial Name[a]	Chinese Name	Botanical Name[a]	TCVM Indications[a,b]	Common Dosage per Formula[a]	Main Chemical Constituents[b–d]	Pharmacological Effects[b,d]
Fructus Lycii	Gou Qi Zi	*Lycium barbarum*	Liver and kidney yin deficiency, dizziness, blurred vision, seminal emission, limpness and aching in the lumbar areas and knees	*Equus & Bos* 15–60 g; *Ovis & Suidae* 10–15 g; *Canis & Felis* 3–8 g	Polysaccharides, carotene, thiamine, betaine, riboflavin, nicotinic acid, ascorbic acid, β-sitosterol, linoleic acid, physalien, atropine, scopoletin, zeaxanthin	Stimulating respiration in rabbits; lowering blood pressure; increasing testosterone concentration; reducing blood lipid levels and blood sugar concentrations; strengthening uterine contractions; improving immunity
Frutus Amomi	Sha Ren	*Amomum villosum, A. longiligulare*	Indigestion due to Qi-stagnation, abdominal and stomach distension, vomiting, upset stomach, diarrhea due to cold deficiency	*Equus & Bos* 15–30 g; *Ovis & Suidae* 3–10 g; *Leporidae & Aves* 1–2 g	Volatile oils (camphor, borneol, bornyl acetate, camphene, linalool, limonene, nerolidol, β-pinene, α-pinene, guaiaol), saponins, flavonoids	Having therapeutic effects for peptic ulcers; inhibiting blood platelet concentration; antibacterial (*Proteus vulgaris, Shigella shigae, Bacterium coli*); promoting movement in the intestinal tract
Herba Epimedii	Yin Yang Huo	*Epimedium brevicornum, E. pubescens, E. wushanense, E. koreanum*	Impotence due to kidney yang deficiency, lack of strength in the lumbar area and knees, rheumatism or numbness, nonestrus of female animals	*Equus & Bos* 15–30 g; *Ovis & Suidae* 10–15 g; *Leporidae & Aves* 0.5–1.5 g	Flavonoids (icariin, epimedosides, icarisoside, sagittatosides, wushanicariin, epimedokoreanoside, anthocyanin, baohuoside), anthraquinones, volatile oils, polysaccharides, ceryl alcohol, phytosterol, palmitic acid, oleic acid, linoleic acid	Strengthening immunological functions; improving sexual ability due to its male hormone-like functions; lowering blood pressure; preventing coronary heart disease; antibacterial effects; relieving and suppressing coughing; calming wheezing; tranquilizing functions

Herba Verbenae	Ma Bian Cao	*Verbena officinalis*	Blood stasis postpartum, water retention, stomach distension, abscesses and sores	*Equus & Bos* 30–120 g; *Ovis & Suidae* 15–30 g	Glycosides (iridoid glycosides, verbenalin, hastatoside, eukovoside, verbascoside), β-sitosterol, ursolic acid, artemetin, triterpenes, lupeol	Anti-inflammatory effects on conjunctivitis in rabbits; analgesic effects; relieving and suppressing coughing
Radix Aconiti Lateralis Preparata	Fu Zi	*Aconitum carmichaeli*	Yang collapse and deficiency, cold limbs, weak pulse, abdominal pain and loose stools due to yang deficiency, impotence, rheumatism, or numbness caused by draught, cold, or damp	*Equus & Bos* 15–30 g; *Ovis & Suidae* 3–9 g; *Canis & Felis* 1–3 g; *Leporidae & Aves* 0.5–1 g	Diterpenoid alkaloids (aconitine, mesaconitine, hypaconitine, talatisamine, benzoylaconine, benzoylmesaconine, benzoylhypaconine), aminophenols, hygenamine, coryneine, salsolinol	Contains high concentrations of toxic alkaloids: oral administration of 1.0–3.0 mg/kg of aconite results in death in mice, LD_{50} of aconitine (intravenous injection; mice) = 0.22 mg/kg; causing general paralysis and respiratory arrest in animals; cardiotonic effects; arrhythmia effects; enlarging cardiovascular and cerebral vessels; preventing coma; tranquilizing function; analgesic effects; anti-inflammatory effects
Radix Angelicae Sinensis	Dang Gui	*Angelica sinensis*	Blood stasis, exterior injuries, abscesses and sores, constipation due to dry bowels	*Equus & Bos* 15–60 g; *Camelus* 35–75 g; *Ovis & Suidae* 5–15 g; *Canis & Felis* 2–5 g; *Leporidae & Aves* 1–2 g	Volatile oils (n-butylide phthalide, ligustilide, α-pinene, β-pinene, camphene, *p*-cymene, myrcene), *p*-methyl-benzoalcohol, ferulic acid, succinic acid, nicotinic acid, vitamin B_{12}, B_1, E	Inhibiting platelet aggregation and serotonin; relaxing effects on animal tracheal smooth muscles; antiasthmatic and spasmolytic activity; improving microcirculation; regulating uterus function; inhibiting blood platelet concentration, arrhythmia, and cardiac ischemia; reducing blood lipid concentrations

(continued)

TABLE 15.2 (CONTINUED)
Traditional Descriptions, Chemical Constituents, and Pharmacological Data of TCVM Herbs Described in This Chapter

Pharmacopoeial Name[a]	Chinese Name	Botanical Name[a]	TCVM Indications[a,b]	Common Dosage per Formula[a]	Main Chemical Constituents[b–d]	Pharmacological Effects[b,d]
Radix Astragali	Huang Qi	*Astragalus membranaceus*	Deficiency of spleen and lungs Qi, exterior weakness, spontaneous sweating, swelling due to Qi deficiency, sores, abscesses, ulcers	*Equus & Bos* 20–60 g; *Camelus* 30–80 g; *Ovis & Suidae* 5–15 g; *Leporidae & Aves* 1–2 g	Astragalus polysaccharides, astragalosides, soyasaponin, daucosterol, β-sitosterol, palmitic acid, kumatakenin, choline, betaine, folic acid, formononetin, calycosin, cycloastragenol	Enhancing immunity; therapeutic effects for chronic hepatitis (preventing decrease of hepatic glycogen contents and raising the total serum protein and album concentrations); anti-inflammatory effects; antibacterial effects (*Shigella shigae, Haemolytic streptococci, Staphylococcus aureus*); reducing free radicals; inhibiting gastric secretion and preventing gastric ulcers; lowering blood sugar
Radix Bupleuri	Chai Hu	*Bupleurum chinense*	Cold and flu, high fever, alternate attack of chills and fevers, diarrhea due to spleen deficiency, prolapse of rectum or uterus	*Equus & Bos* 15–45 g; *Ovis & Suidae* 3–10 g; *Leporidae & Aves* 1–3 g	Saponins (saikosaponins, saikogenin, longispinogenin), volatile oils (bupleurumol, limonene, myrcene), oleic acid, linolenic acid, palmitic acid, α-spinasterol	Antipyretic effects; preventing stress ulcers in rats; inhibiting the histologic damage of liver tissue in rats; moderate analgesic effects in mice; tranquilizing functions; relieving and suppressing coughing; antibacterial and antiviral effects; anti-inflammatory effects

Radix Codonopsis	Dang Shen	*Codonopsis pilosula, C. tangshen*	Deficiency of energy Qi and blood, reduced appetite, coughing and wheezing due to lung Qi deficiency, loose stools, fatigue and weakness	*Equus & Bos* 20–60 g; *Ovis & Suidae* 5–10 g; *Leporidae & Aves* 0.5–1.5 g	Sterols (stigmasterol, α-spinasterol), polysaccharides, phenolic acids, terpenes (atractylenolide), inulin, fructose, choline, caproic acid, enanthic acid, pinene	Inhibiting ulcers and gastric secretion, preventing gastric mucous membrane damage; strengthening cardiac contractions; increasing blood pressure; inhibiting blood platelet concentration; improving immunity
Radix Codonopsis Lanceolataee	Wu Ye Shen/Si Ye Shen[f]	*Codonopsis lanceolata*	Weakness after diseases, agalactia, lung abscess, painful swelling, sores, and abscesses	*Equus & Bos* 30–150 g; *Ovis & Suidae* 15–30 g	Triterpene saponins (oleanane-type saponins, taraxerone), flavonoids (luteolin, apigenin, tectoridin), alkaloids (perlolyrien, norharman), stigmasterol, syringarsinol, α-spinasterol, hexadecane acid, succinic acid[e]	Antioxidant effects; increasing the activity of SOD (superoxide dismutase) enzyme; antifatigue in rats; reducing blood pressure in rabbit; improving immunity in mice; tranquilizing functions; analgesic effects; antibacterial effects; relieving and suppressing coughing in mice; improving memory in mice[e]
Radix et Rhizoma Ginseng	Ren Shen	*Panax ginseng*	Fatigue, lack of strength, cold limbs, weak pulse, spontaneous sweating, palpitations, coughing and wheezing due to lung qi deficiency, reduced appetite	*Equus & Bos* 15–30 g; *Ovis & Suidae* 5–10 g; *Canis & Felis* 0.5–2 g	Ginsenosides (protopanaxadiol, protopanaxatriol), volatile oil (α-guaiene, β-patchoulene, *trans*-caryophyllene, humulene, elemene), panaxynol, panaxydol, linoleic acid, citric acid, panax acid, vitamins, sterols, carbohydrates	Improving immunity; improving memory; decreasing heart rate and central venous pressure in dogs; decreasing blood pressure; preventing cardiac ischemia; dilating blood vessels; preventing blood blotting; improving endocrine functions; delaying anesthesia induction by alcohol in rats and rabbits when pretreated with the herb; antitumor effects; stimulating pituitary-adrenocortical system

(*continued*)

TABLE 15.2 (CONTINUED)
Traditional Descriptions, Chemical Constituents, and Pharmacological Data of TCVM Herbs Described in This Chapter

Pharmacopoeial Name[a]	Chinese Name	Botanical Name[a]	TCVM Indications[a,b]	Common Dosage per Formula[a]	Main Chemical Constituents[b–d]	Pharmacological Effects[b,d]
Radix et Rhizoma Rhei	Da Huang	*Rheum palmatum, R. tanguticum, R. officinale*	Constipation due to excessive heat, stomach distension and pain, abscesses and sores, acute conjunctivitis due to wind-heat, external injuries	*Equus & Bos* 30–120 g; *Camelus* 30–90 g; *Ovis & Suidae* 6–12 g; *Canis & Felis* 3–5 g; *Leporidae & Aves* 1.5–3 g	Anthraquinone derivatives (aloe emodin, chrysaron, chrysophanol, emodin, isoemodin, physcion, rhein), tannins (glucose pyrogallol tannins, catechu tannin, free gallic acid), sennoside, rheinosides, anthranol, anthrone	Purgative effects due to stimulating middle part of the colon; inducing secretion of bile; antibacterial effects (*Bacillus subtilis, Staphylococcus aureus, Escherichia coli, Micrococcus luteus, Candida albicans, Clostridium perfringens, Fusobacterium varium, Bacteriodes fragilis*); antipyretic effects; anti-inflammatory effects; improving immunity; diuretic effects in rabbits; hemostatic activity; preventing gastric ulcers and bleeding
Radix Glycyrrhizae	Gan Cao	*Glycyrrhiza uralensis, G. inflata, G. glabra*	Fatigue, lack of strength, coughing and wheezing, sore throat, sores, and abscesses	*Equus & Bos* 15–60 g; *Camelus* 45–100 g; *Ovis & Suidae* 3–10 g; *Canis & Felis* 1–5 g; *Leporidae & Aves* 0.6–3 g	Triterpenoid saponins (glycyrrhizin, glycyrrhizic acid, glycyrrhetic acid, glabrolide), flavonoids (liquiritin, liquritigenin, isoliquiritin, neoliquiritin, liqurazid, licoricone, isolicoflavonol), urease, sucrase enzymes	Inhibiting gastric secretion; prolonging biological effects of cortisol, aldosterone, and testosterone by delaying clearance of corticosteroids during metabolism; antibacterial effects; antiallergic effects; promoting urination; relieving gastrointestinal spasm; inhibiting hepatitis viruses; relieving coughing and promoting bronchial secretion; lowering blood lipid concentrations
Radix Isatis	Ban Lan Gen	*Isatis indigotica*	Cold and flu due to wind-heat, sore throat,	*Equus & Bos* 30–100 g; *Ovis & Suidae* 15–30 g;	Indoxyl-β-glucoside, isatin, β-sitosterol, γ-sitosterol,	Antibacterial effects (*Staphylococcus aureus,*

			mouth ulcer, sores and abscesses	*Canis & Felis* 3–5 g; *Leporidae & Aves* 1–2 g; fishes 1–2 g/kg	amino acids, epigoitrin	*Escherichia coli, Shigella shigae, Salmonella enteridis*)
Radix Puerariae Lobatae	Ge Gen	*Pueraria lobata*	Fever due to wind-cold or wind-heat, first stage of measles, thirst due to stomach heat, diarrhea due to spleen deficiency	*Equus & Bos* 20–60 g; *Ovis & Suidae* 5–15 g; *Leporidae & Aves* 1.5–3 g	Isoflavones (daidzin, daidzein, pueroside, puerarin), arachidic acid, daucosterol, allantoin	Antipyretic effects; lowering blood pressure in dogs; relaxing coronary vessels and improving coronary circulation
Radix Scutellariae	Huang Qin	*Scutellaria baicalensis*	Cough due to excessive lung heat, diarrhea, jaundice, high fever, thirst, sore and swollen throat, vomiting of blood, uterine, bleeding, hemafecia, abscess, threatened abortion due to internal heat	*Equus & Bos* 20–60 g; *Ovis & Suidae* 5–15 g; *Leporidae & Aves* 1.5–2.5 g; fishes 2–4 g/kg	Flavonoids (baicalin, baicalein, wogonoside, wogonin, neobaicalein, skullcap flavones, oroxylin A, 7-methoxybaicalein, 7-methoxynorwogonin), palmitic acid, oleic acid, benzoic acid, sterols (β-sitosterol, stigmasterol)	Antibacterial effects (*Vibrio comma, Staphylococcus aureus, Shigella shigae, Corynebacterium diphtheriae, Pseudomonas aeruginosa, Bacterioides melamnogenicus intermedius,* and other gram-positive bacteria); antipyretic effects; antihistaminic, anticholinergic, and papaverine-like activity in guinea pigs; reducing blood lipid concentrations
Rhizoma Coptidis	Huang Lian	*Coptis chinensis, C. deltoidea, C. teeta*	Diarrhea, dysentery and vomiting due to damp-heat in stomach and intestines, sore and swollen throat, mouth sores, red and swollen eyes, abscesses and ulcers	*Equus & Bos* 15–30 g; *Camelus* 25–45 g; *Ovis & Suidae* 5–10 g; *Leporidae & Aves* 0.5–1 g; fishes 2–5 g/kg	Isoquinoline alkaloids (berberine, coptisine, worenine, palmatine, columbamine, jatrorrhizine, epiberberine, magnoflorine), ferulic acid, chlorogenic acid	Antibacterial effects (*Staphylococcus aureus, S. hemolyticus, Salmonella typhosa, Shigella dysenteriae, S. paradysenteriae, Escherichia coli, Neisseria gonorrhoeae, Diplococcus pneumoniae*); anti-inflammatory effects; antipyretic, effects; inhibiting the growth of Ehrlich and lymphoma ascites tumor

(*continued*)

TABLE 15.2 (CONTINUED)
Traditional Descriptions, Chemical Constituents, and Pharmacological Data of TCVM Herbs Described in This Chapter

Pharmacopoeial Name[a]	Chinese Name	Botanical Name[a]	TCVM Indications[a,b]	Common Dosage per Formula[a]	Main Chemical Constituents[b–d]	Pharmacological Effects[b,d]
						cells; lowering blood pressure in rats and rabbits; increasing bile bilirubin concentration in rats; decreasing serum cholesterol concentrations
Rhizoma Dioscoreae	Shan Yao	*Dioscorea apposita*	Reduced appetite, loose stools due to spleen deficiency, coughing and wheezing, frequent uresis	*Equus & Bos* 30–90 g; *Ovis & Suidae* 10–15 g; *Leporidae & Aves* 1.5–3 g	Mannan, diosgenin, choline, saponins, starch, glycoproteins, batatasine, phytic acid, allantoin, diastase enzymes, vitamin C	Improving secretion of biliary cholesterol, bile salt and phospholipids; improving digestion; lowering blood sugar concentrations; improving immunity; expelling sputum
Rhizoma Paridis	Chong Lou	*Paris polyphylla*	Cough due to excessive lung heat, sore throat, swollen and toxic abscesses, fright epilepsy, convulsions, snakebites, external injuries	*Equus & Bos* 30–60 g; *Ovis & Suidae* 15–60 g	Saponins (pariphyllin, dioscin, diosgenin, pennogenin), alkaloids, polysaccharides, aspartic acid, asparagine, alanine, amino acids	Relieving and suppressing coughing and wheezing; antibacterial effects (*Staphylococcus aureus, Paracolon bacilli, Shigella shigae*); antivirus (type A Asian influenza virus); tranquilizing function; inhibiting carcinoma

[a] Committee of China Veterinary Pharmacopeia (2005).
[b] Xu and Wang (2002).
[c] Zheng (1999).
[d] Tang and Eisenbrand (1992).
[e] References in notes b–d did not record *Radix Codonopsis Lanceolatae*; therefore, chemical and pharmacological data of this herb were compiled from Wu, Pu, and Han (2005).
[f] Chinese name was recorded as Si Ye Shen in *CVP* 2005.

literature. Most of the articles published in the English literature described studies examining the immunomodulatory and antibacterial or antiviral effects of TCVM. For example, a Chinese herb, Ma Bian Cao (*Herba Verbenae*), at a concentration of 1% (vol/vol) increased bovine blood neutrophil activity by over 40% compared with control in an *in vitro* study (Hu et al., 1992). *Herba verbenae* was recorded in the ancient Chinese book *Ben Cao Gang Mu* as a treatment for equine pharyngitis, and it is also included in the *CVP* (Committee of China Veterinary Pharmacopeia, 2005). The study also demonstrated that herbs traditionally believed to have antipyretic effects (*CVP*, 2005), such as Chai Hu (*Radix Bupleuri*) and Jin Yin Hua (*Flos Lonicerae*), could stimulate bovine milk neutrophil functions and increase phagocytosis.

Other *in vivo* and *in vitro* studies have examined the immunostimulatory effects of active compounds extracted from Chinese herbs for use as adjuvants to animal vaccines, such as Newcastle disease vaccine for chickens and rabbit hemorrhagic disease vaccine (Kong et al., 2006; Sun et al., 2006; Wang et al., 2005). For example, Kong et al. (2006) showed that *Astragalus* polysaccharide (constituent of Huang Qi, *Radix Astragali*), *Isatis* root polysaccharide (constituent of Ban Lan Gen, *Radix Isatis*), and *Epimedium* flavone (constituent of Yin Yang Huo, *Herba Epimedii*) could optimally enhance the cell proliferation of chick embryo fibroblasts in response to the Newcastle disease virus when applied at concentrations of 600, 150, and 20 μg/mL, respectively, compared with the relative controls containing no active herbal constituents ($p < .05$) (Kong et al., 2006). This study also indicated the importance of correct concentrations of active compounds used in developing immune stimulators or adjuvants for the vaccine. In another study (Sun et al., 2006), carried out in zelanian rabbits vaccinated against rabbit hemorrhagic disease, a rapid increase in serum antibody titer (determined by the hemagglutination inhibition test) was observed after injections of a mixture of Chinese herbal compounds containing *Astragalus* polysaccharide and ginsenoside when compared with the control group. Sun et al. suggested that chemical constituents in Chinese herbs may play an important role in improving immune response by enhancing peripheral T-lymphocyte proliferation and serum hemagglutination inhibition antibody titer; therefore, they could be considered for use as immunopotentiators.

Although isolated chemical constituents from some herbs have been shown to have certain pharmacological effects, it is not clear whether the same therapeutic effects will occur if several herbs are used in combination, as is more frequently the case in TCVM. With combination herbal treatments, there is the potential for synergistic or antagonistic interactions to occur during preparation or administration. Some veterinary studies have investigated the effects of TCVM mixtures (Lin et al., 1988; Lin et al., 2000). In a comparative study, a TCVM formula (0.5 g), Koken-Huanglien-Huangchin-Tang, containing Ge Gen (*Radix Puerariae Lobatae*), Huang Qin (*Radix Scutellariae*), Huang Lian (*Rhizoma Coptidis*), and Gan Cao (*Radix Glycyrrhizae*) was administered orally twice a day to 1-day-old piglets with preweaning diarrhea. The results showed that, compared with the control (0.5 g lactose), the formula reduced the incidence of infection ($p < .1$) during the first 10 days of life and significantly reduced the duration of preweaning diarrhea ($p < .01$) (Lin et al., 1988).

A recently developed TCVM formula, Yi-Kang-Kong, was claimed to have potential antiviral activity against transmissible gastroenteritis virus (TGEV) in swine

(Lin et al., 2000). Yi-Kang-Kong is composed of several herbal extracts, including Gou Qi Zi (*Fructus Lycii*), Huang Qi (*Radix Astragali*), Shan Zha (*Fructus Crataegi*), Chong Lou (*Rhizoma Paridis*), Dang Gui (*Radix Angelicae Sinensis*), Bai She Hua Cao (*Ancistrodon acutus*), Shan Yao (*Rhizoma Dioscoreae*), and Wu Ye Shen (*Radix Codonopsis Lanceolatae*). An *in vitro* study found that Yi-Kang-Kong extract at a concentration of 10 μg/mL could decrease TGEV infection in the cultured swine testis (ST) cells by 45% at a temperature of 25°C, and that the infectivity was reduced by around 25% in ST cells pretreated with the extract for over 2 h (Lin et al., 2000).

Although potential synergistic or antagonistic effects may exist in a combined TCVM formula, little research has been done to examine this issue. Sun et al. (2006) also compared the immune-enhancing effects of mixtures of Chinese herbal isolated active constituents (e.g., *Astragalus* polysaccharide and ginsenoside) and that of a mixture of Chinese herbal extracts (e.g., water decoction of herbs, *Astragalus*, and Ren Shen/ginseng). The results showed that, based on the equivalent active constituents, Chinese herbal extracts were less effective than the mixture of isolated constituents, although both types of preparation achieved positive results in the immuno-enhancement trials (Sun et al., 2006). This study may indicate the potential antagonistic effects between different chemical constituents in the herbal mixture. However, in practice, the use of isolated active constituents rather than prepared decoctions of herbal mixtures may not necessarily be considered as a therapeutic priority without understanding the toxicity of both types of preparation. Moreover, there is a lack of toxicity studies on TCVM herbs or their active constituents.

It is beyond the scope of this chapter to review all available studies published in Chinese databases as studies published in the Chinese literature cover a wide range of therapeutic uses. At present, common types of veterinary herbal medicines used in China include herb-only traditional remedies (which may include animal and mineral products as TCVM ingredients), integrated traditional herbs with conventional drugs, active compounds isolated from Chinese herbs, and herbs as additives to animal feeds.

Over 1,500 articles associated with Chinese herbal veterinary treatments have been published since the 1970s and are recorded in the China Academic Journal Database (http://chinanew.eastview.com/). Most clinical articles in the Chinese literature described case reports of effectiveness of TCVM treatments rather than controlled studies assessing efficacy. However, in recent years authors and scientists have applied the principles of evidence-based medicine to veterinary medicine: Evidence-based veterinary medicine (EBVM) has been suggested to assess the quality of scientific evidence for veterinary medicines (Holmes and Cockcroft, 2004; Kastelic, 2006; Schmidt, 2007). EBVM uses a systematic approach to reviewing all published scientific evidence for veterinary treatments and aims to improve standards for veterinary drug treatment trials. However, current published reviews mostly focused on orthodox medical treatments; only a few involved TCVM. A systematic review of 40 studies for canine atopic dermatitis (published in 1980–2002) included several studies assessing Chinese herbal remedies for this condition; the review concluded that there was insufficient evidence to recommend or to advise against the use of Chinese herbal treatments (Olivry, Mueller, and The International Task Force on Canine Atopic Dermatitis, 2003). The study implied that systematic review of the

Chinese literature may be necessary to assess scientific evidence for efficacy and safety of TCVM for specific animal diseases. Such a review of the extensive existing literature is needed to identify areas in which there is adequate evidence of safety or efficacy and those for which more studies are required.

15.7 CURRENT STATUS OF TCVM IN CHINA

At present, in China TCVM is regarded as a mainstream veterinary treatment approach alongside conventional veterinary drugs. TCVM products are regulated under the same regulatory frameworks as are traditional and conventional medicines, and manufacturers of TCVM products are required to comply with the principles of good manufacturing practice (GMP) to demonstrate and ensure product quality. TCVM is also a compulsory subject required for students of veterinary medicine in most Chinese universities offering a 5-year veterinary medicine course. However, unlike traditional Chinese medicine for human health, training in TCVM is offered by few universities in China as an individual degree course; therefore, the number of TCVM practitioners is probably far fewer than that of traditional Chinese medical doctors in China. Nevertheless, TCVM remains a popular approach on many Chinese farms and in veterinary clinics as it is believed to be able to reduce potential adverse effects from conventional drugs, although evidence for this is lacking.

Compared with traditional Chinese medicine for human use, TCVM is a relatively new subject for modern scientific studies in China. Additional rigorous research is needed to investigate the efficacy and safety, including the toxicity of traditional herbs, herbal residues in animals, and herb–drug interactions, of TCVM and ultimately its cost-effectiveness. Moreover, it has been suggested that the exploitation of new medicinal plant resources for veterinary uses in China is important to avoid competition with the medicinal plants used for human treatment (Liang, Zhang, and Li, 2005). This suggestion implies that there is currently competition for supplies of specific herbs that are used for both animals and humans, and that there may not be sufficient supply for both markets.

15.8 VETERINARY FOLK MEDICINES IN CHINA

Although TCVM is the major and best-known form of ethnoveterinary medicine in China, it is not the only ethnoveterinary approach currently used among Chinese societies. Many of the 56 ethnic minority groups in China have their own traditional plants as folk veterinary medicines. Some of these "folk herbs" may not be recorded in Chinese veterinary textbooks or pharmacopoeia. The practice of veterinary medicine by some minority groups may have similarities to that of TCVM. For example, a preliminary investigation of folk veterinary medicine in Bouyei society in southwest China found some similarities with TCVM in diagnostic techniques (e.g., interrogation, observation, and palpation) and therapeutic approaches (e.g., internal herbal decoction, needle stimulation, and cupping) (Lu, 1999). The study identified a few unique practice methods used in this society, including bloodletting and bamboo tube therapy. They found that Bouyei folk veterinary medicine is believed to be effective for bone setting and external injuries, such as snakebites, that plant names are based on the plant's reputed therapeutic function, and that veterinary practice is still influenced by local shamanism. However, the study did not collect specimens of the plant resources used, so no botanical or scientific validation of the plant uses was possible. Although this type of investigation is rarely published in the English or Chinese literature, it still provides a good ethnomedical source for understanding ethnoveterinary medicine in some remote areas in China.

Several other ethnobotanical studies have been carried out among ethnic minority groups in China, although most have explored medicines used for human health (Chen, Chen, and Xiao, 2003; Pei et al., 2006). Information on the medicinal plants identified in these studies is potentially valuable for development of ethnoveterinary plant medicines. However, the current practices and features of ethnoveterinary medicines within ethnic minority societies in China are rarely explored and documented. In addition, some veterinary medicines used in ethnic minority groups in China might evolve under the influence of TCVM and modern (Western) veterinary medicine. Therefore, it is suggested that more ethnoveterinary investigations are needed to explore veterinary practice in remote areas of China and to record their traditional knowledge.

15.9 CONCLUSION

TCVM has developed based on ancient Chinese philosophy. At present, it is an integral part of veterinary medicine in China. However, there have been limited systematic investigations of plants used specifically for TCVM treatments. Further studies investigating efficacy and safety of TCVM and systematic reviews of the existing literature are needed to support traditional veterinary uses, as well as ethnoveterinary studies to record traditional knowledge, particularly in remote areas. Collectively, this approach may help identify safe and effective new veterinary medicines for pharmaceutical development.

REFERENCES

Chan, T.Y., Tomlinson, B., Tse, L.K., Chan, J.C., Chan, W.W., and Critchley, J.A. (1994). Aconitine poisoning due to Chinese herbal medicines: a review. *Veterinary and Human Toxicology*, 36(5), 452–455.

Chen, S.B., Chen, S.L., and Xiao, P.G. (2003). Ethnopharmacological investigations on Thalictrum plants in China. *Journal of Asian Natural Products Research*, 5(4), 263–271.

Chinese Academy of Agricultural Sciences. (1963). *The comprehensive collections of Yuanheng Liaoma Niutuo Jing*. Beijing: China Agriculture Press.

Commission of Chinese Veterinary Pharmacopeia. (2005). *Veterinary Pharmacopoeia of the People's Republic of China 2005*, Vol. 2, in Chinese. Beijing: China Agriculture Press.

Commission of Chinese Veterinary Pharmacopeia. (2008). *Veterinary Pharmacopoeia of the People's Republic of China 2005*, English Edition. Beijing: Chemical Industry Press.

Feng, H.Q. (2000). The investigation and annotation of veterinary formulae in Ben Cao Gan Mu. *Agricultural Archaeology*, 3, 289–292.

Haas, K.B. (2004). Acutherapy over the ages. *Journal of Equine Veterinary Science*, 24(5), 182–183.

He, T.C. (2007). Ancient literature of animal husbandry and veterinary medicine in China (from Han to Qing dynasty). *Shanghai Journal of Animal Husbandry and Veterinary Medicine*, 1, 43–47.

Holmes, M., and Cockcroft, P. (2004). Evidence-based veterinary medicine 1. Why is it important and what skills are needed? *In Practice,* 26(1), 28–33.

Hu, S., Cai, W., Ye, J., Qian, Z., and Sun, Z. (1992). Influence of medicinal herbs on phagocytosis by bovine neutrophils. *Journal of Veterinary Medicine Series A*, 39(8), 593–599.

Jaggar, D. (1992). History and basic introduction to veterinary acupuncture. *Problems in Veterinary Medicine*, 4(1), 1–11.

Kastelic, J.P. (2006). Critical evaluation of scientific articles and other sources of information: an introduction to evidence-based veterinary medicine. *Theriogenology*, 66(3), 534–542.

Kong, X.F., Hu, Y.L., Yin, Y.L., Wu, G.Y., Rui, R., Wang, D.Y., and Yang, C.B. (2006). Chinese herbal ingredients are effective immune stimulators for chickens infected with the Newcastle disease virus. *Poultry Science*, 85(12), 2169–2175.

Li, Z.X. (2008). Preliminary discussion on the trends in development of modern veterinary Chinese medicine. *Science and Technology Information*, 11, 46.

Liang, J.P., Zhang, Y.L., and Li, Z.R. (2005). Prospects of research and development of Chinese herbal veterinary medicine. *Journal of Traditional Chinese Veterinary Medicine*, 4, 21–23.

Lin, J.H., Lo, Y.Y., Shu, N.S., Wang, J.S., Lai, T.M., Kung, S.C., and Chan, W.W. (1988). Control of preweaning diarrhoea in piglets by acupuncture and Chinese medicine. *American Journal of Chinese Medicine*, 16(1–2), 75–80.

Lin, S.H., Cheng, I.C., Weng, C.N., and Chen, C.M. (2000). Effects of a Chinese herbal medicine yi-kang-kong on TGEV infectivity in tissue culture. *Journal of the Chinese Society of Veterinary Science,* 26(2), 110–116.

Lu, Q.L. (1999). Preliminary study of folk veterinary medicines in Bouyei people. *Chinese Journal of Veterinary Medicine*, 25(7), 57–58.

Niu, J.P. (1991). The derivation and development of traditional Chinese veterinary medicine. *Agricultural History of China,* 1, 78–85.

Olivry, T., Mueller, R.S., and the International Task Force on Canine Atopic Dermatitis (2003). Evidence-based veterinary dermatology: a systematic review of the pharmacotherapy of canine atopic dermatitis. *Veterinary Dermatology*, 14(3), 121–146.

Pei, D., Yang, L.M., Huang, J., Zhang, X.M., Zhao, J.H., and Pan, L.T. (2006). Comparative study of ethnodrug among Miao nationality, Shui nationality, Buyi nationality and Gelao nationality in Guizhou Province. *Chinese Journal of Ethnomedicine and Ethnopharmacy,* 3, 14–16.

Schmidt, P.L. (2007). Evidence-based veterinary medicine: evolution, revolution, or repackaging of veterinary practice? *Veterinary Clinics of North America—Small Animal Practice*, 37(3), 409–417.

Sun, J.L., Hu, Y.L., Wang, D.Y., Zhang, B.K., and Liu, J.G. (2006). Immunologic enhancement of compound Chinese herbal medicinal ingredients and their efficacy comparison with compound Chinese herbal medicines. *Vaccine*, 24(13), 2343–2348.

Tang, W., and Eisenbrand, G. (1992). *Chinese drugs of plant origin, chemistry, pharmacology and use in traditional and modern medicine.* Berlin: Springer-Verlag.

Wang, D., Hu, Y., Sun, J., Kong, X., Zhang, B., and Liu, J. (2005). Comparative study on adjuvanticity of compound Chinese herbal medicinal ingredients. *Vaccine*, 23(28), 3704–3708.

Wang, D.G. (2005). *Foundation and clinical practice of Chinese veterinary medicine*. Beijing: China Agricultural University Press.

Wang, Y.L. (1991). The preliminary exploration of administration of experienced formulae documented in Yuanheng Liaoma Collections. *Journal of Traditional Chinese Veterinary Medicine,* 1, 40–42,

Wu, L.H., Pu, H.S., and Han, C.J. (2005). Chemical and pharmacological research progress of *Codonopsis lanceolata. Chinese Journal of Information on TCM*, 12(10), 97–99.

Xu, L., and Wang, W. (2002). *Chinese materia medica: combinations and applications.* Hertfordshire, UK: Donica Publishing Ltd.

Yang, Y.F. (2005). The application of Chinese herbal medicine on veterinary medicine and animal husbandry. *Animal Health,* 3, 35–36.

Zheng, J.H. (1999). *Pharmacognosy,* 3rd ed. Beijing: People's Medical Publishing House.

16 Ethnoveterinary Medical Practice in the European Union (EU)

A Case Study of the Netherlands

Tedje van Asseldonk

CONTENTS

16.1 INTRODUCTION

The Netherlands is a small country (41.526 km^2) in the northwest of Europe, consisting mainly of the fertile deltas of three rivers. Nearly half of the country is situated below sea level. The climate is strongly influenced by the sea. There are moderate yearly temperature fluctuations (January 2°C, July 17°C) and rain all year round (50–80 mm each month). The original vegetation consists of the temperate deciduous forest, partly kept open by big grazers. Of this original landscape, nearly nothing is left. At present, the Netherlands has the highest population density in the Western world (465 persons/km^2, nearly twice as much as the United Kingdom, 17 times as much as the United States). The majority of this population is concentrated in the west part of the country (Holland) and nearest the sea (Anonymous, 2001). In spite of this fact, the country is the world's third largest agricultural export nation, both for vegetable (flowers, vegetables) and for animal products. The agricultural activities use 55% of the land and have been modernized through many technological innovations, leading to intensification. For example, pig farms have an average of 1,400 pigs on 13 hectares of land, with megafarms holding up to 16,000 pigs (Wisman, 2003). The dairy producer has an average of 55 dairy cows on the farm, but the 100 largest farms have 270 cows on 54 hectares of land on average (Pierik, 2001). Apart from the millions of cattle and pigs, about 92 million chickens, 1.6 million small ruminants (sheep/goats), and an increasing number of pets (about 135,000 horses and around 5 million cats and dogs) live in the Netherlands (CBS, 2008). The feed for these animals is mainly imported from other countries. The dense concentration of animals makes contagious diseases a great risk for farmers. Not only innovative management systems but also frequent use of antibiotics are used to control and ameliorate against these risks.

Traditional use of herbs for both humans and animals can be found in handbooks and textbooks from 1500 to 1900 (Wagenfeld, 1844; Numan, 1844; Ludwig, 1996); afterward, chemical synthetic medicine gained popularity.

The purpose of this chapter is to discuss present-day use of herbal medicines for animals in the Netherlands, both the traditional use in folk medicine and the use resulting from modern agricultural innovations and from international trends in animal health care.

16.2 DUTCH ETHNOVETERINARY TRADITIONS

Ethnobotanical and ethnoveterinary research taking place in the Netherlands itself are not big topics in Dutch universities and are mainly provided by the Institute for Ethnobotany and Zoopharmacognosy (IEZ), a small private knowledge institute that performed field studies to make inventories of traditional Dutch folk remedies used for animals (Van Asseldonk, 2001; Van Asseldonk and Beijer, 2006). Wageningen University questioned organic farms in the Netherlands for their actual medicine use, and this revealed some use of alternative remedies (Kijlstra et al., 2004; Van der Meulen et al., 2004a, 2004b; Van der Werf et al., 2004a, 2004b). A lack of knowledge regarding quality, safety, and efficacy of these remedies existed both among farmers and among the researchers.

A field study of 30 organic dairy farms was also done by the Dutch Institute for Food Safety RIKILT (Groot, 2003). Many of these farms used homeopathic remedies (27% of treatments), whereas allopathic medicines consisted of 61% and phytomedicines 10% of the treatments mentioned. However, it is worth noting that in the Netherlands, as in most of Western Europe, many herbal tinctures are registered as homeopathic and were indicated as such by the farmers. Many of them should be categorized as herbal medicines due to the concentrations used.

In the Netherlands, herbal use for both animals and humans is similar to that in Germany, Austria, and Switzerland (Van Asseldonk and Beijer, 2006). IEZ recorded 168 folk remedies for animals, including 63 plant genera (68 plant species) (Van Asseldonk and Beijer, 2006). The animals involved most frequently were horses. There were only a few remedies for pigs, and contact with the association of organic pig farmers confirmed that their members hardly use herbs. Of the 68 plant species (involving 36 families), 92% were indigenous to or naturalized in the Netherlands.

TABLE 16.1
Eight Plant Genera Most Involved in 168 Ethnoveterinary Recipes Reported to IEZ 1998–2004

Genus/Species	No.	Plant Parts	Animals Involved	Health Problem Example
Allium sativum	11	Cloves	Horse, dog, chicken	Worms, cough, insect repellant, general health
Calendula officinalis	6	Flower, Herb	Dog, cat, horse, sheep, guinea pig	External application on wounds
Linum usitatissimum	20	Seed	Horse, cow, sheep	Improving molting; on wounds; prevent colic
Matricaria chamomilla	9	Flower	Horse, dog, cat, ferret	External for prevention of eye infection, cough
Mentha species	6	Leaves	Horse, cow, goat, rabbit	Colic, scour, hard udder
Taraxacum officinalis	8	Herb, root	Rabbit, horse, dog, pigeon	General health improvement
Trigonella foenum graecum	5	Seed	Horse	Cough
Urtica dioica/ urens	15	Herb	Horse, chicken, turkey, pig, cow, goat	Regaining strength and stimulating milk after delivery; *roborans* (tonic to speed convalescence)

No. = Exact reported number of recipes with these genera.

Apart from nettles, there was hardly any harvesting from the wild. The eight remedies mentioned most often are listed in Table 16.1.

The most frequently reported plant (20 times) was *Linum usitatissimum,* commonly called flaxseed. It is used for a range of different health conditions but mainly for skin and digestive problems in horses and cows. In two cases, it was used for sheep. Many of the users are enthusiastic about the influence these seeds have on molting, but they added a warning: The seeds should be boiled before use to neutralize the prussic acid/hydrogen cyanide (HCN). Also, in some cases only the slimy residue was used, and the seeds were left behind.

Several studies have been conducted in Canada on the feeding of flaxseed to dairy cows. This was shown to have a positive influence on the milk quality (Gonthier et al., 2005; Lessard et al., 2003; Petit, 2002; Petit, Germiquet, and Lebel, 2004; Soita et al., 2003). The beneficial influence of flaxseed on the skin of atopic horses also has been confirmed by a Canadian clinical study (O'Neill et al., 2002). Flaxseed has considerable nutritional value due to the presence of α-linolenic acid and soluble and insoluble fiber (Degenhardt et al., 2002). Lignans have been isolated and found to be anticarcinogenic, estrogenic, and antiestrogenic. While the α-linolenic acid may tend to improve milk quality (Dewhurst et al., 2003), the phytoestrogens may have deleterious effects on fertility and milk production if flaxseed is used in excess

(Tou et al., 1998). In the Netherlands, flaxseed is part of calf and piglet feed and of some types of horse feeds. It is not a part of the normal feed for cows or pigs.

The second most popular plant (mentioned in 15 reports) is the nettle. Two *Urtica* species, *U. dioica* and *U. urens,* are available in the Netherlands, and they are both being used. These are among the few plants that are collected from the wild. Their use is very broad, covering all animal species. The use is mostly as a tonic (general strength improvement) and to aid with recovery. One interesting practice was to rub the leaf onto a sow's nipples; the burning sting would make her allow the piglets to suckle more easily. We also feel that the internal use of the nettle around the period of delivery is an application that deserves more research. *Urtica* species contain lignans, lectins, and steroids (Anonymous, 2007). Polysaccharides and caffeic malic acid also isolated from *Urtica* species were found to have anti-inflammatory activity and stimulated T-lymphocyte synthesis in animal experiments. This genus is in fact widely used in humans for rheumatism and inflammation of bruises and sprains (Wang et al., 2008).

The third plant, reported 11 times, was garlic. *Allium sativum* seems to be a worldwide remedy that was reported several times for ethnoveterinary use (Mathias, Rangnekar, and McCorkle, 1998; Pieroni, 1999; Williams and Lamprecht, 2008) and in case studies in Trinidad and Tobago (Lans, 1996). This is not an indigenous plant to the Netherlands, but it is cultivated on a small scale. A lot of the reports concerned commercial preparations sold for horses. Because of the antiparasitic, antibacterial, insect-repellant, antitussive, and antioxidant properties of the organosulfides in garlic, it appears to make a good equine nutraceutical component (Williams and Lamprecht, 2008).

Accidents are likely to happen soon as the market for these animal herbal products is booming without enough traditional herbal knowledge left among animal owners to resist the advertisements. It is known that garlic in large quantities, or its chronic use, can cause anemia in sheep and anemia or urticaria in horses (Miyazawa et al., 1991; Parton, 2000; Pearson et al., 2005). It is also known that the aromatic compounds of garlic pass the ovine placental barrier (Nolte et al., 1992).

The rather popular human remedies *Calendula officinalis* L. and *Matricaria chamomilla* L. were mentioned only for external applications. Other popular plants were several mint species with several different applications and *Trigonella foenum-graecum* L. (Greek hay). This is another herb strongly promoted by the commercial animal herbal industry, mostly for cough. We could not find studies confirming the rationality of this use. It has previously been shown to have larvicidal (Halim and Morsy, 2006) and nematicidal (Zia et al., 2001) activity.

The treatments that were reported most often were alimentary tract and metabolism treatments and dermatological applications. Respiratory, antiparasitic, and reproductive treatments are also common. There was only one cardiac remedy mentioned; this was *Crataegus laevigata* (Poir) DC (hawthorn), also sold as a commercial preparation for humans in Europe and the United States. It was noteworthy that this herb was reported by Zitterl-Egelseer and Franz (1999) as the most prescribed herb by veterinarians in German-speaking countries, mainly for mild cases of heart failure in old dogs. Hawthorne (*Crataegus* species) is one of the most popular European herbal products. Its cardiotonic and cardioprotective effects are subscribed to the oligomeric procyanidins and (–)-epicatechin (Svedström et al., 2002).

16.3 ETHNOVETERINARY TRADITIONS IN EUROPE

Since 1996, there have been several studies on ethnoveterinary practices in Europe. This makes it possible to compare some results. Table 16.2 shows results of several European studies.

Allen and Hatfield (2004) compiled a survey of medicinal plants from British and Irish folk traditions. They presented 105 genera of vascular plants that were used to treat animals. Of these, the IEZ study (Van Asseldonk and Beijer, 2006) shared 23 genera, although they were often applied differently. For example, although the English name for *Artemisia absinthium* L. is wormwood, and IEZ found two reports of its use against worms in the Netherlands, Allen and Hatfield only reported the use for cuts on cow udders. Likewise, it was found that of the 23 genera, only 7 were used in a more or less similar way in both the Netherlands and the British Isles.

A study by Agelet and Vallès (1999) in Spain mentioned 76 remedies, of which IEZ shared 20; again, the indications are quite different. The IEZ study only shared a few remedies with the study of Uncini Manganelli, Camangi, and Tomei (2001) in Tuscany, but of the ones shared, at least one of the indications was quite often similar. This was less the case for the survey Pieroni et al. (2004) conducted in the south of Italy. There were many commonalities between the herbal remedies used in the Netherlands and the herbal remedies reported for Austria (Ludwig 1996).

The Rubia project, a large study coordinated at Wageningen University by Pieroni (Pieroni et al., 2006), described the actual Mediterranean use of herbs, including ethnoveterinary uses. In the survey, 135 veterinary preparations and

TABLE 16.2
Plant Genera Used for Ethnoveterinary Practice in the Netherlands (IEZ Study) Compared to Genera Reported by Other Recent Studies

	No. Genera	Shared	Indication Shared
IEZ study (Van Asseldonk and Beijer, 2006)	63		
United Kingdom/Ireland (Allen and Hatfield, 2004)	105	23	7 (30%)
Spain, Catalonia (Agelet and Vallès, 1999)	76	20	8 (40%)
Italy, Tuscany (Uncini Manganelli, Camangi, and Tomei, 2001)	77	14	9 (64%)
Italy, south (Pieroni et al., 2004)	55	10	5 (50%)
Austria, dairy (Ludwig, 1996)	37	22	13 (59%)

110 plant taxa were recorded, with Asteraceae and Lamiaceae the most mentioned botanical families.

Following requests from several organic pig farmers, the IEZ performed a controlled study of six farms in the Netherlands, testing some traditional Dutch medicinal plants. Dried herbs (nettles, fennel, aniseed, and thyme) were fed to both sows and piglets. They appeared to have a positive impact on piglet survival (Van Asseldonk et al., 2005) given the fact that on average one extra piglet survived in each nest as compared to the untreated controls. However, this result could not be repeated by another researcher (De Vries, Bestman, and Wagenaar, 2006).

16.4 EMERGING MARKETS FOR HERBAL PET PRODUCTS

Zoopharmacognosy (self-medication, or spontaneous use of uncultivated or medicinal plants by animals) is a popular issue among pet owners. Many pet owners think these phenomena are scientifically proven facts, whereas in fact up to now only a few scientists worldwide have addressed this subject (Huffmann, 2005). Advertisements for herb mixes (which sometimes seem to be undefined or irrationally created) for pets use this belief for the purpose of increased sales.

Commercial activities regarding herbs for pets and training of specialized prescribers of these products to become holistic pet therapists in private institutes is a booming business in the Netherlands as well as in the rest of Europe. Some veterinarians are involved because they feel the consumer's need for these new approaches; instead of combating disease to try to improve health, there is the move from treatment-oriented medicine to prevention. But, in the Netherlands the sole veterinary university does not provide an educational background for a critical involvement with these issues. As a result, interested veterinarians are vulnerable to commercial activities. Education for nonveterinarians provided by private parties tends to lack a scientifically sound background and be of a commercial nature, which may cause conflicts of interest.

Many herbal products are offered to owners of horses that are used for recreation and sports, and many do buy these products. However, research into the quality, efficacy, and safety of these products is lacking. A better situation exists in the German-speaking countries (Germany, Switzerland, Austria), where there has been more involvement of science in herbal therapies. For example, the Veterinary University of Vienna (Austria) has a long-standing tradition of researching herbal medicines for animals. However, from the large number of registered herbal medicines for animals that existed in Germany and Austria at the end of the 20th century, next to nothing is left. New European Union regulations demand new and costly (re-)evaluation of these medicines, which is not affordable in a market in which the same herbs continue to be sold as feed items. Many herbal remedies that were used by veterinarians some years ago are now being sold as feed supplements, through veterinarians or directly to the owners. A growing market exists for use of herbs and other natural products (such as prebiotic substances and probiotics: active microorganisms), both for pets and for farm animals, the latter being of great economic importance.

16.5 HERBAL FOOD ADDITIVES FOR FARM ANIMALS

At the request of the organic farmers organization, in 2006 the Dutch Ministry of Agriculture, Nature, and Food Quality initiated the Fyto-V project for the development of phytotherapy as a tool for reducing or preventing diseases in farm animals. Organic farmers should preferably use natural remedies such as phytotherapy to treat animals, but since few are registered as veterinary medicines, information on the quality and the efficacy of herbal products on the market was not easily available. The goal of the project was to increase knowledge about a possible rational use of herbal therapies in organic farming. The project was performed by RIKILT Institute of Food Safety in Wageningen in cooperation with several research institutes: Utrecht University, the Animal Science Group of Wageningen University, the Louis Bolk Institute, IEZ, and the professional agricultural university in Den Bosch (Groot et al., 2008).

The project started with an inventory of the available herbal products for farm animals in 2006. Most products were not registered as veterinary medicines but instead were used as feed additives or feed constituents. Information on all 255 products for pigs, cows, or poultry was compiled in a database (http://www.fyto-v.nl). Literature references to studies regarding the efficacy of the products or their herbal ingredients were included.

A large number of herbal products have been developed partly as a response to the ban on antimicrobial growth promoters in animal feed in the European Union that went in place in January 2006. The replacement of growth-promoting antibiotics by plant extracts and other natural materials was a concern for many scientists and corporations in several EU-(co)financed projects in the period 2001–2007 (Rumen-up, Replace, Feed for Pig Health, Healthy Pig Gut, and Safewastes).

The FytoV project identified, in addition to the product information concerning the herbal products that were on the market in 2007, about 30 related (*in vitro*) studies and 150 clinical studies with herbs that were not confidential. Only 65 (43%) were published in peer-reviewed journals. Table 16.3 shows the number of herbal products and number of relevant publications found in relation to the indication for their use and the target species. Several producers showed their documentation on experiences with their products in confidence, both performed by independent research institutes and results from practice. This indicates that many promising products are available. For certain important indications (e.g., respiratory conditions or parasites in poultry and pigs), there are still only a few products available on the market (Groot et al., 2008). These figures may be biased by the restriction of medicinal claims that can be made for feed substances or feed additives. The herbal products are mostly marketed as aromatic feed additives. Still, many of them are made from herbs that are listed in the European Pharmacopoeia because they are used for medicinal applications in human medicine. The claim is mostly restricted to "increased appetite," sometimes in addition to "better growth." In several cases, documentation existed of the preventive or damage control effect that these herbs will have in animals, for example, in relation to stressful periods such as weaning or transport or during infections with *Escherichia coli* or clostridia. Dermatologicals and external preparations for gynecological problems (mastitis) belong to a second group that is well represented in this market.

TABLE 16.3
Numbers of Herbal Products and Numbers of Relevant Publications[a] Found at the Start of the Fyto-V Project in Relation to the Indication for Their Use

		Products and Target Species				
Therapeutic Uses	**Publications[a]**	**Sum**	**Cattle**	**Pigs**	**Poultry**	**Small Ruminants**
Alimentary tract (against diarrhea and constipation)	5	13	6	6	3	2
Alimentary tract (enhancing appetite and growth)	96	55	22	26	23	6
Alimentary tract (specific organ functions and special diets)	7	24	10	8	9	6
Alimentary tract (prebiotic and dried probiotic)	13	19	8	9	5	2
Dermatological	1	12	11	9	3	9
Genitourinary tract	22	24	20	4	0	10
Immunological	6	2	0	0	0	0
Anti-infective (systemic)	23	0	0	0	0	0
Musculoskeletal	0	1	1	0	0	1
Nervous system	9	5	0	3	4	0
Antiparasitic products, insecticides, and repellents	10	7	2	2	5	2
Respiratory system	0	7	6	6	2	3
Sum	**192**	**168**	**86**	**74**	**54**	**41**

[a] The number of publications concerns mostly clinical studies with the product in one of the target animals and some *in vitro* studies. Some of the products carry more than one indication and some products are used in more than one animal species for the same indication. Therefore the sum of the number of products is not the same as the sum of products for different target species.

In many cases, the plant species used in these feed additives was not public information, and more transparency would be desirable. Therefore, the listing in Table 16.4 of the herbs that are most often used for livestock may be biased. Still, the most important herbs (i.e., oregano, garlic, yeast, cinnamon, and chicory) can be identified. Several studies exist on the veterinary use of these five plants. Oregano oil appeared very promising for the prevention of *E. coli* infections in the pig gut (Jongbloed, Maiorano, and Wagenaars, 2008). Chicory root is sold as such or with enriched content of fructooligosaccharides (FOS) that has a probiotic effect (i.e., stimulates the growth of beneficial microorganisms in the gut). The use of the other plants seems to be based mainly on tradition or is derived from human use, for which extensive documentation exists for these herbs (European Scientific Cooperative on Phytotherapy [ESCOP], 2003).

TABLE 16.4
Plant Species Used in More Than Four Herbal Products for Veterinary Use Identified in the Dutch Fyto-V Project with Numbers of Products and Numbers of Publicized Clinical Animal Studies in Which They Appear

Species (Botanical Name)	English	Products	Publications[a]	Use
Acorus calamus	Sweetflag	6	7	Digestive aid
Allium sativum	Garlic	12	24	Gut health, antibiotic
Aloe vera	Aloe	5	1	External (skin, udder)
Arnica montana	Mountain arnica	5	0	External (bruises)
Artemisia absinthium	Absinth	7	2	Digestive aid
Cichorium intybus	Cichory	8	12	Prebiotic (fiber)
Cinnamomon camphora	Camphor	6	0	External (udder)
Cinnamomon zeylanicum	Cinnamon	7	16	Gut health, antibiotic
Echinacea purpurea	Purple coneflower	6	14	Immune stimulant
Erythraea centaurium	Common centaury	11	0	Digestive aid
Foeniculum vulgare	Fennel	6	7	Digestive aid
Lavendula officinalis	Lavender	5	1	External, skin
Matricaria chamomilla	Chamomile	5	0	Digestive aid
Mentha piperita	Mint	5	4	External (udder)
Origanum vulgare	Oregano	6	37	Gut health, antibiotic
Quercus robur	Oak	9	3	Antidiarrhea
Rosmarinus officinalis	Rosemary	9	2	Gut health, antibiotic
Saccharomyces spp.	Yeast	14	10	Prebiotic (cell walls)
Silybum marianum	Milk thistle	13	0	Digestive aid/liver support
Thymus vulgaris	Thyme	6	22	Gut health, antibiotic
Trigonella foenum graecum	Fenugreek	8	1	*Roborans* (tonic to speed convalescence)
Urtica urens	Nettle	13	3	*Roborans*

[a] Number of publications refers to clinical studies with poultry, pigs, or cows. For all of these herbs, except for oregano and chicory, which seem to be more specific animal remedies, many studies exist, both *in vitro* and animal studies, for the validation of all of these remedies concerning human use (ESCOP, 2003). All of these herbs are also (part of) registered human medicines in Germany. But, clinical use in animals has for some herbal products only been marginal or has not been tested in spite of a long tradition. Moreover, not all studies dedicated to agricultural applications were positive. For example, the use of *Echinacea* has strong evidence based on *in vitro* and clinical studies for humans, and this herb also is used for animals; it has been tested in products for several animal species, but the results were disappointing for poultry, and results for pigs were inconsistent. One of the remedies tested on cows in the Fyto-V project was a herb mixture with *Echinacea* as an important constituent that showed *in vitro* activity against *Staphylococcus aureus*.

16.6 TESTING OF PRODUCTS

Of the hundreds of products summarized, 10 herbal products were selected for quality and efficacy tests in the Fyto-V project. Three herbal preparations were clinically tested on pigs, five on poultry, and three on dairy cattle (one in two animal species). In general, all these herbal products were very well tolerated during the trials.

The products tested on pigs were a simplex of standardized oregano oil on puffed wheat; a fixed combination of oregano, aniseed, and lemon peel oil on powdered chicory root; and a fixed complex mixture of about 10 dried herbs.

All three products showed positive effects as claimed by the suppliers either on growth or on feed conversion in weaners, but due to large variations per animal there was no strong significance ($.05 < p < .13$). Interesting results were found at the time of slaughtering: Differences in carcass and organ assessments and percentages of meat and fat were registered as compared to untreated controls. Oregano oil gave less fat, and the oregano-aniseed-orange oil tended to give on average more meat. A number of livers were collected for *in vitro* research of biomarkers regarding gut health. Promising results were found with gene expression assays for oxidative stress, which clearly showed health improvement for the animals using herbs in their feed (Groot et al., 2008).

The five poultry products that were tested for their effects on coccidiosis infection were developed as feed additives for broilers. They did not cause a reduction of the effect of an acute *Eimeria* infection in young layer hens (Lourens and Jongbloed, 2008). However, the synthetic ionophore coccidiostatic that was used as an internal control for the trial was equally ineffective in this challenge study. On day 28, the effects of the infection on the gut had disappeared; only a slightly retarded growth (15 g) remained that was better conquered by the coccidiostatic than by the herbs. More studies are necessary to determine whether herbal products may perform better in the case of milder and chronic infection at a higher age, preferably tested in a plot that simulates the actual farm problems more accurately.

Three herbal products currently used by several dairy farms in the Netherlands, and claiming to have a beneficial influence on high cell count, were given to dairy cows. None of the products made a significant difference in cell count compared to the control group. Due to an unknown cause, the cell count barely increased at the start of the trial, unlike the beginning of the prior period. Therefore, this test could not conclude whether the effect could have been beneficial in the case of high cell counts.

An important part of the Fyto-V study was the study of quality (in laboratory tests) and of biological activity (in chemical and bioassays) of the herbal products used in the animal trials. All but 2 producers of 10 showed documents that proved their products to be of stable quality. Tests regarding the presence of specific herbs and the absence of antibiotics, ionophores, and hormones revealed no abnormalities.

All herbal products showed antioxidant activity in the ORAC (oxygen radical absorbence capacity) tests, but when comparing the feed preparations with and without herbs added, there was no statistical difference. As a result, clinical effects cannot be accounted for by their antioxidant value.

The three pig products were active against *E. coli*. Two of three herbal products tested for dairy cows showed activity against biofilm formation by *Staphylococcus aureus*, possibly by reducing bacterial growth. One of them was a fixed combination of tinctures of seven herbs; the other was garlic extract, enriched to a high percentage of stable allicin.

These five products were also shown to decrease the functioning of polymorphonuclear granulocytes, thus indicating an immune-modulating effect, without having a toxic effect.

At the end of the Fyto-V project, it was concluded that some herbal products could be beneficial to animal health. However, more literature as well as clinical studies will be necessary to make better judgments regarding the possible benefits and optimal use of these products by farmers. In particular, studies that measure relevant examples of gene expression profiles promise to provide statistical proof for health effects of some herbal products for animals (Groot et al., 2008).

16.7 PROSPECTS OF HERBAL MEDICINES AT THE EU LEVEL

During the course of the 20th century, interest in herbal products for animals was lost in Europe, but it regained importance near the start of the 21st century. Future developments in this field will be strongly influenced by legislation that is harmonized at the level of the European Union. It will be necessary to adapt regulations for herbal veterinary products. The European Medicines Agency (EMEA) has already adopted rules for human herbal medicine and food supplements (Groot et al., 2007). Dutch legislation in the area of animal feed and veterinary medicinal products is based on European legislation and consists of a national implementation of this legislation.

A distinction should be made between the uses of herbs as animal feed, animal feed supplements, animal feed additives or therapeutic products. Currently, herbs without medical claims can be used in animal feed, provided they are safe and not mentioned on the list of undesirable substances of the European Directive 2002/32/EC (European Parliament and Council, 2002) and provided they do not contain toxic substances above permitted levels.

The growing interest in the use of herbs for several health conditions requires both adequate legislation and proper bioassays to value and reward the honest and transparent health claims. This is a challenge science and industry will face in the coming years (Groot, 2008).

16.8 CONCLUSION

The potential of herbal animal remedies is slowly dawning in the Netherlands and all over Western Europe, both in the pet care field, in which scientific involvement is rare, and in the field of animal feed, in which innovative companies are developing new herbal products. The introduction of herbal and related natural products in the field of animal feed has contributed to maintaining the productivity of the livestock after the banning of the use of antibiotics as growth promoters. To gain the benefit of these promising approaches, the scientific community should get more involved and support innovative developments in this area. Legislative constraints for these developments should be evaluated and, if possible, removed. It is of equal importance that science is involved in monitoring and controlling the risks accompanying these new products. Little is known about

- The interaction of natural growth promoters with each other and with veterinary medication (be it negative or, under control, positive).
- The residual or other effects these products have on meat quality. Again, risks must be assessed here, but there are also some positive signals (e.g., feeding on oregano seems to give more unsaturated fatty acids in meat). The impact on the production and quality of milk and eggs of herbal ingestion also needs to be investigated (Gerber, 1997; Wenk, 2005).
- The implementation of natural products within the specific management situation of each farm. Many general health problems are solved by vaccination and balanced feed; the remaining problems are often farm related and require specific additional products (like herbal supplements) to solve them. The choice of natural products is overwhelming, and good monitoring of their results in specific situations will be necessary to make better use of them.

Expertise on these topics needs to be built up by all involved institutions and should be supported by international cooperation.

ACKNOWLEDGMENTS

We thank everyone involved in the studies described. The ethnoveterinary fieldwork for IEZ was performed by 85 students of two professional universities and received no funding. The Fyto-V project involved the research activities of 20 people in seven institutes and two organic farms. M. Groot (RIKILT) coordinated the Fyto-V project. M. N. Zijlstra (IEZ) edited the literature studies that were gathered in a spreadsheet and edited this chapter. The Fyto-V project was financed by the Dutch Ministry of Agriculture, Nature, and Food (LNV) at the request of the Bioconnect organization.

REFERENCES

Agelet, A., and Vallès, J., Vascular plants used in ethnoveterinary in Pallars (Pyrenees, Catalonia, Iberian Peninsula, in *Herbs, humans and animals/Erbe, uomini e bestie*, Pieroni, A., Ed., Experiences, Cologne, Germany, 1999, 14–35.

Allen, D.E., and Hatfield, G., *Medicinal plants in folk tradition, an ethnobotany of Britain and Ireland*, Timber Press, Portland/Cambridge, UK, 2004.

Anonymous, *De Grote Bosatlas*, Wolters-Noordhoff, Groningen, the Netherlands, 2001.

Anonymous, *Urtica dioica*; *Urtica urens* (nettle). Monograph. *Alternative Medicine Review* 12 (3), 2007, 280–284.

CBS, Centraal Bureau voor de Statistiek (Statistics Netherlands), Statistics Netherlands is responsible for collecting and processing data in order to publish statistics to be used in practice, by policymakers and for scientific research. In addition to its responsibility for (official) national statistics, Statistics Netherlands also has the task of producing European (community) statistics. http://www.cbs.nl (accessed September 1, 2008).

Degenhardt, A., Habben, S., and Winterhalter, P., Isolation of the lignan secoisolariciresinol diglucoside from flaxseed (*Linum usitatissimum* L.) by high-speed counter-current chromatography. *Journal of Chromatography A* 943 (2), 2002, 299–302.

De Vries, A., Bestman, M., and Wagenaar, J.P., *Gebruik van kruiden bij de opfok van biggen*, Louis Bolk Instituut, Driebergen, the Netherlands, 2006.

Dewhurst, R.J., Fisher, W.J., Tweed J.K.S., and Wilkins, R.J., Comparison of grass and legume silages for milk production. 1. Production responses with different levels of concentrate. *Journal of Dairy Science* 86, 2003, 2598–2611.

European Parliament and Council, Directive 2002/32/EC of the European Parliament and of the Council of 7 May 2002 on undesirable substances in animal feed. Published in the *Official Journal of the European Communities* 30.05.2002. Available through http://ec.europa.eu/food/food/animalnutrition/contaminants/index_en.htm

European Scientific Cooperative on Phytotherapy (ESCOP), *Monographs—the Scientific Foundation for Herbal Medicinal Products*, 2nd ed., ESCOP, Exeter, UK, 2003.

Gerber, A., *Salvia officinalis* L. als Futteradditiv in der Schweinemast, PhD dissertation, Veterinärmedizinische Universität, Vienna, Austria, 1997.

Gonthier, C., Mustafa, A.F., Berthiaume, R., Petit, H.V., Martineau, R., and Oullet, D.R., Effects of feeding micronized and extruded flaxseed on ruminal fermentation and nutrient utilization by dairy cows. *Journal of Dairy Science* 88 (2), 2005, 748–756.

Groot, M.J., Noordam, M.Y., and Kleter, G.A., *WP3: wettelijke regelingen over het gebruik van kruiden bij landbouwhuisdieren, Rapport 2007.017,* RIKILT, Wageningen, the Netherlands, 2007.

Groot, M.J., *Deskstudie alternatieve gezondheidszorg voor melkvee,* RIKILT, Wageningen, the Netherlands, 2003.

Groot, M.J., Noordam, M.Y., Kleter, G.A., Van Asseldonk, A.G.M., Kleijer-Ligtenberg, E., Halkes, S.B.A., Fink-Gremmels, J. and Van Os, H.H., *Fyto-V Eindrapport.* Rapport 2008.010. RIKILT, Wageningen, the Netherlands, 2008. www.rikilt.wur.nl

Halim, A., and Morsy, T.A. Efficacy of *Trigonella foenum-graecum* (fenugreek) on third stage larvae and adult fecundity of *Musca domestica. Journal of the Egyptian Society of Parasitology* 36 (1), 2006, 329–334.

Huffman, M.A., *A study of primate self-medication*, collection of papers by the CHIMPP Group on Primate Self-Medication (1989–2004), 2005, http://www.pri.kyoto-u.ac.jp/shakai-seitai/seitai/huffman.

Jongbloed, A.W., Maiorano, R., and Wagenaars, C.M.F., *Effect of several plant products on prevention of* E. coli *adhesion in the gastrointestinal tract of weaned piglets*, Rapport 62, Animal Science Group, Wageningen, the Netherlands, 2008.

Kijlstra, A., Van der Werf, J.T.N., and Buitendijk, J. *Inventarisatie diergeneesmiddelengebruik in de biologische geitenhouderij*, CIDC Lelystad, WUR, Wageningen, the Netherlands, 2004.

Lans, C., *Ethnoveterinary practices used by livestock keepers in Trinidad and Tobago*, thesis, WUR, Wageningen, the Netherlands, 1996.

Lessard, M., Gagnon, N., and Petit, H.V., Immune response of postpartum dairy cows fed flaxseed. *Journal of Dairy Science* 86 (8), 2003, 2647–2657.

Lourens, S., and Jongbloed, A.W., *Effect van kruiden en kuikenkwaliteit op de respons van jonge biologische opfokhennen na een coccidiosebesmetting.* ISSN 1570-8616. Rapportnummer 120, Animal Sciences Group, Wageningen University, the Netherlands, 2008.

Ludwig, M., *Phytotherapie beim Rind einst und jetzt—alte Indikationen neu bewertet*, Inaugural dissertation, Veterinary University, Vienna, Austria, 1996.

Mathias, E., Rangnekar, D.V., and McCorkle, C.M., Eds., *Ethnoveterinary medicine: alternatives for livestock development*, proceedings of an international conference held in Pune, India, 1997, BAIF Development Research Foundation, Pune, India, 1998.

Miyazawa, K., Ito, M., and Ohsaki, K., An equine case of urticaria associated with dry garlic feeding. *Journal of Veterinary Medical Science* 53 (4), 1991, 747–748.

Nolte, D.L., Provenza, F.D., Callan, R., and Panter, K.E., Garlic in the ovine fetal environment. *Physiology and Behaviour* 52 (6), 1992, 1091–1093.

Numan, A., *Handboek der genees-en verloskunde van het vee*, Utrecht, the Netherlands, 1844.

O'Neill, W., McKee, S., and Clarke, A.F, Flaxseed *(Linum usitatissimum)* supplementation associated with reduces skin test lesional area in horses with Culicoides hypersensitivity. *Canadian Journal of Veterinary Research*, 66 (4), 2002, 72–77.

Parton, K., Onion toxicity in farmed animals, *New Zealand Veterinary Journal* 48 (3), 2000, 89.

Pearson, W., Boermans, H.J., Bettger, W.J., McBride, B.W., and Lidinger, M.I., Association of maximum voluntary dietary intake of freeze-dried garlic with Heinz body anemia in horses. *American Journal of Veterinary Research* 66 (3), 2005, 457–465.

Petit, H.V., Digestion, milk production, milk composition, and blood composition of dairy cows fed whole flaxseed. *Journal of Dairy Science*, 85 (6), 2002, 1482–1490.

Petit, H.V., Germiquet, C., and Lebel, D., Effect of feeding whole, unprocessed sunflower seeds and flaxseed on milk production, milk composition, and prostaglandin secretion in dairy cows. *Journal of Dairy Science* 87 (11), 2004, 3889–3898.

Pierik, C., *Grote melkveebedrijven groeien minder snel*, CBS (Statistics Netherlands) Webmagazine on www.cbs.nl, June 18, 2001.

Pieroni A., Giusti, M.E., de Pasquale, C., et al., Circum-Mediterranean cultural heritage and medicinal plant uses in traditional animal healthcare: a field survey in eight selected areas within the RUBIA project. *Journal of Ethnobiology and Ethnomedicine* 2, 2006, 16.

Pieroni, A., Ed., *Herbs, humans and animals/Erbe, uomini e bestie*, Experiences, Cologne, Germany, 1999.

Pieroni, A., Howard, P., Volpato, G., and Santoro, R.F., Natural remedies and nutraceuticals used in ethnoveterinary practices in inland southern Italy. *Veterinary Research Communications*, 28, 2004, 55–80.

Soita, H.W., Meier, J.A., Fehr, M., Yu, P., Christensen, D.A., McKinnon, J.J., and Mustafa, A.F., Effects of flaxseed supplementation on milk production, milk fatty acid composition and nutrient utilization by lactating dairy cows. *Archiv für Tierernährung*, 57 (2), 2003, 107–116.

Svedström, U., Vuorela, H., Kostiainen, R., Tuominen, J., Kokkonen, J., Rauha, J.-P., Laakso, I., and Hiltunen, R., Isolation and identification of oligomeric procyanidins from *Crataegus* leaves and flowers. *Phytochemistry* 60 (8), 2002, 821–825.

Tou, J.C.L., Chen, J., and Thompson, L.U., Flaxseed and its lignan precursor, secoisolariciresinol diglycoside, affect pregnancy outcome and reproductive development in rats. *The Journal of Nutrition* 128 (11), 1998, 1861–1868.

Uncini Manganelli, R.E., Camangi, F., and Tomei, P.E., Curing animals with plants: traditional usage in Tuscany (Italy). *Journal of Ethnopharmacology* 78, 2001, 171–191.

Van Asseldonk, A.G.M., Traditional and modern herbalism in the Netherlands, research report presented as a short paper at the ethnobotany conference in Antigua, Guatemala, IEZ, Beek-Ubbergen, the Netherlands, 2001.

Van Asseldonk, A.G.M., et al.,. *Kruiden in de kraamstal—een praktijkonderzoek voor en door biologische varkenshouders*, IEZ, Beek-Ubbergen, the Netherlands, 2005, http://www.ethnobotany.nl.

Van Asseldonk, A.G.M., and Beijer, H., Herbal folk remedies for animal health in the Netherlands, in *Proceedings of the fourth International Congress on Ethnobotany (ICEB 2005),* edited by ZF Ertug, Ege Yayinlan, Istanbul, 2006, 257–263.

Van der Meulen, J., Kijlstra, A., Van den Berg, M., and Spruit, J., *Inventarisatie diergeneesmiddelengebruik in de biologische pluimveehouderij*, CIDC Lelystad, WUR, Wageningen, the Netherlands, 2004a.

Van der Meulen, J., Kijlstra, A., Reijmer, E. and Sielias, W., *Inventarisatie diergeneesmiddelen gebruik in de biologische varkenshouderij*, CIDC Lelystad, WUR, Wageningen, the Netherlands, 2004b.

Van der Werf, J., Kijlstra, A., Buitendijk, J., Klink, M., Munniksma, K. and Van der Schaaf, R., *Inventarisatie diergeneesmiddelengebruik in de biologische melkveehouderij*, CIDC Lelystad, WUR, Wageningen, the Netherlands, 2004a.

Van der Werf, J.T.N., Kijlstra, A., Klink, M.C.M. and Munniksma, K., *Inventarisatie diergeneesmiddelengebruik in de biologische vleesveehouderij*, CIDC Lelystad, WUR, 2004b.

Wagenfeld, dr. L., *De Vakbekwame Veearts*, 1844.

Wang, W., Yan, X. G., Duan, L. X., Zhou, Y., Tang, L., Wang, Y. Q., and Feng, B.M., Two new secolignans from the roots of *Urtica mairei* Levl. *Chinese Chemical Letters,* 19 (10), 2008, 1212–1214.

Wenk, C., *Einsatz von Kräutern und deren Extracten in der Tierernährung: Erwartungen und Möglichkeiten*, Tagungsband 4.BOKU Symposium: Tierernährung ohne antibiotische Leistungsförderer, Vienna, Austria, 2005, 17–27.

Williams, C.A., and Lamprecht, E.D., Some commonly fed herbs and other functional foods in equine nutrition: a review. *The Veterinary Journal* 178 (1), 2008, 21–31.

Wisman, A., *Europese varkenshouderij: Nederland is goede middenmotor*, LEI, Agrimonitor, Augustus, 2003.

Zia, T., Siddiqui, I.A., and Nazrul-Hasnain., Nematicidal activity of *Trigonella foenum-graecum* L. *Phytotherapy Research* 15 (6), 2001, 538–540.

Zitterl-Egelseer, K., and Franz, C., Medicinal herbs of Central Europe used in the therapy of different animal species, in *Herbs, humans and animals/Erbe, uomini e bestie*, Pieroni, A., Ed., Experiences, Cologne, Germany, 1999, 187–195.

17 Herbal Medicines for Pet and Companion Animals

David R. Katerere and Vinny Naidoo

CONTENTS

17.1 INTRODUCTION

A cursory glance at the popular literature or a general Internet search reveals that there is an increasing interest in the use of herbal products for domestic pets and horses. This mirrors the same interest for using, and self-medicating with, complementary and alternative medicine (CAM) among humans. As with humankind, there is evidence that animals have been self-medicating for millennia (Huffman and Hirata, 2004), and in some cases humans may have started to use certain plants on observing animals using them to good effect (Huffman and Seifu, 1989).

In the 1990s, there was an unprecedented growth in the herbal medicine market, with at least 60 million Americans using botanicals (Miller, 1998; Kennedy, 2005). In Germany, the use of "natural remedies" has more than doubled since the 1970s, from 20% in 1970 to 56% of the population in 2002 (Ernst and Dixon, 2006). This reflects a worldwide trend that has probably exponentially increased for various reasons. The reasons most commonly cited are increased cost of medical care (Dubnick, 1986) leading persons, particularly those of low economic status, to seek cheaper alternatives; the need for control by patients over their health care that self-medication appears to satisfy; and globalization, which has led to greater awareness of other medical systems and philosophies, especially those from China and India (Ernst and Dixon, 2006). There is also the general perception that "natural is safe," and this partly arises due to the adverse effects that are associated with the use of synthetic chemicals and pharmaceuticals (Dubnick, 1986).

With specific reference to veterinary pharmaceutical products, a good example is the environmental and human safety concerns arising from the use of topical fipronil (Frontline®), the premier tick control product in dogs (Geary, Conder, and Bishop, 2004). It is now known to be toxic to fish, bees, rabbits, and game birds (Gupta, 2007). Imidacloprid (Advantage®), a neonicotinoid, has been banned in Germany because of its toxicity to honeybees (Benjamin, 2008). In India, Nepal, and Pakistan, the use of diclofenac in cattle was banned because of the threat to the survival of scavenger vultures, which feed on animal carcasses (Swan et al., 2006).

For the foregoing reasons, there is increased use of botanicals in humans that is now filtering through to use in their animal companions for the same reasons. The renewed interest in industrialized countries in traditional veterinary medicine has tracked the revival in interest of CAMs for human use (Schillhorn van Veen, 1997), and this has coincided with new-found global interest in green issues and respect for indigenous cultures and traditional ecological knowledge. Due to globalization, there is also a shift in human–pet relations that started in industrialized countries and now is spreading to emerging markets, in which pets are now considered part of family, and their lifestyle (including food and medicine) has been "humanized" (Reportbuyer, 2009).

Interestingly the acceptance of CAMs by the veterinary establishment appears to have been more enthusiastic than that by the human medical fraternity. In the United States, the American Veterinary Medicine Association (AMVA) has spearheaded the founding of the Veterinary Botanical Medical Association, and professional associations elsewhere (e.g., Britain and South Africa) have followed suit.

Countries with strong Germanic influence (e.g., Germany, Austria, Switzerland, and the Netherlands) have been leading proponents of phytomedicine in humans and more recently in animals. A survey done among veterinarians in Austria, Germany, and Switzerland, for example, found increasing interest in the use of veterinary phytotherapies (Hahn, Zitterl-Eglseer, and Franz, 2005), with three-quarters of the respondents already using herbal products on their patients, often for liver, joint, kidney, heart, and skin diseases. The same study found that herbal drugs were more acceptable to pet owners and veterinarians due to the belief that they had fewer side effects when used for the management of chronic diseases.

This chapter discusses phytotherapies commonly used to treat and maintain health in companion animals, mainly dogs, cats, and horses. Seven common therapeutic areas have been selected and are discussed. An apparently glaring omission is in the treatment of allergies, but because allergies in animals are caused by numerous known and unknown insults, management is symptomatic. The remedies used for skin, immune modulation, and cough can also be used for managing allergies in pets.

17.2 HERBS FOR TICKS

Pets are carriers of ectoparasites (fleas, ticks, and lice) that if untreated cause misery to the pet or are carriers of severe blood-borne parasitic diseases, such as babesiosis and ehrlichiosis. In addition, these ectoparasites pose a zoonotic risk for diseases such as psittacosis, cat scratch disease, and tapeworm (Kayne et al., 2001). Ectoparasites belong to a wide group of parasites that live in or on the skin of animals. The parasites that live on the skin include blood-sucking flies, ticks, fleas, and lice (Horowitz, 2003). In seeking appropriate interventions, it is important to understand the parasite and its life cycle; adult fleas live on the animal, while eggs, larvae, and pupae live in the environment (i.e., carpet, bedding, and soil), so eradication methods should be aimed at both populations (Wynn and Chalmers, 2002). Ticks, on the other hand, may have a one-host, two-host, and three-host cycle and may therefore complete their life cycle over a longer period of time, thereby contaminating the environment for a prolonged period if they carry diseases.

The parasites living within the skin are the mites (e.g., sarcoptic and demodectic mange). Their management is also wide and includes products such as medicated topical washes, spot-ons, collars, dips, and powders.

In the past, organochlorine and organophosphate compounds were used for tick and insect control, but these fell out of favor because of their toxicity and persistence in the environment. They were largely replaced by the pyrethroids (Casida, 1980), which are synthetic analogues of naturally derived pyrethrins and the formidines.

The natural pyrethrins (from chrysanthemums) and D-limonene (from citrus) are widely used and investigated for insecticidal activity. The pyrethrins are found naturally in the pyrethrum flower *Tanacetum* or *Chrysanthemum cinerariifolium* (Casida, 1980; Kiriamiti et al., 2003). The pyrethrum flower has a history dating to its use for body lice in central Europe and Persia in the 1800s. Synthetic analogues, the pyrethroids (e.g., permethrin and cypermethrin), are widely used as domestic, industrial, and agricultural insecticides (Soderlund et al., 2002). Pyrethrins and related compounds affect the nervous system of insects and other arthropods by disrupting ion channels (Townsend, 2005). Pyrethrins are rapidly broken down and have little residual effect, which makes them environmentally safe, while the pyrethroids are more stable, persist in the environment for longer, and hence are more toxic.

Neem oil, derived from *Azadirachta indica,* has a broad spectrum of activity and is safe for all pets, including kittens and puppies. The azadirachtins and salannin are thought to be the insecticidal constituents of neem oil (Wynn and Chalmers, 2002). There is good clinical evidence to suggest that spraying fleas for 3 to 5 days with neem oil gives the best outcome (Wynn and Chalmers, 2002). It has been shown to be safe not only for kittens and puppies but also for humans.

Azadirachtin and related neem compounds exhibit various actions against insects, e.g., antifeedancy, growth regulation, fecundity suppression and sterilization, oviposition repellency or attractancy, changes in biological fitness, and blocked development of vector-borne pathogens. These compounds show broad-spectrum activity against species of insects of medical and veterinary importance (e.g., mosquitoes, flies, triatomines, cockroaches, fleas, and lice) (Mulla and Su, 1999).

Melaleuca alternifolia (the tea tree) is an Australian plant with a long history of use among the Aborigines and with proven potent antimicrobial activity (Carson, Riley, and Cookson, 1998; Allen, 2001). It is the source of tea tree oil (TTO), which is a complex mixture of terpene hydrocarbons and tertiary alcohols distilled mainly from plantation stands of *Melaleuca alternifolia* (Maiden and Betche) Cheel of the Myrtaceae family. Data from individual components suggest that TTO has the potential to be developmentally toxic if ingested at higher doses; however, TTO and its components are not genotoxic (Hammer et al., 2006). TTO has acaricidal properties and has shown activity against nymphs of *Ixodes ricinus* (Iori et al., 2005).

Another Australian native and also a member of the Myrtaceae family, the *Eucalyptus* (gum tree) also possesses essential oil with fungicidal, insecticidal or insect repellent, acaricidal, and nematicidal properties (Batishet, Singh, and Kohli, 2008). The lipophilic terpenoids terpinen-4-ol and 1,8-cineole (eucalyptol), major constituents of TTO and gum tree, are known to possess broad-spectrum activity because of their ability to disrupt cellular membranes and associated receptors and signalling pathways (Camp, 2004).

Mammea Americana, *Pouteria sapota,* and *Manilkara zapota* are used for ectoparasites in dogs in the Caribbean (Lans et al., 2000). *Mammea americana* is a tree from the tropical rain forests of South America that is now domesticated for its edible fruits (*zapote domingo*) (Reyes-Chilpa et al., 2008). In Columbia, powdered seeds are applied in the form of a paste to dogs and humans for the elimination of fleas, lice, and mange mites. Several coumarins (so-called mammeins) and xanthones with insecticidal and antibacterial activity have been isolated from *M. Americana* (Crombie et al., 1970; Yasunaka et al., 2005).

In Canada, fleas and flies on dogs and cats are treated with *Artemisia vulgaris* (mugwort), *Citrus x limon* (lemon), *Juniperus communis* (common juniper), *Lavandula officinalis* (lavender), *Melissa officinalis* (lemon balm), and *Thuja plicata* (western red cedar) (Lans, Turner, and Khan, 2008). All of these plants are aromatic and possess insecticidal activity, probably due to the presence of monoterpenoids. Limonene, for example, is a common constituent of citrus aromatic or essential oils and acts by dissolving the cuticle of the adult flea (Wynn and Chalmers, 2002). However, it causes hypersensitivity in cats.

In central Italy, the leaves of the tomato plant (*Lycopersicon esculentum*) are placed in dog kennels to repel fleas and other parasites (Guarrera, 1999). Tomato plants synthesize the glycoalkaloids dehydrotomatine and α-tomatine, possibly as a defense against bacteria, fungi, viruses, and insects (Kozukue et al., 2004).

Also in Italy, an infusion, leaves, or flowers of the dwarf elder (*Sambucus ebulus*) were rubbed on the animal's body or placed on bedside tables to remove mosquitoes or other insects (Guarrera, 1999). Pallets were also made with dwarf elder branches to prevent fleas in kennels (Guarrera, 1999; Ebrahimzadeh, Mahmoudi, and Salimi, 2006). *Sambucus ebulus* contains lectins called ebulitins (de Benito et al., 1995), which are ribosome-inactivating proteins (RIPs) and inhibitors of proteolytic enzymes and glycohydrolases (Ryan, 1990).

In the United States, the aromatic herbs pennyroyal (*Mentha pulegium*), peppermint (*Mentha x piperita*), or spearmint (*Mentha spicata*) are used in natural flea collars (Horowitz, 2003). There is no literature that could be found that showed investigation of the antitick or insecticidal efficacy of the mint. However, it can be postulated that the role of the monoterpenes is central, and as Wynn and Chalmers (2002) have suggested, the mechanism is by disrupting the cell wall.

As a direct result of ethnoveterinary information, a herbal preparation containing oil extracts of plants *Cedrus deodare* (deodar cedar) and *Pongamia glabra* was successfully used in the management of dogs heavily infested with *Sarcoptis scabiei* in a controlled clinical study (Das, 1996). The results showed this combination to be highly effective, with even the most severely infested dog recovering by day 14 of treatment. Most importantly, the cure was clinically comparable with a commercial amitrax formulation. This new formulation was based on previous ethnoveterinary information suggesting that the individual plants were effective against animal parasites on production animals (Sharma et al., 1997).

17.3 HERBS FOR SKIN DISEASES

As with treatments for ticks, pet owners are concerned about the long-term effects of the use of pharmaceutical products, especially corticosteroids, for skin diseases (Wynn and Chalmers, 2002). Various products containing oatmeal, TTO, *Aloe vera*, and hammamelis have been used, mainly as shampoo to control pruritus.

Aloe vera is one of the most widely known herbal products and is used for all manner of purported therapeutic and cosmetic reasons. It has been reported to assist wound healing by increasing epithelialization in trials on dogs (Farstvedt, Stashak, and Othic, 2004). Extracts of *Aloe vera* also showed reduced leukocyte adhesion, tumor necrosis factor α (TNF-α), and interleukin (IL) 6 in an *in vivo* study on burn wounds. An *in vitro* study by Naidoo, Zweygarth, and Swan (2006) also showed the potential benefits of the related *Aloe marlotthii* extract in the management of ehrlichiosis.

Sarcoptic mange is one of the most important veterinary skin diseases. Conventional therapy and control of sarcoptic mange involves the use of chemotherapeutics. Most of the chemical acaricides currently used have several drawbacks (e.g., resistance, toxicity to the patient and the ecosystem) (Du et al., 2008). In Estonia, various plant species are used to treat mange, such as *Heracleum sosnowskyi*, *Artemisia vulgaris*, *Tanacetum vulgare* (common tansy), *Artemisia absinthium*, *Allium sativum* (garlic), *Piper nigrum* (black pepper), *Juniperus communis*, and *Cymbopogon nardus* (citronella grass) (Magi, Jarvis, and Miller, 2006). They have mainly been shown to have activity against ectoparasites.

Azadiracta (neem), which is widely used for tick repellency, also features as a treatment for sarcoptic mange. Petroleum ether extracts of neem oil were found effective against *Sarcoptes scabiei* var. *cuniculi* larvae (Du et al., 2008).

For the treatment of mange in Trinidad and Tobago, dogs are bathed in water with leaves of bamboo (*Bambusa vulgaris*), black sage (*Cordia curassavica*), and Congo lala (*Eclipta alba*) (Lans et al., 2000). *Bambusa vulgaris* is, however, toxic to horses (Barbosa et al., 2006). *Cordia curassavica* is used in Central America and the Caribbean islands for humans with respiratory and dermatological disorders (Hernandez et al., 2007). The leaf branch of *C. curassavica* is used for grooming horses in Trinidad and Tobago (Lans et al., 2006). Extracts of this species have been shown to have antibacterial, larvicidal, and antifungal activity (Hernandez et al., 2007; Ioset et al., 2000). Biological activity has been ascribed to meroterpenoid naphtoquinones called cordiaquinones (Ioset et al., 2000). In Suriname, neonates are bathed with black sage and a mixture of *Lantana camara* and *Renealmia* species to protect from a variety of illnesses (Ruysschaert et al., 2009), which reinforces the purported efficacy of black sage. *Eclipta alba* (false daisy) is used in India as a topical antiseptic and has been shown to have broad-spectrum antibacterial and antifungal activity (Kumar et al., 2006).

Lans et al. (2000) have suggested that the treatment of mange mirrors that of various human skin conditions; for example, *Scoparia dulcis* (licorice weed) is used to treat impetigo and herpes, *Bixa orellana* (achiote) seeds have been used for burns and wounds; the sap of *Musa* spp. is used for burns and wounds, while *Eclipta alba* has been used as a hemostatic and for bruises and wounds in cattle. These species are now being used for dogs.

In Tuscany (Italy), numerous remedies are recommended for topical wound management, such as *Pinus halepensis* (Aleppo pine), *Equisetum telmateia* (giant horsetail), and *Stachys officinalis* (Bishop's wort), among others (Uncini Manganelli, Camangi, and Tomei, 2001). *Pinus halepensis* resin is warmed in olive oil or lard and used in dogs; dried and chopped cauli of *E. telmateia* may be applied to small wounds, especially those caused by hematophagous insects. *Stachys officinalis* is applied locally to infected and purulent wounds of cats, dogs, and horses. It is rich in polyphenolic compounds and phenolic acids, which gives it good free-radical scavenging properties (Vundać et al., 2007), and it may play a part in wound healing. The phenylethanoid glycosides have also been isolated from it (Miyase, Yamamoto, and Ueno, 1996). *Nerium oleander* (oleander) is applied locally on the skin of dogs and cats affected by fungal disease. *Nerium oleander* is well known for the presence of cardenolides, which are toxic to man and animals; therefore, its application should be done carefully (Barbosa et al., 2008).

For horse hoof lesions, wonder of the world (*Kalanchoe pinnata*), young banana leaves (*Musa* species), castor bean leaves (*Ricinus communis*), or turmeric root (*Curcuma longa*) may be crushed and stuck onto affected areas with soft candle wax or epsom salts (Lans et al., 2006). These treatments are used in the horse-racing industry in Trinidad and British Columbia but have yet to be validated.

17.4 HERBS FOR ARTHRITIS AND JOINT SUPPORT

Bone and joint disorders are a frequent occurrence in both dogs and horses. The etiologies of the diseases are different, however, and numerous. In dogs, the typical cause of joint problems results from the developmental diseases, such as osteochondrosis dystrophy and degenerative joint disease (DJD). Of these, the latter tends to be caused by incorrect diet during the growing phase of bone. In horses, the cause could be developmental and may also be due to excessive training. Symptomatic treatment and the alleviation of pain are the common treatment modalities. With bone disorders, the only manner of managing permanent changes is surgically, or through the implantation of prosthesis. Infection, trauma, and inflammation may also cause joint disease.

In Canada, horses with injuries have rachette (*Nopalea cochenillifera*) applied directly to the injured area or as a mixture with *Aloe vera*, flour, or Epsom salts (Lans et al., 2006). Another remedy was castor bean leaves (*Ricinus communis*) quickly passed over a flame and wrapped around clay, which is used as a dressing on the injured area. Alternatively, leaves of wonder of world (*Kalanchoe pinnata*) were used to remove the "heat" from the injured leg.

The leaves, root, and seed oil of castor (*Ricinus communis*) are used as an antirheumatic externally as a poultice in the Middle East (Marc et al., 2008). Extracts of cayennne pepper, turmeric, horsetail, and Chinese skullcap are also widely used in the Middle East, Asia, and the Mediterranean for arthritis and joint pain (rheumatism). Cayenne pepper (*Capsicum* sp.) is mixed with olive oil to form a paste, which could be rubbed onto the affected area. The capsaicinoids, acid amides isolated from the fruits of *Capsicum* species, are commonly formulated into topical analgesics for clinical use (Li et al., 2009). Turmeric (*Curcuma longa*), which is closely related to another popular spice, ginger (*Zingiber officinale*), has been widely used either topically or orally in traditional Chinese medicine and ayurveda for the relief of pain and inflammation (Anonymous, 2001). The curcuminoids, isolated from *Curcuma longa,* have been shown to inhibit cyclooxygenase (Cox) enzymes (Cox-1 and Cox-2) (Ramsewak, DeWitt, and Nair, 2000). In animal studies, curcumin was shown to reduce levels of the inflammatory glycoprotein GpA-72, with a concomitant reduction in paw inflammation (Vitetta, Cicuttini, and Sali, 2008). Ginger, on the other hand, has been known for its anti-inflammatory effects for centuries. This has now been validated by research, which showed that gingerols inhibit prostaglandins and leukotrienes (Ali et al., 2008). Ginger has been found to modulate some biochemical pathways activated in chronic inflammation (Grzanna, Lindmark, and Frondoza, 2005). It was found to inhibit the induction of several genes involved in the inflammatory response, and some of these genes encode cytokines, chemokines, and the inducible enzyme Cox-2 (Grzanna, Lindmark, and Frondoza, 2005). Ginger extracts have potent Cox-1 inhibitor effects (Williams and Lamprecht, 2008), but they also have the tendency to increase proinflammatory cytokines. Ginger has been used in horses and been shown to reduce postexercise recovery times.

Scutellaria baicalensis (Chinese skullcap) was found to have activity comparable to the standard treatment in the arachidonic acid (AA)-induced mouse ear edema on topical application (Cuéllar et al., 2001). It is an important medicine in the Chinese pharmacopoeia used for inflammation and as an antiviral, anticancer preparation (Li, Jiang, and Chen, 2004). It contains several constituents (e.g., phenylethanoids, amino acids, and a large variety of flavonoids). The flavones baicalein and baicalin were shown to be responsible for the anti-inflammatory activity via free-radical scavenging or antagonizing ligand-initiated Ca^{2+} influx leading to inhibited leukocyte adhesion (Shen et al., 2003).

Harpagophytum procumbens (devil's claw) has validated anti-inflammatory activity that has been attributed to various iridoid glycosides and acetylated phenolic glycosides (Williams and Lamprecht, 2008). Harpagoside, one of the glycosides, has been shown to suppress Cox-2 and inducible nitric oxide synthase (iNOS) *in vitro*, but not Cox-1 activity. It also has chrondroprotective effects. Most of the human clinical studies reported reduced pain intensity and increased joint flexibility, and these studies have been used to justify the use of devil's claw preparations in animals, particularly horses. Pearson, McKee, and Clarke (1999) reported an anti-inflammatory effect in horses due to a reduction of the prostaglandin E_2 (PGE_2) synovial fluid content.

Flaxseed (*Linum usitatissimum*) has also been evaluated for use as an anti-inflammatory and found to reduce both Cox-1 and Cox-2 levels (Williams and Lamprecht, 2008). It is high in essential free fatty acids, particularly α-linolenic acid and lignan-type phytoestrogens, of which the main component is secoisolariciresinol diglucoside (SDG) (Degenhardt, Habben, and Winterhalter, 2002). Flaxseed extracts have been shown to have anti-inflammatory and antioxidant activity due to the presence of lignins and α-linolenic acid (Saleem et al., 2005). The primary use of flaxseed is to improve hair and coat quality, which is particularly important in show pets. It is commonly added into the newer scientifically formulated pet foods because of the high concentration of α-linolenic acid and soluble and insoluble fiber, which make it of high nutritional value.

Yucca schidigera, native to the southwestern United States and northern Mexico, is traditionally used for treating pain (Cheeke, Piacvente, and Oleszek, 2006). It has phenolics that inhibit nuclear factor kappa B (NFκB), resulting in suppression of nitric oxide, an inflammatory agent. Yucca extracts are currently used in feed additives for livestock and poultry in the United States for their growth-promoting, nematocidal, and antiprotozoal effects and in horses and dogs for antiarthritic indications (Cheeke, Piacvente, and Oleszek, 2006).

Arnica montana, native to central Europe and Scandinavia, is a popular homeopathic remedy that has shown good effect in acute inflammation (Macêdo et al., 2004). The sesquiterpene lactones were found to inhibit NFκB (Klaas et al., 2002). Arnica preparations, however, should not be used on broken skin as it contains formic acid and arnicine, which may cause cardiac damage and hepatoenteric inflammation (Harman, 2002).

17.5 HERBS FOR INTESTINAL WORMS

Helminths are an important cause of disease in both dog and horse, and infection may result from in utero and in milk exposure in puppies or from environmental exposure. For the latter, it may involve the active penetration of the immature larvae into the skin of the animal (e.g., hookworm in dogs) or through the ingestion of immature larvae during normal feeding or, in the case of dogs, through the ingestion of beetles or lizards carrying the parasite. While most worms will be present in the gastrointestinal tract, this is not always the case as some worms live in the blood, the airways, and the trachea. Clinical signs in affected animals will vary according to the location and type of parasites; for example, blood-sucking parasites tend to cause severe anemia, while other worms cause loss of appetite, weight, and general health. Lung worms cause coughing, while esophageal worms may cause nausea and vomiting, among other symptoms.

The allopathic treatment of these animals is simply to deworm them with one of many available commercial remedies. These remedies may be specific not only for the parasite but also for the animal being treated.

Neem (*Azadirachta indica*) belongs to the Meliaceae family and widely used in traditional medicine to treat various diseases including elimination of nematodes (Biswas et al., 2002). It has also been shown to be effective against ectoparasites such as *Bovicola ovis* (Heath et al., 1995) and *Hyalomma anatolicum excavatum* (Abdel-Shafy and Zayed, 2002). Leaf extracts of *A. indica* were shown to have ovicidal and larvicidal actions against *Haemonchus contortus* (Costa et al., 2008), while azadirachtin, a chemical compound extracted from *A. indica* seeds, inhibited 68% of *H. contortus* egg hatchi. Ground leaves of neem are put into food or water for dogs in Trinidad and Tobago for deworming purposes (Lans et al., 2000).

Various other herbs have been cited. In Canada, wormwood (*Artemisia absinthium*), rue (*Ruta graveolens*), peppermint (*Mentha piperita*), sage (*Salvia officinalis*), and mugwort (*Artemisia vulgaris*) are recommended for the control of endoparasites in pets, while worm grass (*Chenopodium ambrosioides*) is used as an anthelmintic in horses (Lans et al., 2006; Lans et al., 2007). Other remedies used in horses include aerial parts of the elecampane (*Inula helenium*) or wormwood powdered and added to the feed daily for 1 week. Alternatively, wormwood was given in equal combination with elecampane (*Inula helenium*) and thyme (*Thymus* sp.). Artemisia species have a long history of use from the time of the Roman Empire, when they were used for *Ascaris*, *Enterobius*, and tapeworm infections (Grove, 1990). The sesquiterpene santonin from *Artemisia* has been shown to be active against these worms but not against *Oxyuris* spp. and cestodes.

Papaya (pawpaw) preparations have a long history of use as human anthelmintic remedies originating from South America (Behnke et al., 2008). Seeds of papaya are used for human and farm animal intestinal worm treatment in the West Indies and India (Quinlan, Quinlan, and Nolan, 2002), and extracts have shown *in vitro* activity against *Caenorhaditis elegans* and other helminths (Wilson et al., 2002). The anthelmintic activity of papaya, ficus (fig), and pineapple (*Ananas comosus*) extracts is attributed to the actions of cysteine proteinase (CP) enzymes (Behnke et al., 2008). Kermanshai et al. (2001) also suggested that benzyl isothiocyanate may be the phytoconstituent responsible for activity.

A popular anthelmintic in livestock in Nordic countries stretching back to Greek times is the powdered rhizome of the male fern (*Dryopteris flix-mas*) for the treatment of tapeworm infections (Waller et al., 2001). Filicic acid has been identified as the active compound.

Garlic extracts have shown broad antiparasitic activity against human and nonhuman parasites. Allicin, diallyl trisulfide, and ajoene have activity against trophozoite, protozoan, plasmodium, trypanosome, coccidian, and flagellate species, and various mechanisms have been elucidated to explain this broad-spectrum effect (Williams and Lamprecht, 2008).

Commonly consumed vegetables (e.g., carrot, brassica, pumpkin, and cucumber seeds) also have a long history of use against intestinal worms (Waller et al., 2001). Tanniferous plants have shown anthelmintic activity that has been attributed to improved host resistance from better protein absorption as a result of tannin ingestion or direct nematodicidal activity in both the animal and on the free-living fecal stage.

Feeding culinary herbs with medicinal properties to pets is also feasible, particularly if they are herbivores, for example, caraway (*Carum carvi*), parsley (*Petroselinium crispum*), chicory (*Cichorium intybis*), chervil (*Anthriscus cerefolium*), and dill (*Anethum graveolens*) (Waller et al., 2001) are used. Some of the spices (e.g., caraway, thyme, and mint) have shown activity against *Trichostronglus* larvae *in vitro* and in sheep. Their use in pets still has to be scientifically documented.

17.6 HERBS FOR IMMUNOMODULATION

The immune system is divided into B cell (which includes circulating antibodies such as immnoglobulins [Igs] IgG, IgM, and IgE) and T cell or cell mediated (which involves the lymph nodes, thymus, and spleen and operates via the helper, cytotoxic, and suppressor cells) (Dodds, 2002). Immune modulators may have beneficial effects in the prevention and treatment of cancer, infections, and inflammatory diseases (Plaeger, 2003; Patwardhan and Gautam, 2005). There are various natural products offered for immune support, such as *Panax ginseng* (ginseng), *Artemisia annua* (annual wormwood), *Glycyrrhiza glabra* (licorice), *Uncaria tomentosa* (cat's claw), *Viscum album* (common mistletoe), and *Echinacea* species, among others (Patwardhan and Gautam, 2005). There is increasing interest in the use of herbal products for this purpose in human medicine, but there is little or no research for equivalent veterinary application.

Various compounds have been shown to influence immunostasis, among them the phytosterols, lectins, polysaccharides, alkaloids, flavonoids, and terpenoids (Reis et al., 2008; Mishra, Singh, and Dagenais, 2000; Shao et al., 2004; Chen et al., 2004). The phytosterols have been shown to improve T-lymphocyte and natural killer cell actions. In a clinical trial with cats infected with the feline immunodeficiency virus (FIV), phytosterol-rich supplementation slowed the progression of the disease (Bouic, 1997).

Flavonoids are also known to play a part in immune modulation, although different ones may be immunosuppressive or immunostimulatory (Ielpo et al., 2000). Quercetin, for example, has shown potent inhibition of the release of lysosomal enzymes in polymorphonuclear leukocytes (PMNs), which are also called neutrophils or granular leukocytes. The immunomodulatory effects of flavonoids are linked to their antioxidizing effects (Ielpo et al., 2000)

The best-known herb for immune modulation is *Echinacea* (purple coneflower), and it is used for the treatment of colds, coughs, and upper respiratory infections in humans. Polysaccharides from *Echinacea* have been shown to increase phagocytosis, chemotaxis, and the oxidative burst in neutrophils and macrophages in both healthy and immunosuppressed animals (Dodds, 2002; Barrett, 2003). The use of *Echinacea* in horses is now widespread and has been covered in Chapter 9.

Some mushroom species, such as *Grifola frondosa* (maitake), *Ganoderma lucidum* (reishi), and *Lentinula edodes* (shitake), enhance macrophage and T-cell function (Dodds, 2002; Shao et al., 2004). This immunostimulatory activity has been attributed to polysaccharides and lectins (Singh, Singh, and Singh, 2004; Chen et al., 2004).

As discussed, the use of botanical immunodrugs is of immense interest in human medicine but is hardly applied at all at the moment in veterinary medicine.

17.7 HERBS FOR BEHAVIORAL AND COGNITIVE DISORDERS

Behavioral diseases in dogs and horses are wide ranging. They include disorders such as excessive barking, aggression, noise phobia, separation anxiety, boredom (wind sucking and crib biting), and so on. The management of these diseases usually involves the blunting of the behavior of the animal through the use of specific sedatives, tranquilizers, or antidepressants. The role of behavioral modification through nonmedical means has also been used as an alternate to medical management of behavioral problems in patients.

One commonly used commercial remedy in the management of behavioral problems in cats and dogs is the homeopathic Rescue Remedy®, formulated by Edward Bach in 1930 by mixing the essence of floral extracts of rock rose (*Cistus* spp.), star of Bethlehem (*Ornithogalum umbellatum*), impatiens, cherry plum (*Prunus cerasifera*), and clematis. At present, this commercial remedy is recommended for the management of various behavioral problems, which include excessive barking in dogs, stress due to being left alone, shock, trauma or mistreatment, anxiety when adapting to new surroundings, and the fear of loud noises. The product is not specifically formulated for animals and as such is also sold for use in humans.

Canine cognitive dysfunction syndrome describes a neurological condition of older dogs (>7 years) that is characterized by a gradual decline in learning, memory, perception, and awareness and is akin to Alzheimer's disease in humans (Landsberg, 2005). A number of different remedies have been recommended for the management of this condition, extending from commercial ethical products to the use of dietary supplements or complementary medicines.

Clinical trials have shown that nutraceutical supplementation can ameliorate the effects of canine cognitive dysfunction (Heath, Barabas, and Craze, 2007; Cotman et al., 2002). The protective effects of flavonoids and polyphenolic compounds against neurodegenerative disease have been demonstrated in numerous experimental animal models (e.g., dietary supplementation with spinach or berry fruit have been reported to reduce neurological deficits in animal models) (Ramassamy, 2006).

A specific commercial diet containing supplements is currently marketed by a leading pet nutrition company following clinical trials indicating the value of the diet in managing the condition. This formulation makes use of antioxidants and mitochondrial cofactors as a means of improving antioxidant defenses and possibly decreasing the toxic damage induced by free radicals specifically in the brain (Cotman et al., 2002). The latter is important as the brain is particularly susceptible to free-radical damage due to a combination of its rapid metabolism and its high lipid content. Their diet is currently fortified with vitamins E and C, beta-carotene, selenium, α-lipoic acid, flavonoids, and carotenoids of plant origin as antioxidants. In addition to exerting an antioxidant action, flavonoid-rich extracts also lower levels of NFκB and hence influence apoptosis and the regulation of cell signaling (Ramassamy, 2006). In addition, L-carnitine and α-lipoic are included to enhance mitochondrial function. In experiments with aged dogs, an antioxidant-rich diet was shown to significantly reduce cognitive decline (Osella et al., 2007; Zicker, 2005).

A number of complementary therapies have also been recommended and include melatonin, valerian root, phosphatidylserine, *Gingko biloba*, pyridoxine, and vitamin E, vitamin B, ginseng, bilberry, α-lipoic acid, folic acid, n-acetylcysteine, grapeseed, rosemary, sage, choline, lecithin, and essential fatty acids. As mentioned, vitamin E, flavonoids, and carotenoids are most likely effective due to their antioxidant effect, while the B vitamins may have a neuroprotective effect (Zicker, 2005).

Ginkgo biloba extracts (mainly the standardized form called EGb 761) are a mixture of sesquiterpenes and flavonoids and possess a combination of antioxidant, antiamyloidogenic, and antiapoptotic effects that reduce cerebral ischemia, protect dopaminergic neurons, and prevent amyloid β peptide toxicity (Ramassamy, 2006). Vitamin E is important in maintaining neuronal integrity and preventing cell loss (Cantuti-Castelvetri, Shukitt-Hale, and Joseph, 2000). However, studies are still ongoing to establish whether this effect leads to improved behavioral and cognitive function.

α-Lipoic acid, a dithiol that exists in two enantiomeric forms, the naturally occurring form (R) and the synthetic S, is a good scavenger of hydroxyl, peroxy radicals, and nitric oxide (Cantuti-Castelvetri, Shukitt-Hale, and Joseph, 2000). It has been shown to improve memory in aged mice and protect neurons *in vitro* against hypoxic, glutamate-, and iron-induced insults. Phosphatidylserine, a phospholipid, has been reported to improve memory and learning and to offer neuroprotection (Osella et al., 2008). The fatty acids are required for maintenance of normal brain cell function, most likely due to their structural importance for cell membrane integrity (Youdim, Martin, and Joseph, 2000; Calon and Cole, 2007).

17.8 HERBS FOR COUGHING

Coughing may be defined as the forced expiration of air from the lungs against a closed glottis. Coughing in animals, as in humans, is controlled by the cough center in the brain after a stimulus is conducted there through afferent fibers. The stimulus for coughing arises from receptors in the pharynx, larynx, trachea, bronchi, and smaller airways through chemical or mechanical irritation (Anderson-Wessberg, 2005). The stimulation of these receptors can be vast and includes airway inflammation, mechanical irritation from congestive heart failure, allergic response, trauma, and even parasitic diseases.

Allopathic management is diverse and depends on the etiology of the disease and whether the cough is productive or dry. Airway inflammation in pets is most commonly caused by infectious agents such as viruses, fungi, bacteria, or mycoplasma. The treatment of this relies on the use of pathogen-specific antimicrobial agents if available. In addition, the degree of airway irritation may be eased through the use of specific demulcents (e.g., pectin, glycerin, and honey) that soothe the irritated mucosa and blunt the stimulus to cough. Common herbal demulcents include comfrey (*Symphytum officinale*), corn silk (*Zea mays*), couchgrass (*Agropyrum repens*), flaxseed (*Linum usitatissimum*), mallow (*Malva sylvestris*), oatmeal (*Avena sativa*), and liquorice (*Glycyrrhiza glabra*).

Congestive heart failure may result in a persistent dry cough due to the increased heart placing pressure on the aorta. Management of the diseased heart tends to alleviate the respiratory irritation. Central active antitussives, which inhibit the cough center directly and include drugs such as diphenhydramine and codeine, may also be of use.

Allergies in all animal species but notably the cat and horse may manifest as coughs and sneezes. The management of allergies involves therapy with drugs with various modes of action. This includes blunting of the histamine pathway by using drugs such as antihistamines and mast cell stabilizers (the latter tend to produce better results in horses than dogs), decreasing local inflammation by using corticosteroids, and stimulating bronchodilation by using bronchodilators such as the β2-agonists or the methylxanthines (e.g., theobromine and theophylline from green tea *Camellia sinensis*).

Parasitic disease in the dog may result in coughing. The solution would be to control the offending parasite with drugs.

In general, with coughs, the dry coughs should be treated with antitussive agents, which act to inhibit the cough reflex in the cough center, while for productive coughs, in which animals produce large qualities of phlegm, the cough can be managed by allowing for better expression of the inspissated mucus. Treatments include the expectorants, mucolytics, and simple nebulization with plain water.

There is little in the literature that addresses the use of medicinal plants in treating coughs and related conditions in domestic animals and pets. Lans et al. (2000) documented some products used in Canada. A commercial herbal product containing lungwort (*Pulmonaria officinalis*), bioflavonoids, and vitamin K was recommended for the management of exercise-induced pulmonary congestion, while elecampane (*Inula helenium*) mixed with cranberry bush (*Virbunum opulus*), root of liquorice, and thyme (*Thymus* sp.) was used for the management of stable cough. Another remedy for coughing included cloves of crushed garlic (*Allium sativum*) added to the feed. Treatment for a runny nose included pure garlic powder and mullein (*Verbascum thapsus*) fed in grain until recovery (Lans et al., 2000).

17.9 PITFALLS IN THE USE OF HERBS FOR COMPANION ANIMALS

The use of herbs is not without problems, as discussed in greater detail in the previous chapters of this book. One of the biggest problems is that, despite their widespread adoption, there are few clinical studies to validate herbal use in domestic animals. Much of the use and evidence of use is adopted from anecdotal evidence and human uses. For example, products containing aloe may be marketed variously for "detoxifying" or constipation. This use is based on the human use of aloe powders as laxatives and purgatives. Similarly, devil's claw (*Harpagophytum procumbens*) is marketed in combination with glucosamine and chondroitin for arthritic and joint pain. Similar products have been in use by humans for many decades. The lack of clinical studies presents problems with selecting a rational and safe dosing regimen in the various target species.

With the burgeoning of the market in herbal products for companion animals, there is increasing interest in clinical studies. One such study was described by Wynn and Chalmers (2002). It was an intriguing clinical trial of a traditional Chinese medicine formula patented for atopic eczema in adult humans. When the product showed good effect and few side effects, some of its constituent species were formulated into a similar product for the same condition in dogs. It was found to be effective when tested in a randomized clinical trial in dogs and has now been commercialized. This example gives credence to those like Mathias and Perezgrovas (1997), who argued that because of the overlap between medicinal plants used for humans and animals, the pharmacognosy of many ethnoveterinary plants is already well studied in the human context and therefore does not have to be repeated in animals.

The adverse effects of commonly used herbs should also be of concern, and there is a need to generate good pharmacovigilence data in this regard. Herbal products with chamomile, echinacea, garlic, ginger, and so on have all been shown to have some adverse effects in humans (Miller, 1998), but few studies have documented such effects in animals. The use of garlic may affect spermatogenesis and decrease platelet aggregation, ginkgolide B from *Ginkgo biloba* is a potent inhibitor of platelet-activating factor, and ingestion has resulted in hematoma in an elderly man, while the persistent use of echinacea may increase the risk of hepatotoxicity (Miller, 1998). Herb–drug interactions are also important considerations that may limit the use of herbal products. Ginger, garlic, and gingko have been shown to enhance the anticoagulant effects of aspirin and warfarin, while tannin-rich herbs such as chamomile, hawthorn, and green tea may reduce iron absorption (Miller, 1998).

One of the greatest pitfalls in the use of veterinary botanicals (and herbal medicine in general) is the ever-present risk of opportunistic quackery (Schillhorn van Veen, 1997). This results in the marketing of products that may be unsafe or without any proven anecdotal or scientific efficacy. Much of the marketing peddles the mysticism that is associated with traditional herbal use in humans (e.g., ayurverda or homeopathy). However, evidence to support these uses remains elusive. Unscrupulous botanical manufacturers and traders abound because the regulatory regime of commercial herbal products is still in its infancy in many territories, and many phytomedicines are classified as dietary supplements.

This lack of regulation of the CAM sector has led to confusion. In the United Kingdom, herbal remedies are classified as either licensed veterinary "herbal medicines" or "herbal supplements," and their advertising, labeling, and selling are proscribed by law. Herbal medicines are licensed by the Veterinary Medicines Directorate, a government agency, and treated the same as allopathic drugs; herbal supplements, on the other hand, cannot make a therapeutic claim.

There are also other potential problems, such as the identity and authenticity of herbs used not being verified or confirmed. Further adulteration, especially of high-value raw materials, has been documented in several studies. Such adulteration renders the product inefficacious or, worse, toxic (Ernst, 2002).

Contamination of herbs with heavy metals, bacteria, mycotoxins, and pharmaceuticals have all been reported in the past. Contamination with arsenic, cadmium, and lead have been reported in African (Obi, Akunyili, and Orisakwe, 2006), European (Arpadjan et al., 2008), and Asian (Ernst, 2002) herbal products. Undeclared pharmaceuticals (e.g., antibiotics, anti-inflammatories, etc.) have also been shown to be common contaminants in botanicals. The adulteration may be intentional and aimed at increasing potency or accidental (Ernst, 2002). Contamination with mycotoxins, which are potent mutagenic and carcinogenic compounds or which may be estrogenic and have neurotoxic effects, is also of major concern (Katerere et al., 2008). Animals are particularly prone to acute mycotoxicosis.

17.10 CONCLUSION

Ethnoveterinary research in the pet industry will become increasingly important in the future. This is mainly because of a shift in human–pet relations (the so-called humanization trend) that started in the United States and Europe and now is occurring in the emerging markets of China and India, where the nouveau riche regard pets as part of the family and spending and self-medication habits for pets increasingly mirror those of the owners (Reportbuyer, 2009). The pet industry is a major segment of the U.S.

economy and is estimated to be worth US$45 billion in 2009, having doubled in value since 1998 (American Pet Products Association, 2009). Nearly 90% of this money is spent on food and medicines, and as with humans, there is a trend toward functional, condition-specific, or novel ingredient types of foods.

While commonality of use in animals mirrors that of herbal drug therapy in people, their use needs to be scientifically validated. This is important to rule out factors such as species-specific toxicity and to eradicate noneffective medication and quackery. There is widespread acceptance among veterinarians and pet owners regarding the use of herbal products, but there is a need for greater regulatory oversight to ensure safety and quality.

REFERENCES

Abdel-Shafy S., Zayed A.A. In vitro acaricidal effect of plant of neem seed oil (*Azadirachta indica*) on egg, immature, and adult stages of *Hyalomma analicum excavatum* (Ixodoidea: Ixodidae), *Veterinary Parasitology* 2002; 106: 89–96.

Ali B.H., Blunden G., Tanira M.O., Nemmar A. Some phytochemical, pharmacological and toxicological properties of ginger (*Zingiber officinale* Roscoe): a review of recent research. *Food and Chemical Toxicology* 2008; 46 (2): 409–420.

Allen P. Tea tree oil: the science behind the antimicrobial hype. *The Lancet* 2001; 358 (9289): 1245.

American Pet Products Association (APPA). 2009. Industry Statistics & Trends. http://americanpetproducts.org/press_industrytrends.asp (accessed April 30, 2009).

Anderson-Wessberg K. Cardiorespiratory. In: Ettinger S.J., Feldman E.C. (Eds.), *Textbook of veterinary internal medicine,* 6th ed., Elsevier, St. Louis, MO, 2005, Chapter 53, 189.

Anonymous. *Curcuma longa* (turmeric). Monograph. *Alternative Medicine Review* 2001: 6; Suppl. S62–S66.

Arpadjan S., Çelik G., Taşkesen S., Güçer Ş. Arsenic, cadmium and lead in medicinal herbs and their fractionation. *Food and Chemical Toxicology* 2008; 46 (8): 2871–2875.

Barbosa J.D., de Oliveira C.M.C., Duarte M.D., Riet-Correa G., Peixoto P.V., Tokarnia C.H. Poisoning of horses by bamboo, *Bambusa vulgaris. Journal of Equine Veterinary Science* 2006; 26 (9): 393–398.

Barbosa R.R., Fontenele-Neto J.D., Soto-Blanco B. Toxicity in goats caused by oleander (*Nerium oleander*). *Research in Veterinary Science* 2008; 85 (2): 279–281.

Barrett B. Medicinal properties of *Echinacea*: a critical review. *Phytomedicine* 2003; 10 (1): 66–86.

Batish D.R., Singh H.P., Kohli R.K. Eucalyptus essential oil as a natural pesticide. *Forest Ecology and Management* 2008; 256 (12): 2166–2174.

Behnke J.M., Buttle D.J., Stepek G., Lowe A., Duce I.R. Developing novel anthelmintics from plant cysteine proteinases. *Parasites and Vectors* 2008; 1: 29.

Benjamin A. Pesticides: Germany bans chemicals linked to honeybee devastation (published May 23, 2008). http://www.guardian.co.uk/environment/2008/may/23/wildlife.endangeredspecies (accessed January 2009).

Biswas K., Chattopadhyay I., Banerjee R.K., Bandyopadhyay U. Biological activities and medicinal properties of neem (*Azadirachta indica*). *Current Science* 2002; **82:** 1336–1345.

Bouic P.J.D. Immunomodulation in HIV/AIDS: The Tygerberg/Stellenbosch University Experience. *AIDS Bulletin* 1997; 6: 18–20.

Calon F., Cole G. Neuroprotective action of omega-3 polyunsaturated fatty acids against neurodegenerative diseases: evidence from animal studies. *Prostaglandins, Leukotrienes and Essential Fatty Acids* 2007; 77 (5–6): 287–293.

Camp R. The Wizard of Oz, or the intriguing tale of the tea tree. *Journal of Investigative Dermatology* 2004; 23: xviii–xix.

Cantuti-Castelvetri I., Shukitt-Hale B., Joseph J.A. Neurobehavioral aspects of antioxidants in aging. *International Journal of Developmental Neuroscience* 2000; 18 (4–5): 367–381.

Carson C.F., Riley T.V., Cookson B.D. Efficacy and safety of tea tree oil as a topical antimicrobial agent. *Journal of Hospital Infection* 1998; 40 (3): 175–178.

Casida J.E. Pyrethrum flowers and pyrethroid insecticides. *Environmental Health Perspectives* 1980; 34: 189–202.

Cheeke P.R., Piacvente S., Oleszek W. Anti-inflammatory and anti-arthritic effects of *Yucca schidigera*: a review. *Journal of Inflammation* 2006; 3: 6–13.

Chen H.-S., Tsai Y.-F., Lin S., Lin C.-C., Khoo K.-H., Lin C.-H., Wong C.-H. Studies on the immuno-modulating and anti-tumor activities of *Ganoderma lucidum* (Reishi) polysaccharides. *Bioorganic and Medicinal Chemistry* 2004; 12 (21): 5595–5601.

Costa C.T.C., Bevilaqua C.M.L., Camurça-Vasconcelos A.L.F., Maciel M.V., Morais S.M., Castro C.M.S., Braga R.R., Oliveira L.M.B. In vitro ovicidal and larvicidal activity of *Azadirachta indica* extracts on *Haemonchus contortus*. *Small Ruminant Research* 2008; 74 (1–3): 284–287.

Cotman C.W., Head E., Muggenburg B.A., Zicker S., Milgram N.W. Brain aging in the canine: a diet enriched in antioxidants reduces cognitive dysfunction. *Neurobiology of Aging* 2002; 23 (5): 809–818.

Crombie L., Games D.E., Haskins N.J., Read G.F. Isolation and structure of insecticidal components from *Mammea americana* L. *Tetrahedron Letters* 1970; 11 (3): 251–254.

Cuéllar M.J., Giner R.M., Recio M.C., Máñez S., Ríos J.L. Topical anti-inflammatory activity of some Asian medicinal plants used in dermatological disorders. *Fitoterapia* 2001; 72 (3): 221–229.

Das S.S. Effect of a herbal compound for treatment of sarcoptic mange infestations on dogs. *Veterinary Parasitology* 1996; 63: 303–306.

de Benito F.M., Citores L., Iglesias R., Ferreras J. M., Soriano F., Arias J., Méndez E., Girbés T. Ebulitins: a new family of type 1 ribosome-inactivating proteins (rRNA *N*-glycosidases) from leaves of *Sambucus ebulus* L. that coexist with the type 2 ribosome-inactivating protein ebulin 1. *FEBS Letters* 1995; 360 (3): 299–302.

Degenhardt A., Habben S., Winterhalter P. Isolation of the lignan secoisolariciresinol diglucoside from flaxseed (*Linum usitatissimum* L.) by high-speed counter-current chromatography. *Journal of Chromatography A* 2002; 943 (2): 299–302.

Dodds W.J. Complementary and alternative veterinary medicine: the immune system. *Clinical Techniques in Small Animal Practice* 2002; 17 (1): 58–63.

Du Y.-H., Jia R.-Y., Yin Z.-Q., Pu Z.-H., Chen J., Yang F., Zhang Y.-Q., Lu Y. Acaricidal activity of extracts of neem (*Azadirachta indica*) oil against the larvae of the rabbit mite *Sarcoptes scabiei* var. *cuniculi* in vitro. *Veterinary Parasitology* 2008; 157 (1–2): 144–148.

Dubick M.A. Historical perspectives on the use of herbal preparations to promote health. *Journal of Nutrition* 1986; 116 (7): 1348–1354.

Ebrahimzadeh M.A., Mahmoudi M., Salimi E. Antiinflammatory activity of *Sambucus ebulus* hexane extracts. *Fitoterapia* 2006; 77 (2): 146–148.

Ernst E. Toxic heavy metals and undeclared drugs in Asian herbal medicines. *Trends in Pharmacological Sciences* 2002; 23 (3): 136–139.

Ernst E., Dixon A. Alternative medicines in Europe. In: Mossialos E., Mrazek M., Walley T.M. (Eds.), Regulating pharmaceuticals in Europe: striving for efficiency, equity and quality. 2006, Chapter 18. http://www.euro.who.int/document/E83015.pdf (accessed April 22, 2009).

Farstvedt E., Stashak T.S., Othic A. Update on topical wound medications. *Clinical Techniques in Equine Practice* 2004; 3 (2): 164–172.

Geary T.G., Conder G.A., Bishop B. The changing landscape of antiparasitic drug discovery for veterinary medicine. *Trends in Parasitology* 2004; 20 (10): 449–455.

Grove D.I. *A history of human helminthology*. Oxford University Press, New York, 1990.

Grzanna R., Lindmark L., Frondoza C.G. Ginger—an herbal medicinal product with broad anti-inflammatory actions. *Journal of Medicinal Food* 2005; **8:** 125–132.

Guarrera P.M. Traditional antihelmintic, antiparasitic and repellent uses of plants in central Italy. *Journal of Ethnopharmacology* 1999; 68 (1–3): 183–192.

Gupta R.C. (Ed.). Veterinary toxicology—basic and clinical principles. Elsevier–Academic Press, Oxford, UK, 2007.

Hahn I., Zitterl-Eglseer K., Franz C. Phytomedizin bei Hund und Katze: Internetumfrage bei Tierarzten und Tierarztinnen in Österreich, Deutschland und der Schweiz [Phytomedicine in dogs and cats: Web-based survey among veterinarians in Austria, Germany and Switzerland]. *Schweizer Archiv für Tierheilkunde* 2005; 147 (3): 135–141.

Hammer K.A., Carson C.F., Riley T.V., Nielsen J.B. A review of the toxicity of *Melaleuca alternifolia* (tea tree) oil. *Food and Chemical Toxicology* 2006; 44 (5): 616–625.

Harman J. The toxicology of herbs in equine practice. *Clinical Techniques in Equine Practice* 2002; 1 (2): 74–80.

Heath A.C.G., Lampkin N., Jowett J.H. Evaluation of non-conventional treatments for control of the biting louse (*Bovicola ovis*) on sheep, *Medical and Veterinary Entomology* 1995; 9: 407–412.

Heath S.E., Barabas S., Craze P.G. Nutritional supplementation in cases of canine cognitive dysfunction—a clinical trial. *Applied Animal Behaviour Science* 2007; 105 (4); 284–296.

Hernandez T., Canales M., Teran B., Avila O., Duran A., Garcia A.M., Hernandez H., Angeles-Lopez O., Fernandez-Araiza M., Avila G. Antimicrobial activity of the essential oil and extracts of *Cordia curassavica* (Boraginaceae). *Journal of Ethnopharmacology* 2007; 111 (1): 137–141.

Horowitz S., Integrative health care for pets: an overview of today's treatments. *Alternative and Complementary Therapies* 2003, 9(6): 317–321.

Huffman M.A., Hirata S. An experimental study of leaf swallowing in captive chimpanzees: insights into the origin of a self-medicative behavior and the role of social learning. *Primates* 2004; 45: 113–118.

Huffman M.A., Seifu M. Observations on the illness and consumption of a possibly medicinal plant *Vernonia amygdalina* (DEE.), by a wild chimpanzee in the Mahale Mountains National Park, Tanzania. *Primates* 1989; 30 (1): 51–63.

Ielpo M.T.L., Basile A., Miranda R., Moscatiello V., Nappo C., Sorbo S., Laghi E., Ricciardi M.M., Ricciardi L., Vuotto M.L. Immunopharmacological properties of flavonoids. *Fitoterapia* 2000; 71 (S1): S101–S109.

Iori A., Grazioli D., Gentile E., Marano G., Salvatore G. Acaricidal properties of the essential oil of *Melaleuca alternifolia* Cheel (tea tree oil) against nymphs of *Ixodes ricinus*. *Veterinary Parasitology* 2005; 129 (1–2): 173–176.

Ioset J-R., Marston A., Gupta M.P., Hostettmann K. Antifungal and larvicidal cordiaquinones from the roots of *Cordia curassavica. Phytochemistry* 2000; 53 (5): 613–617.

Katerere D.R., Stockenström S., Thembo K.M., Rheeder J.P., Shephard G.S., Vismer H.F. A preliminary survey of mycological and fumonisin and aflatoxin contamination of African traditional herbal medicines sold in South Africa. *Human and Experimental Toxicology* 2008; 27: 793–798.

Kayne S., Shakespeare M., Jepson M., Cockbill S. New opportunities in veterinary pharmacy. *The Pharmaceutical Journal* 2001; 267: 159–166.

Kennedy J. Herb supplement use in the U.S. adult population. *Clinical Therapeutics* 2005; 27: 1847–1858.

Kermanshai R., McCarry B.E., Rosenfeld J., Summers P.S., Weretilnyk E.A., Sorger G.J. Benzyl isothiocyanate is the chief or sole anthelmintic in papaya seed extracts. *Phytochemistry* 2001; 57 (3): 427–435.

Kiriamiti H.K., Camy S., Gourdon C., Condoret J.S. Pyrethrin exraction from pyrethrum flowers using carbon dioxide. *The Journal of Supercritical Fluids* 2003; 26 (3): 193–200.

Klaas C.A., Wagner G., Laufer S., Sosa S., Della Loggia R., Bomme U., Pahl H.L., Merfort I. Studies on the anti-inflammatory activity of phytopharmaceuticals prepared from *Arnica* flowers. *Planta Medica* 2002; 68 (5): 385–391.

Kozukue N., Han J.S., Lee K.R., Friedman M. Dehydrotomatine and alpha-tomatine content in tomato fruits and vegetative plant tissues. *Journal of Agricultural Food Chemistry* 2004; 52 (7): 2079–2083.

Kumar V.P., Chauhan N.S., Padh H., Rajani M. Search for antibacterial and antifungal agents from selected Indian medicinal plants. *Journal of Ethnopharmacology* 2006; 107 (2): 182–188.

Landsberg G. Therapeutic agents for the treatment of cognitive dysfunction syndrome in senior dogs. *Progress in Neuro-Psychopharmacology and Biological Psychiatry* 2005; 29: 471–479.

Lans C., Harper T., Georges K., Bridgewater E. Medicinal plants used for dogs in Trinidad and Tobago. *Preventive Veterinary Medicine* 2000; 45: 201–220.

Lans C., Turner N., Brauer G., Lourenco G., Georges K. Ethnoveterinary medicines used for horses in Trinidad and in British Columbia, Canada. *Journal of Ethnobiology and Ethnomedicine* 2006; 2: 31.

Lans C., Turner N., Khan T. Medicinal plant treatments for fleas and ear problems of cats and dogs in British Columbia, Canada. *Parasitology Research* 2008; 103(4): 889–898.

Lans C., Turner N., Khan T., Brauer G. Ethnoveterinary medicines used to treat endoparasites and stomach problems in pigs and pets in British Columbia, Canada. *Veterinary Parasitology* 2007; 148: 325–340.

Li F., Lin Y., Wang X., Geng Y., Wang D. Preparative isolation and purification of capsaicinoids from *Capsicum frutescens* using high-speed counter-current chromatography. *Separation and Purification Technology* 2009; 64 (3): 304–308.

Li H., Jiang Y., Chen F. Separation methods used for *Scutellaria baicalensis* active components. *Journal of Chromatography B* 2004; 812 (1–2): 277–290.

Macêdo S.B., Ferreira L.R., Perazzo F.F., Tavares Carvalho J.C. Anti-inflammatory activity of *Arnica montana* 6cH: preclinical study in animals. *Homeopathy* 2004; 93 (2): 84–87.

Magi E., Jarvis T., Miller I. Effects of different plant products against pig mange mites. *Acta Veterinaria* 2006; 75: 283–287.

Marc E.B., Nelly A., Annick D.-D., Frederic D. Plants used as remedies antirheumatic and antineuralgic in the traditional medicine of Lebanon. *Journal of Ethnopharmacology* 2008; 120 (3): 315–334.

Mathias E., Perezgrovas R. Applications of ethnoveterinary medicine: where do we stand? In: Mathias E., Rangnekar D.V. (Eds.), *Ethnoveterinary medicine: alternative for livestock development*. Proceedings of an International Conference held in Pune, India, November 4–6, 1997. http://www.vetwork.org.uk/pune14.htm (accessed May 12, 2006).

Miller L.G. Herbal medicinals—selected clinical considerations focusing on known or potential drug-herb interactions. *Archives of Internal Medicine* 1998; 158: 2201–2211.

Mishra L., Singh B.B., Dagenais S. Scientific basis for the therapeutic use of *Withania somnifera* (Ashwagandha): a review. *Alternative Medicine Review* 2000; 5(4): 334–346.

Miyase T., Yamamoto R., Ueno A. Phenylethanoid glycosides from *Stachys officinalis. Phytochemistry* 1996; 43 (2): 475–479.

Mulla M.S., Su T. Activity and biological effects of neem products against arthropods of medical and veterinary importance. *Journal of the American Mosquito Control Association* 1999; 15(2): 133–152.

Naidoo V., Zweygarth E., Swan G.E. Determination and quantification of the in vitro activity of *Aloe marlothii* (A. Berger) subsp. *marlothii* and *Elephantorrhiza elephantina* (Burch.) skeels acetone extracts against *Ehrlichia ruminantium. Onderstepoort Journal of Veterinary Research* 2006; 73(3): 175–178.

Obi E., Akunyili D.N., Orisakwe B.E.O.E. Heavy metal hazards of Nigerian herbal remedies. *Science of the Total Environment* 2006; 369 (1–3): 35–41.

Osella M.C., Re G., Badino P., Bergamasco L., Miolo A. Phosphatidylserine (PS) as a potential nutraceutical for canine brain aging: a review. *Journal of Veterinary Behavior: Clinical Applications and Research* 2008; 3 (2): 41–51.

Osella M.C., Re G., Odore R., Girardi C., Badino P., Barbero R., Bergamasco L. Canine cognitive dysfunction syndrome: prevalence, clinical signs and treatment with a neuroprotective nutraceutical. *Applied Animal Behaviour Science* 2007; 105 (4): 297–310.

Patwardhan B., Gautam M. Botanical immunodrugs: scope and opportunities. *Drug Discovery Today* 2005; 10 (7): 495–502.

Pearson W., McKee S., Clarke A.F. Effect of a proprietary herbal product on equine joint disease. *Journal of Nutraceuticals, Functional and Medical Foods* 1999; 2 (2): 31–46.

Plaeger S.F. Clinical immunology and traditional herbal medicines. *Clinical and Diagnostic Laboratory Immunology* 2003; 10 (3): 337–338.

Quinlan M.B., Quinlan R.J., Nolan J.M. Ethnophysiology and herbal treatments of intestinal worms in Dominica, West Indies. *Journal of Ethnopharmacology* 2002; 80 (1): 75–83.

Ramassamy C. Emerging role of polyphenolic compounds in the treatment of neurodegenerative diseases: a review of their intracellular targets. *European Journal of Pharmacology* 2006; 545 (1): 51–64.

Ramsewak R.S., DeWitt D.L., Nair M.G. Cytotoxicity, antioxidant and anti-inflammatory activities of curcumins I-III from *Curcuma longa. Phytomedicine* 2000; 7(4): 303–308.

Reis S.R.I.N., Valente L.M.M., Sampaio A.L., Siani A.C., Gandini M., Azeredo E.L., D'Avila L.A., Mazzei J.L., Henriques M.G., Kubelka C.F. Immunomodulating and antiviral activities of *Uncaria tomentosa* on human monocytes infected with dengue virus-2. *International Immunopharmacology* 2008; 8 (3): 468–476.

Reportbuyer. The pet industry: food, accessories, health products and services (published January 2006). http://www.reportbuyer.com/consumer_goods_retail/pet_products/pet_industry_food_accessories_health_products_services.html (accessed April 30, 2009).

Reyes-Chilpa R., Estrada-Muñiz E., Vega-Avila E., Abe F., Kinjo J., Hernández-Ortega S. Trypanocidal constituents in plants. 7. Mammea-type coumarins. *Memórias do Instituto Oswaldo Cruz* 2008; 103: 5.

Ruysschaert S., van Andel T., Van de Putte K., Van Damme P. Bathe the baby to make it strong and healthy: plant use and child care among Saramaccan Maroons in Suriname. *Journal of Ethnopharmacology* 2009; 121 (1): 148–170.

Ryan C.A. Proteinase inhibitors in plants: genes for improving defenses against insects and pathogens. *Annual Review of Phytopathology* 1990; 28: 425–449.

Saleem M., Kim H.J., Ali M.S., Lee Y.S. An update on bioactive plant lignans. *Natural Product Reports* 2005; 22: 696–716.

Schillhorn van Veen, T. W. Sense or nonsense? Traditional methods of animal parasitic disease control. *Veterinary Parasitology* 1997; 71: 177–194.

Sharma D.K., Saxena V.K., Sanil N.K., Singh N. Evaluation of oil of *Cedrus deodara* and benzyl benzoate in sarcoptic mange in sheep. *Small Ruminant Research* 1997; 26 (1–2): 81–85.

Shen Y., Chiou W., Chou Y., Chen C. Mechanisms in mediating the anti-inflammatory effects of baicalin and baicalein in human leukocytes. *European Journal of Pharmacology* 2003; 465 (1–2): 171–181.

Singh J., Singh J., Singh K.S. A novel mitogenic and antiproliferative lectin from a wild cobra lily, *Arisaema flavum. Biochemical and Biophysical Research Communications* 2004; 318 (4): 1057–1065.

Soderlund D.M., Clark J.M., Sheets L.P., Mullin L.S., Piccirillo V.J., Sargent D., Stevens J.T., Weiner M.L. Mechanisms of pyrethroid neurotoxicity: implications for cumulative risk assessment. *Toxicology* 2002; 171 (1): 3–59.

Swan G., Naidoo V., Cuthbert R., Green R.E., Pain D.J., et al. Removing the threat of diclofenac to critically endangered Asian vultures. *PLoS Biology* 2006; 4 (3): 0395–0402.

Townsend L. External parasite control for horses. *Journal of Equine Veterinary Science* 2005; 25 (9): 411.

Uncini Manganelli R.E., Camangi F., Tomei P.E. Curing animals with plants: traditional usage in Tuscany (Italy). *Journal of Ethnopharmacology* 2001; 78: 171–191.

Vitetta L., Cicuttini F., Sali A. Alternative therapies for musculoskeletal conditions. *Best Practice and Research Clinical Rheumatology* 2008; 22 (3): 499–522.

Vundać V.B., Brantner A.H., Plazibat M. Content of polyphenolic constituents and antioxidant activity of some *Stachys* taxa. *Food Chemistry* 2007; 104 (3): 1277–1281.

Waller P.J., Bernes G., Thamsborg S.M., Sukura A., Richter S.H., Ingebrigtsen K., Höglun J. Plants as de-worming agents of livestock in the Nordic countries: historical perspective, popular beliefs and prospects for the future. *Acta Veterinaria Scandinavica* 2001; 42; 31–44.

Williams C.A., Lamprecht E.D. Some commonly fed herbs and other functional foods in equine nutrition: a review. *The Veterinary Journal* 2008; 178: 21–31.

Wilson R.K., Kwan T.K., Kwan C-Y., Sorger G.J. Effects of papaya seed extract and benzyl isothiocyanate on vascular contraction. *Life Sciences* 2002; 71 (5): 497–507.

Wynn S.G., Chalmers S. Alternative therapies for pruritic skin disorders. *Clinical Techniques in Small Animal Practice* 2002; 17 (1): 37–40.

Yasunaka K., Abe F., Nagayama A., Okabe H., Lozada-Pérez L., López-Villafranco E., Estrada Muñiz E., Aguilar A., Reyes-Chilpa R. Antibacterial activity of crude extracts from Mexican medicinal plants and purified coumarins and xanthones. *Journal of Ethnopharmacology* 2005; 97 (2): 293–299.

Youdim K.A., Martin A., Joseph J.A. Essential fatty acids and the brain: possible health implications. *International Journal of Developmental Neuroscience* 2000; 18 (4–5): 383–399.

Zicker S.C. Cognitive and behavioral assessment in dogs and pet food market applications. *Progress in Neuro-Psychopharmacology and Biological Psychiatry* 2005; 29 (3): 455–459.

Index

Page references in **boldface** type refer to tables.
Page references in *italics* refer to figures.

B

C

D

E

I

M

Q

R

U

Y

Z